V. W. Cairns, P. V. Hodson, and J. O. Nriagu
Contaminant Effects on Fisheries — Volume 16 1984

J. O. Nriagu and C. I. Davidson
Toxic Metals in the Atmosphere — Volume 17 1986

A. H. Legge and S. V. Krupa
Air Pollutants and Their Effects on the Terrestrial Ecosystem — Volume 18 1986

Air Pollutants and Their Effects on the Terrestrial Ecosystem

Volume

18

in the Wiley Series in

Advances in Environmental Science and Technology

JEROME O. NRIAGU, Series Editor

Air Pollutants and Their Effects on the Terrestrial Ecosystem

Edited by

ALLAN H. LEGGE

Kananaskis Centre for Environmental Research
The University of Calgary
Calgary, Alberta

SAGAR V. KRUPA

University of Minnesota,
St. Paul, Minnesota

A WILEY-INTERSCIENCE PUBLICATION

JOHN WILEY & SONS

New York • Chichester • Brisbane • Toronto • Singapore

Library of Congress Cataloging in Publication Data:

Main entry under title:

Air pollutants and their effects on the terrestrial ecosystem.

(Advances in environmental science and technology; v.18.)

"A Wiley Interscience Publication."
Includes index.
1. Air—Pollution—Environmental aspects. 2. Air—Pollution. I. Legge, Allan H. II. Krupa, Sagar V. III. Series.
TD180.A38 vol. 18 628 s [575.5′222] 85-6461
[QA545.A3]
ISBN 0-471-08312-7

Printed in the United States of America

10 9 8 7 6 5 4 3 2 1

CONTRIBUTORS

A. C. CHAMBERLAIN, Atomic Energy Research Establishment, Harwell, England

CARL W. CHEN, Systech Engineering, Inc., Lafayette, California

E. E. CUDBY, Chevron Standard Ltd., Calgary, Alberta, Canada

KENNETH L. DEMERJIAN, Environmental Sciences Research Laboratory, U.S. Environmental Protection Agency, Research Triangle Park, North Carolina

A. A. ELSEEWI, Department of Soil and Environmental Sciences, University of California, Riverside, California

R. L. FINDLAY, Amoco Canada Petroleum Company Ltd., Calgary, Alberta, Canada

J. H. B. GARNER, Environmental Criteria and Assessment Office, U.S. Environmental Protection Agency, Research Triangle Park, North Carolina

S. GODZIK, Polish Academy of Sciences, Institute of Environmental Engineering, Zabrze, Poland

ROBERT A. GOLDSTEIN, Electric Power Research Institute, Palo Alto, California

GLEN E. GORDON, Department of Chemistry, University of Maryland, College Park, Maryland

R. GUDERIAN, University of Essen, Essen, West Germany

JEREMY M. HALES, Battelle Pacific Northwest Laboratories, Richland, Washington

G. STEVE HART, Environment Canada, Ottawa, Ontario, Canada

WALTER W. HECK, U.S. Department of Agriculture/North Carolina State University, Raleigh, North Carolina

B. B. HICKS, National Oceanic and Atmospheric Administration, Atmospheric Turbulence and Diffusion Laboratory, Oak Ridge, Tennessee

G. M. HIDY, Environmental Research and Technology Inc., Newbury Park, California

R. P. HOSKER, Jr., Air Resources, Atmospheric Turbulence and Diffusion Laboratory, National Oceanic and Atmospheric Administration, Oak Ridge, Tennessee

R. B. HUSAR, Washington University, St. Louis, Missouri

W. B. JOHNSON, Atmospheric Science Center, SRI International, Menlo Park, California

RONALD N. KICKERT, Division of Biological Control, University of California, Berkeley, California

ALLAN H. LEGGE, Kananaskis Centre for Environmental Research, The University of Calgary, Calgary, Alberta, Canada

STEVEN E. LINDBERG, Environmental Sciences Division, Oak Ridge National Laboratory, Oak Ridge, Tennessee

S. N. LINZON, Ontario Ministry of the Environment, Toronto, Ontario, Canada

JAMES R. MCCLENAHEN, Ohio Agricultural Research and Development Center, Wooster, Ohio

W. A. MCCLENNY, U. S. Environmental Protection Agency, Research Triangle Park, North Carolina

D. C. MCCUNE, Boyce Thompson Institute for Plant Research at Cornell University, Ithaca, New York

S. B. MCLAUGHLIN, Environmental Sciences Division, Oak Ridge National Laboratory, Oak Ridge, Tennessee

PAUL R. MILLER, Forest Service, U. S. Department of Agriculture, Riverside, California

P. K. MUELLER, Electric Power Research Institute, Palo Alto, California

LEONARD NEWMAN, Environmental Chemistry Division, Department of Applied Science, Brookhaven National Laboratory, Associated Universities Inc., Upton, New York

A. L. PAGE, Department of Soil and Environmental Sciences, University of California, Riverside, California

W. R. PIERSON, Scientific Laboratory, Ford Motor Company, Dearborn, Michigan

T. H. RISBY, The Johns Hopkins University, Baltimore, Maryland

V. C. RUNECKLES, Department of Plant Science, University of British Columbia, Vancouver, British Columbia, Canada

P. J. W. SAUNDERS, Natural Environment Research Council, Swindon, Wiltshire, England

J. SHINN, Lawrence Livermore Laboratory, Livermore, California

DAVID S. SHRINER, Environmental Sciences Division, Oak Ridge National Laboratory, Oak Ridge, Tennessee

I. R. STRAUGHAN, Research and Development, Southern California Edison Co., Rosemead, California

M. A. TABATABAI, Iowa State University, Ames, Iowa

M. UNSWORTH, University of Nottingham, Sutton Bonington, Loughborough, England

T. CRAIG WEIDENSAUL, Laboratory for Environmental Studies, Ohio State University, Ohio Agricultural Research and Development Center, Wooster, Ohio

INTRODUCTION TO THE SERIES

The deterioration of environmental quality, which began when mankind first congregated into villages, has existed as a serious problem since the industrial revolution. In the second half of the twentieth century, under the ever-increasing impacts of exponentially growing population and of industrializing society, environmental contamination of the air, water, soil, and food has become a threat to the continued existence of many plant and animal communities of various ecosystems and may ultimately threaten the very survival of the human race. Understandably, many scientific, industrial, and governmental communities have recently committed large resources of money and human power to the problems of environmental pollution and pollution abatement by effective control measures.

Advances in Environmental Sciences and Technology deals with creative reviews and critical assessments of all studies pertaining to the quality of the environment and to the technology of its conservation. The volumes published in the series are expected to service several objectives: (1) stimulate interdisciplinary cooperation and understanding among the environmental scientists; (2) provide the scientists with a periodic overview of environmental developments that are of general concern or that are of relevance to their own work or interests; (3) provide the graduate student with a critical assessment of past accomplishment which may help stimulate him or her toward the career opportunities in this vital area; and (4) provide the research manager and the legislative or administrative official with an assured awareness of newly developing research work on the critical pollutants, and with the background information important to their responsibility.

As the skills and techniques of many scientific disciplines are brought to bear on the fundamental and applied aspects of the environmental issues, there is a heightened need to draw together the numerous threads and to present a coherent picture of the various research endeavors. This need and the recent tremendous growth in the field of environmental studies have clearly made some editorial adjustments necessary. Apart from the changes in style and format, each future volume in the series will focus on one particular theme or timely topic, starting with Volume 12. The author(s) of each pertinent section will be expected to critically review the literature and the most important recent developments in the particular field; to critically evaluate new concepts, methods, and data; and to focus attention on important unresolved or controversial questions and on probable future trends. Monographs embodying the results of unusually extensive and well-rounded investigations will also be published in the series. The net result of the new editorial policy should be more integrative and comprehensive volumes on key environmental issues and pollutants. Indeed, the development of realistic standards of environmental quality for many pollutants often entails such a holistic treatment.

JEROME O. NRIAGU, Series Editor

PREFACE

The physical and chemical nature of the global atmosphere and its impact on our environment as a result of human activities is of major concern. One of the most important issues to arise out of this concern is the current and future potential effects of atmospheric pollutants on terrestrial and aquatic ecosystems. To address this complex issue, one must characterize and quantify pollutant emissions and determine the dispersion and deposition of these emissions that lead to observed and/or predicted ecosystem effects. Within the terrestrial ecosystem, specific receptors, such as plants, soil, or water, can react directly with atmospheric deposition of pollutants as well as indirectly through interactions with receptors. It is clear that, because of the diversity of sciences required to investigate the total (integrated atmosphere–biosphere) system, there is a need for interdisciplinary cooperation and communication. This integrated approach requires an understanding and an exchange of ideas between investigators in the physical sciences and in the biological sciences, at both the basic and applied levels.

Consequently, this book covers a wide range of subjects, from government and industry regulatory perspectives, atmospheric chemistry, meteorological processes and pollutant deposition, pollutant measurement technology, vegetation and soil effects of air pollutants, and data acquisition strategies, to total ecosystem research.

The preparation of this book has been an experience that will not be soon forgotten. Like all endeavors of this magnitude, a number of people made this publication possible. We would like to thank Mrs. K. Seto, Department of Communications Media, The University of Calgary, for the preparation of the excellent final graphics for the book. A very special note of thanks is extended to Linda Jones, Julie Lockhart, and Della Patton of the Kananaskis Centre. They tirelessly edited, typed, and retyped the review and synthesis papers until they were letter perfect.

A. H. LEGGE
S. V. KRUPA

Calgary, Alberta
St. Paul, Minnesota
February 1985

ACKNOWLEDGEMENTS

This unique International Conference would not have been possible without the financial sponsorship, which is gratefully acknowledged, of the following organizations:

Alberta Environment, Edmonton, Alberta, Canada;

Kananaskis Centre for Environmental Research, The University of Calgary, Calgary, Alberta, Canada;

Electric Power Research Institute, Palo Alto, California, U.S.A.;

Oil Sands Environmental Study Group, Calgary, Alberta, Canada;

Environmental Protection Agency, Research Triangle Park, North Carolina, U.S.A.;

The University of Calgary, Calgary, Alberta, Canada;

Northern States Power Company, Minneapolis, Minnesota, U.S.A.;

U.S. Department of Interior, Fish and Wildlife Service, Washington, D.C., U.S.A.;

Southern California Edison, Rosemead, California, U.S.A.;

Petroleum Association for the Conservation of the Canadian Environment, Toronto, Ontario, Canada.

CONTENTS

Air Pollutants and Their Effects on the Terrestrial Ecosystem

PART ONE

INDUSTRY VIEWPOINTS AND GOVERNMENTAL AIR QUALITY CRITERIA DEVELOPMENT: TERRESTRIAL IMPACTS BY AIR POLLUTANTS

Industrial abatement strategies relative to the loading of the atmosphere by air pollutants and the corresponding governmental air quality criteria have been changing continuously in recent years. Cost–benefit ratios as well as the problems of energy shortages and energy distribution play an extremely important role. Further, pollution control strategies are highly complex and, according to some, may be comparatively ineffective relative to sources such as the transportation (mobile sources) situation. Alternative and/or new energy technologies pose questions relative to every new pollutant and/or its product generation. These considerations have implications in agriculture and forestry.

This section addresses both mobile and stationary source–originated pollutants. This section is the major point of reference for all subsequent sections.

Air Quality Criteria Development

J. H. B. GARNER
Environmental Criteria and Assessment Office
U.S. Environmental Protection Agency Research Triangle Park, North Carolina

1. BACKGROUND

Air quality criteria are an up-to-date compilation of the scientific knowledge of the relationship between various concentrations of air pollutants and their adverse effects on man and his environment. Air quality criteria constitute the basis for ambient air quality standards. The presentation which follows explains the procedures and guidelines used by the U.S. Environmental Protection Agency (EPA) in preparing criteria documents. Since the terms *air quality criteria* and *air quality standards* will be used frequently in this presentation, definitions of these terms may be helpful. *Air quality criteria* are descriptive; that is, they describe the effects that have been observed when the concentration of a pollutant in the ambient air has reached or exceeded a specific figure for a specific period of time (Hueter et al., 1970; Stopinski and Boksleitner, 1973). They describe the latest scientific knowledge and serve as the scientific basis for the EPA administrator's regulatory decision as provided in Section 108 of the Clean Air Act.*

Air quality standards are prescriptive (Hueter et al., 1970; Stopinski and Boksleitner, 1973). They prescribe pollutant concentrations that cannot legally be exceeded during a specific span of time in a specific geographic area. National Ambient Air Quality Standards (NAAQS) are mandated by law (Section 109 of the Clean Air Act). They are based on technical reviews that compile and assess the published scientific evidence which identifies associations between exposures to airborne pollutants and the resulting effects on humans, as well as the effects on crops, materials, and natural ecosystems. In addition to utilizing the scientific criteria, ambient air standards are also based on "best judgments" and "adequate margins of safety" necessary to protect the segment of the human population that is at greatest risk. The scientific reviews upon which standards are based are issued by EPA's Office of Research and Development as air quality criteria documents.

The initial criteria documents were published in 1969–1970 before the EPA was established by the Department of Health, Education and Welfare. As part of the 1977 Clean Air Act Amendment, however, Congress mandated that the scientific evidence supporting ambient air quality standards be reviewed prior to the end of 1980 and updated every five years thereafter. Depending on the volume and import of the additional research evidence, this could mean anything from a brief amendment for an existing

*Clean Air Act (42 U.S.C. 1857 et seq.) includes the Clean Air Act of 1963 (P.L. 88-206), and amendments made by the Motor Vehicle Air Pollution Control Act—P.L. 89-272 (October 20, 1965), the Clean Air Act Amendments of 1966—P.L. 89-675 (October 15, 1966), The Air Quality Act of 1967—P.L. 90-148 (November 21, 1967), the Clean Air Amendments of 1970—P.L. 91-604 (December 31, 1970), the Comprehensive Health Manpower Training Act of 1971—P.L. 92-157 (November 18, 1971), and the Energy Supply and Environmental Coordination Act of 1974—P.L. 93-319 (June 22, 1974).

document, to an extensive rewrite and publication of a new criteria document.

It has now been about 10 years since the initial air quality criteria documents were published. Therefore, completely rewritten documents are being prepared for each of the existing criteria pollutants—namely, sulfur oxides, particulate matter, oxidants, oxides of nitrogen, and carbon monoxide. A document for lead was issued in late 1977, and revision was not required until the end of 1982.

The task of preparing the revised criteria documents is under the guidance of Environmental Criteria and Assessment Office (ECAO), situated in Research Triangle Park, North Carolina.

Before considering the details of document preparation, let us consider how the criteria document is used by the EPA in the standard setting process. The preparation of the criteria document is only the first of three separate steps in the possible regulation of an ambient air pollutant.

1. The first step is preparation of the air quality criteria document as specified by Section 108(c) of the Clean Air Act. This step, as mentioned before, is carried out by the EPA's Office of Research and Development (ORD). It includes preparing a scientific document and identifying the spectrum of air pollution effects caused by specific concentrations of the pollutant in the atmosphere. Based on the scientific information, the administrator decides whether to regulate and, if necessary, the extent of regulation.
2. Step two, as specified in Section 109 of the Clean Air Act, is to propose—if necessary—an ambient air quality standard based upon findings in the criteria document. Within the EPA, the Office of Air, Noise, and Radiation is responsible for proposing the standard. The object of the standard is to specify a concentration of the pollutant in the atmosphere which, if not exceeded, will prevent the effects to public health and welfare (the nonhuman environment) identified in the criteria document. The ambient exposure concentration associated with the effects, plus a margin of safety, comprise the core of the standard setting rationale. The proposed standard is announced for public review and comment, and the process culminates in the promulgation of a new or revised standard.
3. The third step is carried out at the time a standard is promulgated. It consists of issuing regulations and air quality management information which can be used by the states in preparation of their implementation plans for the attainment and maintenance of the standard. The air quality management information which is issued at this third step is treated as a separate subject from that of identifying the effects brought on by the pollutant. The Office of Air, Noise, and Radiation is again responsible for providing this information to the states for developing implementation plans.

2. THE OBJECTIVE OF THE CRITERIA DOCUMENT

As stated in Section 108 of the Clean Air Act, the purpose of a criteria document is to identify the *effects* on public health and welfare (nonhuman environment) brought about by exposure to ambient concentrations of a pollutant. The focus of the document is not the information on air quality management necessary for the attainment and maintenance of possible standards. Instead, the objective in preparing a criteria document is to develop a scientific document that focuses on effects occurring in the ambient environment which might, if detrimental, be prevented by an ambient air quality standard. Air quality management information, such as modeling, emission inventories, measurement methodologies, and atmospheric transport, are included in the document only to the extent necessary to put the air pollution effect in perspective with reality.

Air quality criteria should be objective statements of the available scientific evidence. They should be free from bias, from speculation stated as fact, and from economic considerations relative to the protection of public health. They should not make statements regarding concentrations of pollutants in a way that can be construed as recommending standards.

Criteria documents deal with substances that have been identified clearly as pervasive pollutants that pose threats to health and the environment. Hence, when the EPA begins preparing a criteria document, a sizeable body of literature on each "criteria pollutant" already exists. Therefore, the treatment of a pollutant in a criteria document is usually rather comprehensive. An air quality criteria document usually includes the following information:

1. An extensive review of the literature dealing with the health and welfare effects of the pollutant or class of pollutants in question—specifically, the effects on humans and our environment, including plants, animals, aquatic and terrestrial ecosystems, and materials both natural and manmade.
2. Condensed chapters dealing with the chemical and physical properties of the substance, methods of measurement and analysis, sources and emissions, geographic and temporal distribution of sources of exposures, ambient air concentrations, and the global cycle of the pollutant in all its forms.
3. A technical summary chapter.
4. Summary and conclusions of the document, which concentrate on interpreting the data cited therein.
5. Demonstrations wherever possible of the relationship of ambient air concentrations to effects—"no observed effects" or "least-observed effects." (Such relationships do *not* point out a concentration that should be adopted as a standard, but indicate a range of concentrations that is useful in deriving a standard.)

6. "Effects" chapters, in which the report attempts to concentrate on those studies that are most critical in the development of National Ambient Air Quality Standards (NAAQS).
7. Lists of *all* reference material used in the development of the document.
8. Information on the relationship between emission concentrations and ambient air concentrations, especially if such relationships have direct bearing on the explanation of effects sustained by receptors.
9. Risk assessment information (included wherever possible), especially for the at-risk segment of the population.

3. HUMAN HEALTH ASSESSMENTS

Human health assessments are usually considered the most important part of a criteria document. The main purpose of the health assessment is to answer the critical question: How much exposure does what to whom?

The most important component of the health assessment is the exposure analysis. Some of the specific topics relating human exposure to airborne pollutants are:

1. Delineation of external exposure parameters in terms of sources, media, routes, and levels of exposure.
2. Methodologic validity and reliability of exposure measurement techniques.
3. Pharmacokinetics of adsorption, distribution, retention, and excretion of pollutants.
4. Definition of chemical markers and other indices of internal exposure levels.

Of primary concern in the health assessment is movement of the pollutant through the environment—from sources through various mechanisms of physical transport and chemical transformation. This, in turn, results in exposure to various concentrations, depending on meteorological variables and an individual's activity pattern. As a key part of an air quality criteria document, respiratory exposure must be set into context with other possible exposures to the same or related pollutants via food, water, and skin contact. In some instances, such as with lead or cadmium, the criteria document cannot concentrate on air alone, but becomes multimedia when considering the routes of exposure.

The validity and reliability of the measurement techniques used to quantify exposure must be described. This involves questions of accuracy, precision, reproducibility, and interferences. The methodologies used both in measuring pollutants in the ambient air (e.g., routine monitoring and community health studies), as well as in measuring the controlled concentrations presented to human subjects or animals in clinical experiments, are discussed. Usually an entire chapter is devoted to assessing the comparative merits of the various measurement principles and instruments used to quantify the pollutant in question.

The physiological course of the pollutant and its metabolites through the body must also be described, and identifiable symptoms or indicators of incurred doses must be listed. Small amounts of a pollutant may appear to be successfully neutralized, excreted, or sequestered (e.g., in bone). In certain susceptible individuals, however, small increments or chronic low-level doses may stress organs, interfere with enzyme functions, or impair resistance to infection, resulting in symptoms quite different from the classical syndrome associated with acute exposures.

The health effects and risk assessment information should include both qualitative descriptions and quantitative data. The qualitative information is usually derived from human epidemiological studies and from animal toxicological studies. The animal studies, although usually yielding carefully quantified exposure and effects data, cannot be simplistically translated to the human context. These studies, nevertheless, can provide invaluable insight to susceptible tissues and organs and suggest mechanisms of pathology and possible recovery. A crucial question, and one that rarely evokes a consensus, is whether certain effects are merely adaptive responses within an organism's normal range of accommodation or are adverse effects that temporarily or permanently impair functions or vitality.

The quantitative information includes dose–effect relationships (usually derived from clinical studies with volunteer subjects), identification of the most sensitive groups in the population, and finally definition of the dose–response relationships for these sensitive groups.

The human health assessment is used in deriving a primary national air quality standard.

4. WELFARE EFFECTS

The effects of the ambient air concentrations of a particular pollutant on plant life (crops and natural vegetation), animals (domestic and wildlife), terrestrial and aquatic ecosystems, and materials (natural and manmade) are assessed under welfare effects. Social and aesthetic effects (e.g., visibility) are also discussed. As in the case of public health, the criteria document attempts to provide as much insight as possible regarding injurious effects to the nonhuman sector of the environment to provide the basis for a secondary standard if one is needed. In making an assessment of the effects of a pollutant on the nonhuman environment, questions similar to the following are asked:

TABLE 1. Lines of Inquiry

Ambient Air Measurements	Vegetational or Ecosystem	Air Quality Criteria
Measured concentrations	Identification of effects	Ambient air concentrations at which effects occur
Validity of measurements	Establishment of dose–response concentrations	Significance for species populations or ecosystems effects correlated with concentrations
Source identification	Identification of species, populations, or ecosystems at risk	

1. What is the potential for environmental damage?
2. What concentration of the pollutant contributes to the observed effects?
3. Can dose–response or dose–yield functions or relationships be expressed?
4. What species are being injured, and what regions are most affected?
5. What is the relationship between environmental damage and the pollutant concentration and emissions?
6. What are the dollar costs associated with the injury or damage? With crop loss or materials damage?
7. What courses do the pollutants or their transformation or degradation products follow through the ecosystem? Where do they tend to accumulate? What are the possible long-term consequences?

Criteria documents are essentially effects documents. As such, they are a synthesis of two major lines of inquiry (see Table 1).

5. THE DOCUMENTATION PROCESS

The preparation of a criteria document takes place in two major steps:

1. The preparation and review of a working EPA draft.
2. External review and revision leading to a final document.

The initial draft is usually prepared under the direction of the Environmental Criteria and Assessment Office through the assistance of a group of expert consultants. These consultants are usually individuals from academia or the private sector who have had extensive experience in the field about which they are writing. These consultants also assist in subsequent revisions and in evaluations of public comment.

The process of preparing the internal draft can be subdivided into three phases (see Table 2). *Phase* 1 consists of assigning a project manager for the document, selecting other EPA staff members for the document management team, and recruiting both EPA and other government agency experts for an internal review task force. In addition, it involves recruiting consultants from outside the government, developing an outline, planning the timetable for the document, and beginning literature searches.

Phase 2 consists of reading the literature and analyzing the pertinent references, writing first drafts of the chapters, and the informal review and the polishing of these drafts.

Once prepared, the preliminary review copy is circulated to the EPA task force and to selected outside technical reviewers, who will also be invited to the workshop (a working meeting of writers and reviewers) that begins *Phase* 3. The remainder of Phase 3 involves assessing and incorporating additions and revisions recommended at the workshop and production of the external review draft. This concludes the three phases of the internal step.

The subsequent external step can also be described in three phases. To emphasize that these phases build on the results of the first three phases, they are numbered 4, 5, and 6.

In *Phase* 4, beginning the external step of the process, a *Federal Register* notice is published announcing that the external review draft is available for public review and comment. This external draft is circulated to other federal and state government agencies, to EPA's Clean Air Science Advisory Committee (CASAC), and to the general public. A meeting of the CASAC, open to the public, is scheduled and held to review the external draft, and the CASAC submits its report evaluating the document's strengths and weaknesses.

In *Phase* 5 the evaluation provided by the CASAC plus the comments

TABLE 2. Criteria Document Process

	Phase	Activity	Time (days)
Internal step	1	Planning and initiation	30–40
	2	Preparation of working draft	60–200
	3	Review and revision of working draft	60–120
External step	4	Public/CASAC review of external draft	60–120
	5	Post CASAC meeting document revision	60–100
	6	CASAC closure on document status	60–90
			1–2 years

from the public sector and from other agencies are weighed by the Environmental Criteria and Assessment Office staff and the contributing authors and reviewers. The responses are documented and kept on file in a public docket.

Consideration of public comments requires significant time and very careful scrutiny in order to separate the scientific facts from opinion not supported by scientific evidence. The many suggestions received from outside the agency can lead to inclusion of new material in the document. Also, there is a legal requirement for a detailed summary (docket) of the EPA's responses to all comments to be available for review by anyone. The docket is maintained in the EPA's offices in both North Carolina and Washington, D.C. The end result of this phase is a revised draft, which is normally resubmitted to the Scientific Advisory Committee.

Phase 6 consists of a second review by the CASAC. If no problems exist, then final editing and printing of the document begin. Formal release of the document is coincident with the administrator's announcement concerning the need for a new or revised standard.

6. LIMITATIONS OF AIR QUALITY CRITERIA

There are several general areas of knowledge where information is usually inadequate; because of the nature of their shortcomings, these areas merit attention only to point out that our knowledge of them may never be complete.

1. Effects related to exposures to ambient air can never take into account the presence of all the pollutants which may be causing the net result. Some of the pollutants may be intermediate or unstable products of known substances, while the presence of others may be unsuspected and thus not even monitored.
2. Laboratory experiments using simulated air pollution cannot totally and simultaneously replicate the actual ambient air in composition, temperature, and humidity. This is both because of the constant fluctuations of these variables in the ambient air and because of the presence of other pollutants (some not routinely measured), which may be contributing to the effects observed.
3. Dose–response relationships between the ambient air and observed effects can only be estimated. The actual dosage is only an estimate or average of an exposure, which is often extremely variable over relatively short periods of time.
4. It is extremely difficult—if possible at all—to state minimum or threshold levels for a particular pollutant with reference to a particular effect. Long-term exposures permit too many variables to exert

an additional influence on the outcome of a particular observation. Short-term exposures are frequently difficult to measure and if the response is not a visible effect, the time and conditions under which it occurred may not be determined.

5. Most studies of effects are not directly comparable with each other because of nonparallel exposure times or conditions and because of variation in measurement technique and averaging times.
6. Air quality data often do not reflect the actual exposure of the subjects being studied.

7. NEEDED RESEARCH

The shortcomings of air quality criteria as listed above indicate the need for research in those areas where information is usually lacking or inadequate. Those planning research should keep in mind the ecological paradigm that the whole is more than the sum of its parts. This fact is common knowledge, but it is not usually practiced, because the reductionist approach of looking at one aspect of the whole is simpler. It is easier to study the effects of an air pollutant on a single tree than on the forest. It is also easier to conduct research in the laboratory, where the parameters can be more easily circumscribed than in the field, where variables—many of which may be unknown or uncontrollable—exist.

Research in the following areas will aid in filling the gaps in knowledge usually found in the air quality criteria:

1. Effects of air pollutants on vegetation and terrestrial ecosystems in which the exposures are related to conditions in the ambient air. Attempts are being made through the use of open-top field chambers, zonal air pollution systems (ZAPS), and other exposure systems to conduct research which fills this requirement.
2. Studies using standardized conditions so that the results of the different studies can be compared. The National Crop Loss Assessment Network is a step in this direction.
3. Research on dose–response and dose–yield functions to determine the concentrations of a pollutant at which vegetational injury or response is severe enough to cause suppression of yield.
4. Economic assessment methodologies that relate dollar costs to vegetational injury, damage, crop loss, or suppression of yield.
5. The synergistic effects of SO_2 and O_3, O_3 and CO_2, and SO_2 and NO_2.
6. Studies delineating the chronic effects of combinations of the above-listed pollutants on growth and yield.

7. Studies delineating the effects of acid precipitation on vegetation, soils, and terrestrial ecosystems.
8. Studies explaining the cellular and biochemical changes (plant mechanisms) involved in plant response to air pollutants.

The protection of human welfare (natural and agroecosystems), as well as public health, requires action based upon the best evidence of causation available.

The accuracy and precision of data concerning the effects of air pollution on health and welfare often leave something to be desired. Until better measurements are possible, our actions must be based upon the best knowledge we have and must be guided by the principle of enhancing the quality of the environment and, thus, of human life.

REFERENCES

Hueter, F. G., J. C. Romanovsky, and D. S. Barth. Considerations in developing air quality criteria. In *Proceedings of the 1st Annual Conference on Environmental Toxicology*. September, Fairborn, OH. Aerospace Medical Research Laboratory. Air Force Systems Command Report No. AMRL-TR-70-102. Available from NTIS, (National Technical Information Service), Springfield, VA; AD 727022 (1970). pp. 287–294.

Stopinski, O. W. and R. P. V. Boksleitner. Air Quality Criteria. In *Energy Needs and the Environment*. R. L. Seale and R. A. Sierka (eds.). University of Arizona Press, Tucson, AZ (1973), pp. 149–156.

Industry Overview of Air Emission Regulations

E. E. CUDBY
Chevron Standard Limited
Calgary, Alberta, Canada

1. INTRODUCTION

It is a distinct pleasure to be given the opportunity to present a review paper on industry's viewpoints on government regulatory policies, particularly as they apply to the Canadian and provincial scenes where I am most at home. I shall, therefore, discuss the topic primarily in provincial terms rather than in the much broader worldly sense. I trust the contents will encompass worldly concerns in spite of the provincial origins of my approach.

2. SUMMARY

There is no doubt that serious questions are posed when attempting to marry regulations to the environment in a scientifically sound union. From the sour gas plants of the province of Alberta to the nickel beneficiating plants of Ontario, the Canadian air pollution scene probably epitomizes the dilemma between a rational ambient air pollution emissions policy on the one extreme, and a repeatable, reputable monitoring system on the other. Somewhere in between, a scientifically unassailable biomonitoring system seems worthy of serious attention, to credibly tie these pollution control tools together.

(It should be noted here that the term sour gas applies to a natural gas containing sulfur in the form of H_2S in excess of an amount specified as the maximum allowable for home consumption. When such excess H_2S is removed, the treated gas, for domestic or industrial consumption, is then said to have been "sweetened.")

One can see the validity in the two extreme positions, but also note the apparent difficulty in perceiving, let alone reaching, an acceptable operating standard for an air-polluting industrial complex. Surely a blending of the best monitoring systems adapted to the site-specific issue offers us a reachable, worthy, and defensible goal.

At the extremes, however, the present monitoring systems are not without their problems. Ambient air monitoring instrumentation still does not provide the level of accuracy or repeatability required for the regulatory policing action being administered. Unfortunately, regulatory agencies continue to rely heavily on the perceived accuracy of records to judge a company in violation. This precipitates emotional and defensive reactions. At the established levels of readings beyond which violations begin, the accuracy range of the instrumentation (or tool) is often greater than the range of the reading in apparent violation.

We note the unsuccessful attempt of government to set laboratory standards for the analysis of monitoring records (SO_2 and H_2S cylinders, plates, strips, etc.), thus pointing out how truly quixotic regulations can be. The ability to compare historical monitoring analytical results, as between laboratories, is all but impossible and questionable at best.

The present monitoring systems' weaknesses are also observed at the judicial level where the public seeks redress of both apparent or real grievances in the courts, usually unsuccessfully, and industry is in the position of always looking over its shoulder to see who is "shadowing" it, and for what reason. In many ways we are all losers.

Some effort is necessary to establish a level at which each air pollutant in the ambient air represents a clear and dangerous threat to both the human population and ecosystems impacted. The level does not have to be final. It must, however, be reasonable and based on the scientific facts at hand. As additional scientific data become available these levels should be revised accordingly.

In eastern Canada, the ultimate sink for airborne sulfur emissions is predominately the Canadian (Precambrian) Shield, which essentially is incapable of neutralizing acid-forming compounds (oxides of both sulfur and nitrogen). As a result, this specific environment has been physically and biologically changed for the worse. Some very strict emission controls appear to be central to an overall solution to the acid rain problem. But beyond that, what?

In western Canada, by and large, the emission control level of a specified pollutant (SO_2 particularly) into the atmosphere is presently established at a value sufficiently safe to preclude vegetative or related ecological damage. Therefore, ambient air monitoring should perhaps be recognized as a surveillance, operating, and research tool providing long-term (chronic versus acute abuse) utility.

As a summation of the many unresolved air quality related issues that should be addressed, the following represent a number of dilemmas worthy of attention. There are dilemmas:

in the setting of regulations for which no authority appears to exist, or which are scientifically indefensible;

with industry and public input into the regulatory decision-making process;

in the administration of justice, where the regulations often exhibit no legal force, which makes the "scientific" basis for the regulations rather suspect;

about technical difficulties of applying "reasonable scientific measurement" in a court of law;

with reputable scientific data being available in the event that class action litigations are approved for our courts;

in the air pollution monitoring systems,

as to the validity of the instrumentation as well as its utilization and accuracy,

as to the interpretation of results;

as to the quality of surveillance systems in terms of acceptable emission levels that yield unacceptable ambient air-emission records;

in systems that provide for both antidegradation policies and the impetus for development of (good) control technology;

with varying threshold levels for pollutants which are based on scientific, medical, or socioeconomic considerations;

in setting maximum acceptable levels of pollution for protection, and qualifying these levels with regard to "desirable" and "tolerable" levels which are also "acceptable";

where an administrative permission (that is, approval to do something which is not necessarily in conflict with a regulation) can override regulation (that is not based on scientific evidence);

of government not wishing to prosecute violators of clean-air regulations because of the element of scientific contradiction;

of more effective systems to monitor ambient air pollutants, such as biomonitoring;

of the sensibilities of a half-hour ambient air-emission allowable level;

in the move towards environmental trade-offs in terms of socioeconomic and/or scientific soundness;

in linking biological monitoring with atmospheric dispersion modeling to develop a sensitive biological forecasting capability;

in designing large-scale air dispersion studies to trace regional pollution patterns under various ambient air conditions;

in relating various types of ecological sinks to ambient weather models;

in developing damage risk criteria;

in determining cost-effectiveness of developing equitable air emission guidelines from meaningful research;

in quantifying environmental protection;

in documenting criteria which relate to social acceptability and benefit;

of identifying source and routes of air pollutants, in economic terms, as a prerequisite to fate and effect studies;

of establishing probability of a bad event and the consequences;

of producing detailed research with the assurance that the regulations will be revised to reflect the results; and

of research finding the answers from an applications viewpoint, and recognizing that complexities and interrelationships of biological systems cannot be considered in isolation.

The current viewpoint of industry in the matter of airborne pollutants is that industry has attempted to adopt a holistic approach to any environmentally oriented concern. However, government agencies attempting to monitor industry have seemed to zero in on one pollutant concern in apparent isolation from everything else where a process understanding is essential, if a reasonable solution to an environmental problem is to be found.

For the most part the research community has also tackled environmentally oriented research problems in isolation from other aspects of the system. If we are to make scientific sense of our research results, we must tackle the solutions to air pollutant emission problems on a holistic basis, as we do many engineering problems. This concept recognizes the continuing need for single-element research. However, such environmentally oriented research should always be in the context of the ecosystem within which the singular study exists. One such study that exemplifies this approach is the one undertaken by the Whitecourt Environmental Study Group (WESG) (Legge et al., 1977, 1978, 1981; Legge, 1980). I discuss this study briefly later. In terms of a systems study, it is unique and is worthy of detailed review and assessment by those scientists who would relish a fresh approach, where the research laboratory is in the boreal forest to record the impact of industrial SO_2 on the ecosystem.

The following paper attempts to put in context the dilemmas, as summarized above, for useful discussion at this conference.

3. CONTENT BACKGROUND AND ISSUES

3.1 Federal/Provincial (Alberta) Air Emission Regulations

In 1973, following a substantial increase in the number of sour gas plants being built, Alberta made, persuant to Section 10 of the Clean Air Act of the Province of Alberta (S.A. 1971, c. 16), some new clean air regulations. However, even these were promulgated before the recommendations of a public hearing on sulfur emissions into the atmosphere were completed. (Refer to Table 1 for details related specifically to the matter of air emissions.)

Subsequent to these new provincial regulations, a subcommittee of provincial and federal representatives, under the sponsorship of the Canadian Council of Resource and Environmental Ministers (CCREM), recommended model environmental conservation regulations. After much deliberation, the federal government promulgated National Ambient Air Quality Objective No. 1 in 1974, followed by additional Objectives 2 and 3, adding tolerable levels to the existing desirable and acceptable levels, plus standards for NO_x. These are shown in Table 2.

An Environment Canada paper (Swan, 1976) outlined the Department's concept of air quality objectives. I would like to quote and comment on several interesting aspects as follows:

(*i*) *The Federal Clean Air Act allows the setting of emission standards only if the pollutant in question could constitute a significant danger to health.*

TABLE 1. Section 10 of the Clean Air Act of the Province of Alberta

10. *Regulations*

The Lieutenant Governor in Council may make regulations

(a) governing applications for and issue of
 (i) permits and amendments thereto, and
 (ii) licences and amendments thereto, and prescribing the form for any application and the form of the permit and licence;

(a1) prescribing terms and conditions attached to all permits and all licences or any class of either;

(a2) prescribing different types of permits and licences, the length of time for which they are issued and permitting the Director of Standards and Approvals to issue permits and licences for a shorter period of time than prescribed in the regulations;

(b) exempting any plant, structure or thing or class of plants, structures or thing from the operation of Section 4, 4.1 and 4.2;

(c) specifying a plant, structure or thing or a type of class of plants, structures or things that are subject to Section 4, 4.1 and 4.2;

(d) prescribing the procedure relating to any proceedings under Section 7 and of inquiries held under that section;

(e) providing for any procedure relating to any proceedings under this Act or the regulations;

(f) authorizing the payment of compensation by the Crown to any person for loss or damage to that person as a result of the application to him of any provision of this Act or the regulations or as a result of an order directed to him under this Act, prescribing the cases in which the compensation shall be paid and the loss or damage for which the compensation shall be paid and the loss or damage for which the compensation is to be paid, and conferring jurisdiction on the Supreme Court of Alberta, a district court or the Public Utilities Board in connection with settlement of the compensation to be paid;

(g) authorizing the Minister to expropriate on behalf of the Crown any estate or interest in land if he considers it necessary to do so for the purpose of enforcing or carrying out the provisions of this Act or the regulations;

(h) requiring the submission to the Director of Pollution Control or some other person of any returns or reports by any person pertaining to the construction or operation of any plant, structure or thing or any class thereof that is or may be the source of an air contaminant;

(i) prescribing a tariff of fees payable to the Provincial Treasurer
 (i) pertaining to applications under Section 4, 4.1, 4.2, 4.4 and 4.5,

TABLE 1 (*Continued*)

(ii) for the filing of any returns, reports or other documents that are required or permitted to be filed under this Act or the regulations, and

(iii) for any other service provided by the Department under this Act or the regulations;

(j) referring to, incorporating or adopting, in whole, in part or with modifications, documents that set out standards of or relate to air quality, the prevention or control of air pollution or the design, construction, maintenance or operations of a plant, structure or thing that may be a source of air pollution; (1976, c. 65, s. 1(8));

(k) for the control, restriction or prohibition of the manufacture, sale or use of any equipment, device or service designed or provided for any purpose related to the control or elimination of air pollution;

(l) providing, with respect to any provision of the regulations, that its contravention constitutes an offence;

(m) providing, in respect of an offence provided for pursuant to Clause (l), for penalties by way of fine or imprisonment or both for which the offender is liable on summary conviction therefore;

(n) generally, for the prevention, control or prohibition of air pollution and the regulation of sources of air pollution;

(o) setting the date required for Section 11.1;

(p) concerning the submission of monitoring reports, establishing forms and the times at which reports must be submitted to the Director of Pollution Control;

(q) to permit the preparation and publication by the Department of the Environment of guidelines for the construction of any plant, structure or thing;

(r) prohibiting the release of toxic air contaminants to the atmosphere unless the approval of the Director of Standards and Approvals has been first obtained;

(s) prohibiting or regulating the removal or rendering ineffective any device or thing that reduces or prevents or is intended to reduce or prevent the emission of any air contaminant attached or connected to or forming part of any plant, structure, motor vehicle or thing. (1972, c. 19; 1974, c. 15, s. 4)

TABLE 2. Clean Air Regulations ($\mu g/m^3$ unless otherwise stated)

	Average Concentration					Arithmetic Mean
	½ hour	1 hour	8 hours	24 hours	30 days	Annual
Alberta[a]						
SO_2	525 (0.20)[b]	450 (0.17)[b]		150 (0.06)[b]		30 (0.010)[b]
H_2S	17 (0.012)[b]	14 (0.010)[b]		4 (0.003)[b]		
NO_2		400		200		60 (0.032)[b]
CO (mg/m^3)		15	6			
Oxidants (Ozone)		100		30		
Particulates (mg/m^3)				100		60
Dustfall ($mg/100\ cm^2$)					53[c]/158[d]	
Canada[e]						
SO_2 #1 Desirable		0–450		0–150		0–30
#2 Acceptable		450–900		150–300		30–60
#3 Tolerable				300–800		

H_2S				
NO_2 #1 Desirable				0–60
#2 Acceptable	0–400 (0.213)[b]		0–200 (0.106)[b]	0–100 (0.053)[b]
#3 Tolerable	400–1000		200–300	
CO #1 Desirable	0–15	0–6		
#2 Acceptable	15–35	6–15		
#3 Tolerable		15–20		
Oxidants #1 Desirable	0–100		0–30	
#2 Acceptable	100–160		30–50	0–30
#3 Tolerable	160–300			
Particulates (mg/m^3) #1 Desirable				0–60
#2 Acceptable			0–120	60–70
#3 Tolerable			120–400	
Dustfall ($mg/100\ cm^2$)				

[a] Alberta Clean Air Act (1975).
[b] Parts per million (ppm).
[c] Residential and Recreation.
[d] Industrial.
[e] Canada Clean Air Act (1978a).
Canada Clean Air Act (1978b).
Canada Clean Air Act (1978c).
Canada Clean Air Act (1978d).

One might ask: Who determines this, and on what basis does a pollutant constitute a significant danger to health?. There is no answer to date.

(*ii*) *The Canadian National Ambient Air Quality Objectives* (*NAAQO*) *have no legal force.*

If so then, "why does the Federal Clean Air Act call for their (the NAAQO) establishment?" The government answers: "It is impossible to effectively plan, organize or control any endeavor if you do not know what your objective is."!

Alberta Environment (AE) suggests,

. . . this obvious fact tends to be overlooked in the protracted arguments of those favoring an "ambient standard" approach to control of specific sources, and those (including the Federal Government of Canada) favoring a performance or emission standard approach to control the same sources.

Swan states,

National Air Quality Objectives provide the basic management focus to the programs of responsible control agencies that lead to overall control strategies and tactics based upon a prior awareness of the effectiveness of such strategies and tactics whether these involve land use planning, control of mobile sources, limitation of sulphur contents of fuels, or, the imposition of emission standards on particular industries.

In the agency's own words, here is the key to how the regulations should be applied.

In assessing options, the managing agency must have as its first concern the total effect on ambient air in a region that will result from a proposed action.

This approach is commendable; however, there is still no reference base for backup in decision making that includes appropriate scientific criteria, socioeconomic impact, or the state of existing technology.

The Canadian System of NAAQO notes the following requirements if they are to be used as a management tool. I quote:

The system must be scientifically defensible.

The system should be readily understood and accepted by the general public.

The system should accommodate the varying control activities involving either or both provincial and federal agencies, and

it should be linked to the Federal/Provincial National Air Pollution Surveillance System,

it should assist in the setting of priorities,

it should assist in the determination of the effectiveness of programs,

it should accommodate "episode" conditions and the development of episode control indices.

The system should provide a basis for an anti-degradation policy and provide impetus for the continuing development of control technology.

None of these requirements can be considered unique. What may be unique, however, is combining these needs in a single system related to air quality objectives. In the government's view, the key element in the combining process was the recognition that there are, in fact, a number of threshold levels for each pollutant (or combination).

These varying threshold levels are dependent upon not only scientific or medical concerns but also on the socioeconomic concerns of the country and the administrative concerns of control agencies. The concept is very desirable if understood clearly.

However, if the "Maximum Acceptable Level is intended to provide adequate protection against effects on soil, water, vegetation, materials, animals, visibility, personal comfort and well being," and if, "when this level is exceeded, control action by a regulatory agency is indicated," then what is the significance of the "desirable" and "tolerable" levels—in this context?

It is in this dilemma that the agencies (provincial and federal) monitor industry, and it is industry which finds itself without proper understanding or direction. What is needed is an Air Pollution Hazard document that includes:

1. Air pollutant analysis methodology that is acceptable and desirable for comparison studies; and
2. Air pollutant toxicity levels for humans, as well as many wildlife or vegetative species (presently there is much information which can be added to or revised as available).

Such a document would form the basis of any Air Pollution Standards document that would include the appropriate standards criteria as a backup for any political or ecological jurisdiction.

It is also in this framework that the "Canada Alberta Accord for the Protection and Enhancement of Environmental Quality 1975*" was written (see Table 3 and Table 4, accord objectives). What is essential to achieving these accords and objectives is a clear strategy statement that includes reference to input by industry (the regulated part) or the public (the subjective pollutant sink). Industry has often initiated joint discussions concerning regulatory control with the express purpose of reaching the best practical regulations. All things considered, this process is proving to be increasingly successful.

*This is a written agreement between the Minister for Environment Canada and the Minister for Alberta Environment. The accord is renewed every five years.

TABLE 3. Alberta Accord for the Protection and Enhancement of Environmental Quality

---

WHEREAS management of the quality of the natural environment involves maintaining or enhancing the ability of the biosphere to produce a wide variety of resources and conditions useful to man; and

WHEREAS an understanding of the biophysical relationships of ecosystems is fundamental to successful attainment of environmental quality objectives; and

WHEREAS institutional systems established to govern man's activities including his impacts on the natural environment, are superimposed upon natural systems; and

WHEREAS both Canada and the provinces have jurisdictions and responsibilities in the field of environmental quality, including pollution prevention, control and abatement;

THEREFORE, the Governments of Canada and Alberta,

RECOGNIZING that programs aimed at achieving environmental objectives should be planned and undertaken in such a way as to ensure comprehensiveness and eliminate duplication,

AGREE to adhere to the principles and practices hereinafter stated in the development and maintenance of complementary programs with each government acting within its constitutional jurisdiction; and

AGREE to develop new coordinating mechanisms and new complementary programs so that they are in harmony with existing cooperative or complementary arrangements in related fields flowing either from legislation or administrative practice; and

AGREE to (certain stipulated) principles and practices relating to the protection and enhancement of environmental quality.

TABLE 4. Objectives of the Canada Alberta Accord for the Protection and Enhancement of Environmental Quality

---

(a) to provide a more effective overall effort on the solution of pollution problems through better coordination of the activities of Canada and the Province; and

(b) to provide a broad framework within which specific agreements can be designed to cope with particular problems.

3.2 Some Legal Aspects

3.2.1 Federal Aspects

"Class action" legislation as a new public force is highlighted in a *Financial Times* item (Whittington, 1976).

> *Since the 1960's, the public confidence in the quality of air impacted by industry has been considerably eroded. This trend, coupled with a recognition that it is possible to fight city hall and sometimes win, has produced increasing pressure on governments to provide the public, who feel they have been unfairly treated by business, with a means of legal redress.*
>
> *Of the legal tactics available to the public, the class action suit has attracted the most attention, particularly in the past year, particularly as the issues affect consumers.*
>
> *It enables a group of consumers to bring a suit for damages against a manufacturer on behalf of other consumers, solely on the basis of their shared interest as purchasers of a product. Though such suits have been in use in the United States for more than two decades, the courts in Canada have been more restrictive about the circumstances under which such suits could be mounted.*
>
> *It is now only a matter of time before class action litigation is allowed throughout Canada. If not through the courts, the legal basis for its use will probably come through legislation.*

Although the present direction is mostly consumer oriented, cases for "quality of air" negative-impact class action suits are also noted, and the following provincial incident emphasizes this.

3.2.2 Provincial Aspects

The remarkable state of relevant legislation at the provincial level, particularly as it existed in Alberta in 1977, is best exemplified in a brief (Elder, 1977) quoted essentially verbatim as follows:

> *An activist environmental protection group took both the Alberta Minister of Environment and the Attorney General to court over a case of nonprosecution of an industry violation of the Clean Air Regulations. When the Environment Minister ignored the activist group's request for punitive action, they tackled the Attorney General.*
>
> *Here was a government whose one minister (Environment) had stated publicly that the Company "on trial" appeared to have breached the law on several occasions, yet another (Attorney General) was solemnly asking a public interest group to give him a* prima facie *case of the breach.*
>
> *A word of explanation is in order here. The Clean Air Act (Alberta Statute* 1971, c. 16, *as amended) does not create any general offence of polluting the air. Instead, it empowers the Minister of the Environment to make various air*

pollution regulations proscribing maximum concentrations of air contaminants both in the ambient air and in an exhaust stream (5.3(1) *as amended by Alta. Stat.* 1974, *c.* 15). *The main regulation made for this purpose does several things. It creates ambient standards (micrograms per cubic metre/period time) for several contaminants (mainly sulphur dioxide, hydrogen sulphide, nitrogen dioxide, carbon monoxide) without creating any offences for their breach. It also creates particulate emission standards (pounds per* 1000 *pounds of effluent) from all but specifically excepted stationary sources of air pollution. Thirdly, it creates density standards for visible emissions on a* 5 *point scale where* 100% *blackness is density no.* 5.

Also, under the Clean Air Act, *no one may construct or operate a "plant, structure or thing" of the types specified either in the Act or Regulations, without a "permit or licence", permission for which is by application.*

After weighing the application for such permission, the Director of Standards and Approvals may issue a permit or licence either with or without conditions, and the conditions must be at least as stringent as those contained in the regulations (Alta. Stat. 1976, *c.* 65 *S.*5*(a)). No* SO_2 *emission regulations exist, but the Company operating licence created (inter alia) a maximum concentration of* SO_2 *in the flue gases of* 16,000 *parts per million by volume at a half hour average level. The Company was charged with exceeding this limit under S.*4.1(8)*(b) of the Act, which makes it an offence to contravene a term in a licence.*

Although the provincial court judge acquitted the company on both charges, the SO_2 emissions side of the issue is worthy of some comment for some very important reasons.

The argument surrounded the question that since the emissions control level as a condition of the plant operating licence was exceeded, the company should incur some punitive action. What was not then (nor not now) taken into account was the cause and effect relationship. That is, was the cause serious and/or negligent and was the effect damaging?

The regulations provide for maximum air levels of pollution but not for pollutant emissions. To meet these regulations with some additional degree of assurance, considering the physical character of atmospheric dispersion as an uncontrollable variable, the government Standards and Approvals Division also put some operational constraints into operating licences by limiting emissions.

The intent, of course, should be to provide a monitoring balance between emissions and sink as a control "team." That is, if emissions exceed limits, but ambient air levels are not exceeded, then no problem exists in the short term. Conversely, if emissions are within requirements, but ambient levels are exceeded, there is also no problem in the short term—at nominal levels. Unfortunately, that is not the case.

The foregoing case, however, emphasizes that the objectives of regulatory control are lost in both the regulatory and public arenas, as other cases addressed herein will confirm. If class action cases are to be a future way of life, there must be some sound bases on which a case can be laid; otherwise the process will remain a mockery. How do we resolve this dilemma?

3.3 Industry Experience

Sometime between 1971 and 1972, eight companies which operated 12 sour gas plants in a 4300 square mile triangle in northwest Alberta, and by permit were allowed to emit some 900 tons/day of SO_2 into the atmosphere, got together to determine the answers to two questions:

1. Is the cumulative effect of 12 concurrent sour gas plant operations cause for any alarm over the life of the projects?
2. Is biomonitoring of an ecosystem a responsible and effective alternative to ambient air monitoring techniques?

Out of the initial meeting, the Whitecourt Environmental Study Group (WESG) was born. The intervening years have witnessed a comprehensive program of research and evaluation, which has recently culminated in a series of conclusions and a film. These have stimulated some rethinking with regard to the whole matter of air pollution transport and deposition.

Perhaps this conference is the true extension of the aforementioned success of a new approach to research in this complex area of concern. Success, however, only existed at the research endorsement level, not in any structural change at the regulatory level—not yet, that is.

Although the initial objective of the WESG was to determine the cumulative effects of the specified "concentrated" sour gas operations, it became more than that. It was soon realized that all member companies were in agreement that there was a better way to monitor the airborne pollutants than by the methods then, and still, employed. SO_2 and H_2S continuous-monitoring trailers, cylinders, as well as sulfur dust collectors, were strategically located around each facility in varying numbers to record the level of emissions at the site-specific receptor. In spite of all the research, these methods are still a condition and requirement of any sour gas operating permit to a greater or lesser degree. I shall come back to this matter later.

Concurrent with the study group deliberations at that time, the Alberta Government Air Pollution Control Section attempted to set standards for laboratory analysis of SO_2 and H_2S cylinders. The substantial difference that prevailed between government laboratory control sample analysis and commercial laboratories' analyses caused the government to abort its efforts. The inconsistencies were so great that the government did not know who was right, themselves included. This situation highlighted another as yet unresolved problem: the need for comparable laboratory analytical standards.

Unfortunately, there were then, and still are, companies which are not concerned about laboratory analysis quality controls or accuracies, and a number of laboratories have treated their SO_2 and H_2S cylinder analyses with equal disdain. The consequence is that a government regulatory control has been turned into a requirement without substance, a cost without value, a historical event without utility. There remain the questions of analytical integrity, competence, and chemical standardizations to be effectively answered.

A better way to monitor the environment, the WESG members felt, would be to largely replace mechanical and electronic instrumentation with biomonitoring techniques, based on research and surveillance related to the study area, and leave mechanical and related monitoring tools largely as an operating control role (e.g., accidental occurrences identification).

As an unrelated followup to this, a subcommittee of the Canadian Petroleum Association (CPA) was formed in the late 1970s, the mandate of which was to assess the validity of existing monitoring techniques and to lessen the use of costly monitoring devices that produce mounds of records that do nothing to improve our understanding when they are not effectively utilized.

By the end of 1979, two Alberta provincial agencies, Energy Resources Conservation Board (ERCB) and Alberta Environment (AE), jointly issued a new Air Monitoring Directive (IL 79-22), which appeared to recognize to some degree the problems with the argument that "more is better." However, the determination of need for air monitoring remains elusive, as the following observations confirm.

With respect to static exposure station monitoring networks, the new directive allows for companies having five years of data to apply for network reductions and station relocations, provided that abnormal occurrences, such as wild wells or pipeline breaks that may have caused short-term, high-exposure levels, are taken into consideration in the network modification application. The criteria for allowing these network reductions is based on whether the past records meet established exceedence allowables, provided that such reductions in number do not exceed 50% of original. In no case should the minimum number be less than 4. No factor value is given to either the level of exceedence, duration, season, location, or socioeconomic consequences. This remains a contentious issue with industry.

Similar reasoning was directed toward reduction in the number of continuous ambient air monitoring trailers required for each plant. There is no scientifically sound basis for the revisions, making the merit of the revised Air Monitoring Directives somewhat questionable.

Perhaps the only real gain in understanding the air pollution sink problem was the directive dealing with sulfur dust. The practice of deploying sulfur dust canisters was replaced by soil pH analysis programs to determine better the level and extent of any sulfur contamination problem. Even here, pending the availability of more sophisticated data-results-oriented systems, sulfur dust canisters can still serve as a useful operating tool, when properly deployed and evaluated.

A number of other examples testify to the dilemma that falls on industry with regard to meaningful air emission and monitoring regulations and guidelines, particularly in Alberta.

EXAMPLE 1

Several years ago the CPA, on behalf of the petroleum and natural gas industry in Alberta, responded to the request for input into public hearings on the subject of "SO_2 Emissions into the Alberta Environment" (ECA, 1972). The resulting CPA study incorporated the observations of highly regarded international scientists experienced in atmospheric diffusion. This study was designed to prove that the regulations on "acceptable ground level concentrations" for specified airborne contaminants were already excessively penalizing (in their arbitrariness) and that there was no evidence to suggest that the regulations should be made even more severe.

However, before the results of the public hearings were released in print, Alberta Environment promulgated by Order in Council new standards for allowable ground level concentrations of certain pollutants—at levels approximately half of what they were. The reasons cited by Alberta Environment, although not based on scientific fact, were that the new concentration levels were considered safe. In addition, whereas other regulatory jurisdictions have established as a minimum acceptable limit either one- or three-hour acceptable average concentrations, Alberta established half-hour average allowable limits. The belief, of course, is that, with a half-hour maximum ambient air pollution level within which industry must operate on a continuous basis, no environmental damage will occur.

It should also be noted that at the plant operations center, operators must also maintain critical surveillance to ensure that the sour gas facility is operated within narrowly defined operating limits as well as severe pollution emission standards. Although the revised regulations have no scientific validity and sophisticated monitoring directives are applicable in only a very strict sense, the regulations remain politically expedient, although often legally obtuse and operationally invalidated on a site-specific basis.

EXAMPLE 2

Some years ago, an Alberta ERCB Board of Inquiry delved into the records of a gas plant operation experiencing a high number (some 500) of H_2S violations occurring in a deep V-cut valley adjacent to the sour gas plant, which was perched on a high escarpment some 600 feet above. Evidence indicated that the sulfur slating facilities in the valley were to blame, and a year was allowed to resolve the problem.

Concurrent with the research effort to model the air dispersion characteristics of the valley, a pilot project to degasify the sulfur of its live H_2S content to a level at which the H_2S emission levels (at various points of emission in the sulfur slating process) would no longer exceed the maximum half-hour allowable ground level concentrations, was also undertaken. In the modeling process, H_2S isopleths (ppm) indicated that the whole valley would exceed the provincial half-hour maximum H_2S allowable *ground level concentrations* (GLCs) under specifically identified atmospheric conditions. Therefore, even under the best operating conditions, exceeding limiting values could not always be avoided. The concurrent WESG research, however, suggested that at the levels of H_2S GLCs recorded there should be no environmental damage whatever.

The present state of regulatory control and agency policy is to meet the air quality standards, thereby safeguarding the environment. Unfortunately, this is an end in itself rather than a means of evaluating the record of environmental damage, if any. Fortunately, the provincial government agencies, in administering the regulations, rarely take punitive action against those operators who only seldom exceed the air quality standards. Considering the potential area for ongoing conflict, industry will continue to be uncomfortable as long as the environmental impact record is not properly taken into account as an operating tool.

Two observations serve to frustrate the environmental management process:

1. Governing agencies tend to let political solutions to environmental problems prevail.
2. Government directives do not always address the environmental problems scientifically and, therefore, often result in questionable requirements.

We are continually confronted with the comment that adequate solutions are neither easy nor simple.

What is missed in all this is that each sour gas plant's historical record indicates the level of pollutant impact under all synoptic conditions. Therefore, a real-time dispersion model could not only indicate what operating levels will produce what levels of ground level pollution concentrations under varying climate states, but also where the pollution is likely to occur.

Once an atmospheric dispersion model has been developed for a facility, all conditions of operation could then be "pollution predictable" and physical monitoring techniques restricted to database information gathering, applicable to further model validation, and refinement. Exactness is not the immediate goal; utility is.

Ecosystem-based research also notes that effective biomonitoring programs are able to utilize on-site environmental indicators to show whether significant damage could or would occur, under anticipated operating conditions.

In spite of any combination of irregular terrain and variable climatic conditions, environmental damage would not necessarily accrue from any H_2S emissions at the levels described, given the historical record and assurances of good operating practices for the future. Therefore, would it not be more responsible to monitor the operating record, and, if such attention is warranted at the operating level, make the necessary corrections? The issue of violations on a recurring basis at nondestructive levels should be put to bed once and for all.

What is the message that I should attempt to get across? One that would be a clear statement for positive and immediate action by industry and government in Canada, relative to the problem of atmospheric pollutant emission? This is a dilemma that ranges from overkill to evasion, depending on which side of the enigma one happens to be on.

On the one hand, since government does not know, in the scientific sense, where to set acceptable limits for emissions from pollution generators, it establishes maximum levels low enough to secure a public oriented, noncontroversial position at best. On the other hand, industry does not know where the acceptable limits should be either. It does know, however, that many of the standards, and the basis for them, are essentially wrong.

In the United States, one particular corporation recently adopted a three-pronged approach to this dilemma to protect both its interest and integrity. The key words are:

1. COMPLIANCE
2. VARIANCE
3. "DEFIANCE"

(I use the word defiance in the poetic sense here. In truth, however, the action really denotes a struggle of conviction.)

The first step is to make every effort to comply with the regulations. If they prove onerous and unrealistic, a second course of action is then indicated. A case is then made for seeking relief from the regulations—having due regard for the specific circumstances—by injecting a cost–benefit analysis for establishing a rational basis for decision making. If this does not prove acceptable, and the company feels strongly that the original regulation (or strict interpretation thereof) is logically and scientifically indefensible, it will take the offending agency to court over the issue.

If one examines these three alternatives, it is clear that it would be far better if it were not necessary to initiate legal action, which, regardless of the case decision, nobody really wins.

When there is a breakdown between the important question of the law or the environment, inevitably someone is going to get hurt, sometimes with long-term drawbacks which are too damaging to recoup. Fortunately, to date in Alberta, we have been able to negotiate a mutually acceptable compromise in the best interests of both industry and government on specific issues. Unfortunately, this takes time, effort, and expense that could be avoided.

3.4 Industry/Government Observations

To put the responsibility where it should be (that is, placing the burden of proof on governments in establishing the controls they feel are essential, while leaving industry to show that its operations are not damaging to the environment) is, of course, only the tip of the pollutant emission iceberg. Further, it is not sufficient simply to suggest that a more disciplined approach to the regulatory-making process is the answer, using cost–benefit and effectiveness techniques. There are other considerations as well.

As an extension of the previously mentioned defiance stage of a regulatory ruling, there is a concerted effort by the industry toward deregulation (particularly in the United States where industry is developing the power to achieve its objectives). Industry is beginning to reflect a serious concern for the overregulation syndrome by becoming directly and knowledgeably involved in the research process necessary to assure an acceptable basis for sensible change.

To write documents which justify the basis for tradeoffs and offsets in setting a specific emission limit, it will be necessary to understand the difference between, and application of, related basic and applied research. In a recent article, Robinson (1979) stated that "... a highly desirable regulatory philosophy should emerge, one that is aimed at maximizing social good through a mixture of health, economic, aesthetic and resource conservation elements." He suggests that in Canada, where the degree of public participation in the governmental decision-making process tends to be less than in the United States, it is of special significance when we consider public input commenting on tradeoffs and exposure of the regulatory decision-making process to public view.

Moving toward a more scientific method of sulfur pollution analysis is to be applauded; but the holistic approach to monitoring systems objectives has not yet been properly addressed.

First, it is suggested that the present method of continuous monitoring of the ambient air produces a staggering amount of information that is useless unless the repository agency evaluates its significance.

Second, it is clear that there is the danger that, as we try to marry regulations and the environment, a hybrid, combining the informational inadequacies of both, could result.

Third, there is the danger, from an applications viewpoint, that scientific research will never find the answers because of the inherent complexities and interrelationships of biological and environmental processes.

I referred earlier to the research of the WESG, to which I would now like to return. Results of some studies showed rather conclusively that SO_2 uptake by pine trees in the boreal forest resulted in reduced basal area increment, or decreased wood production. The suggestion might be that the impact is all negative (in the forest management sense), representing a loss of timber wealth. I wonder, however, if something positive may be happening!

Perhaps a better grain and harder wood is being developed, or even a better looking, hardier, healthier tree as one looks at the change from the tree's usual expected characteristics.

Let us now suggest a solution-oriented scenario relevant to making a decision on where we could go from here. We should review the data, establishing actual relationships between recorded ground level concentrations of pollutants, with the emission volumes and weather, to see what the cumulative data tell us. They may well indicate that for each site-specific case, one can predict with considerable accuracy expected ground level concentrations under any given set of atmospheric conditions. If we use this predictive capability, we would be in a position to revise or modify operating procedures accordingly. This should eliminate the need for an ambient air monitoring capability designed to satisfy, at best, questionable regulations. Of course, technical control via feedback is workable only if industry will agree to apply the techniques.

Before I conclude, allow me to add the following observations made by a federal minister of the environment in his speech to a Canadian government affairs seminar (Hon. J. Fraser, 1979) held in Ottawa, Canada.

The Hon. John Fraser addressed the problem of the long range transport of airborne pollutants (LRTAP) and achieving a balance between good science and appropriate regulation. In prioritizing the national concern over the effects of acid rain, he noted that information is less well-known for land than it is for water ecosystems and that research investigations should relate to:

1. the source of air pollutant emissions;
2. how prevailing winds transport the pollutants;
3. what chemical reactions take place during airborne movement of pollutants;
4. monitoring the deposition of airborne pollutants; and
5. the impact assessment of airborne pollutants
   a. in sensitive areas on land or water, and
   b. from long-term exposure to low level concentrations.

He went on to say that

> *. . . the international and interprovincial dimension of the "acid rain" problem demands action. This has led to the formation of a Canada/United States research group to coordinate research and develop a scientific data base as the foundation for development of control strategies. It reflects the shared judgments of scientists in both countries.*
>
> *As a followup, a policy for scientific investigation and cooperation is being drafted where a scientific appraisal as prerequisite to a sound policy is presently under way.*

The minister also stated that the

> *... development and implementation of control strategies to alleviate [airborne pollution] damage cut right to the heart of many vital issues in North-America today—energy, inflation and industrial development.*

It is interesting to note that the minister recognized that the problem could not be solved unilaterally when he clearly stated in his conclusion that

> *... it cannot be left to private enterprise to solve this problem alone. The benefits of cleanup will accrue to society at large and so eventually will the costs; these costs must, therefore, be understood and accepted by society if an effective program is to be implemented.*

4. CONCLUSIONS

This brings me to the area where research can be effectively utilized to answer some nagging questions. I see two levels of research need—applied and pure—against which criteria can be designed to reflect the protective levels required under various environmental conditions and operating practices. The place for pure and applied research should then stand out more clearly. Consider:

1. Macro dispersion studies should be designed to show where airborne pollution will go under ambient atmospheric conditions, using regional point sources and regional climatic conditions (air drainage basins).
2. Relate the various types of environmental sinks to the ambient climatic models to determine the significance of the potential hazard, given the pollutant.
3. Analyze the information and develop damage risk criteria to any given operation on any given day:
   a. stack SO_2, NO_x emissions,
   b. sulfur loading and handling, including the type of bulk carriers acceptable for long transportation hauls, as well as storage and transfer.
4. If relevant research designed to provide the criteria were an objective, the results would permit solution of the practical aspects of the problem exhibited in items 1 to 3.
5. Long-term monitoring of the results of potential airborne pollution could then be effectively managed by using the appropriate biomonitoring techniques most suitable to any case under study.
6. Cost and effectiveness guidelines would then provide the proper balance necessary to meet both society's assurance for adequate

environmental protection and industry's assurance that its survival is not unduly or unfairly jeopardized.

A further list of industry concerns to meet research "needs" only are:

1. cost and benefits of research,
2. ability to tie risk analysis to assimilative capacity,
3. ability to quantify environmental protection and forget BACT and BAT (best available control technology and best available technology),
4. relationship to site-specific controls,
5. ability of air quality control (AQC) to reflect the level at which the public or the environment should be protected.

   NOTE: One should be certain that expenditures in health-oriented research are valid, that is, do the statistics support the type of research proposed?

Industry should also ensure:

1. That practical research priorities are set.
2. That criteria documentation are related to
   a. social acceptability, and
   b. benefit to society.
3. The rigorous documentation of scientific fact.
4. The rigorous evaluation of qualitative aspects of regulations and the establishment of reasonable regulations.
5. That the long-term problem of transport of air pollutants is considered in tandem with cost and benefit.
6. That fate and effect studies are seriously considered.

   NOTE: It was suggested that before the adequate regulation and control of hazardous substances could be actualized objectively, the source and routes into the market place must be surveyed, including fate and effect studies.

7. That consideration be given to risk estimation in determining the safety factor based on human, animal, and other biologic factors. Also,
8. What is the probability of a bad event having bad results.
9. Social evaluation of risk is a political responsibility, is it not? Our involvement should be an interconnected process of investigation, research, criteria, standards, regulations, and enforcements.

   NOTE: What is the use of engaging in detailed, expensive research if the regulations do not reflect the results?

One can talk about biological resiliency, but few government monitoring agencies seem to appreciate its significance. One should compare short-term damage with long-term damage and even with the effects occasioned by not doing anything (e.g., use or nonuse of dispersants in an impacted marine or vegetated environment).

There should be a balance between free and controlled research at three levels: applied, pure, and social.

Data could be translated into "gutsy" language for public consumption and include risk analysis lists (e.g., primary risk versus alternative risk). I read somewhere that it was a French statesman (unknown to me) who said "government regulations replace chance with error." Must it be this way? In doing something constructive let us then also recognize:

1. The cost of strict compliance to regulations may nullify a project.
2. There is a public lack of confidence in government regulatory processes.
3. Regulations should not be ends in themselves.
4. There are some who believe the public should be involved in regulation and control of industry.
5. Emission reductions versus environmental consequences could be used to preserve levels of technical capability.

There are some who believe that we should institute a control strategy group who would direct the control process to:

1. Avoid damage (penalty should be based on damage).
2. Obtain more collection apparatus information.
3. Define the extent of the problem.
4. Assess level of impact.
5. Balance industry, that is, activity, versus control.
6. Find realistic ways to facilitate a solution.

Notwithstanding the foregoing, the ultimate objective should be, of course, to reduce ignorance around which judgments will be made. As we continue to meet first the regulatory requirements presently in place, we should also strive to seek regulatory relief (variance through research and study), and only as a last resort prepare court cases to redress grievances.

As a final observation, may I suggest:

1. Producing a biological balance model (namely CO_2, NO_2 cycles), like a process flow chart for a gas plant applicable for every project, showing source, quantity, and sink could prove beneficial.

2. Project task teams to develop strategy documents (including the multidisciplinary inputs of engineers, lawyers, biologists, toxicologists, pharmacologists, etc.) that may represent a starting link in the solution chain.

It is hoped that we shall not be at cross purposes insofar as biologically oriented research objectives may affect the business of industry at the micro versus macro levels.

The task of developing pollutant control and research strategies has already begun. It is therefore timely that this conference has been convened to appraise the status of international research efforts and, hopefully, set guidelines for the future which will be beneficial to the research communities everywhere. I wish everyone well in meeting the challenges before us.

REFERENCES

Alberta Clean Air Act, Alberta Regulations 218/75, *The Alberta Gazette Part II* **71**(16), 612–618 (1975).

Canada Clean Air Act, Ambient Air Quality Objectives Order No. 1, Clean Air Act, Chapter 403. *Consolidated Regulations of Canada* **IV**: 2869–2871 (1978a).

Canada Clean Air Act, Ambient Air Quality Objectives Order No. 2, Clean Air Act, Chapter 404. *Consolidated Regulations of Canada* **IV**: 2873–2874 (1978b).

Canada Clean Air Act, Ambient Air Quality Objectives No. 3, Clean Air Act, Registration SOR/78-74. *Canada Gazette Part II* **112**(3), 467–468 (1978c).

Canada Clean Air Act, Amendment, Ambient Air Quality Objectives No. 3, Clean Air Act Registration SOR/78-812. *Canada Gazette Part II* **112**(21), 3929–3930 (1978d).

Elder, P. S. (ed.) Alberta's Clean Air Act fails the test. *Canadian Environmental Law Association Newsletter* (*CELA*) **2**(5), 67–69 (1977).

Energy Resources Conservation Board. Revised environmental monitoring requirements for plants processing sour gas (IL 79–22), Information Letter, Calgary, Alberta, December 13 (1979).

Environmental Conservation Authority. Environmental effects of the operations of sulphur extraction gas plants. Canadian Petroleum Association submission to the Environment Conservation Authority, September (1972).

Fraser, Hon. J. A. Proposed Solution to the "Acid Rain" Problem. An address in the Second Canadian Government Affairs Seminar, Ottawa, September 23–26 (1979).

Legge, A. H. Primary productivity, sulphur dioxide, and the forest ecosystem: An overview of a case study. In *Proc. of Symposium on Effects of Air Pollutants on Mediterranean and Temperate Forest Ecosystems, Riverside, California, June 22–27, 1980*, pp. 51–62. Available as Gen. Tech. Rep. PSW–43, 256 pp. Pacific Southwest Forest and Range Exp. Stn., Forest Serv., U.S. Dept. Agric., Berkeley, CA (1980).

Legge, A. H., D. R. Jaques, R. G. Amundson, and R. B. Walker. Field studies of pine, spruce and aspen periodically subjected to sulphur gas emissions. *Water, Air, Soil Pollut.* **8**, 105–129 (1977).

Legge, A. H., D. R. Jaques, H. R. Krouse, E. C. Rhodes, H. U. Schellhase, J. Mayo, A. P. Hartgerink, P. F. Lester, R. G. Amundson, and R. B. Walker. Sulphur gas emissions in the boreal forest: The West Whitecourt case study, II. Final Report. Publication Number 78–18, 615 pp. Submitted to Whitecourt Environmental Study Group, Kananaskis Centre for Environmental Research, University of Calgary, Calgary, Alberta (1978).

Legge, A. H., D. R. Jaques, H. R. Krouse, E. C. Rhodes, H. U. Schellhase, J. Mayo, A. P. Hartgerink, P. F. Lester, R. G. Amundson, and R. B. Walker. Sulphur gas emissions in the boreal forest: The West Whitecourt case study, I. Executive Summary. *Water, Air, Soil Pollut.* **15**, 77–85 (1981).

Robinson, R. M. Socio-economic impact analysis: Bureaucratic exercise or essential tool? *J. Air Pollut. Cont. Assoc.* **29**, 1026–1027 (1979).

Swan, H. R. From criteria to standards—The Canadian federal government view, Environment Canada. Mr. L. Edgerworth, Assistant Deputy Minister (1976).

Whittington, D. L. Class action suits add new impetus to consumer movement. *Financial Times*, **64**(50), 8 (1976).

SYNTHESIS

INDUSTRY VIEWPOINTS AND GOVERNMENTAL AIR QUALITY CRITERIA DEVELOPMENT

R. L. Findlay
Amoco Canada Petroleum Company Limited
Calgary, Alberta, Canada

J. H. B. Garner
U.S. Environmental Protection Agency
Research Triangle Park, North Carolina

G. Steve Hart
Environment Canada
Ottawa, Ontario, Canada

BACKGROUND

Industrial abatement strategies relative to the atmospheric loading of air pollutants and the corresponding governmental air quality criteria have been changing continuously in recent years. Cost–benefit ratios as well as the current energy shortage play an extremely important role. Further, pollution control strategies are highly complex and, according to some, may be comparatively ineffective relative to sources such as the automobile. Alternative and/or new energy technology poses questions relative to new pollutant and/or product generation. These considerations have implications in agriculture and forestry.

DISCUSSION

Based on air quality criteria, ambient air quality standards are used as the basis for controlling ambient air pollution. Air quality criteria documents are up-to-date compilations of the scientific knowledge of the relationships between the concentrations of various air pollutants and their effects on man and his environment. Air quality criteria documents attempt to describe the effects observed when the concentration of a specific pollutant in the ambient air reaches or exceeds a specific value for a specified time period.

Ambient air quality standards imply that a certain amount of air pollution is acceptable, and that it is possible to determine the pollutant concentrations at which human beings and the ecosystems in which they live are harmed. When writing criteria documents many questions need to be considered. Some of these questions are:

1. *How accurate are the pollutant measurements?*
2. *How well can the pollutant pathways be determined?*
3. *Are the scientific criteria used for standard setting developed in the laboratory or in the field?*
4. *If the criteria are developed in the laboratory, are they applicable to the real world?*
5. *Does the standard actually reflect real world situations and conditions?*
6. *How accurate is the monitoring equipment?*
7. *How accurately can receptor response be determined?*
8. *If the present methods are not accurate or valid, are there better methods which may be used to set standards?*

A. The Scientific Basis for Standards

There is more than one objective in studying the effects of air pollutants on vegetation. Criteria and standards are not the only reason for research; also important is the understanding of any environmental stress on crop production.

The development of criteria that will be used in the control of pollution involves the use of pollutant effects data. Not all data are comparable because the studies have been conducted by many different scientists in different laboratories and frequently in different countries under a variety of experimental conditions. If standardized methods for conducting research were available, criteria documents would not be encyclopedic, but could be more selective and disregard studies that were not done by the standardized or comparable methods.

B. Biomonitoring

Biomonitoring through the use of sensitive plants can be an aid in establishing criteria if concentrations, exposure duration, and effects are measured. In the Netherlands an ambient biomonitoring network has been established. An attempt is being made to determine the relationships between pollutant concentrations and effect-intensity levels. However, this is confounded by environmental conditions. It is necessary to know not only the concentration and the duration of the pollutant, but also the condition of the plant. The stage of growth and the genetics of the plant are important in determining its response. In addition, specific physiological parameters of the plants must be measured. The concentrations of a pollutant on the plant surface, the concentrations entering the plant, and the mechanism of action within the plant must be related to the effects.

If biomonitoring is to be a viable tool, sensitive plants in each geographic region must be identified. Results obtained by the techniques used in Europe, Canada, and some parts of the United States indicate that biomonitoring has potential as a tool for measuring the environmental impacts of air pollutants. However, at present, in general only the qualitative but not the quantitative aspects of pollutant impacts can be determined by these methods.

C. Type of Standards Needed

Very low concentrations of pollutants affect plants physiologically, resulting in subtle injury (not visibly detectable) and reduce plant productivity and quality. A pollutant may appear to produce no effects over a short period of time but may be ecologically unsafe over a long period of time. Mixtures of pollutants occur under natural conditions, and the total load of pollutants should be considered when setting standards.

In Canada, air quality objectives are used rather than air quality standards. Three pollutant regimes are prescribed: maximum desirable, maximum acceptable, and maximum tolerable. The lowest level, maximum desirable, sets long-term controls as part of the policy of nondegradation of the environment. Maximum acceptable levels are intended to provide maximum protection from pollutant effects on soil, water, vegetation, visibility, personal comfort and well being, as well as human health. Maximum tolerable levels are the highest permitted. These objectives have been established by the federal government, and the provinces are expected to establish standards compatible with these.

Flexible standards reflect the real world situation better than fixed ambient air standards. One examines maximum sensitivity and then tries to set an ambient air quality standard that will be an acceptable and equitable standard across the continent. Averages are usually used to determine the acceptable concentration when it may be peak concentrations of short duration that are causing the effect.

No one standard can really satisfy all land resource uses. All receptors, be they plants, lakes, soils, or humans, are not equally sensitive relative to space and time. Temporal and spatial variations should be considered when setting standards. Different amounts of pollutants can be tolerated at different times of the year. When human populations are at risk, a strict primary standard must be enforced; however, different types of vegetation (forest and agricultural plants) are not equally sensitive at different times of the year. Modeling that considers the various populations at risk—be they plant, animal, or human—could be a satisfactory approach to establishing a reasonable standard. Another approach is to indicate a range of air quality guidelines that specifies the no-effect level, a median, and a tolerable range.

The only way to reduce air pollution is with emission standards. It will never be possible to establish the kind of links between emissions and ambient standards as we currently conceive them that will be satisfactory to the public, the regulators, and the industry.

Emission standards would also have to vary. In areas with very critical ecosystems, they would have to be quite low. In other places, such as the middle of Nevada where population pressures are low and there are few industrial emissions, the emission standards could be relaxed.

In the province of Ontario, there are point-of-impingement standards. Unless there is a single source in the area, it is difficult to determine which source is responsible for increasing the ambient air pollution. Therefore, it is very difficult to enforce ambient air quality standards. It is easier to enforce an impingement standard which is based on emissions. This is a calculated standard based on the stack height, velocity of emission, and a number of other factors. The concentration is calculated for a half-hour period, and if this point is exceeded, or calculated to be exceeded, then the industry must install corrective measures to meet the standard.

D. Modeling

Modeling depends on the accuracy that is required in relating sources to ambient concentrations for control purposes. If any accuracy factor of 2 is adequate, then one can relate source to ambient concentrations on a scale of 500–600 km. Where change is made from ambient standards to emission standards using modeling that is accurate only to a factor of 2, then with an error of 1%, the source exceeding the emission standard will be penalized, most likely by the actions of federal or local governments.

Air quality models have improved in the last 10 years, particularly relative to long range and regional transport. They may not be totally satisfactory, but they are good models relative to the cost involved in their development and application compared with the costs of emission control.

A difficulty in using models lies in the conceptual difference between using averages and means. Models do not predict what is observed. When a half-hour standard is used, there are no models that will adequately predict concentrations within 40–50%. Therefore, to use models, one has to use an averaging time of at least 3 hours.

E. Joint Effects (Pollutant Mixtures)

Standards that are set to control a single pollutant, for example sulfur dioxide, neglect the fact that at the same time there may be elevated concentrations of other pollutants in the atmosphere. Standards in the future will have to consider that mixtures may have a greater impact than single pollutants.

Pollutant mixtures often cause effects at lower concentrations than do pollutants occurring singly. Such effects may be additive, more than additive, and less than additive. To address the problem of pollutant mixtures, the researchers need to know what mixtures occur, and in what concentrations in the atmosphere, so experiments may be designed to determine the effects.

Another complicating factor when considering multiple pollutants is the impact of biotic diseases, pests, and different climatic conditions, such as frost. These factors, in combination with pollutants, can result in more serious effects.

F. Cost–Benefit and Socioeconomic Concern

If the results of mixed pollutant studies are used, standards will have to be reduced drastically. This will have quite an economic effect on industry. The polluter-pay principle should be considered. "Best available technology" (BAT) is often used with regard to emissions. Best available technology can change rapidly and be improved if pressure is applied to the profit margin. This does not mean that industry alone should pay. Within our economic structure, it is common to pass this on to the consumer. It seems inherently fair that a bigger portion of the cost be borne by the industry responsible (via the consumer) rather than by distributing it equally throughout society.

The question of tradeoffs needs to be considered; for example, when certain types of controls are used, new pollutants are formed. Under those circumstances the disposal of the new pollutant may result in water pollution; therefore, the holistic concept and a better systems approach need to be used in air pollution control and regulation. When control technology is used, studies need to be performed to determine whether the equipment is performing satisfactorily. Regardless of the technology or controls used, short-term economic tradeoffs should not be at the expense of long-term ecological and economic considerations. Good long-term economics should be synonymous with good long-term ecological management as the economy depends on ecology for its existence.

Better dose–response or dose–yield data are needed so that yield loss and crop damage can be quantified economically. Better cost–benefit models are needed. If crop losses were quantified in economic terms, standard setting would be easier and more meaningful. However, economics should not be the sole basis for setting a standard. The environment should be considered in its entirety, not just those aspects that can be expressed monetarily. It must be recognized that the environment is a cost factor that needs to be included in the equation.

There is a need for:

1. *Development of standards that are more closely related to relevant scientific data.*
2. *Continued development on a crop-by-crop basis of techniques for measuring economic loss.*
3. *Establishment, by the regulatory agencies, of criteria for the types of experimental work they require to fill perceived gaps in information.*
4. *Further development of the concept of flexible standards to reflect real world situations.*
5. *The development of more precise and comprehensive criteria for determining the joint effects of multiple pollutants on the terrestrial ecosystem.*
6. *More research into the interaction between pollutants and other factors such as biotic disease, pests, and climatic problems (e.g., frost).*
7. *Research into the economic implications of the subtle-injury concept.*
8. *Better models to predict the social and economic impact of pollution-control regulations.*
9. *Ecosystems analyses to ensure that tradeoffs between short-term economic considerations are not to the detriment of long-term environmental and economic requirements.*

10. *Economic, biological, and chemical research into appropriate waste-disposal technology and the choice of hazardous waste-disposal sites.*

11. *Further development of biological process modeling.*

12. *Instrumentation and techniques for measuring pollutant concentration at the leaf surface.*

13. *Instrumentation and techniques for measuring pollutant concentrations entering the plant.*

14. *Better understanding of physiological processes of plants in the presence of pollutants.*

15. *Characterization of atmospheric pollutants, including their residence time, long-distance transport mechanisms, and formation of secondary pollutants.*

PART TWO

ATMOSPHERIC CHEMISTRY OF AIR POLLUTANTS

Air pollution–terrestrial ecosystem effects involve source–emission–transport–transformation–deposition–reception phenomena. Atmospheric chemistry of air pollutants is a critical factor. Biologists need to be made aware of the complexities of atmospheric chemistry of pollutants. Complicated questions still remain to be fully answered, such as the residence time, long-distance transport, as well as tropospheric injection of ozone. Additionally, we do not fully understand the catalytic oxidation and production of ozone. Sulfur chemistry poses similar problems. Homogeneous and heterogeneous oxidation of sulfur dioxide, relationships of these to the acid rain phenomenon, biogenic versus man-originated atmospheric loading of sulfur, formation and role of sulfur compounds such as stable sulfite species, dithionate, and organic sulfur volatiles require further study. Sulfur compounds and particulates play an integral role. In addition, one needs to understand atmospheric organics as well as other pollutants.

This section addresses these questions and others of current and future concern.

The Sulfur Oxide–Particulate Matter Complex

G. M. HIDY†
P. K. MUELLER*
Environmental Research and Technology, Incorporated
Newbury Park, California

†*Present address*: Desert Research Institute, Reno, Nevada.
**Present address*: Electric Power Research Institute, Palo Alto, California.

1. INTRODUCTION

The term *sulfur oxide–particulate matter complex* (*SPC*) apparently was coined a few years ago by Nelson and his colleagues (Nelson et al., 1975). The term has been used as a generic description of the diverse chemical contributors to the polluted atmospheric aerosol, which include sulfur dioxide (SO_2), particulate sulfate (SO_4^{2-}), and soot associated with the aerosols occurring during fuel burning. Sulfur dioxide, combined with total suspended particulate matter (*TSP*), is the best observed of industrial air pollution in a historical sense. The early criteria documents supporting U.S. National Ambient Air Quality Standards (NAAQS) for SO_2 and TSP take note of the suspected health effects and other influences of these pollutants (USHEW, 1969, 1970). By the middle 1960s, information was available to show that ambient SO_2 concentrations and urban exposures in themselves were not sufficient to explain adverse health effects attributed to air pollution. However, in combination with particulate matter, there was evidence that SPC was implicated in community health deterioration from air contamination. The historic pollution incidents of Donora, Pennsylvania, in 1948 and London in 1952, involving SO_2 and black particulate matter, are the dramatic evidence perhaps most often cited in the literature.

In the early 1970s, workers in the U.S. Environmental Protection Agency (USEPA) believed that new epidemiologic and other toxicological studies, partly associated with the Community Health Environmental Surveillance Study (CHESS), implicated particulate sulfate as a major factor in adverse health effects (USEPA, 1974). This evidence has proven to be largely unsupportable on closer examination. However, as a result of this and other pressures, considerable effort has been devoted recently to the investigation of sulfur oxide behavior in the atmosphere. The emphasis of the new work shifted from SO_2 in association with TSP to specific consideration of the atmospheric chemistry of sulfur compounds to form end products, including sulfate. This component of TSP is believed to come mainly from the oxidation of SO_2 and other sulfurous gases in the atmosphere and is sometimes called a *secondary pollutant*, as contrasted with *primary emissions* of material.

Particulate sulfate in its common atmospheric form of free acid and alkaline metal and ammonium salts has not been demonstrated to have adverse effects on human health at ambient air concentrations found in North America. Therefore, workers have extended their investigations to other components of the SPC in search of their origins and evolution in polluted air. Other sulfur oxide constituents may include sulfite–metal complexes, thiosulfates, and mixed salts such as zinc ammonium sulfate. Nonsulfurous components include particulate nitrate, carbon compounds, and metal oxides or salts in the presence of gaseous products of atmospheric chemical reactions. These include ozone, peroxyacyl nitrates or other nitro compounds, and partially oxidized hydrocarbon vapors.

At the same time that the complexity of chemical composition has emerged, investigations have begun to show the potential importance of pollutant behavior spread over great geographical distances from their origins. The interaction with clouds and precipitation contributes to acidity in rain; widespread impairment in visibility also seems to be a symptom of large-scale effects of the SPC (Husar et al., 1978).

This survey focuses on the atmospheric behavior of SO_2 and associated constituents of airborne particulate matter. Photochemical oxidation processes involving nitrogen oxides (NO_x) and hydrocarbons as precursors are discussed separately in the following chapter by Demerjian. This review begins with consideration in detail of the constituents of the SPC to better define the concept. Spatial and temporal variations are described, and the interactions of various rate processes are surveyed. The discussion then turns to atmospheric processes as part of geochemical cycling; the budgeting of sulfur in eastern North America is examined as an example of the cycle concept. Some available information on long-term trends is discussed, and some globally sensitive areas to pollution are hypothesized. In completion, several unknowns about SPC behavior that are important in environmental assessment are identified and summarized.

2. DESCRIPTION OF THE SULFUR OXIDE–PARTICULATE COMPLEX

With each year, the list of identified trace constituents in the atmosphere becomes longer and more diverse. As knowledge of chemical processes develops and methods of detection improve, there seems to be no end to the identification of compounds that may be a factor in the SPC. As an example, the gaseous sulfur and nitrogen constituents are listed in Table 1. The most common sulfurous gas found in the urban or near-urban atmosphere is SO_2. In some rural areas, however, hydrogen sulfide is found at locally comparable concentrations. Of the nitrogen oxides, nitrous oxide evidently is the most common gas in rural areas. However, nitric oxide and nitrogen dioxide are the more abundant in urban air. Ammonia concentrations generally are low, in the parts per billion (ppb) to parts per trillion (ppt) range, because of the chemical reactivity of this gas. There are several nitrogen oxide species found in the atmosphere that are products of chemical reactions, including nitric acid and peroxyacyl nitrates.

Paralleling the variety of sulfurous and nitrogenous gases are the organic vapors, which are listed generically in Table 2. Although methane is by far the most common carbonaceous gas in the air, it is of least interest in its environmental impact. The olefinic and aromatic compounds, with the oxygenated materials, represent potential sources of reactive species and some are believed to adversely influence human health. For example, aromatics, such as benzene, are believed to be carcinogens. Although the

TABLE 1. Sulfurous and Nitrogenous Gases Known to Be in the Lower Atmosphere

		Range of Concentration (ppb)		
Constituent	Formula	Urban	Rural	Origins
Sulfur				
Sulfur dioxide	SO_2	5–500	1–50	Fossil fuel combustion
Hydrogen sulfide	H_2S	1–50	1–50	Biogenic; geothermal
Carbonyl sulfide	COS	0.1	0.1	Biogenic
Dimethyl sulfide	$(CH_3)_2S$	0.1	0.1	Biogenic
Carbon disulfide	CS_2	0.1	~0.1	Biogenic
Methyl mercaptan	CH_3SH	—	0.1	Biogenic
Nitrogen				
Nitrous oxide	N_2O	1–300	1–300	Biogenic
Nitric oxide	NO	10–1000	0.1–100	Fossil fuel combustion
Nitrogen dioxide	NO_2	1–500	0.1–100	Fossil fuel combustion Atmospheric oxidation
Nitric acid	HNO_3	0.1–20	0.02–0.3	Atmospheric reactions of NO_x
Ammonia	NH_3	1–80	0.1–10	Biogenic; fuel combustion
Peroxyacetyl nitrate (PAN)	CH_3COONO_3	0.1–60	0.1–1	Atmospheric reaction product

TABLE 2. Organic Vapors Identified in Ambient Air

Constituent	Range of Concentration (ppb)		Origins
	Urban	Rural	
Methane	20,000	1000	Biogenic; geogenic
Alkanes (ethane, propane, . . .)	100	10	Gasoline, fuels, solvents, etc., partially burned hydrocarbon
Alkenes (olefins)	100	10	Gasoline, fuels, solvents, etc., partially burned hydrocarbon
Cyclic alkanes	1	0.1	Gasoline, fuels
Aromatic (benzene, toluene, . . .)	1	0.1	Gasoline, fuels
Polycyclic aromatics (e.g., benzopyrene)	<0.1	<0.01	Combustion of fuel
Terpenes	0.1	0.1	Biogenic
Aldehydes (formaldehyde, . . .)	1–10	0.1–1.0	Biogenic; atmospheric reaction products
Other oxygenates (ketones, alcohols, acids)	<0.1–1.0	< 0.1(?)	Biogenic; atmospheric reaction products
Alkynes (acetylene)	1–10	0.1–1.0	Partially burned fuel
Cyclic alkenes	0.1	0.01	Gasoline, fuels

polycyclic aromatic compounds generally are very low in ambient concentrations, they have been a concern for many years as potential carcinogens.

2.1 Particle Chemical Composition

Suspended in this complex mixture of gases are the aerosol particles, whose chemical makeup is equally diverse and perhaps far more poorly characterized. Typical estimated compositions of collected particulate material in North American cities and rural areas are listed in Table 3. The components of particulate matter are listed as major and minor ones by mass concentration. One readily sees in this list that the most common constituents are identified as ammonium sulfate and nitrate, lead halides, and elemental and organic carbon material. Additional components include material with composition similar to the earth's crust, such as fly ash and wind-blown soil dust, as well as ubiquitous quantities of water.

If sulfites and nitrites are present, they generally are in small quantities; they are likely to be unstable as ions, but may exist as organic materials or metal–ion complexes. The sulfate and nitrate components are associated with the atmospheric oxidation of gaseous precursors and represent an atmospheric end product for such gases. Sulfate is widely present in quantities above 0.5–2.0 $\mu g/m^3$. There are large variations from place to

TABLE 3. Chemical Composition of Particulate Matter Sampled in Different North American Locations (Concentrations in $\mu g/m^3$)

	Location						
Constituent	New York City[a] (Aug 10–16, 1976)	Montague MA[b] (Oct 21–22, 1977)	Duncan Falls OH[b] (Oct 21–23, 1977)	Allegheny Mtn[c] (Jul 23–Aug 11, 1977)	Pulaski TN[b] (Oct 21–23, 1977)	Denver CO[d] Dec 18, 1978)	Los Angeles[e] (Aug 14, 1977)
Secondary—Anthropogenic							
Ammonium	3.5	2.2	2.3	2.3	1.5	10.1	11.2
Sulfate	22.4	12 (8.0)[f]	13 (8.4)[f]	14.1	11 (9.8)[f]	15.0	24.3
Nitrate	8.6	0.05	0.37	0.53	0.40	14.0	12.7
Carbon/organic							
Elemental C	—	—	—	—	—	33.2	—
Organic C	—	—	—	—	—	39.0	—
Total Noncarbonate C	14.0	5.7	1.5	0.78	1.6	—	27.1
Automotive							
Lead and Halide (Pb + Br + Cl)	1.6[g]	0.41	0.39	0.2	0.52	13.3	4.3[g]
Light metals							
Sodium	0.93	0.09	0.36	0.2	0.22	—	1.25
Magnesium	0.31	0.07	0.27	0.07	0.17	—	0.61
Aluminum	0.82	0.25	0.91	0.7	0.87	8.7	2.75
Silicon	—	0.50	3.0	1.0	3.3	25.4	10.5
Potassium	0.31	0.11	0.43	0.07	0.27	3.14	0.942
Calcium	—	0.16	0.77	0.33	2.2	—	1.55

Heavy metals							
Titanium	0.084	<0.05	0.09	0.04	0.09	0.48	0.161
Vanadium	0.046	<0.05	<0.05	0.003	<0.05	0.06	0.074
Chromium	0.032	<0.05	0.05	—	0.05	0.054	—
Manganese	0.034	<0.05	0.05	0.009	0.05	0.041	0.076
Iron	1.48	0.23	0.82	0.32	0.71	7.0	2.74
Nickel	0.018	—	—	—	—	0.05	0.048
Copper	—	<0.03	<0.25	—	<0.25	0.164	0.042
Zinc	0.44	0.09	0.07	0.02	0.03	0.453	0.375
Arsenic	3.9×10^{-3}	—	—	—	—	4×10^{-3}	—
Selenium	7.1×10^{-3}	—	—	0.003	—	—	—
Total mass concentration	90	30	50	—	61	290	170

[a]Bernstein and Rahn (1979); Lioy et al. (1979); Daisey et al. (1979); Kleinman et al. (1979).

[b]Sulfate Regional Experiment (Mueller et al., 1980); Metal Data Florida St. University (Winchester, personal communication).

[c]From Pierson et al. (1980).

[d]Denver Haze Study (Heisler et al., 1980).

[e]Los Angeles Aerosol Study (Heisler et al., 1979), <15-μm diameter, 0800–1200 PST.

[f]Parentheses refer to <2.5-μm diameter data.

[g]Coastal cities may have significant fraction of Cl from sea salt.

place in nitrate concentration, ranging from <0.1 to >10 $\mu g/m^3$. Part of this variability appears to be related to unresolved sampling problems created by the volatility and instability of nitrates and losses after filter collection and storage. Ammonium nitrate, for example, shows a strong temperature dependence on vapor pressure, and susceptibility to decomposition which creates special problems for sampling (Stelson et al., 1979). High concentrations of particulate nitrate generally are found in areas of excess atmospheric ammonia and cold air. Volatile nitric acid vapor then can be absorbed at low temperatures in particles and neutralized to form the ammonium salt. In warm air, nitrates appear to be less likely to form from nitric acid vapor or to remain stable in the condensed phase. Thus, nitrate in particles often appears as a winter maximum concentration, while sulfate has a summer maximum with an SO_2 minimum and a winter maximum accompanying an SO_2 maximum. Where large quantities of ammonia are available to neutralize acid (e.g., Los Angeles, Denver, or the San Joaquin Valley of California) high levels of nitrate have been observed (Heisler et al., 1979; Grosjean et al., 1979a).

The acidity of the SPC is an important issue in relation to the chemical nature of the mixture of contaminants. Work initiated by Brosset (1978) and others (Charlson et al., 1978; Stevens and Dzubay, 1978) has indicated that significant hydrogen ion is present in some particulate samples. Stoichiometric arguments have been applied to assign the acidity of the particulate material. Such calculations indicate that aerosol particle mixtures can be accounted for as ammonium salts of sulfate and nitrate. If particulate nitrate is present as a neutralized salt and chloride concentrations are small, the sulfur oxide mixtures frequently correspond to ammonium bisulfate and ammonium sulfate. Some examples of conditions where hydrogen ion content is in excess of ammonium bisulfate have been reported (Mueller et al., 1980; Pierson et al., 1980). Generally, insufficient observations of alkaline metal concentrations have been available to determine if sulfuric acid or other acids are actually present in appreciable quantities. However, the observations of Pierson et al. (1980) indicate that acidity in particles from western Pennsylvania air could not be eliminated by considering measured alkaline metal concentrations. A combination of observations in rural areas of Pennsylvania by infrared spectroscopy, Gran titration, and stoichiometry suggests that a mixture of ammonium sulfates and free sulfuric acid can coexist in some heavy pollution conditions. Data taken in the Allegheny Mountains near Pittsburgh during July and August 1977 indicate a high H^+ to SO_4^{2-} correlation, with mole ratios approaching 2 (Pierson et al., 1980). In this case, acidity was especially strong under conditions of high sulfate levels. During conditions of regional pollution events in the East during July 1978, it was found that the calculated mass ratio of sulfate to ammonium or aerosol samples can vary over periods of hours from 2.6 to more than 6, suggesting the presence of free acids as well as neutralized ammonium salts (Mueller et al., 1980). In contrast with the

Allegheny Mountain results, these investigations found high acidity with low sulfate concentrations.

Recently, Lazrus (personal communication) proposed potentially useful parameters for characterizing the total acidity of air. The acidity of air can be defined as

$$\text{Acidity} = \text{Equivalents of protonic acid} - \text{Equivalents of base} \quad (1)$$

where sulfate, nitrate, and chloride are included as acid contributors and bases include ammonium and alkali or alkaline earth metals. The total potential acidity of the air, taking into account the major gas phase constituents, is

$$\text{Potential acidity} = \text{Acidity} + \text{Equivalents of } NO_x \text{ and } SO_2 - \text{Equivalents of ammonia} \quad (2)$$

Lazrus has attempted to estimate the acidity and potential acidity from samples taken during the first phase of the Acid Precipitation Experiment (APEX). For conditions in summer 1978 over the Ohio River Valley and Atlantic Coast areas, the acidity or potential acidity corresponded to calculated pH values from an ion balance roughly in the range between 2.5 and 4.

The presence of water in aerosol particles has been measured directly only in California (Hidy, 1975). It is known, however, that atmospheric particles are hygroscopic in nature so that they will require water to equilibrate to given humidity. Even at relative humidities less than 50%, particles are expected to contain on the order of 5–10% water, as noted by Meyer et al. (1973).

The common crustal elements contained in particles are probably alkaline metal oxides or salts, either found in soil dust or contained in fly ash from emissions of fossil fuel combustion. There are traces of certain toxic elements in fly ash that are of concern in conjunction with the sulfate and nitrate species. Other possible sources of metals include petroleum and metal-refining operations.

Lead in the form of oxide or lead bromochloride also is found in aerosol particles. The origin of this material is most likely automobile exhaust, except in the vicinity of smelting operations.

The carbonaceous material found in aerosol particles is made up of soot, including elemental carbon and organics. These may come from natural sources, incomplete combustion of fossil fuels, vegetative debris, or atmospheric chemical reactions. The organic fraction generally has been poorly characterized. Only recently have simple measurements segregating total elemental and/or organic carbon become available from different locations (Grosjean et al., 1979b). The diversity of organic material found in particles is illustrated in Tables 4 and 5. The data in Table 4 are analyses of samples taken in Belgium (Cautreels and van Cauwenberghe, 1978). The variety of compounds given are attributed to primary emissions from combustion

TABLE 4. Concentrations of Organic Compounds Formed in Air Samples Taken in Belgium[a]

Compound	Concentrations ng/m^3 Particulate Samples (P)	Gas Phase Samples (G)[b]	Distribution Factor[c] (P)/(G)
Aliphatic Hydrocarbons			
n-Nonadecane	0.80	15.1	0.053
n-Eicosane	0.85	7.55	0.113
n-Heneicosane	1.08	4.12	0.262
n-Docosane	2.33	4.23	0.551
n-Triosane	4.75	3.38	1.41
n-Tetracosane	8.15	4.63	1.76
n-Pentacosane	9.50	5.74	1.66
n-Hexacosane	9.73	8.70	1.12
n-Heptacosane	11.1	9.03	1.39
n-Octasane	8.10	7.80	1.04
n-Nonacosane	15.8	7.32	2.41
n-Tricontane	5.75	4.87	1.18
n-Hentriacontane	11.2	3.99	2.81
Polyaromatic Hydrocarbons			
Phenanthrene and anthracene	1.21	44.7	0.027
Methylphenanthrene and methylanthracene	0.90	10.2	0.088
Fluoranthene	2.22	8.52	0.261
Pyrene	3.17	3.36	0.488
Benzofluorenes	2.33	1.87	1.246
Methylpyrene	0.93	—	P
Benz[*a*]anthracene and chrysene	12.2	3.87	3.15
Benzo[*k*]fluoranthene and benzo[*b*]fluoranthene	23.1	2.01	11.5
Benzo[*a*]pyrene, benzo[*e*]pyrene, and perylene	20.1	2.69	7.47
Phthalic Acid Esters			
Di-isobutyl phthalate	1.73	32.8	0.053
Di-n-butyl phthalate	101.0	353.0	0.286
Di-2-ethylhexyl phthalate	54.1	127.0	0.426

TABLE 4. (*Continued*)

Compound	Concentrations ng/m^3 Particulate Samples (*P*)	Gas Phase Samples (*G*)[b]	Distribution Factor[c] (P)/(G)
	Miscellaneous		
Anthraquinone	1.59	5.66	0.281
	Fatty Acid Esters		
Lauric acid	0.01	30.3	0.0003
Myristic acid	1.39	7.58	0.183
Pentadecanoic acid	3.60	5.35	0.673
Palmitic acid	29.0	4.77	6.08
Heptadecanoic acid	2.84	5.71	0.497
Oleic acid	2.06	—	P
Stearic acid	35.7	2.27	15.73
Nonadecanoic acid	1.91	1.02	1.87
Eicosanoic acid	9.04	3.00	3.01
Heneicosanoic acid	2.56	1.65	1.55
Docosanoic acid	13.7	—	P
Tricosanoic acid	3.23	—	P
Tetracosanoic acid	10.7	—	P
Pentacosanoic acid	2.55	—	P
Hexacosanoic acid	9.12	—	P
	Aromatic Acids		
Pentachlorophenol	2.43	—	P
	Basic Compounds		
Acridine, phenantridine, and benzoquinolines	0.94	—	P
Benzacridines	0.85	—	P

[a]From Cautreels and van Cauwenberghe (1978).

[b]Low-molecular-weight organics found only in the gas phase were not included in this extract of the author's original table.

[c](P) listed in this column means that essentially all of the material is contained in the particles.

TABLE 5. Organic Constituents of Aerosol Particle Samples from Photochemical Smog, Attributed to Atmospheric Chemical Reactions[a]

Compounds Identified	Possible Gas-Phase Hydrocarbon Precursors
Aliphatic Multifunctional Compounds	
1. $X-(CH_2)_n-Y$ ($n = 3, 4, 5$):	1. Cyclic olefins
X — Y	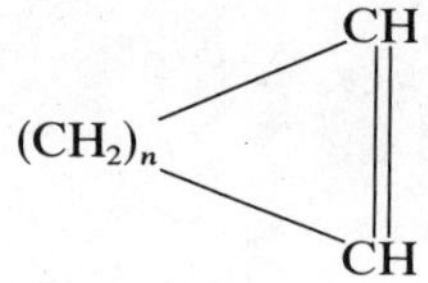
COOH — CH_2OH	
COOH — COH	
COOH — COOH	
COOH — CH_2ONO	
or COH — CH_2ONO	
COH — CH_2OH	
COH — COH	
COOH — COONO	and/or diolefins
or COH — $COONO_2$	
COH — COONO	$>C=CH-(CH_2)_n-CH=C<$
COOH — $COONO_2$	
COOH — CH_2ONO_2	
2. Others:	2. Not known; possibly from aromatic ring cleavage
$CH_2OH-CH=C(COOH)-CHO$	
$CH_2OH-CH_2-CH=C(COOH)-CHO$	
$CHO-CH=CH-CH(CH_3)CHO$	
$CH_2OH-CH=CH-CH=C(CH_3)CHO$	
$C_5H_8O_3$ isomers[b]	
Nitrocresols	
$C_6H_6O_2$ isomers[b]	
Aromatic Monofunctional Compounds	
3. $C_6H_5-(CH_2)_n-COOH$ ($n = 0, 1, 2, 3$)	3. Alkenylbenzenes $C_6H_5-(CH_2)_n-CH+CHR$; also toluene for C_6H_5COOH
4. $C_6H_5-CH_2OH$	4. Toluene, styrene, other monoalkylbenzenes?
C_6H_5CHO	
Hydroxynitrobenzyl alcohol	
Terpene-Derived Oxygenates	
5. Pinonic acid	5. α-Pinene
Pinic acid	
Norpinonic acid	
6. Isomers of pinonic acid	6. Other terpenes?
$C_9H_{14}O_2$ isomers	
$C_{10}H_{14}O_3$ isomers	
$C_{10}H_{14}O_2$ isomers	

[a]From CMBEEP (1977).

[b]Isomers not resolved by mass spectrometry.

sources, or from natural processes. The distribution factor shown is the ratio of material found in the particles (P) to that found in the gas phase (G). One notes that considerable quantities of some of the materials exist in the vapor phase; some were observed by these investigators only in the gas phase. Roughly speaking, the lower the vapor pressure, the more material in the condensed phase. Table 5 gives a list of organic compounds identified by high-resolution mass spectroscopy in particles sampled from photochemical smog in the Los Angeles area (CMBEEP, 1977). These compounds are suggested as condensable products of atmospheric reactions with different hydrocarbon precursors. The precursors are listed for each of the groups of compounds in the table.

2.2 Particle Size Distribution

Another dimension of the SPC is the range of aerosol particle size found in the atmosphere. This is shown schematically in Figure 1. The range of diameter covers particles from less than 0.001 micrometers (μm) to giant particles exceeding 10 μm. The current picture of the mass versus particle size distribution suggests a multimodal distribution. The coarse particles larger than a few micrometers in diameter are dominated by material from comminution or breakup processes, such as blowing soil dust or sea spray, or by spores and pollen. The very finely divided particles, less than 0.1 μm in diameter, are identified with nucleation or combustion processes. The range roughly between 0.1 and 1.0 μm diameter is associated with atmospheric chemical reaction, accumulation of condensable species, including combustion, and the aging by coagulation of the very small particles.

The chemical composition of aerosol particles tends to separate into unique parts according to the origin of the material. Thus, the soil dust elements, for example, are concentrated in the coarse particles, while the combustion-related components and secondary particles are in the finely divided material (Figure 1). Depending on the conditions of production or evolution of sulfate, for example, the mass median diameter can vary substantially. A striking difference between photochemical conditions in coastal southern California, where moist air prevails, and in the dry Southwest has been found using a low-pressure impactor for taking observations (Hering, personal communication). The desert samples have a mass median diameter of approximately 0.2 μm, compared with 0.5 μm found in the Los Angeles area. This difference may be explainable from the chemical reactions of sulfate formation. The process may be influenced by heterogeneous reactions in California, with moist, gas-absorbing particles. In the desert, homogeneous oxidation followed by nucleation may be most important. On the other hand, the desert aerosol also may be a "dried" residue of larger sulfate particles. In any case, these differences fail to support universal, simplistic theories about sulfate formation in the atmosphere.

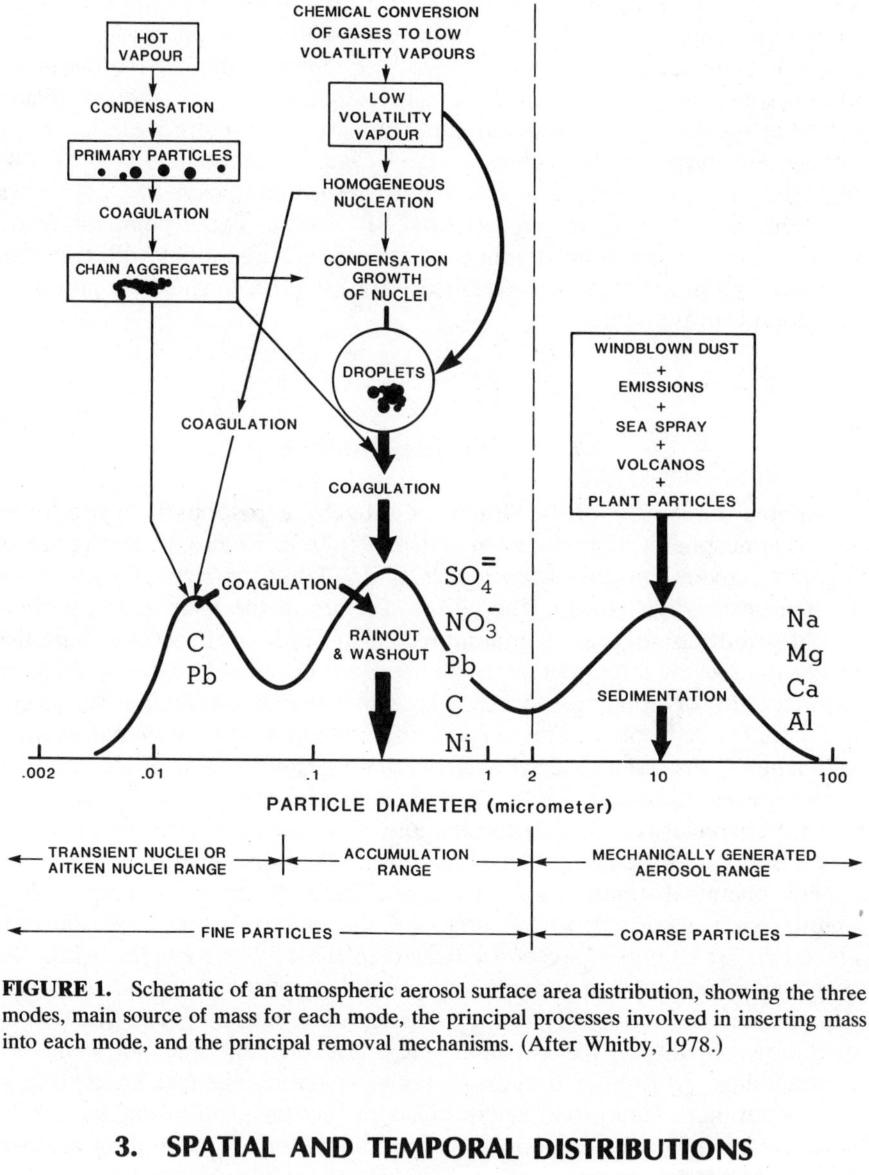

FIGURE 1. Schematic of an atmospheric aerosol surface area distribution, showing the three modes, main source of mass for each mode, the principal processes involved in inserting mass into each mode, and the principal removal mechanisms. (After Whitby, 1978.)

3. SPATIAL AND TEMPORAL DISTRIBUTIONS

3.1 Ambient Concentrations and Exposure Levels

A rational assessment of SPC impact depends on the capability to estimate pollutant exposures in a region of interest. This can be done either directly from available measurements, or by calculation from simulation models. It is impractical to count on extensive measurements in every location because of costs and time to accumulate data. Modeling, on the other hand, suffers from a lack of credibility, resulting from (1) limitations in accounting quantitatively for certain atmospheric processes, and (2) uncertain reliability in simulating observed distributions of pollutants in different geographical areas. Given the drawbacks of modeling, measurements remain a principal basis for SPC impact assessment.

In North America, the emphasis of measurements has focused on airborne concentrations of pollutants rather than dosage (a product of concentration and time) or accumulation by deposition. In recent years, considerable effort has been made to develop a uniformly comparable air monitoring system in the United States. Except for hydrocarbon vapors, the so-called criteria pollutants (SO_2, NO_2, O_3, HC, CO and TSP) are measured routinely on a continuing basis in most parts of the United States and Canada.

With the exception of areas under the local influence of large sources, the current levels of SO_2 in urban and rural areas of North America are well below the annual average, with levels of 60–80 $\mu g/m^3$ and 24-h concentrations of 365 $\mu g/m^3$. The trends of ambient SO_2 levels have shown decreases through the middle 1970s in the United States (USEPA, 1978).

Perhaps the largest systematic air monitoring in the United States has involved TSP sampling in the National Air Surveillance Network (NASN). This system has provided data for over 20 yr in some cities. Observations range from a few $\mu g/m^3$ in remote areas to mg/m^3 over 24-h intervals in dusty or heavily industrialized areas. Over a period of approximately 15 yr, between 1958 and 1977, national trends have shown a steady decrease in TSP levels. From 1962 to 1972 a nationwide improvement of 2%/yr has been reported.

Although ambient SO_2 levels in the United States have tended to decrease in the past several years (Tables 6A and 6B), regional sulfate levels have remained relatively stable (Altshuller, 1976). In the Midwest, east of the Mississippi River, and in the Southeast, there is a suggestion of an actual increase in long-term averaged sulfate levels. Accepting the historical data as self-consistent, with no systematic error, workers have been puzzled by the decreases in precursor SO_2 emissions and ambient concentrations without accompanying decrease in sulfate levels. It is believed that this difference in long-term trends is associated with redistribution of sources relative to the monitoring concentrated in cities and the influence of long-range transport of fine particles without atmospheric removal relative to gases, partly on account of taller smokestacks.

Certain geographical differences in distributions of SO_2, TSP, sulfate, and nitrate are illustrated in recent data reported for rural sites in the northeastern United States. The composite geometric means from four representative season months from the Sulfate Regional Experiment (SURE) in 1977 and 1978 are shown in Figure 2. Taken together, they give an estimate of annual mean distributions. The observations taken over a large region show the rather localized nature of TSP and nitrate, centered

TABLE 6A. Three-Year Running Averages for Sulfur Dioxide and Sulfate (in $\mu g/m^3$) by Geographical Region for Urban Sites in the United States (Altshuller, 1976)

Region	Pollutants	1963–65	1964–66	1965–67	1966–68	1967–69	1968–70	1969–71	1970–72
East Coast[a]	SO_2	147	146	132	119	100	74.0	66.0	ND[g]
	SO_4^{2-}	18.4	18.4	17.7	16.9	17.2	16.6	15.7	15.7
Southeast[b]	SO_2	ID[h]	ID	ID	ID	28.0	26.0	ID	ID
	SO_4^{2-}	10.1	9.8	10.0	10.0	10.4	10.6	10.3	10.2
Midwest–east of	SO_2	ID	ID	75.0	71.0	63.0	54.0	46.0	ID
Mississippi River[c]	SO_4^{2-}	14.8	14.9	13.7	14.6	15.2	15.6	15.2	15.0
Midwest of	SO_2	ID	ID	22.0	20.0	19.0	18.0	ID	ID
Mississippi River[d]	SO_4^{2-}	ID	6.5	6.5	6.2	7.4	7.5	ID	ID
Western interior[e]	SO_2	ID	ID	ID	15.0	15.0	14.0	ID	ID
	SO_4^{2-}	ID	ID	ID	4.9	5.5	6.0	6.7	ID
West Coast[f]	SO_2	ID	ID	ID	ID	24.0	24.0	21.0	ID
	SO_4^{2-}	ID	ID	ID	ID	9.3	9.3	9.1	ID

[a] Providence, RI; Hartford, CT; New Haven, CT; New York City, NY; Jersey City, NJ; Newark, NJ; Camden, NJ; Philadelphia, PA; Wilmington, DE; Baltimore, MD; Washington, DC.

[b] Norfolk, VA; Atlanta, GA; New Orleans, LA; Nashville, TN; Chattanooga, TN; Louisville, KY.

[c] Pittsburgh, PA; Charleston, WY; Cincinnati, OH; Columbus, OH; Dayton, OH; Youngstown, OH; Cleveland, OH; Detroit, MI; Indianapolis, IN; East Chicago, IN; Chicago, IL; St. Louis, MO; Milwaukee, WI.

[d] Minneapolis, MN; Des Moines, IA; Omaha, NE; Dubuque, IA; Kansas City, MO.

[e] Denver, CO; Salt Lake City, UT; Casper, WY; Tulsa, OK; Oklahoma City, OK; Houston, TX; Pasadena, TX; Phoenix, AZ.

[f] Long Beach, CA; San Bernardino, CA; San Francisco, CA; Portland, OR; Seattle, WA.

[g] No data.

[h] Insufficient data available to compute averages for region (ID).

TABLE 6B. Three-Year Running Averages for Sulfur Dioxide and Sulfate (in $\mu g/m^3$) by Geographical Region for Nonurban Sites in the United States (Altshuller, 1976)

Region	Pollutants	1965–67	1966–68	1967–69	1968–70	1969–71	1970–72
East Coast[a]	SO_2	ND[h]	ND	ID[i]	(9)[j]	(10)[j]	ND
	SO_4^{2-}	8.1	8.2	8.4	(7.0)[j]	(7.1)[j]	(7.7)[j]
Southeast[b]	SO_4^{2-}	5.4	6.1	7.2	ID	8.4	ID
Midwest—east of Mississippi River[c]	SO_4^{2-}	7.3	7.8	8.3	9.5	9.8	10.4
Midwest—east of Mississippi River[c]	SO_4^{2-}	3.1	2.7	3.2	ID	3.7	ID
Southwest[e]	SO_4^{2-}	3.5	3.3	4.2	5.2	5.8	5.6
Mountain[f]	SO_4^{2-}	1.7	1.2	1.5	1.8	2.1	ID
West Coast[g]	SO_4^{2-}	3.0	3.0	3.2	3.7	3.8	ID

[a] Arcadia Natl. Park, ME; Coos Co., NH; Orange Co., VT; Washington Co., RI; Kent Co., DE; Calvert Co., MD.

[b] Shenandoah Park, VA; Cape Hatteras, NC; Richland Co., SC; Jackson Co., MS; Montgomery Co., AK.

[c] Jefferson Co., NY; Clarion Co., PA; Monroe Co., IN; Parke Co., IN.

[d] Shannon Co., MO; Thomas Co., ND; Black Hills Natl. Park, SD.

[e] Cherokee Co., OK; Matagovda Co., TX; Grand Canyon Park, AZ.

[f] White Pine Co., NV; Yellowstone Natl. Park, WY; Glacier Natl. Park, MT; Butte Co., ID.

[g] Humboldt Co., CA; Curry Co., OR.

[h] No Data (ND).

[i] Insufficient data available to permit computation of averages for region (ID).

[j] Computations based on smaller number of sites.

around major metropolitan areas of industrial complexes. Sulfur dioxide, on the other hand, has a broad distribution with elevated concentrations along a zone of the Midwest–Ohio River Valley extending into Pennsylvania, New Jersey, and New York. This zone is linked closely with the distribution of emission density of SO_2. Superimposed on the SO_2 geography is a broad sulfate distribution oriented in a similar way. The region of elevated sulfate concentration above approximately 5 $\mu g/m^3$ extends from west to east over approximately 1000 km, with a north–south span of 500 km or more according to the SURE data. The extent of coverage of the zone of elevated sulfate has been associated with long-range transport of pollution. Taking into account a correlation between SO_2 and sulfate concentrations, it can also be explained by the localized influence of very large areas of high SO_2 emission density (Mueller et al., 1980). Such a correlation emerged from analysis of the SURE data but did not appear in the Allegheny Mountain observations in summer 1977 (Pierson et al., 1980). The role of long-range

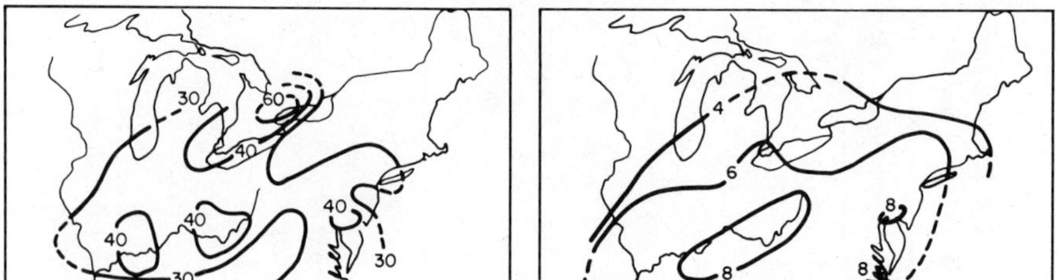

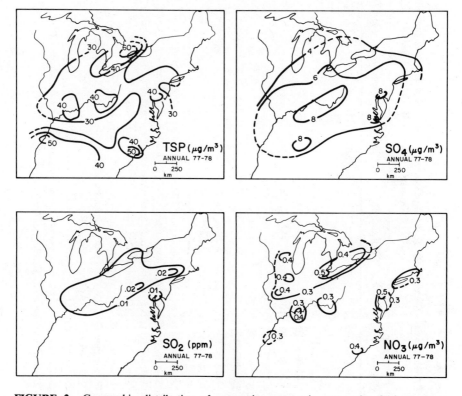

FIGURE 2. Geographic distribution of composite geometric means for 24-hour mass, sulfate, and nitrate, and 1-hour SO_2 values based on four months' data (August 1977, October 1977, January–February 1978, and April 1978) from 54 rural stations. (After Mueller et al., 1980.)

transport has been addressed in detailed analyses of certain kinds of pollution events in this region (Mueller and Hidy, 1983; Samson, 1980).

The frequency of occurrence of elevated sulfate levels is shown in Figure 3. The average distribution from data of the National Aerometric Data Bank (NADB) in 1977–1978 and the SURE data taken in these years indicate that sulfate concentrations above 20 $\mu g/m^3$ occur 5% of the time or less. But concentrations of sulfate >10 $\mu g/m^3$ may occur half the time based on the NADB data or 22% of the time in rural areas according to the SURE. The median of the 1977–1978 SURE data from nine rural stations is 6 $\mu g/m^3$, while the NADB set of combined urban and rural stations (mostly urban) is about 9 $\mu g/m^3$. We believe that the differences are related to the location of sampling sites as well as sampling artifact associated with the SO_2-adsorbing glass fiber filters of the NADB versus a nonadsorbing substrate used in the SURE. The two data sets are remarkably similar, on the average, in the range of high concentrations.

A suggestion of the sulfate concentration trends on a regional scale covering the northeastern United States is found by comparing the SURE data with earlier 1974–1975 observations taken at rural stations similar to those of the SURE (Hidy et al., 1976), and in the New York Metropolitan

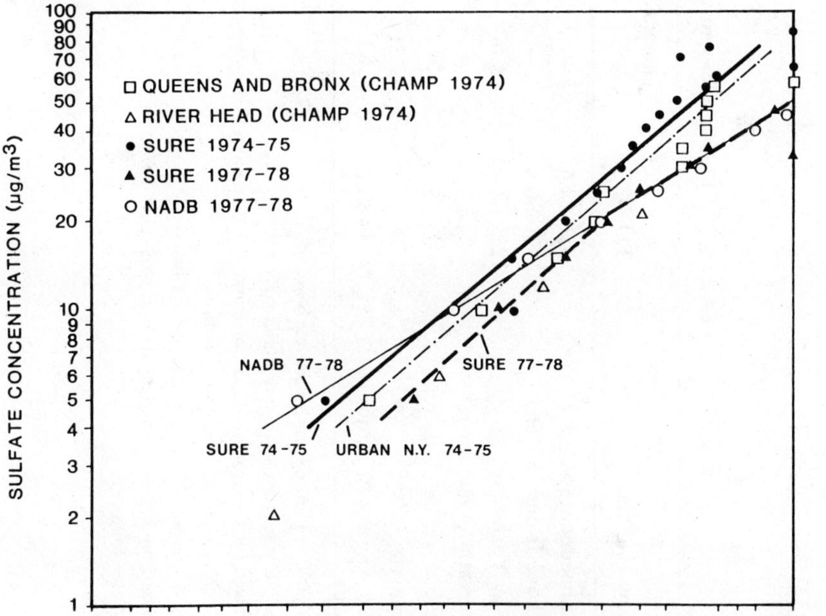

FIGURE 3. Frequency of occurrence of sulfate concentrations in the SURE region as shown by various data sets. (After Mueller et al., 1980.)

TABLE 7. Carbon Content of Atmospheric Aerosols (in $\mu g/m^3$)

New England	TONC[a]	Elemental Carbon	SSOC[b]	TONC % Total Mass	Reference
Connecticut (5 sites)	3.7–14.4	—	2.0–13.0	3.0–7.7	Grosjean et al. (1979b)
Rhode Island (3 sites)	1.8–30.4	—	—	2.8–18.8	Grosjean et al. (1979b)
New York City	12.5[c]	5.6	—	18.3	Countess et al. (1980)
Houston (4 stations)	—	—	3.9–7.3[d]	15.3–33[e]	Grosjean et al. (1979b)
Rockport, IN (1 station)	0.5–15	—	—	—	Grosjean et al. (1979b)
Denver	17.4[f]	7.3	—	21.4	Countess et al. (1980)
Los Angeles (2 stations)	5.9–50.2	—	1.0–44.6[g]	—	Grosjean et al. (1979b)

[a]TONC = total noncarbonate carbon.
[b]SSOC = solvent soluble organic carbon.
[c]Based on 170 organic C samples and 145 elemental C samples (24-h average).
[d]Dichloromethane–methanol soluble.
[e]% of particles <2.5 μm diameter only.
[f]Based on 260 4-h samples taken in Nov–Dec 1978.
[g]Water soluble.

area (Queens, Bronx, and Riverhead, New York). The latter data are from the USEPA Community Health Monitoring Program (CHAMP). Comparison of these data suggests that there may be a difference in occurrence of elevated concentration frequency, with a downward trend between 1974 and 1978. Perhaps more likely, the differences are attributable to the urban influence of New York and a bias in the early SURE-related data associated with a rural station in the upper Ohio River Valley. We suggest from these results that no trend in regional sulfate levels can be identified over this 4-yr period. Over the same period, SO_2 emissions over the greater Northeast were said to be constant or decreasing.

Effort has been spent on tracing the geographical distribution and trends in metal components of particles. The available data have concentrated on fuel-related elements, lead, vanadium, nickel, and titanium, and heavy-industry-related materials, cadmium, chromium, copper, iron, and manganese. These metals have been monitored by the NASN in the United States in some cases since the early 1960s. The results of a regional trend analysis (Faoro and McMullen, 1977) suggest a distribution relating to specific source categories and urban areas. Basically, trends from the mid-1960s to 1975 appear to be consistent with changes in particulate emission patterns associated with fuel use and industrial control.

The history of widespread carbon measurement for spatial and temporal variations is rather short. Monitoring methods for determining total carbon and organic carbon are still under development. However, some observations have become available recently in several areas, showing at least the magnitude of concentrations in different areas. Selected data are listed in Table 7. Particulate carbon was found to be the most abundant element not only in the Los Angeles aerosol, but also in some of the other areas listed. Such data, in combination with the complex nature of the material suggested in Tables 4 and 5, indicate the importance of gathering more data on this part of the SPC.

3.2 Deposition of Material

Until recently in North America the distributed accumulation of (non-radioactive) pollutants by surface deposition has been of less interest than ambient concentrations. However, concern over the impact of deposited pollutants on ecological processes stemming from the Scandinavian experience has created increased interest in deposition patterns (Ottar, 1979). Of particular concern are precipitation distributions in polluted areas where the influence of rain acidity is a major issue. It is plausible that constituents of the SPC are assimilated in precipitation during droplet formation and growth as a result of nucleation and scavenging process. The resulting change in acidity probably comes from such an influence. Little information exists on the distribution of chemical constituents in precipitation over North America beyond the early studies from the National

Precipitation Network (Junge and Werby, 1958; Lodge et al., 1968). Only recently has a major effort been made to construct an international precipitation chemistry–monitoring program in North America under the auspices of the U.S. Department of Agriculture, the U.S. Department of Energy, and Environment Canada.

From limited measurements the distribution of acidity in precipitation was estimated by Cogbill (1976). The situation for pH in the eastern United States and southeastern Canada for 1972–1973 is shown in Figure 4. Comparison of this map with distributions shown in Figure 2 suggests a

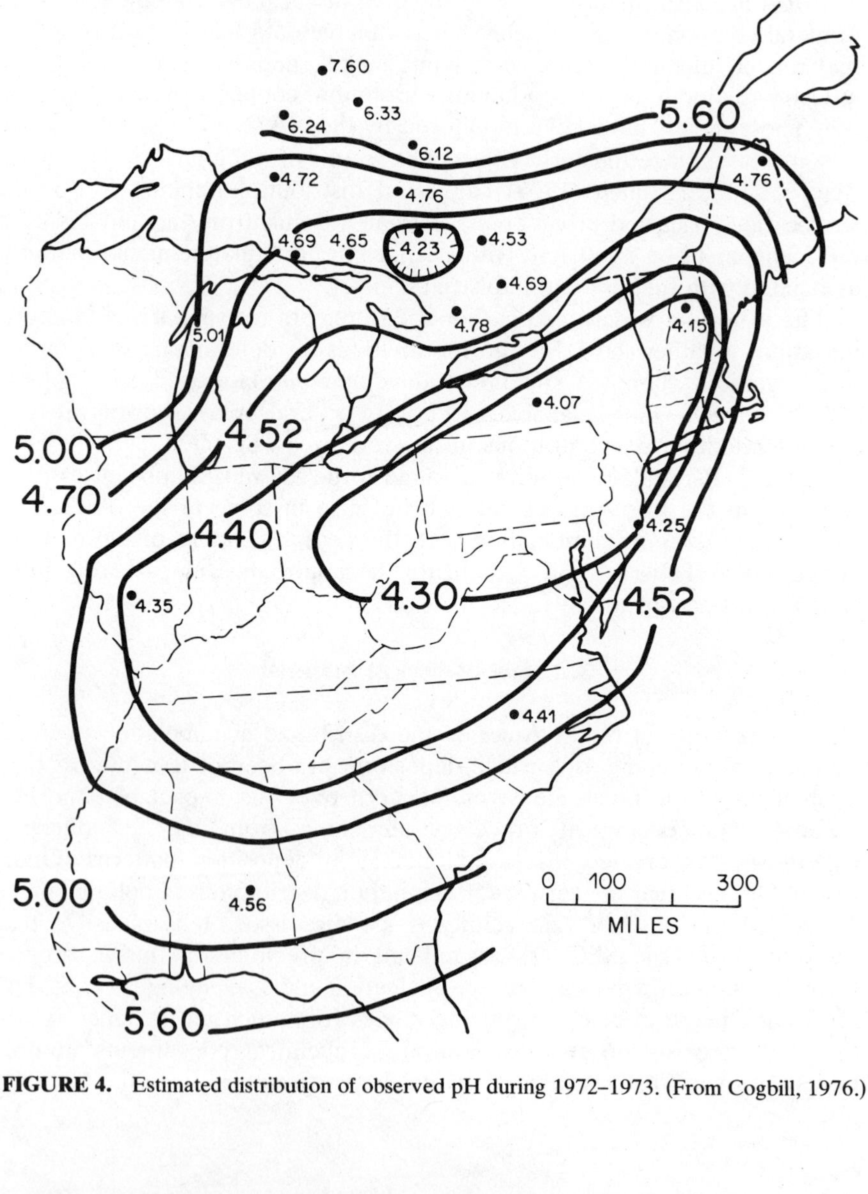

FIGURE 4. Estimated distribution of observed pH during 1972–1973. (From Cogbill, 1976.)

similar pattern between rain acidity and sulfur oxides, but not necessarily (particulate) nitrate. The chemical contribution to precipitation acidity in the eastern states is suggested in the results reported by Likens (1976). The acidity appears to be dominated by sulfate as indicated in Table 8, but the nitrate contribution appears to have been increasing recently as has been the case in Scandinavia. The cause of this is uncertain, but is speculated to be related to increasing NO_x emissions. Liljestrand and Morgan (1978) have reported that nitrate dominates precipitation acidity in the Los Angeles area. This work is one of the few extensive studies on rain chemistry in polluted areas of the western United States, so it is uncertain how widely applicable it is.

Precipitation water contains a variety of materials, including metals and organics, which are likely to be in the SPC. At present, it is uncertain what the influence of these materials may be on ecological processes, separate from, or in combination with, increased acidity.

Although wet deposition chemistry is becoming better appreciated, there is still little information on the chemical nature and distribution of deposition from dry processes. However, these presumably will follow qualitatively the ground level concentration patterns, since the local deposition flux is taken to be proportional to ambient concentrations (see also Section 5).

TABLE 8. Sulfuric and Nitric Acids as Major Sources of Acidity in Precipitation

Substance	Concentration in Precipitation (mg/L)	Contribution to Free Acidity[b] (μeq/L)	Contribution to Total Acidity[c] (μeq/L)
H_2CO_3	0.62[d]	0	20.0
NH_4^+	0.92	0	51.0
Al, dissolved	0.05[e]	0	5.0
Fe, dissolved	0.04[e]	0	2.0
Mn, dissolved	0.0005[e]	0	0.1
Total organic acids	0.34	2.4	4.7
HNO_3	4.40	39.0	39.0
H_2SO_4	5.10	57.0	57.0
Total		98.0	179.0

[a] Based on a sample of rain collected at Ithaca, NY, on October 23, 1975 (Likens, 1976).

[b] At pH 4.01.

[c] In a titration to pH 9.0.

[d] Equilibrium concentration.

[e] Average value for several dates.

4. SOURCES OF GASEOUS AND PARTICULATE CONSTITUENTS

The concentration patterns of material found in the atmosphere depend on their sources, in combination with air transport, turbulent diffusion, chemical transformations, and deposition processes. In general, one finds from available evidence that atmospheric processes create a variability in ambient pollution conditions far in excess of that estimated from variations in emissions from human activities. For example, the daily variation in SO_2 emissions in the northeastern United States is about ±15%, while ambient SO_2 and sulfate concentrations vary by more than an order of magnitude (Mueller et al., 1980). This may or may not be true of natural emissions. In the case of atmospheric sulfur, the natural contributions come from sea salt and blowing soil dust, as well as volcanic and biogenic emissions. Sulfurous gases are emitted and are subsequently oxidized to sulfate. Of the natural sources of sulfur, biogenic emissions are believed to be persistent, major contributors to atmospheric sulfur on a global scale. However, they are probably weak relative to anthropogenic contributions in urbanized, heavily populated regions. The emissions distribution by source category for SO_2 for the eastern United States and southern Canada is given in Table 9. This region covers an area whose perimeter is shown in Figure 2. These data are taken from a current survey associated with the SURE (Klemm and Brennan, 1979). The emissions of SO_2 amount to an average of 29.5×10^6 tons/yr in this region as compared with an estimated biogenic contribution of about 1% of this total (Adams, 1980). The eastern geographical distribution of major SO_2 emissions is associated with the large metropolitan areas combined with the heavily industrialized Ohio River Valley and western Pennsylvania. The areas of high emission density are roughly coincident with the regionally high ambient SO_2 and sulfate concentrations. The tonnage reported in the SURE region accounts for more than the total SO_2 emissions of 27.4×10^6 tons/yr given for the continental United States (USEPA, 1978). This difference is unresolved, but may be related to the inclusion of large sources in Canada in the SURE region.

In the Northeast, as in much of North America, SO_2 emissions are dominated by electric power plants. In other parts of North America, like the southwest United States and Ontario, metal smelting is the major feature of SO_2 sources.

The SURE estimates of emissions of particulate matter and nitrogen oxides by major source category are listed with the SO_2 values in Table 9. The total emitted particles (TEP) are 15×10^6 tons/yr and NO_x emissions are 12.0×10^6 tons annually. These are compared with the nationwide U.S. TEP of 12.4×10^6 tons/yr, and NO_x of 23.1×10^6 tons/yr from the USEPA (1978).

The TEP are approximately 50% of SO_2 emissions in the greater northeastern United States. The NO_x emissions are an average of about

TABLE 9. Aggregated Sulfur Dioxide and Nitrogen Oxide Emissions within the SURE Study Area by Season (thousands of tons/day)[a]

	Winter	Spring	Summer	Fall
SO_2				
Total	99.0	86.6	86.8	88.5
Utility	52.5	48.7	52.5	50.1
Industrial	36.5	33.0	31.6	33.5
Commercial	6.7	2.7	0.9	2.7
Residential	2.2	0.7	0.05	0.7
Transportation	1.3	1.5	1.6	1.5
NO_x				
Total	36.6	34.8	36.1	35.4
Utility	11.4	10.6	11.6	10.8
Industrial	7.5	6.5	6.0	6.6
Commercial	2.1	0.8	0.2	0.8
Residential	1.6	0.5	0.0	0.5
Transportation	14.0	16.5	18.2	16.7
Total Emitted Particulate (TEP)				
Total	46.3	44.5	45.0	45.2
Utility	5.8	5.4	5.7	5.1
Industrial	31.4	33.5	35.1	34.5
Commercial	4.1	1.6	0.4	1.5
Residential	2.0	0.6	0.025	0.6
Transportation	2.9	3.4	3.7	3.4

[a]Based on data of Klemm and Brennan (1979) taken in 1977.

40% of the SO_2 emissions in this region, based on the SURE inventory. Interestingly, the ratio of nitrate and NO_x in the air to total sulfur oxides in the air over the SURE region sometimes is significantly less than the emission ratio. The reason for the apparent loss of NO_x relative to sulfates in the atmosphere is an important question which remains unanswered, but may be related to gaseous and particulate deposition rates. That is, most of the nitrogen oxides in the atmosphere are gases, in contrast to sulfate, and may be lost faster as a result of dry deposition rate differences (see also Section 5.2).

Natural emissions of particles are associated with blowing dust, pollens and spores, sea salt, and volcanic disturbances. Even in highly urbanized or industrialized areas, these sources can contribute significantly to the total particle concentrations in the atmosphere, as currently defined by the high volume sampler measurement. Natural NO_x emissions are believed to be largely biogenic N_2O, which is chemically inert in the troposphere. However, soil exhalations of NO have been observed, and lightning can produce NO and NO_2.

Comparison of the distribution by source category with the SO_2 values indicates different groupings of contributions for each of these pollutants. Utilities, for example, account for approximately 60% of the SO_2 emissions according to data in Table 9. The next largest contributor is industrial sources, with about 34%. TEP, in contrast, is identified largely with the non-power-plant industry, while NO_x is associated with a 46% contribution from transportation, 19% from industry, and 31% from electricity generation. On a tonnage basis, products of SO_2 emissions, at least in eastern North America, probably represent the largest contributor to the SPC, but NO_x and TEP are of similar magnitude. The addition of reactive organic vapors of approximately 18×10^6 tons/yr adds another dimension to this complex mixture. The geographic distribution of major TEP and NO_x emissions over the eastern United States differs somewhat from the SO_2 distribution. These pollutants are concentrated heavily in the major metropolitan or industrial areas of the Midwest and Atlantic Coast, but to a lesser extent along the Ohio River Valley where SO_2 emissions are large.

Comparison between the area SO_2 emission densities of eastern North America and western Europe (Semb, 1978) indicates that the North American emission densities equal or exceed the European estimates when the latter are compared with the Ohio River Valley, Pennsylvania, and elsewhere. Thus the general northeastward migration of air across the area of very high emission density in North America creates the situation analogous to western Europe and Scandinavia.

A concern in defining the SPC is the relatively large contribution by mass of soil dust and other fugitive dust emissions. In many rural areas of North America, especially the arid Midwest and West, the SPC actually may be principally wind blown dust by mass concentration. As noted previously, soil dust influences the coarse particle fraction of airborne particulate matter. It is not clear at this time what significance to adverse environmental impact blowing dust has as a constituent of the SPC, along with other materials.

From this discussion and other information, it is apparent that the emissions of all of the pollutants in North America, with the possible exception of volatile organic carbon (VOC), are concentrated on the eastern half of the continent.

The link between geographic regions experiencing heavy air pollution and the consumption of energy is obvious and can be used to identify areas of potential health or ecological impact (Hidy et al., 1978). As an example, a map of annual energy flux density in kilograms of coal equivalents utilized per square kilometer for 1970 is shown in Figure 5. Using such a map, one can readily see that the zones where heavily industrialized populations are located can be expected to suffer regionally from different symptoms of pollution, ranging from widespread events of photochemical smog and haze to acidic precipitation. In addition to eastern North America and western Europe, central Europe, the western United

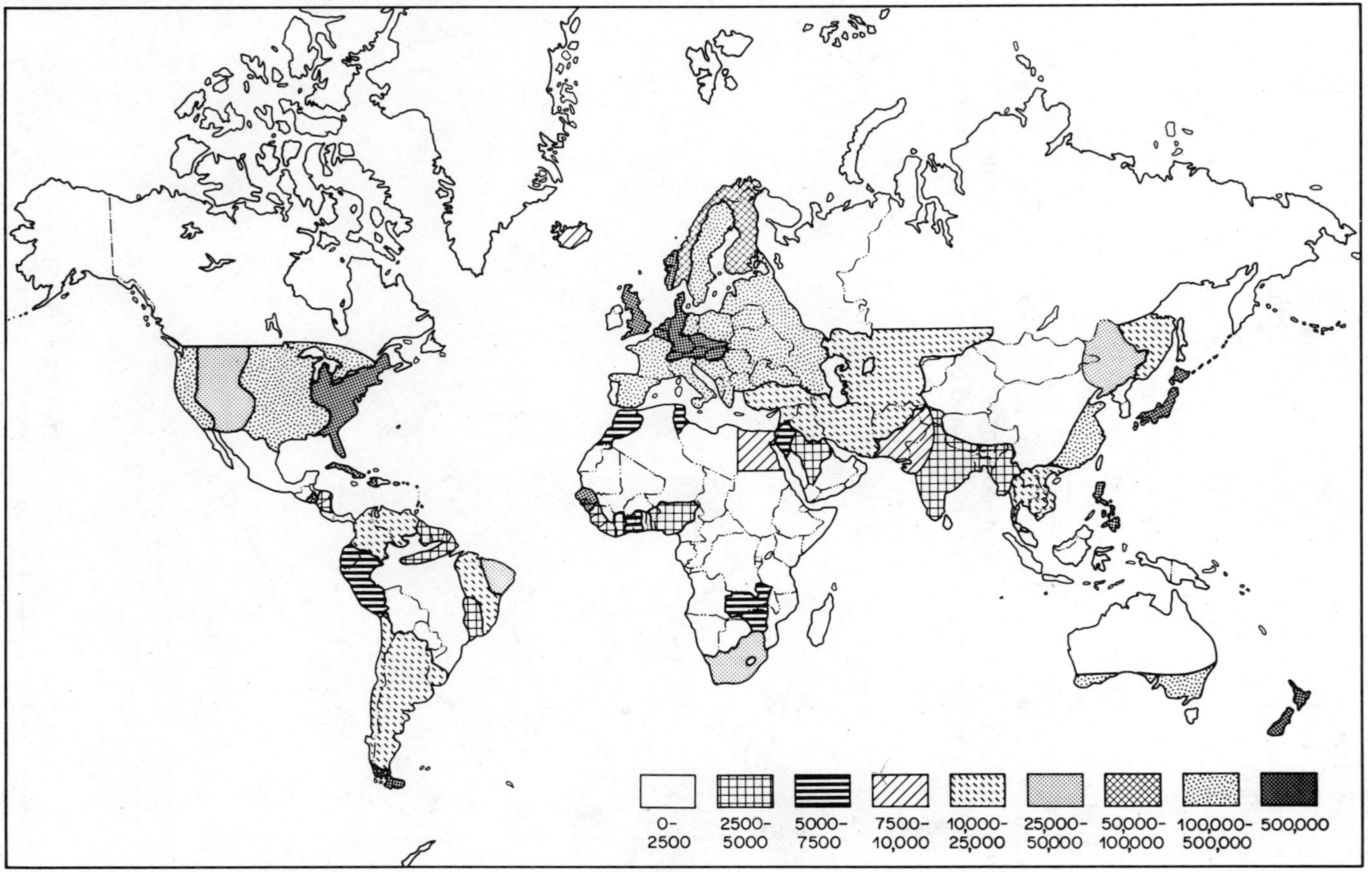

FIGURE 5. World energy flux density (kg of coal/km^2-year), 1970. (GSC, 1977, p. 108.)

States, western China, southern Australia, and Japan use similar quantities of energy. Hence, they are highly susceptible regions for air pollution impact. This qualitative argument is used later in Section 6 to speculate about the broadening influence of effects with increased worldwide energy consumption over the next 50 years.

In the United States, the history of emission control from 1970 to 1977 (Figure 6) has been moderately encouraging, with dramatic reductions in TEP and modest reductions in HC(VOC) and SO_x. However, NO_x emissions have increased, as noted previously, in conjunction with nitrate in rain. It also should be noted that TEP is defined in terms of total particulate matter without respect to size. Because of changing technology, there is no guarantee that the emission of finely divided particles has decreased proportionally to TEP. Also, it is important to note that production of fine particles by atmospheric reactives of SO_2 and NO_x may have increased with changes in emission distribution of these gases. The prospect for future emissions is not encouraging for SO_x and NO_x with increasing pressure to convert to coal use for electric power generation and eventually for space

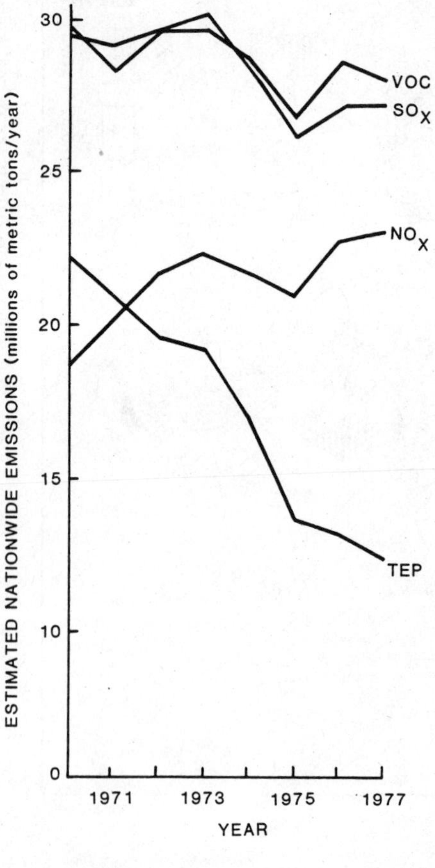

FIGURE 6. Estimates of U.S. nationwide annual emissions for TEP, SO_x, NO_x, and VOC (volatile organic compounds) for 1970–1977. (Source: U.S. EPA, 1978.)

heating. Energy consumption probably will increase with North American (and worldwide) industrialization, and these areas will continually rely on coal as a major contributor (Geophysical Study Committee, 1977).

5. ATMOSPHERIC RATE PROCESSES

The ambient concentration patterns of the SPC components are determined as a net effect of dynamic processes that are highly variable in time. An example of large changes on an event basis estimated for sulfate concentrations is shown in Figure 7. These data were calculated from the

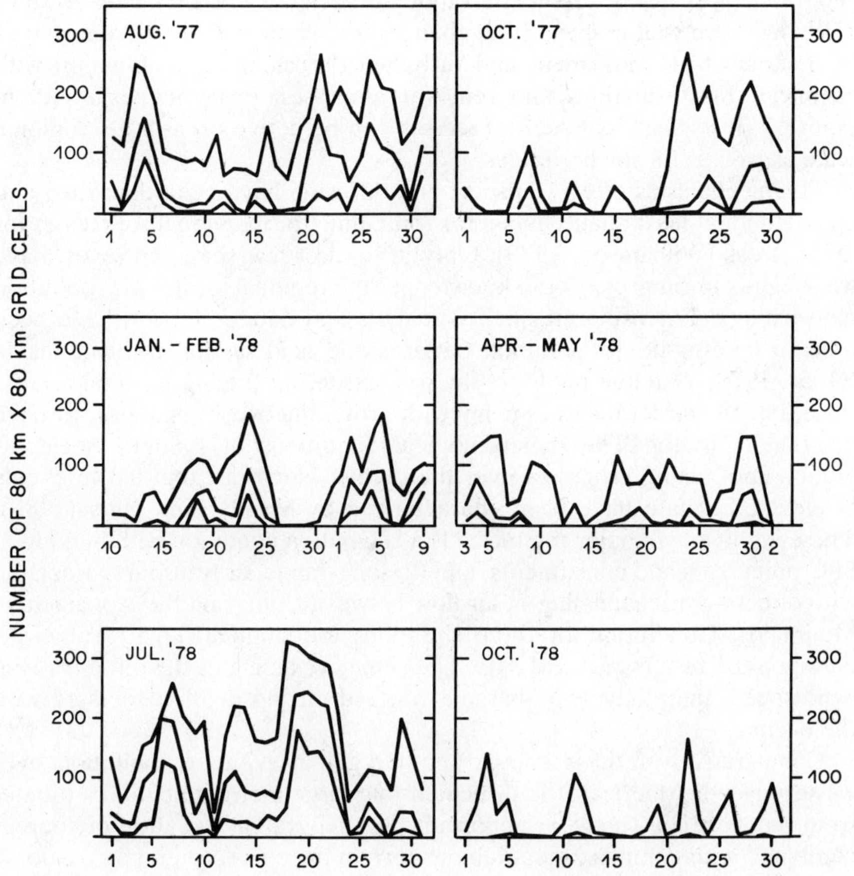

FIGURE 7. Number of 80 by 80 km grid cells experiencing 24-hour average sulfate concentrations of 10 (top line), 15 (middle line), and 20 $\mu g/m^3$ (bottom line) during intense months. A $1/r^2$ ($r = 240$ km) interpolation was employed for these estimates using available data (total possible coverage is about 350 cells from 45 stations).

SURE measurements over the northeastern United States for several months in 1977 and 1978. The variations are given in terms of the numbers of 80 × 80 km grids estimated to have a sulfate concentration exceeding a stated level. The SURE area included a total of 650 grids, 350 of which were over land. When a coverage of grids over the greater Northeast exceeds 200 for 10 $\mu g/m^3$ sulfate concentrations of 40% of the total land area, we have assigned a regional scale event (see also Mueller and Hidy, 1983). During such events, there are some locations centered in the area of greatest pollutant accumulation which experience 24-h sulfate levels up to 20 $\mu g/m^3$ and above. The regional-scale sulfate events take place over periods of 3 to 11 days and involve excursions in ground-level concentrations from a few $\mu g/m^3$ to excesses of 30 $\mu g/m^3$ and more. This variability is far greater than the estimated variability in emissions. On geographical scales of urban size (approximately 50 × 50 km), similar kinds of changes are well known. Such changes are associated with meteorological factors of air movement and turbulent diffusion, in combination with chemical transformations and removal processes. From the examples in Figure 7, it is clear that such processes can be active on a large, regional scale as well as on smaller scales.

The mechanisms of air transport and diffusion have been described and understood, at least qualitatively, for some time for individual sources or for urban areas (Williamson, 1973). Only in the last few years, however, have we begun to improve our knowledge of regional scale air pollution meteorology. Extensive air quality analysis and data collection have been used to investigate air pollution patterns and acid rainfall in Scandinavia (Ottar, 1979). Satellite photography has assisted in tracing haze on a large scale, but the direct measurements with ground networks and aircraft have been most useful. The meteorological conditions of synoptic scale air motion and mixing processes over the greater Northeast that led to events of elevated sulfate have been characterized by Mueller and Hidy (1983). These involve two major regimes: (1) a stagnation condition with buildup in SPC precursors and constituents, and (2) long-range air transport associated with directing or channeling of air flow between fronts and the Appalachian Mountains. Conditions for these situations are summarized in Table 10, based on the results of the SURE. The summary indicates the differences in wind speed, thermodynamic parameters, and mixing depth associated with the events.

Comparatively little is known about regional events of pollution over western North America. It is suspected that large metropolitan area plumes from the west coast may be important, but geographic isolation of sources combined with complex, mountainous terrain appears to segregate regional airsheds more than in the East. The southwestern states, the Great Plains, and the northwestern states may be separate in their regional pollution climatology.

TABLE 10. Meteorological Regimes Conducive to Regional Sulfate Accumulation and Transport in the Northeastern United States

Type	Synoptic Pattern	Seasonal Preference	Transport Winds	Afternoon Mixing Heights	Zone of Influence	Thermodynamics
Stagnation	Dominant high-pressure cell surface and aloft	Any season	Light (~5 m/s)	Low <1000 m	100–400 km	Any T, generally moist
Channeling	Stationary front along Great Lakes Bermuda High circulation Begins with stagnation	Summer	Moderate SW winds (7–12 m/s)	Moderate (1000–1500 m)	Long-range transport (>500 km)	Warm, moist

5.1 Chemical Transformations

Over the past few years a considerable improvement has taken place in knowledge of concentration distributions of pollutants and their (easily identified) chemical end products. A similar experience in understanding of the rates of chemical change in the atmosphere has emerged from the study of homogeneous gas phase processes. However, heterogeneous chemistry involving suspended particles of hydrometeors remains uncertain. The production of a variety of noncondensable species in the gas phase is believed to be largely associated with photochemical processes. These are discussed separately in the following chapter by Demerjian. Products in the condensed phase are identified as sulfur and nitrogen compounds, as well as organics. Reactions forming such products can be either homogeneous or heterogeneous in nature, or a combination of the two.

The oxidation reactions of SO_2 that are believed to be important in the troposphere are listed in Table 11. The first grouping is associated with homogeneous processes. Available laboratory data on rate constants indicate that the dominant reaction is that of the $^{\cdot}OH$ radical. The $RO_2^{\cdot}$ reaction ($R \equiv$ organic alkyl group) may be important in some circumstances, but the $HO_2^{\cdot}$ reaction recently has been found to be slower than originally thought.

The heterogeneous reactions believed to be important in the troposphere are grouped into aqueous and nonaqueous processes (Table 11). The aqueous reactions begin with the absorption of SO_2 to form bisulfite and sulfite (S^{IV}). These undergo oxidation by dissolved oxygen, O_3, or dissolved hydrogen peroxide. The reaction involving O_2 and O_3 is catalyzed by metal ions, including Mn^{2+} and Fe^{2+} (Beilke and Gravenhorst, 1978), and perhaps suspended elemental carbon (Chang et al., 1979). A particularly interesting feature of the latter reaction is that it is zero order in SO_2 over a range in concentration, in contrast with some of the other reactions. The SO_2 oxidation process is enhanced in an acid (ammonium) buffered solution (Beilke and Gravenhorst, 1978; Hegg and Hobbs, 1978) but suppressed with high acidity. Under acidic conditions, the H_2O_2 reaction appears to become important (Penkett et al., 1979), and perhaps dominant. Nonaqueous oxidation of SO_2 may take place on metal oxides such as MnO_2, Fe_3O_4, or on elemental carbon particles. A combination of the heterogeneous reactions may lead to SO_2 oxidation rates approaching 1%/h or more.

Review of available observations for oxidation reactions in plumes suggests a range of rates from <0.1%/h to approximately 10%/h. Heterogeneous processes may be taking place, but the information collected has not established conditions where they may be prevalent (Newman, 1980). In conditions typical of urban smog or summer pollution, evidence from atmospheric studies in Los Angeles and St. Louis suggests maximum oxidation rates from homogeneous processes between 1%/h and 13%/h

TABLE 11. Theoretically Possible Oxidation Paths for SO_2 in the Troposphere

---

Homogeneous Gas Phase Reactions

Inorganic reactions forming SO_3:

$$SO_2 + \tfrac{1}{2}O_2 + \text{sunlight} \rightarrow SO_3$$
$$O(3P) + SO_2 + M \rightarrow SO_3 + M$$

Organic reactions forming SO_3:

$$\overset{\diagup O_3 \diagdown}{CH_2{-}CH_3} + SO_2 \rightarrow SO_3 + 2CH_2O$$
$$^{\cdot}CH_2OO^{\cdot} + SO_2 \rightarrow SO_3 + CH_2O$$
$$CH_2 + O + O + SO_2 \rightarrow SO_3 + CH_2O$$
$$HO_2 + SO_2 \rightarrow HO + SO_3 \quad (a)$$
$$\rightarrow HO_2SO_2^{\cdot} \quad (b)$$
$$CH_3O_2 + SO_2 \rightarrow CH_3O + SO_3 \quad (a)$$
$$\rightarrow CH_3O_2SO_2^{\cdot} \quad (b)$$

Reactions forming $HOSO_2^{\cdot}$ or $ROSO_2^{\cdot}$ radical:

$$HO^{\cdot} + SO_2 \rightarrow HOSO_2^{\cdot}$$
$$CH_3O^{\cdot} + SO_2 \rightarrow CH_3OSO_2^{\cdot}$$

Heterogeneous Reactions

Aqueous:

$$SO_2 + H_2O(l) \rightleftarrows H_2SO_3$$
$$H_2SO_3 \rightleftarrows H^+ + HSO_3^-$$
$$HSO_3^- \rightleftarrows H^+ + SO_3^{2-}$$
$$2SO_3^{2-} + O_2(aq) \rightarrow 2SO_4^{2-}$$
$$SO_3^{2-} + O_3(aq) \rightarrow SO_4^{2-} + O_2$$
$$SO_3^{2-} + H_2O_2(aq) \rightarrow SO_4^{2-} + H_2O$$

Nonaqueous:

$$\left.\begin{array}{r} SO_2(ads) \\ H_2(ads) \\ O_2(ads) \end{array}\right\} + \text{carbons(s)} \xrightarrow{H_2O(?)} H_2SO_4(ads)$$

(Roberts and Friedlander, 1975; Henry and Hidy, 1979, 1982; Chang, 1979). Henry and Hidy (1979, 1982) have found, for example, that sulfate concentrations are strongly dependent on the *combined* levels of O_3 and absolute humidity in Los Angeles, St. Louis, and New York. These results suggest the importance of homogeneous reactions in the three cities. To the extent that the homogeneous reactions are photochemically related, this rate should show a strong diurnal dependence. Diurnal differences appear in the Gillani et al. (1978) observations of sulfur oxidation in a power plant plume. Their work shows a negligible oxidation rate at night, with a maximum of approximately 3%/h at midday for summer conditions near St. Louis.

Observational evidence that heterogeneous SO_2 reactions take place at appreciable rates in the troposphere is difficult to obtain. Indirect evidence comes from statistical relationships between sulfate behavior and aerometric variables. Henry and Hidy's (1979, 1982) analysis indicates that a combination of aerometric variables containing relative humidity influences the variability of sulfate in Los Angeles, St. Louis, New York, and Salt Lake City. Unlike absolute humidity, relative humidity is a measure of liquid water content for moist particles, and may reflect aqueous, heterogeneous processes. Sulfate concentrations in Salt Lake City are not related to O_3 and absolute humidity, but depend strongly on meteorological dispersion conditions, SO_2, and relative humidity. This difference suggests a distinction in mechanisms responsible for sulfate production. Comparing the investigations of urban data with rural data, statistical analysis of rural data from the SURE does not show sulfate dependence on absolute or relative humidity, which varies by season (Mueller, 1980). Pierson et al. (1980) also found little correlation between sulfate and relative humidity at Allegheny Mountain for summer days.

Recent observations confirm that considerable quantities of sulfate can be found in cloud water in polluted air (Lazrus, personal communication). Sulfate in cloud water may come from scavenging of aerosol particles or from oxidation of absorbed SO_2. Lazrus' observation of the presence of H_2O_2 at levels exceeding 0.1 ppb in cloudy air containing acidic droplets also points to the importance of peroxide in the oxidation mechanism.

Regardless of the mechanism, evidence has accumulated that 24-h average SO_2 oxidation rates of approximately 1%/h explain available observations taken from different locations in Europe and the eastern United States (Eliasson, 1978; Mueller and Hidy, 1983).

Like SO_2, NO_x oxidation may occur by homogeneous gas phase processes as well as by heterogeneous reactions. Less is known about NO_x reactions, which produce particulate nitrate, than about sulfate-forming processes. However, reactions that are believed to be relevant to the troposphere are listed in Table 12. The pathway from homogeneous reactions appears to be via nitric acid formation with the $^{\cdot}OH$ radical again playing an important role. Since sulfuric acid has a low vapor pressure, it

TABLE 12. Reactions Potentially Involved in Nitrate Formation

Homogeneous Reactions

Nitrogen oxides:

$$O_3 + NO \rightarrow NO_2 + O_2$$
$$O + M + NO \rightarrow NO_2 + M$$
$$RO_2 + NO \rightarrow NO_2 + RO^{\cdot}$$
$$O_3 + NO_2 \rightarrow NO_3 + O_2$$
$$NO_3 + NO_2 \rightarrow N_2O_5$$

Volatile acids:

$$N_2O_5 + H_2O \rightarrow 2HONO_2$$
$$HO^{\cdot} + NO_2 + M \rightarrow HONO_2 + M$$
$$NO + NO_2 + H_2O \rightarrow 2HONO$$
$$HOSO_2O + NO \rightarrow HOSO_2ONO + H_2O \rightarrow H_2SO_4 + HONO$$
$$HOSO_2O + NO_2 \rightarrow HOSO_2ONO_2 + H_2O \rightarrow H_2SO_4 + HONO_2$$

Gaseous nitrates:

$$NH_3 + HONO_2 \rightarrow NH_4NO_3$$
$$RO_2^{\cdot} + N_2O_5 \rightarrow R'{-}C({=}O)ONO_2 + NO_2 + R'{-}C({=}O)ONO + \cdots$$

Aqueous Reactions

$$N_2O_5 + H_2O(l) \rightarrow 2H^+ + 2NO_3^-$$
$$NO + NO_2 + H_2O(l) \rightarrow 2H^+ + 2NO_2^-$$
$$2NO_2 + H_2O(l) \rightarrow H^+ + NO_3^- + HONO$$
$$HONO + OH^- \rightarrow H_2O + NO_2^-$$
$$2NO_2^- + O_2(aq) \rightarrow 2NO_3^-$$
$$NO_2^- + O_3(aq) \rightarrow NO_3^- + O_2$$
$$2NO_2 + H_2SO_4 \rightarrow HNOSO_4 + HNO_3$$
$$HNOSO_4 + H_2O(l) \rightarrow HONO_2 + H_2SO_4$$
$$3HONO \rightarrow HNO_3 + 2NO + H_2O$$
$$RONO_2 + H_2O(l) \rightarrow H^+ + NO_3^- + R'OH$$

will condense at atmospheric concentrations after formation in the gas phase. However, nitric acid is quite volatile and will not condense at tropospheric concentrations. It then appears necessary to require a base donor such as NH_3 to produce a condensable salt of nitrate. Ammonium nitrate itself can evaporate or decompose so that the presence of particulate nitrate depends on temperature (Stelson et al., 1979) as well as perhaps on nitrate salt solubility in complicated acidic (partially) liquid aerosol particle mixtures.

Heterogeneous processes for the oxidation of NO_x are speculated to be analogous to those for SO_2 (Hidy and Burton, 1980). However, only the homogeneous reactions have been explored to any extent. The pathway of absorption of NO_x in water droplets to form nitrite is logical; oxidation in this case takes place in the presence of dissolved O_2, O_3, or H_2O_2. Little is known of the details of this process, but a basic ion is required to prevent evaporation of the nitrate product after its formation.

There are essentially no measurements of atmospheric NO_x oxidation rates to form nitric acid available in the literature. Limited observations of nitric acid concentrations have been attempted by EPA-sponsored investigators in large power plant plumes in Tennessee and Arizona. Nitrogen oxide conversion rates have not yet been reported from these experiments but may become available shortly.

According to the theoretical models of photochemical processes, the production of nitric acid can be rapid, exceeding SO_2 oxidation. Rodhe et al. (1979) have used this result to explain certain features of nitrate behavior in western Europe, as contrasted with sulfate precipitation distributions. In particular, their work leads to a hypothesis that differences in oxidation rate may account for high nitrate levels near major source regions in Europe, as compared with sulfate distributions which extend great distances from source-intense areas.

The possibility for production of organic aerosol particles from chemical reactions involving hydrocarbon vapors was first recognized by Haagen-Smit and co-workers many years ago. Recent reviews by Hidy and Burton (1980) and CMBEEP (1977) indicate that the probable reactions include O_3 or $^{\cdot}OH$ radical attachment to olefinic compounds. Hypothetically, these reactions produce a similar diversity of products, including the observed variety of oxygenated materials (Table 5). In the case of reactions involving hydrocarbon vapors of carbon number twelve or less, polymerization and oxidation are probably required to generate condensable material at the greater than parts per billion concentrations of reactants found in the atmosphere.

So far, no unambiguous method has been found to estimate the amount of organic carbon produced by atmospheric processes compared with primary emissions. Assessment by compound identification is difficult because it is believed that combustion and photochemical reactions in the atmosphere will generate similar products. Grosjean and Friedlander (1980) have speculated that certain organic nitrates and dicarboxylic acids are likely candidates for secondary organic aerosols. Friedlander (1973) attempted a carbon balance for photochemical smog conditions and showed circumstantial evidence for a large fraction of carbon being of secondary origin in Los Angeles air. This kind of analysis has not been done in other situations.

The production rate of organic aerosols can be rapid under conditions approaching those in the atmosphere. In laboratory experiments, naturally occurring terpenes and cyclic olefins react within seconds to form aerosols where ozone is present. Experiments reported by Hidy and Burton (1980) for linear olefins suggest rates of tenths of a microgram per cubic meter per hour for atmospheric concentrations. This appears to be an order of magnitude lower than projected from observation of buildup in organic aerosol in Los Angeles smog. Grosjean and Friedlander (1980) have reported rates an order of magnitude higher if cyclic or diolefins are present in urban air.

Because of the lack of knowledge about individual organic compounds and their secondary contributions, including production rates, considerable attention needs to be devoted to this part of the SPC. Such knowledge seems to be essential if systematic, precursor-specific, particulate controls are to be established for reduction in particles of secondary origin.

5.2 Dry Deposition Processes

The removal of gases and particles at the earth's surface is a major cleansing process for air. Gases are removed by absorption in soil, buildings, or vegetation, while particles are scavenged by diffusion processes, interception, or impaction on surfaces resulting from inertial forces. The removal of gases like SO_2 occurs at a rate roughly equivalent to its average oxidation rate in the atmosphere. The deposition of NO_x should occur at a rate similar to that for SO_2. Particulate matter is depleted at the surface at a rate lower than that for gases. Some evidence indicates that the particulate removal rate is an order of magnitude lower than removal rates for gases, but recent observations suggest conditions where the two rates are nearly equivalent (Wesely et al., 1977).

Conventionally, the dry deposition rate is given in terms of a deposition velocity v_d, or its reciprocal, the resistance to deposition, r. The flux of material to the surface F then is

$$F\left(\frac{\mu g}{m^2-s}\right) \equiv Cv_d = \frac{C}{r} \tag{3}$$

where C is the concentration of contaminant at a reference height, say 10 m.

The resistance r is given in terms of the sum of an aerodynamic component r_a, a surface term r_s, and a canopy stomatal resistance r_c (Wesely and Hicks, 1977). The first term depends on the turbulent air conditions near the surface, and the second is associated with diffusion processes through thin layers of air adjacent to the surface. The third term depends on the physiological properties of vegetation ($r_c = 0$ for clean soil or building materials). The total resistance at the surface is often taken as 1 s/cm for reactive gases but larger for nonreactive species. The surface resistance is inversely proportional to the friction velocity at the surface, and the stomatal resistance for SO_2 is approximately 0.7 s/cm as a minimum (Garland, 1979). Gaseous removal rates depend not only on the aerodynamics of the air surface layers and physiology of vegetation, but also on the nature and wetness of the surface. These factors are separated for SO_2 and O_3 but have been little studied for other gases.

The deposition of particles depends on the surface roughness as well as on particle size. Observations of deposition ratios from turbulent media in laboratory situations indicate that the variation in deposition velocity v_d is such that a minimum exists in the 0.1 to 1.0 μm diameter range (see Fig. 8). For smooth surfaces, this minimum is rather sharp, with an increase in v_d with large particles associated with inertial deposition and an increase with

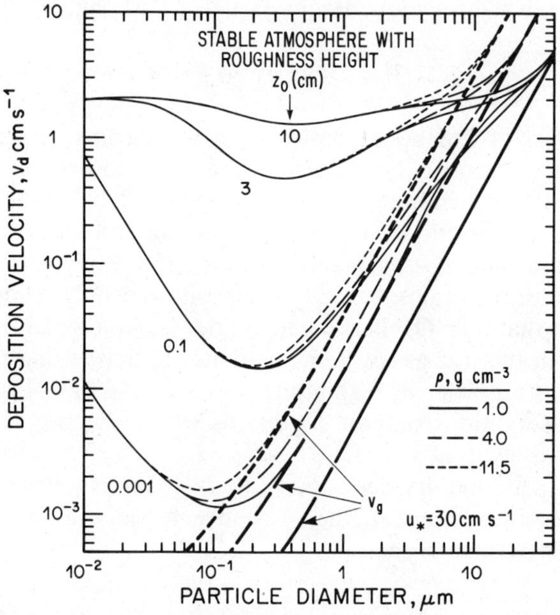

FIGURE 8. Correlation of experimental data by Sehmel and Hodgson (1974) for dry deposition of particles to various solid surfaces. The data encompassed roughness heights only up to about 0.1 cm; therefore, the extrapolation to greater roughness heights is tentative. The deposition velocity plotted is the flux divided by the extrapolated concentration at 1 m; v_g is the gravitational setting speed for particles of indicated density (ρ); the case with $u_* =$ 30 cm/s is shown ($\bar{u} \simeq 10$ m/s).

small particles as a result of Brownian diffusion. As the surface roughness increases, the minimum broadens so that a wide range of particles experience deposition velocities of as high as 1 cm/s for large roughness heights. A value of 0.1 cm/s is often adopted for air quality prediction in the absence of a better value.

Dry deposition may account for a substantial fraction of the apparent variation in concentration of airborne material from day to day and is a factor in the geographical distribution of pollutant impact on the biome. Since relatively little is known about patterns of deposition, or even how to measure them quantitatively, considerable work is needed in this area if impact assessment is to progress.

5.3 Wet Deposition

When precipitation occurs, a substantial increase in deposition of airborne material is expected. The wet deposition of sulfur compounds has been studied with other materials found in rainwater. The mechanism of accumulation in droplets involves in-cloud processes (rainout) and scavenging during fallout from clouds (washout). In both rainout and washout, gases are absorbed in accordance with their solubility, which depends on temperature, pH, and other factors. Particles are collected either by (1) acting as condensation nuclei during droplet formation, or (2) scavenging by collisional processes, including Brownian diffusion and inertial impaction of interception. Both sulfur and nitrogen oxide gases should be absorbed at a reasonable rate in rain, with NO_2 being more soluble than SO_2. Sulfate and nitrate particles should be good nuclei for cloud droplet production since they are generally hygroscopic.

Garland (1978) has calculated the contributions of different mechanisms for wet removal of sulfur. His results are summarized in Table 13. One sees that the solution and oxidation of SO_2 is a large factor but may be overshadowed by (sulfate) nuclei uptake in the droplet-forming process. The absorption of SO_2 by washout also may be significant. Impaction or interception of coarse particles by falling hydrometeors is an inefficient collection process but cannot be eliminated a priori. The latter is likely to be most intense in the first fallout of rain. The investigations of the U.S. Department of Energy's Multistate Atmospheric Power Production Pollution Study (MAP3S) are beginning to show a distinction in sulfur chemistry between summer convective cloud systems and winter stratus clouds. In summer, the absorption of SO_2 from local sources appears to be a significant contributor to sulfate in cloud water (Hales and Dana, 1979). Winter studies over the Great Lakes area suggest that the uptake of sulfate as hydrometeor nuclei is a principal mechanism for sulfate in cloud waters (MacCracken, 1979). Such studies point to the complexities of aqueous chemistry which can be expected in the troposphere.

The contribution of wet processes to the overall deposition rate of

TABLE 13. Mechanisms Contributing to Sulfur in Rainwater (Garland, 1978)[a]

Mechanism	Pollutant Lifetime during Rain (h)		Sulfate Concentration in Rainwater (mg/L)	
	Composition (a)	Composition (b)	Composition (a)	Composition (b)
Particulate sulfate				
Diffusophoresis	—	—	10^{-2}	10^{-3}
Brownian diffusion to cloud droplets	100	4000	0.2	3×10^{-3}
Brownian diffusion to raindrops	10^4	2×10^{-5}	10^{-3}	10^{-5}
Impaction and interception by raindrops	400	1	0.02	1.2
Rainout of condensation nuclei	—	—	3–10	3–10
SO_2				
Solution and oxidation in cloud droplets	—	—	3	3
Uptake by falling raindrops	10	1	1	1

[a] Assumptions: (i) 10 $\mu g/m^3$ of sulfate of Composition (a)—80% submicron particles of typical diameter 0.2 μm and Composition (b)—20% larger particles of typically 4-μm diameter; (ii) 10 $\mu g/m^3$ of SO_2; (iii) rain falling at 1 mm/h as 1-mm drops from a cloud containing 100 drops/cm^3, each of 10 μm radius.

TABLE 14. The Magnitude of the Sulfur Budget Terms for Eastern North America (Tg S/y)[a]

Term	Eastern Canada	Eastern United States	Eastern North America
Inputs			
Man-made emissions	2.1	14.0	16.1
Natural emissions	0.36	0.47	0.76
Inflow from oceans	0.04	0.2	0.2
Inflow from west	0.1	0.4	0.5
Transboundary flow	2.0	0.7	—
Totals	4.6	15.8	17.4
Outputs			
Transboundary flow	0.7	2.0	—
Wet deposition	3.0	2.5	5.5
Dry deposition	1.2	3.3	4.5
Outflow to oceans	0.4	3.9	4.3
Totals	5.3	11.7	14.3

[a]Tg sulfur per year; Tg = 10^6 metric tons. From Whelpdale and Galloway (1979).

material will depend on the precipitation occurrence as well as the chemistry of assimilation into rain. For sulfur fallout over northern Europe, Eliasson (1978) has estimated that the annual wet deposition of sulfur is about half that of dry deposition. For eastern North America, Whelpdale and Galloway (1979) have calculated that annual wet deposition rates and dry deposition rates are approximately the same (Table 14).

6. BUDGETING

6.1 Eastern North America

Given the estimates of residence times and "reservoir concentrations" from rate processes for atmospheric contaminants, their geochemical budgets can be constructed. Global budgets for the flow of elements through the environment are instructive for exposure studies and ecological cycles. Semiquantitative budgets for components of the SPC have been reported in the literature (Garrels et al., 1975). They have indicated the significance of human activities in contrast to nature in many circumstances. Perhaps most extensively studied in the atmosphere is sulfur. Granat et al. (1976) have prepared a regional sulfur budget for northwestern Europe showing a balance in sulfur transport and deposition. A similar calculation for eastern North America has been prepared by Whelpdale and Galloway (1979), and Shinn and Lynn (1979). The former authors' results are shown in Table 14. Unlike many atmospheric sulfur budgets, these calculations are based on "independent" estimates of each term. The inputs are derived from man-made and natural emissions. As expected, human activities account for most of the atmospheric sulfur in this geographical region. The transboundary flow of sulfur from the United States to Canada is about the same as Canada's emissions. Total deposition of sulfur on eastern Canada is approximately the same as that on the eastern United States. It is estimated that a significant amount of sulfur is also transported eastward over the Atlantic Ocean from the continent.

The calculations of Whelpdale and Galloway underscore the issues of the international aspects of pollution involving the sulfur in the SPC, as has the situation in Scandinavia. With intensive industrialization in parts of the world, it appears inevitable that the pollution components with atmospheric residence longer than a few days will have a multinational influence. New cases will arise where pollution will be transported across borders, imparting adverse effects to neighbors who may have little "home-produced" pollution. The impact assessment technology for such situations has only recently begun to emerge.

Sulfur is essentially the only component of the SPC which has been examined for regional budgets. Similar analyses are needed for nitrogen oxides and eventually for carbon compounds. Taking into account differences in chemistry and deposition behavior, an improved perspective will be obtained in geographical areas which are sensitive to large dosages of the SPC.

6.2 A Future Picture

Even without detailed budgets, it is possible to obtain a qualitative picture of regions where SPC pollution will be a factor over the next 20 years. There is little doubt that continuing urbanization and industrialization will give rise to an increase in emissions contributing to the SPC. The acidic or organic gases involved are known to have residence times less than 3 days in duration, while the particulate residence times are in the order of 5 days or less. Increased pollution basically can be tied to corresponding increases in energy consumption. For the next 45 years, energy production will be heavily influenced by fossil fuel combustion, with increasing contributions from coal utilization. Estimates of coal consumption have been made and suggest intensification of SPC pollution in the Northern Hemisphere as well as in parts of the Southern Hemisphere (Husar and Patterson, 1979).

Taking into account the geographical distribution of increased energy production, the global prevailing winds, and residence times for SPC components, Hidy et al. (1978) have estimated certain regional zones of influence over the world.

As suggested in Figure 9a, areas of high-density energy consumption will extend to the regional and larger scale in the Northern Hemisphere and also will become more prevalent in the Southern Hemisphere (compare with Fig. 5). Emerging areas of regional pollution of potential concern are the east coast of South America, the Middle East, eastern Europe, and the USSR, as well as southern China, Manchuria, Korea, and Australia.

Within the projected pattern of urbanization over the next 45 years, it appears that international issues focusing on pollution will involve domestic and transboundary problems in regions of high energy consumption and heavy industrialization. For example, emissions of particles (TEP) and SO_2 are expected to increase substantially during that period (Table 15). According to these estimates, the United States, western Europe, the USSR, and Asia will continue to account for most of the pollution associated with SO_2 and TEP.

The qualitative assessment of regions of susceptibility to pollution can be done for the year 2025 following the arguments used by Hidy et al. (1978).

TABLE 15. Estimated Emissions of Pollutants for the Year 2025 Based on Lists from the Geophysics Study Committee (1977)[a]

	Pollution in Tg[b]	
Region	TEP	SO_2
United States	383 (284)	245 (182)
North America less U.S.	15 (4)	4.9 (1.5)
Western Europe	240 (69)	94 (27)
Oceania	18 (18)	6 (5.3)
Latin America	171 (5)	219 (14)
Japan	136 (–)	107 (–)
Communist Asia	124 (124)	72 (72)
Other Asia	175 (49)	88 (41.3)
Africa	63 (34)	20 (11.4)
USSR	342 (193)	304 (181)
Communist E. Europe	152 (11)	202 (14.4)
Total	1820 (791)	1360 (549)

[a]Data without parentheses are for regional shortfalls supplied by coals, data in parentheses are for regional shortfalls supplied by nonpolluting fuels.

[b]$Tg = 10^6$ metric tons.

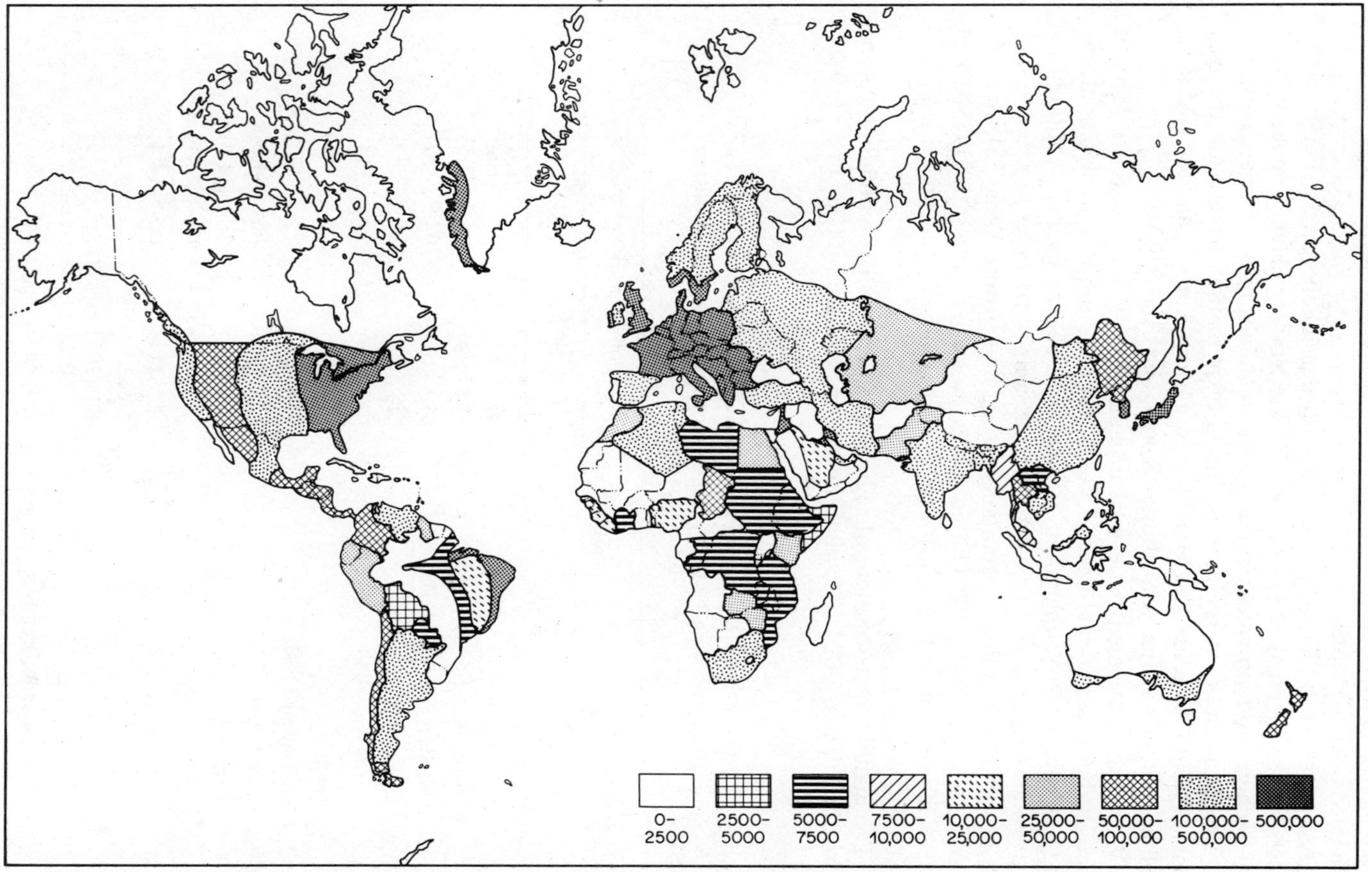

FIGURE 9*a*. World energy flux density (kg coal/km^2-yr) for the year 2025. (GSC, 1977.)

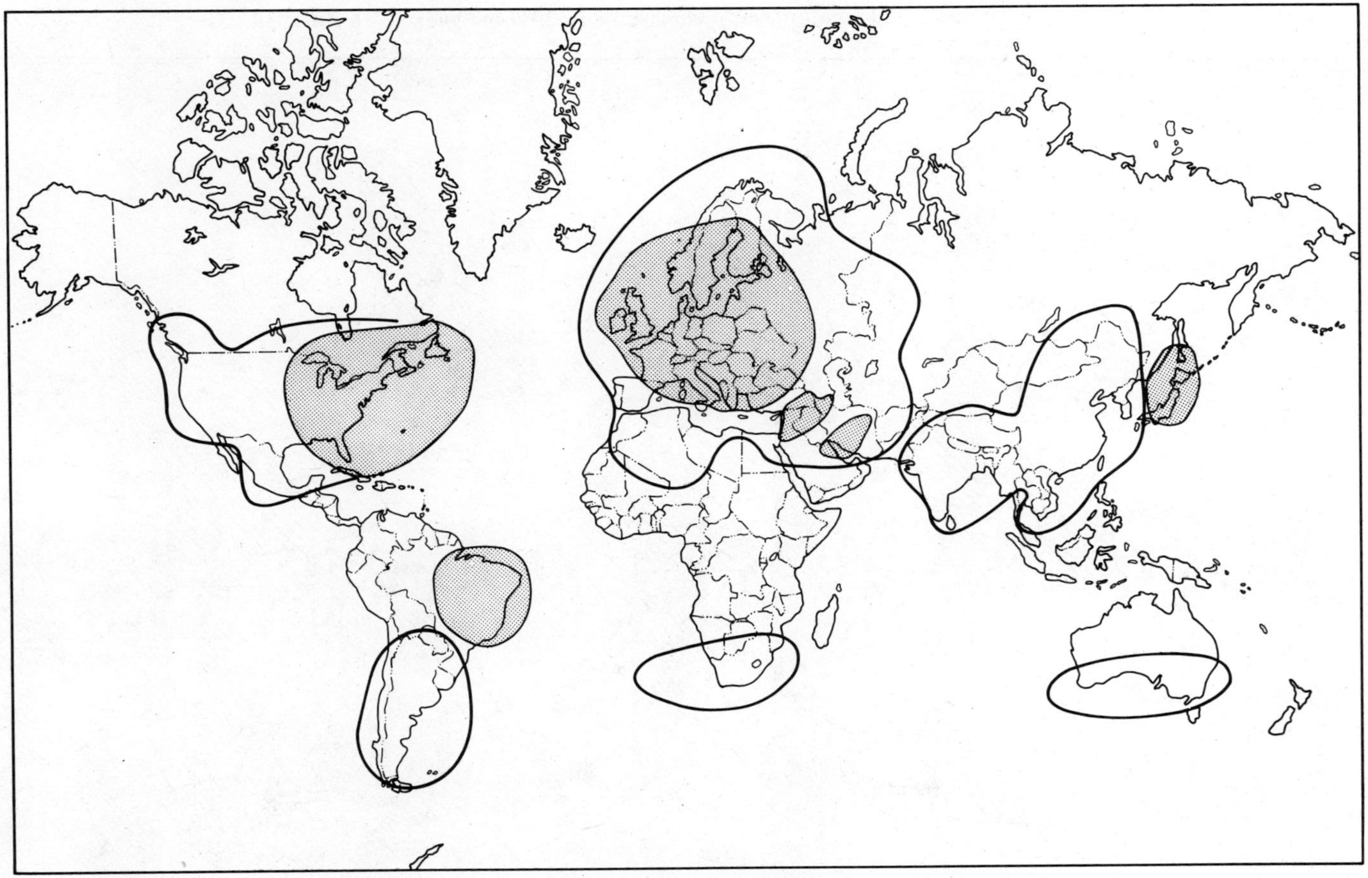

FIGURE 9*b*. Regions of exposure to pollutants with turnover times of less than a few days. (Hidy et al., 1978.)

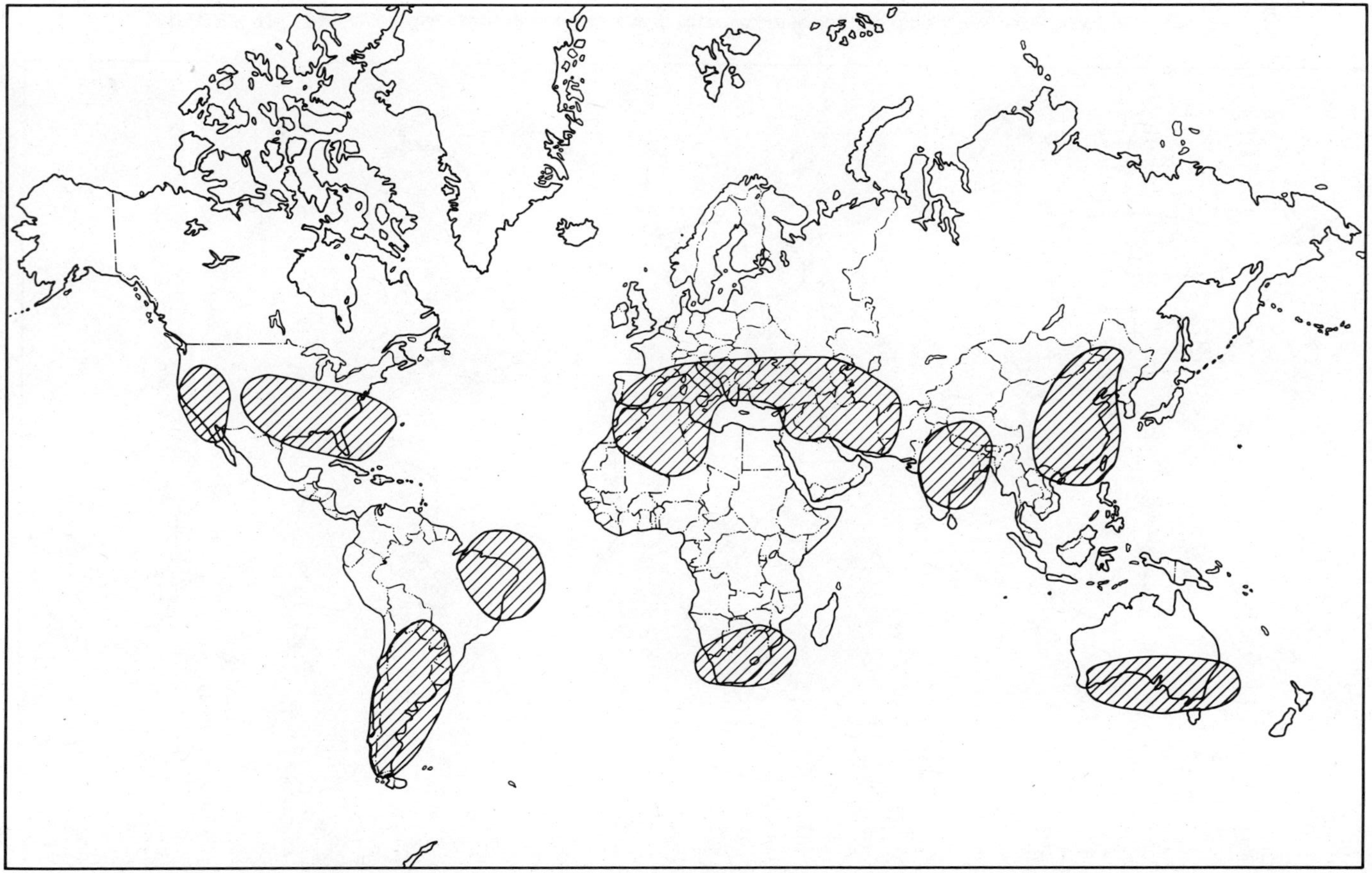

FIGURE 9c. Regions of high susceptibility to photochemical smog. (Hidy et al., 1978.)

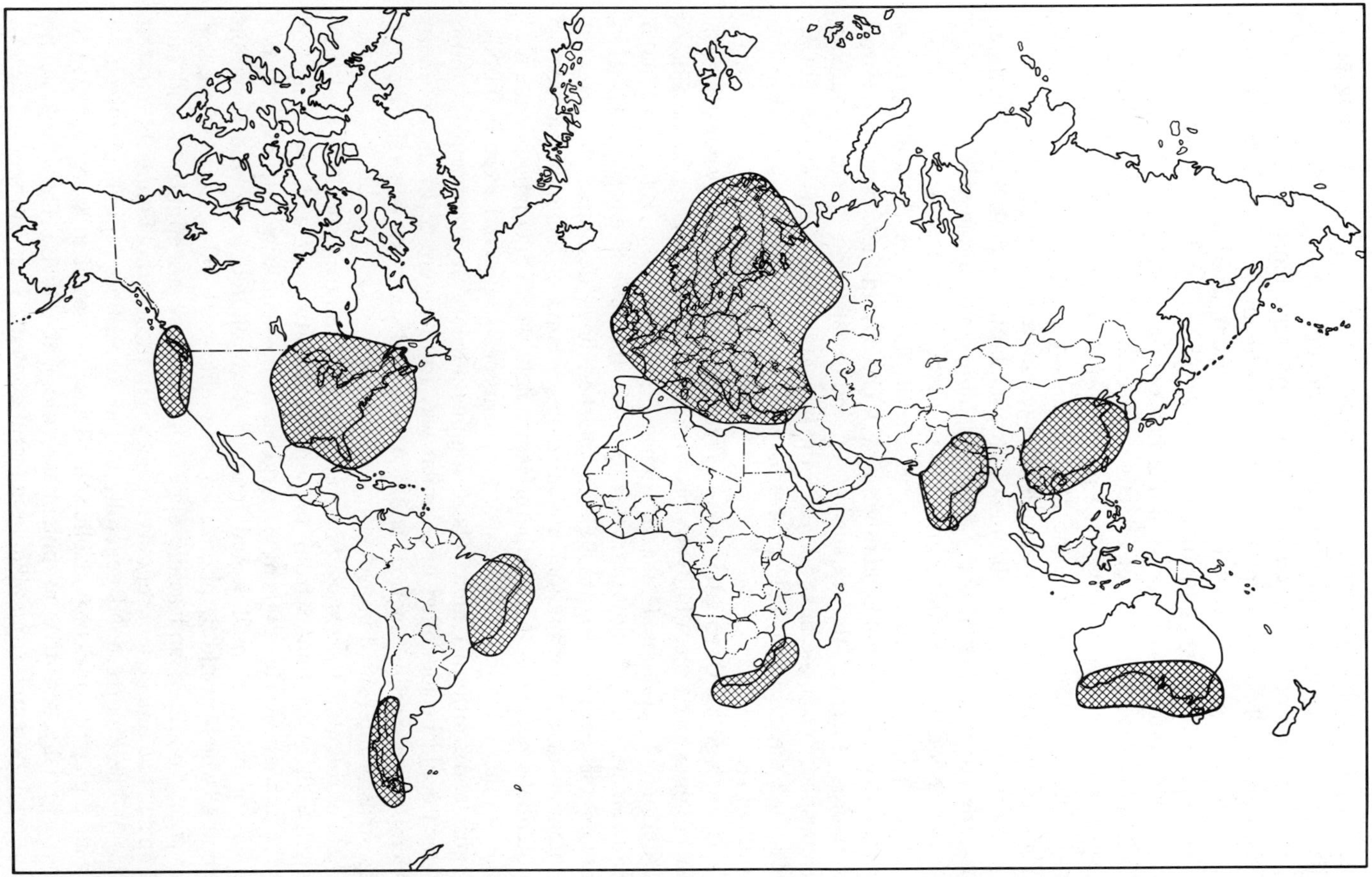

FIGURE 9*d*. Regions of high susceptibility to acid precipitation or pollution-contaminated precipitation. (Hidy et al., 1978.)

Essentially, three maps overlaying the energy density distribution are of interest: the pollution transport estimates, the photochemical smog estimate, and the acid precipitation impact estimate.

Let us assume that a regional impact of large emission density zones ($\geq$100,000 kg/km^2 yr) is in the order of 1000 km downwind, based on a pollutant residence time of 1 to 3 days (as exemplified by SO_x). Then the potential for regional pollution effects as suggested by transport of reactive, short-lived pollutants is shown in the first overlay on Figure 9a, Figure 9b. This indicates that, without effective implementation of controls, we can expect a widespread expansion of regional pollution occurrence in most of the United States, Europe, the Mediterranean area, the USSR, and Asia. The regional exposure may actually merge into a zonal band of pollution in the Northern Hemisphere throughout the midlatitudes, with the ocean areas and the northern USSR as the only remaining sinks of material.

In the Southern Hemisphere, there may be a buildup of regional-scale pollution in Argentina, Chile, and Brazil, as well as in South Africa. Southern Australia will be another area of concern in the absence of control measures. Despite the expansion of industrialization in the Southern Hemisphere over the next 50 years, the impact of zonal transport should be less, and the regional scale problems will remain relatively isolated in the midlatitudes.

Using a global distribution of solar radiation and the energy density map, the regions of susceptibility to photochemical smog can be identified. A criterion of combined energy density greater than 100,000 kg coal equivalents/km^2 yr and solar flux of 300–400 cal/cm^2 day is applied to obtain the overlay, Figure 9c. The slashed lines designate the areas of potential for intense photochemical smog. A band of photochemical smog potential emerges in the Northern Hemisphere which encompasses much of the southern United States, southern Europe, south and western USSR, India, and southern China. These regions are largely expansions of those which could be identified in a similar way from Figure 5. In the Southern Hemisphere, South America, South Africa, and southern Australia emerge as areas of intensified impact.

The potential for exposure to acid gas pollutants, especially acid precipitation, is estimated from the energy-density map and the zones of precipitation of 100–200 cm/yr. Arid areas have little wet deposition, and areas of heavy rainfall would tend to have diluted, dissolved pollution content which would be percolated away efficiently. Also, areas where soil is poorly buffered and acidic appear to be most susceptible to acid rain impact. The antecedent regions are shown in the overlay Figure 9d (crossed lines). These correspond generally to the susceptible regions identified by using a map of zones of podzolic soils. These classes of soils are expected to be especially sensitive to potential ecological effects. They lack acidic buffering potential from alkaline mineral content. They are seen in eastern

North America, Europe (Scandinavia), and India, as well as in Chile, Argentina, and South Africa.

In a general way, the qualitative assessment of regional impact tends to identify essentially overlapping regions, with heaviest effects predicted to continue for eastern North America and Europe. Thus, the future regional pollution patterns will be similar to those known now, but the problems may become increasingly widespread and severe without intensification of control measures. Again, as Figure 9 illustrates, the potential influence of the SPC in the future extends well beyond national boundaries. Consideration of the issues will take on an increasing international flavor in the future.

7. SUMMARY AND CONCLUSIONS

Over the past few years, the information describing the nature, origins, and spatial or temporal distribution of components of the sulfur oxide–particulate complex (SPC) has expanded substantially but remains incomplete. A variety of specific compounds have been identified by direct or indirect methods, but a material balance by constituent has not been possible. The mechanisms for production of constituents by atmospheric chemical processes are best known for sulfate and less well understood for nitrates and organic compounds. Chemical interactions of the SPC with clouds and precipitation result in changes in atmospheric composition and wet deposition of material. Relatively little is known about these effects compared with dry processes.

Because the mechanisms for removal are relatively slow, SPC components can be distributed regionally, in excess of 1000 km downwind of sources. Although such distributions are regional in nature, their maximum influence will be at distances less than a few hundred kilometers because of the strong dilution ability of the atmosphere. Regional influence of the SPC on tropospheric concentration occurs infrequently on an event basis. These pollution events can occur any season of the year and may involve widespread air mass stagnation, sometimes followed by persistent conditions of forced advection of polluted air masses over periods of 3 to 5 days. The exposure or dosage of elevated SPC concentrations as measured by SO_x can vary widely, depending on the frequency of occurrence of pollution events and the association of precipitation with them. The regional influence of SPC in wet deposition from precipitation processes has only recently been explored in detail both in Europe and North America.

Despite the progress in characterizing the nature and occurrence of the SPC, there remain several aspects that require continued investigation. These include the following areas:

1. The SPC has a major component of sulfur oxide, but is a complicated mixture of material including both inorganic and organic constituents. As yet, adverse environmental effects potentially related to this mixture have not been defined very well.

2. Determination of the chemical composition, and its distribution by particle size, is a continuing requirement for impact assessment, especially for identification of specific compounds.

3. Classification of SPC compounds by concentration and potential toxicity is needed.

4. Improved observations of the spatial and temporal variability in chemical transformations and deposition are required.

5. Dry and wet deposition patterns remain poorly determined in most areas of the world.

6. Characterization of regional-scale air quality meteorology in western North America is required to complete the spatial distribution picture of this continent.

7. Nitrate chemistry needs detailed consideration to extend knowledge of its variability in the troposphere.

8. A nitrogen oxide budget should be constructed for North America to compare its behavior with the sulfur budget.

9. The characterization of carbonaceous material is required to distinguish its significance in rural and urban areas, and to identify primary versus secondary origins.

10. Cloud chemistry and interactions require detailed investigation to establish their importance in regional processes.

11. The role of partially soluble organics in changing gas absorption rates may be a factor.

12. The total acidity of the air relative to precipitation for a "natural" baseline and for polluted conditions needs to be established.

13. Chemistry in clouds may be a dominant factor in nitrate behavior.

14. The trends in surrogate measures of the SPC should be measured systematically for several years to verify expectations of the impact of increasing pollution emissions.

ACKNOWLEDGMENTS

This work was sponsored in part by the Electric Power Research Institute under its Contract No. 862. Thanks go to Dr. W. Pierson for his comments and suggestions to improve this manuscript.

REFERENCES

Adams, D. F., S. O. Farwell, E. Robinson, and M. R. Pack. *Biogenic Sulfur Emissions in the SURE Area*. Report EPRI-EA-1516, 170 pp. (1980).

Altshuller, A. P. Regional transport and transformation of sulfur dioxide to sulfates in the U.S. *J. Air Poll. Contr. Assoc.* **26**, 318 (1976).

Beilke, S., and G. Gravenhorst. Heterogeneous SO_2 oxidation in the droplet phase. *Atmos. Environ.* **12**, 231–240 (1978).

Bernstein, D., and K. Rahn. Trace element concentrations as a function of particle size. *Ann. N.Y. Acad. Sci.* **322**, 87–98 (1979).

Brosset, C. Water soluble sulfur compounds in aerosols. *Atmos. Environ.* **12**, 25–38 (1978).

Cautreels, W., and K. van Cauwenberghe. Experiments on the distribution of organic pollutants between airborne particulate matter and the corresponding gas phase. *Atmos. Environ.* **12**, 1133–1142 (1978).

Chang, S. G., R. Brodzinsky, R. Toossi, S. S. Markowitz, and T. Novakov. Catalytic oxidation of SO_2 on carbon in aqueous solutions. In T. Novakov (ed.) *Proc. Symp. on Carbonaceous Particles in the Atmosphere, March* 20–22, 1978. Report LBL-9037, pp. 123–130. Lawrence Berkeley Laboratory, Carbonaceous Part. Atmos. (1979).

Chang, T. Y. Estimate of the conversion rate of SO_2 to SO_4 from Da Vinci flight data. *Atmos. Environ.* **13**, 1663–1664 (1979).

Charlson, R. J., D. S. Covert, T. V. Larson, and A. P. Waggoner. Chemical properties of tropospheric sulfur aerosols. *Atmos. Environ.* **12**, 39–53 (1978).

Cogbill, C. V. The history and character of acid precipitation in Eastern North America. *Water Air Soil Pollut.* **6**, 407–414 (1976).

Committee on Medical and Biologic Effects of Environmental Pollutants (CMBEEP). *Ozone and Other Photochemical Oxidants*, 51 pp. National Academy Sci., Washington, D.C. (1977).

Countess, R. J., G. T. Wolff, and S. H. Cadle. Denver winter aerosol—A comprehensive chemical characterization. *J. Air Poll. Contr. Assoc.* **30**, 1194–1200 (1980).

Daisey, J. M., M. E. Leyko, M. T. Kleinman, and E. Hoffman. The nature of the organic fraction of the New York City summer aerosol. *Ann. N.Y. Acad. Sci.* **322**, 125–142 (1979).

Eliasson, A. The OECD study of long range transport of pollutants: Long range transport modeling. *Atmos. Environ.* **12**, 479–487 (1978).

Faoro, R., and T. McMullen. National Trends in Trace Metal in Ambient Air 1965–1974. EPA-450/1-77-003. U.S. Environmental Protection Agency, Office of Air Quality Planning and Standards, Research Triangle Park, NC (1977).

Friedlander, S. K. Chemical element balances and identification of air pollution sources. *Environ. Sci. and Technol.* **7**, 235–240 (1973).

Garland, J. A. Dry and wet removal from the atmosphere. *Atmos. Environ.* **12**, 349–362 (1978).

Garland, J. A. Dry deposition of gaseous pollutants. In *Proc. WMO Symp. on the Long-Range Transport of Pollutants and Its Relation to General Circulation*. WMO No. 538, p. 95. World Meteorol. Org., Geneva, Switzerland (1979).

Garrels, R. M., F. T. MacKenzie, and C. Hunt. *Chemical Cycles and the Global Environment*. William Kaufman, Inc., Los Altos, CA, 1975.

Geophysics Study Committee (GSC). *Energy and Climate*. National Acad. Sci. Washington, D.C. (1977).

Gillani, N. V., R. B. Husar, J. D. Husar, D. E. Patterson, and W. E. Wilson, Jr. Project MISTT: Kinetics of particulate sulfur formation in a power plant plume out to 300 km. *Atmos. Environ.* **12**, 589–598 (1978).

Granat, L., R. O. Hallberg, and H. Rodhe. The global sulfur cycle. In B. H. Svenson, and R. Soderlund (eds.) *Nitrogen, Phosphorus and Sulfur–Global Cycles*, pp. 89–134. SCOPE Report 7 (1976).

Grosjean, D., and S. K. Friedlander. Formation of organic aerosols from cyclic olefins and diolefins. In G. M. Hidy et al. (eds.) *The Character and Origins of Smog Aerosols*, p. 435–473. Wiley, New York (1980).

Grosjean, D., G. M. Hidy, and E. Irvine. *Relation Between Sulfur, Nitrogen Oxides and Suspended Particulates in the Southern San Joaquin Valley*. Report No. PA019-100. Prep. for Western Oil and Gas Association, Los Angeles, CA by Environ. Res. and Technol., Inc., Westlake Village, CA (1979a).

Grosjean, D., K. K. Fung, P. K. Mueller, S. L. Heisler, and G. M. Hidy. Particulate organic carbon in urban air: Concentrations, size distribution and temporal variations. Presented at AICHE Symp. on Sampling and Analysis of Particulate Matter, 72nd Annual Meeting, San Francisco, CA (1979b); *AICHE Symp. Ser.* **76**(201), 96–107 (1980).

Hales, J. M., and M. T. Dana. Precipitation scavenging of urban pollutants by convective storm systems. *J. Appl. Meteor.* **18**, 294–316 (1979).

Hegg, D. A., and P. V. Hobbs. Oxidation of sulfur dioxide in aqueous systems with particular reference to the atmosphere. *Atmos. Environ.* **2**, 241–254 (1978).

Heisler, S. L., R. C. Henry, P. K. Mueller, G. M. Hidy, and D. Grosjean. *Aerosol Behavior Patterns in the South Coast Air Basin with Emphasis on Airborne Sulfate*. Report No. P5618 Prep. for Southern Calif. Edison Co., Rosemead, CA by Environ. Res. and Technol., Inc., Westlake Village, CA (1979).

Heisler, S. L., R. C. Henry, and J. G. Watson. The origins and nature of the Denver winter haze. Proc. Annual Meeting Air Poll. Contr. Assoc., Paper 80-58.6, 24 pp. 1980.

Henry, R. C., and G. M. Hidy. Multivariate analysis of particulate sulfate and other air quality variables by principal components—Part I, Annual data from Los Angeles and New York. *Atmos. Environ.* **13**, 1581–1596 (1979).

Henry, R. C., and G. M. Hidy. Multivariate analysis of particulate sulfate and other air quality variables by principal components—Part II, Salt Lake City, Utah and St. Louis, Missouri. *Atmos. Environ.* **16**, 929–943 (1982).

Hidy, G. M. (principal investigator). *Characterization of Aerosols in California, Final Report*. Report No. SL524.25FR, Contract 358, California Air Resources Board (NTIS No. PB-247947) (1975).

Hidy, G. M., and C. S. Burton. Atmospheric aerosol formation by chemical reactions. In G. M. Hidy et al. (eds.) *The Character and Origins of Smog Aerosols*, pp. 385–483. Wiley, New York (1980).

Hidy, G. M., E. Y. Tong, P. K. Mueller, S. Rao, and I. Thompson. *Design of the Sulfate Regional Experiment*. Report EC-125. Electric Power Research Institute, Palo Alto, CA. Vol. I, NTIS PB-251701, 293 pp., 1976; Vol. II, NTIS PB-251702, 120 pp., 1976.

Hidy, G. M., J. R. Mahoney, and B. J. Goldsmith. *International Aspects of the Long Range Transport of Air Pollutants*. Final Report, U.S. Dept. of State. Environ. Res. and Technol., Inc., Westlake Village, CA (1978).

Husar, R. B. and D. E. Patterson, *Synoptic-Scale Distribution of Man-Made Aerosols*. Proc. WMO Symp. on Long-Range Transport of Pollutants and Its Relation to General Circulation. WMO No. 538, Supplement. World Meteorol. Org., Geneva, Switzerland (1979).

Husar, R. B., J. P. Lodge Jr., and D. J. Moore. Report of Workshop—Sulfur in the Atmosphere. Proc. Int'l Symp., Dubrovnik, Yugoslavia, September 7–14, 1977. *Atmos. Environ.* **12**, 3–23 (1978).

Junge, C. E. and R. T. Werby. The concentration of Cl, Na, K, Ca, and SO_4 in rain water over the U.S. *J. Meteorol.* **15**, 417–425 (1958).

Kleinman, M. T., C. Tomczyk, B. P. Leaderer, and R. L. Tanner. Inorganic nitrogen compounds in New York City air. *Ann. N.Y. Acad. Sci.* **322**, 115–124 (1979).

Klemm, H., and R. Brennan. The SURE emissions inventory. Final Report to Electric Power Research Institute, GCA Corp., Bedford, MA (1979).

Likens, G. E. Acid precipitation. *Chem. Eng. News* **54**(48), 29–44 (1976).

Liljestrand, H. M. and J. Morgan. Chemical composition of acid precipitation in Pasadena, California. *Environ. Sci. and Technol.* **13**, 1271–1273 (1978).

Lioy, P. J., G. T. Wolff, and B. P. Leaderer. A discussion of the New York summer aerosol study 1976. *Ann. N.Y. Acad. Sci.* **322**, 153–165 (1979).

Lodge, J. P., Jr., J. B. Pate, W. Basbergill, G. S. Swanson, K. C. Hill, E. Lorange, and A. L. Lazrus. *Chemistry of U.S. precipitation: Final Report on the National Precipitation Sampling Network*. National Center for Atmos. Res., Lab. of Atmos. Science, Boulder, CO (1968).

MacCracken, M. C. (ed.) *The Multistate Atmospheric Power Production Pollution Study—MAP3S*. Progress Report for FY 1977 and 1978. DOE/EV-0040. U.S. Dept. of Energy, Office of Health and Environ. Res., Washington, D.C. (1979).

Meyer, R., G. M. Hidy, and J. H. Davis. Determination of water and volatile organics in filter-collected aerosols. *Environ. Lett.* **4**, 9–20 (1973).

Mueller, P. K., and G. M. Hidy. *Sulfate regional experiment: Report of findings*. Report No. CA-1901. Electric Power Research Institute, Palo Alto, CA (1983).

Mueller, P. K., G. M. Hidy, K. Warren, T. F. Lavery, and R. L. Baskett. The occurrence of atmospheric aerosols in the northeastern United States. Proc. Conf. on Aerosols: Anthropogenic and Natural-Sources and Transport. *Ann. N.Y. Acad. Sci.* **338**, 463–482 (1980).

Nelson, N. (Chairman). *Report of Ad Hoc Sulfate Review Panel*. U.S. Environmental Protection Agency, Office of the Administrator, Washington, D.C. (1975).

Newman, L. Atmospheric oxidation of sulfur dioxide. In *Proc. Symp. on Potential Environmental and Health Effects of Atmospheric Sulfur Deposition, Gatlinburg, Tennessee, October* 14–18, 1979. Ann. Arbor. Science, Ann. Arbor, MI, pp. 131–143 (1980).

Ottar, B. An assessment of the OECD study on the long range transport of air pollutants (LRTAP). *Atmos. Environ.* **12**, 445–454 (1979).

Penkett, S. A., B. M. R. Jones, K. A. Brice, and A. E. J. Eggleton. The importance of atmospheric ozone and hydrogen peroxide in oxidizing sulfur dioxide in clouds and rain water. *Atmos. Environ.* **13**, 123–127 (1979),

Pierson, W., W. W. Brachacze, T. J. Truex, J. W. Butler, and T. J. Korniski. Ambient sulfate measurements on Allegheny Mountain and the question of atmospheric sulfate in the northeastern United States. In Proc. Conf. on Aerosols: Anthropogenic and Natural-Sources and Transport. *Ann. N.Y. Acad. Sci.* **338**, 145–173 (1980).

Roberts, P., and S. K. Friedlander. Conversion of SO_2 to sulfur particulate in the Los Angeles atmosphere. *Environ. Health Perspective* **10**, 103–108 (1975).

Rodhe, H., P. Crutzen, and A. Vanderpol. Formation of sulfuric and nitric acid in the atmosphere during long range transport. In *Proc. WMO Symp. on the Long-Range Transport of Pollutants and Its Relation to General Circulation*, pp. 165–172. WMO No. 538. World Meteorol. Org., Geneva, Switzerland (1979).

Samson, P. J. Trajectory analysis of summertime SO_4 concentrations in the northeastern United States. *J. Appl. Meteorol.* **19**, 1382–1394 (1980).

Sehmel, G., and W. Hodgson. *Atmospheric–Surface Exchange of Particle and Gaseous Pollutants*. R. Engleman and G. Sehmel (eds.). U.S. Atomic Energy Symp., Series No. 38. U.S. Government Printing Office, Washington, D.C. (1974).

Semb, A. Sulfur emissions in Europe. *Atmos. Environ.* **12**, 455–460 (1978).

Shinn, J. H., and S. Lynn. Do man-made sources affect the sulfur cycle of northeastern states? *Environ. Sci. and Technol.* **13**, 1063–1067 (1979).

Stelson, A. W., S. K. Friedlander, and H. H. Seinfeld. A note on the equilibrium relationship between ammonia and nitric acid and particulate nitrate. *Atmos. Environ.* **13**, 369–372 (1979).

Stevens, R. K., and T. G. Dzubay. Sampling and analysis of atmospheric sulfates and related species. *Atmos. Environ.* **13**, 55–68 (1978).

U.S. HEW (Dept. of Health, Education and Welfare). *Air Quality Criteria for Particulate Matter*. Publ. No. AP-49. U.S. Public Health Service, Washington, D.C. (1969).

U.S. HEW (Dept. of Health, Education and Welfare). *Air Quality Criteria for Sulfur Oxide*. Publ. No. AP-50. U.S. Public Health Service, Washington, D.C. (1970).

U.S. Environmental Protection Agency. *Health Consequences of Sulfur Oxides: A Report from CHESS* 1970–1971. Publ. No. EPA-650/11-74-004. U.S. EPA, Health and Environmental Research Lab., Research Triangle Park, NC (1974).

U.S. Environmental Protection Agency. *Position Paper on Regulation of Atmospheric Sulfates*. Publ. No. EPA-450/2-75-007. U.S. EPA, Office of Air and Waste Management, Research Triangle Park, NC (1975).

U.S. Environmental Protection Agency. *National Air Quality Monitoring and Emissions Trends Report*, 1977. Publ. No. EPA-450/2-78-052. U.S. EPA, Office of Air Quality Planning and Standards, Research Triangle Park, NC (1978).

Wesely, M. L., and B. B. Hicks. Some factors that affect the deposition rates of SO_2 and similar gases on vegetation. *J. Air Poll. Contr. Assoc.* **27**, 1110–1116 (1977).

Wesely, M. L., B. Hicks, W. Dannevik, S. Frisella, and R. Husar. Eddy-correlation measurement of particulate deposition from the atmosphere. *Atmos. Environ.* **11**, 561–563 (1977).

Whelpdale, D., and J. N. Galloway. *An Atmospheric Sulfur Budget for Eastern North America*. Proc. WMO Symp. on the Long-Range Transport of Pollutants and Its Relation to General Circulation. WMO No. 538. World Meteorol. Org., Geneva, Switzerland (1979).

Whitby, K. T. The physical characteristics of sulfur aerosols. *Atmos. Environ.* **12**, 135–139 (1978).

Williamson, S. *Fundamentals of Air Pollution*. Addison-Wesley, Reading, MA (1973).

Atmospheric Chemistry of Ozone and Nitrogen Oxides

KENNETH L. DEMERJIAN*
Environmental Sciences Research Laboratory
U.S. Environmental Protection Agency
Research Triangle Park, North Carolina

*On assignment from the National Oceanic and Atmospheric Administration, U.S. Department of Commerce.

1. INTRODUCTION

The human impact on the environment continues to be of increasing concern with respect to human health and welfare. It would seem that our continued pollution of the waters, air, and land cannot proceed without serious consequences to the environment. The true wonder of the environment has been its resilience to these insults. The closed cyclical systems, even as extensive as those acting in the environment, cannot continue unaltered given the persistent and ever-increasing perturbations introduced by mankind. The consequences of upsetting the complex and delicate balances of the biosphere are not totally understood, but the implications are enormous and could impact not only our generation but many generations to come. (Recent examples in air environment-related issues include the depletion of stratospheric ozone by long-lived halocarbons emitted in the troposphere and the climatic effects of increased CO_2 in the troposphere as a result of increased fossil fuel combustion and changing land use.) We are all part of this biosphere and intimately dependent on the physical, chemical, and biological surroundings which make it up. It is important that we understand and learn to live within the limits prescribed by our environment so that we and future generations can enjoy its beauty and thrive within it.

The scientific community is charged with the mission of elucidating the complex physical, chemical, and biological processes acting and interacting in the biosphere. Our role is not only to expand upon the knowledge of these systems but to educate the public to the threat that human excesses pose to the environment. I have been asked to address the chemistry of ozone and nitrogen oxides in the atmosphere. As you may well imagine, these compounds represent only one part of a very large puzzle which makes up the complex air chemistry system of our environment. A companion paper in this book on the chemistry of sulfur oxides in the atmosphere is presented by Dr. George Hidy and P. K. Mueller.

There is an extremely large amount of published material on ozone and nitrogen oxides in the air environment. This results, in part, from the fact that ozone and $NO_x = (NO + NO_2)$ are criteria pollutants identified under the Clean Air Act and its amendments. The scope of this chapter precludes discussing this subject area at the level of detail contained in several comprehensive documents (National Research Council, 1976a, 1976b, 1977; U.S. Environmental Protection Agency, 1978, 1980), but the interested reader is referred to them for more extensive discussions. It also has been indicated that, in addition to a state-of-the-art review, the presentation should identify the critical scientific issues yet unresolved in the subject area and outline technological needs and avenues of research that would help resolve these issues.

2. NATURE OF THE CLEAN TROPOSPHERE

Our understanding of the chemistry of ozone and nitrogen oxides in the air environment must begin with a review of our knowledge of the chemical state of the clean, unpolluted troposphere. This in turn provides the basis for assessing the nature and magnitude of effects that human activities have on atmospheric processes. Since it is impossible to address the chemistry of oxides of nitrogen and ozone without considering the chemistry of carbon (because of the interactive nature of their chemical processes), carbon chemistry has been incorporated into this discussion.

This chapter provides a brief review of the nature of the clean atmosphere and the chemical processes acting within it.

2.1 Trace Gas Concentrations in the Clean Troposphere

The background concentrations of trace gases in the clean troposphere are essentially determined by the competitive physical, chemical, and biological processes occurring within atmospheric and geological systems. In urban polluted atmospheres, by contrast, concentrations are determined predominantly by anthropogenic sources acted upon by physical and chemical processes on limited time and space scales.

Table 1 presents typical ranges of concentration for trace gases within the clean troposphere and for typical urban polluted atmospheres during daylight hours on an hourly averaged basis.

2.2 Clean Tropospheric Chemistry

The chemistry of the clean troposphere and the mathematical simulation of the physical and chemical processes therein have been studied extensively by Levy (1971), Wofsy et al. (1972), Crutzen (1974), Fishman and Crutzen (1977), Chameides and Walker (1973, 1976), and Stewart et al. (1977).

The photochemistry of the unpolluted troposphere develops around a chain reaction sequence involving NO, CH_4, CO, and O_3. A brief discussion of the more important reaction steps involved in the mechanistic sequence is presented below. For the interested reader a detailed mechanism, typical state-of-the-art tropospheric photochemical model, has been reproduced from Chameides (1978) in Table 2. The photochemical reaction chain sequence in the troposphere is initiated by hydroxyl radicals (HO) formed from the interaction of $O(^1D)$, the product of photolysis of ozone in the short-end portion of the solar spectrum, with water.

$$O_3 + h\nu(\lambda \leq 310\ \text{nm}) \rightarrow O(^1D) + O_2 \tag{1}$$

$$O(^1D) + H_2O \rightarrow 2HO \tag{2}$$

TABLE 1. Trace Gas Concentrations in the Clean Troposphere and Typical Urban Polluted Atmospheres

Species	Clean Troposphere (ppb)	Polluted Atmosphere (ppb)	References
O_3	30	100–(200)	National Research Council (1976); Singh et al. (1978)
HO	1.0×10^{-5}–2.0×10^{-4}	4.1×10^{-4}–2.4×10^{-3}	Wang et al. (1975); Davis et al. (1976); Campbell et al. (1979)
HO_2	10^{-4}–10^{-2}	0.1–0.2	Calvert and McQuigg (1975); Cox et al. (1976)
N_2O	330	—	Cicerone et al. (1978)
NO	0.01–0.05	60–740	Drummond (1977); Ritter et al. (1978); Air Quality Criteria for Oxides of Nitrogen (1980)
NO_2	0.1–0.5	40–220	Ritter et al. (1978); Noxon (1978); Air Quality Criteria for Oxides of Nitrogen (1980)
HONO	10^{-3}–10^{-1}	4–21	Nash (1974)
$HONO_2$	0.02–0.3	6–20	Huebert and Lazrus (1978); Doyle et al. (1979)
PAN[a]	<1	10–65	Lonneman et al. (1976)
NH_3	0.1–1	20–80	Dawson (1977); Doyle et al. (1979)
NH_4NO_3	0.03–0.5	8–30	Doyle et al. (1979)
H_2	500	—	Schmidt (1974)
H_2O_2	0.1–1	5–40	Bufalini et al. (1972); Kok et al. (1978)
CH_4	1.6–1.7×10^{3}	2–3×10^{3}	Fink et al. (1964); Altshuller et al. (1973)
NMHC[b]	5–10	10^{2}–10^{3}	Robinson et al. (1973); Leonard et al. (1976)
H_2CO	0.1–1	10–40	Altshuller and McPherson (1963)
CO	50–200	10^{2}–10^{4}	Seiler (1974); E.P.A. National Air Quality and Emissions Trends Report (1977)
CO_2	3.3×10^{5}	—	Lowe et al. (1979)

[a]PAN = peroxyacetyl nitrate.

[b]NMHC = non-methane hydrocarbons.

TABLE 2. Reactions and Rate Coefficients

	Reaction	Coefficient[a]	References
(R1)	$CH_4 + OH \rightarrow CH_3 + H_2O$	$K_1 = 2.35 \times 10^{-12} \exp(-1710/T)$	Hudson (1977)
(R2)	$CH_3 + M + O_2 \rightarrow CH_3O_2 + M$	$K_2 = 3 \times 10^{-32}$	Chameides and Stedman (1977)
(R3a)	$CH_3O_2 + HO_2 \rightarrow CH_3OOH + O_2$	$K_{3a} = K_{19}$	Chameides and Stedman (1977)
(R3b)	$2CH_3O_2 \rightarrow 2CH_3O + O_2$	$K_{3b} = 2.6 \times 10^{13}$	Chameides and Stedman (1977)
(R4a)	$CH_3OOH + OH \rightarrow CH_3O_2 + H_2O$	$K_{4a} = K_{21}$	Chameides and Stedman (1977)
(R4b)	$CH_3OOH \rightarrow$ Rainout	$K_{4b} = 1 \times 10^{-6}$	Assumed
(R5)	$CH_3OOH + h\nu \rightarrow CH_3O + O$	$K_5 = K_{20}$	Chameides and Stedman (1977)
(R6)	$CH_3O_2 + NO \rightarrow CH_3O + NO_2$	$K_6 = 3.3 \times 10^{-12} \exp(-550/T)$	Chameides and Stedman (1977)
(R7)	$CH_3O + O_2 \rightarrow H_2CO + HO_2$	$K_7 = 1.6 \times 10^{-13} \exp(-3300/T)$	Hudson (1977)
(R8)	$H_2CO + OH \rightarrow HCO + H_2O$	$K_8 = 3 \times 10^{-11} \exp(-250/T)$	Hudson (1977)
(R9)	$H_2CO + h\nu \rightarrow HCO + H$	$J_9(z = 0) = 7.1 \times 10^{-6}$ [b]	Chameides and Stedman (1977)
(R10)	$H_2CO + h\nu \rightarrow CO + HO_2$	$J_{10}(z = 0) = 2.2 \times 10^{-5}$ [b]	Chameides and Stedman (1977)
(R11)	$H + O_2 + M \rightarrow HO_2 + M$	$K_{11} = 3 \times 10^{-32}(273/T)^{1.3}$	Chameides and Stedman (1977)
(R12)	$HCO + O_2 \rightarrow CO + HO_2$	$K_{12} = 1 \times 10^{-13}$	Chameides and Stedman (1977)
(R13)	$CO + OH \rightarrow CO_2 + H$	$K_{13} = 2.1 \times 10^{-13} \exp(-115/T) + 7.3 \times 10^{33} n_M$	Sie et al. (1976); Cox et al. (1976); Chan et al. (1977)
(R14)	$H_2 + OH \rightarrow H + H_2O$	$K_{14} = 6.8 \times 10^{-12} \exp(-2020/T)$	Chameides and Stedman (1977)
(R15)	$O(^1D) + H_2O \rightarrow 2OH$	$K_{15} = 2.3 \times 10^{-10}$	Chameides and Stedman (1977)
(R16)	$O(^1D) + H_2 \rightarrow H + OH$	$K_{16} = 1.3 \times 10^{-10}$	Chameides and Stedman (1977)
(R17)	$HO_2 + OH \rightarrow H_2O + O_2$	$K_{17} = 3 \times 10^{-11}$	Hudson (1977)
(R18)	$OH + OH \rightarrow H_2O + O$	$K_{18} = 1 \times 10^{-11} \exp(-550/T)$	Chameides and Stedman (1977)
(R19)	$HO_2 + HO_2 \rightarrow H_2O_2 + O_2$	$K_{19} = 3 \times 10^{-11} \exp(-500/T)$	Chameides and Stedman (1977)
(R20)	$H_2O_2 + h\nu \rightarrow 2OH$	$J_{20a}(z = 0) = 6.2 \times 10^{-7}$ [b]	Chameides and Stedman (1977)
(R20b)	$H_2O_2 \rightarrow$ Rainout	$K_{20b} = 1 \times 10^{-6}$	Assumed
(R21)	$H_2O_2 + OH \rightarrow HO_2 + H_2O$	$K_{21} = 1 \times 10^{-11} \exp(-750/T)$	Hudson (1977)
(R22)	$HO_2 + NO \rightarrow OH + NO_2$	$K_{22} = 8 \times 10^{-12}$	Howard and Evenson (1977)

TABLE 2. (Continued)

	Reaction	Coefficient[a]	References
(R23)	$OH + NO_2 + M \rightarrow HNO_3 + M$	$K_{23} = (1.25 \times 10^{-11})/n_M$	Hudson (1977)
(R24)	$OH + NO + M \rightarrow HNO_2 + M$	$K_{24} = (2 \times 10^{-12})/n_M$	Chameides and Stedman (1977)
(R25a)	$HNO_3 + h\nu \rightarrow OH + NO_2$	$J_{25a}(z=0) = 1.1 \times 10^{-7}$ [b]	Chameides and Stedman (1977)
(R25b)	$HNO_3 + OH \rightarrow NO_3 + H_2O$	$K_{25b} = 8 \times 10^{-14}$	Hudson (1977)
(R26)	$HNO_3 \rightarrow$ Rainout	$K_{26} = 1 \times 10^{-6}$	Assumed
(R27)	$HNO_2 + h\nu \rightarrow NO + OH$	$J_{27}(z=0) = 1.5 \times 10^{-4}$ [b]	Chameides and Stedman (1977)
(R28)	$NO_2 + O_3 \rightarrow NO_3 + O_2$	$K_{28} = 1.1 \times 10^{-13} \exp(-2450/T)$	Chameides and Stedman (1977)
(R29)	$NO + O_3 \rightarrow NO_2 + O_2$	$K_{29} = 2.1 \times 10^{-12} \exp(-1450/T)$	Hudson (1977)
(R30)	$NO_2 + h\nu \rightarrow NO + O$	$J_{30}(z=0) = 2.6 \times 10^{-3}$ [b]	Chameides and Stedman (1977)
(R31)	$NO_3 + NO_2 + M \rightarrow N_2O_5 + M$	$K_{31} = (3.8 \times 10^{-12})/n_M$	Chameides and Stedman (1977)
(R32)	$NO_3 + NO \rightarrow 2NO_2$	$K_{32} = 8.7 \times 10^{-12}$	Chameides and Stedman (1977)
(R33a)	$NO_3 + h\nu \rightarrow NO_2 + O$	$J_{33a}(z=0) = 2.3 \times 10^{-3}$ [b]	Chameides and Stedman (1977)
(R33b)	$NO_3 + h\nu \rightarrow NO + O_2$	$J_{33b}(z=0) = 2.3 \times 10^{-3}$	Chameides and Stedman (1977)
(34)	$NO + NO_2 + H_2O \rightarrow 2HNO_2$	$K_{34} = 6 \times 10^{-37}$	Chameides and Stedman (1977)
(R35)	$N_2O_5 + H_2O \rightarrow 2HNO_3$	$K_{35} = 1 \times 10^{-20}$	Chameides and Stedman (1977)
(R36)	$N_2O_5 \rightarrow NO_3 + NO_2$	$K_{36} = 5.7 \times 10^{14} \exp(-10600/T)$	Chameides and Stedman (1977)
(R37)	$O_3 + h\nu(\text{visible}) \rightarrow O_2 + O$	$J_{37}(z=0) = 1.3 \times 10^{-4}$ [b]	Chameides and Stedman (1977)
(R38)	$O + O_2 + M \rightarrow O_3 + M$	$K_{38} = 1.1 \times 10^{-34} \exp(510/T)$	Chameides and Stedman (1977)
(R39)	$O_3 + h\nu(\text{ultraviolet}) \rightarrow O(^1D) + O_2$	$J_{39}(z=0) = 3.8 \times 10^{-6}$ [b]	Chameides and Stedman (1977)
(R40)	$O(^1D) + M \rightarrow O + M$	$K_{40} = 3.2 \times 10^{-11}$	Chameides and Stedman (1977)
(R41)	$OH + O_3 \rightarrow HO_2 + O_2$	$K_{41} = 1.5 \times 10^{-12} \exp(-1000/T)$	Hudson (1977)
(R42)	$HO_2 + O_3 \rightarrow OH + 2O_2$	$K_{42} = 1 \times 10^{-13} \exp(-1250/T)$	Chameides and Stedman (1977)

[a] The rate coefficient units are s^{-1} for one-body reactions, cm^3/s for two-body reactions, and cm^6/s for three-body reactions, and T is temperature (K).

[b] Quantities shown are the daily averaged photodissociation frequencies for equinoctial conditions at 45°N.

The HO produced reacts with CH_4 and CO present in the clean troposphere, resulting in the generation of peroxy radical species.

$$HO + CH_4 \rightarrow CH_3 + H_2O \tag{3}$$

$$HO + CO \rightarrow H + CO_2 \tag{4}$$

$$CH_3 + O_2 + M \rightarrow CH_3O_2 + M \tag{5}$$

$$H + O_2 + M \rightarrow HO_2 + M \tag{6}$$

The peroxy radicals in turn participate in a chain-propagating sequence, which converts nitric oxide (NO) to nitrogen dioxide (NO_2) and in the process produces additional hydroxyl and peroxy radical species.

$$CH_3O_2 + NO \rightarrow CH_3O + NO_2 \tag{7}$$

$$HO_2 + NO \rightarrow HO + NO_2 \tag{8}$$

$$CH_3O + O_2 \rightarrow HO_2 + H_2CO \tag{9}$$

$$H_2CO + h\nu\ (\lambda \leq 370\ \text{nm}) \rightarrow H + HCO \tag{10}$$

$$HCO + O_2 \rightarrow HO_2 + CO \tag{11}$$

The major chain terminating steps include

$$HO + NO_2 + M \rightarrow HONO_2 + M \tag{12}$$

$$HO_2 + HO_2 \rightarrow H_2O_2 + O_2 \tag{13}$$

$$H_2O_2 + HO \rightarrow H_2O + HO_2 \tag{14}$$

The role that the chemistry of the clean troposphere plays in controlling the abundance of tropospheric ozone has been a subject of considerable debate. The model calculations indicate that photochemical processes can produce and destroy tropospheric O_3 at rates equivalent to those estimated for stratospheric injection and depositional losses at the earth's surface. The reaction sequence for O_3 production involves converting NO to NO_2 at a rate sufficiently high to maintain a NO_2/NO ratio to sustain the observed background levels of O_3.

$$HO_2 + NO \rightarrow HO + NO_2 \tag{8}$$

$$NO_2 + h\nu\ (\lambda \leq 430\ \text{nm}) \rightarrow NO + O \tag{15}$$

$$O + O_2 + M \rightarrow O_3 + M \tag{16}$$

$$NO + O_3 \rightarrow NO_2 + O_2 \tag{17}$$

$$HO + CO \rightarrow H + CO_2 \tag{4}$$

In general, reactions (15) through (17) govern the ozone concentration levels present in the sunlight-irradiated atmosphere at any instant and to a first approximation the steady state relationship (Leighton, 1961)

$$(NO_2)K_{15}/(NO)K_{17} = (O_3)$$

provides a good estimate of ozone given the ratio $(NO_2)/(NO)$ and K_{15}/K_{17}. The photolytic rate constant K_{15} is directly related to the integrated actinic solar flux over the wavelength range 290–430 nm.

The paths for ozone destruction in the troposphere include the reaction sequence

$$HO_2 + O_3 \rightarrow HO + 2O_2 \quad (17)$$

$$HO + O_3 \rightarrow HO_2 + O_2 \quad (18)$$

Hydroxyl radical abundances predicted by the tropospheric photochemical models, 10^5–10^6 moles/cm^3, are in qualitative agreement with measurements by Davis et al. (1976), Perner et al. (1976), and Campbell et al. (1979) and with inferred HO levels based on measured trace gas abundances in the troposphere by Singh (1977). The apparent agreement between predicted and observed concentrations of hydroxyl radical in the troposphere, a key species in the photooxidation cycle, provides some credence for the proposed reaction mechanism. What seems to be at issue is not particularly the photooxidation reaction sequence per se but its ability to account for background tropospheric ozone levels. Singh et al. (1977, 1978) have argued that the ozone reservoir present in the troposphere is predominantly of stratospheric origin and that the photochemical oxidation processes discussed above do not contribute significantly to the net O_3 balance in the tropospheric reservoir. The bases for Singh's et al. (1979) conclusions are as follows:

1. Reliable measurements of NO_x in the free troposphere are limited, with most recent measurements indicating NO_x levels approaching 0.01 ppb. Such levels are insufficient for synthesizing tropospheric ozone in its observed abundances via the proposed photochemical reaction mechanisms.
2. Seasonal variations in tropospheric O_3 are out of phase with solar flux levels, contrary to photochemical theory predictions.
3. Observed vertical gradients of O_3 are not consistent with a dominant photochemical source of O_3 within the troposphere.
4. The contention that measured abundances of tropospheric background O_3 in the northern and southern hemisphere cannot be supported based on observations of stratospheric/tropospheric exchanges in the two hemispheres is no longer valid. Reliable experimental measurements of ^{90}Sr and ^{7}Be indicate that stratospheric/tropospheric exchange rates are two to three times larger in the northern hemisphere (NH) than in the southern hemisphere (SH), thus explaining the apparent contradiction in levels due to higher destruction rates in the NH as a result of its greater land to water ratio.

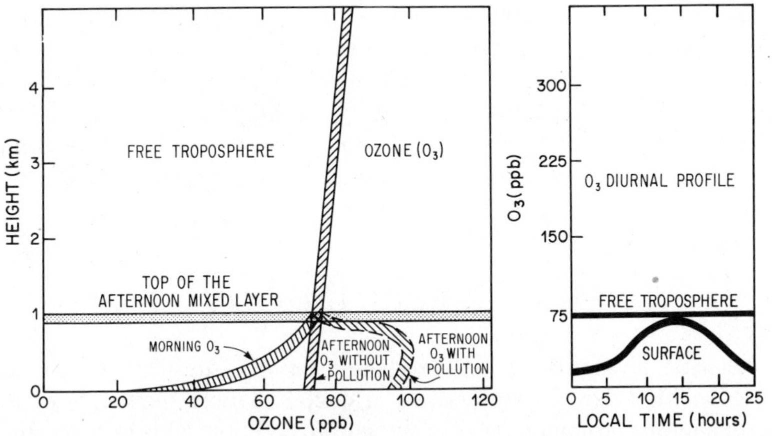

FIGURE 1. A schematic of ozone variations at the surface and in the free troposphere. (Singh et al., 1978.)

To explain the behavior of O_3 in remote locations, Singh et al. (1977, 1978) have developed a schematic representation of ozone variations at the surface and in the free troposphere (Figure 1) and ozone variations by season (Figure 2). Figure 1 depicts a large O_3 reservoir with no average

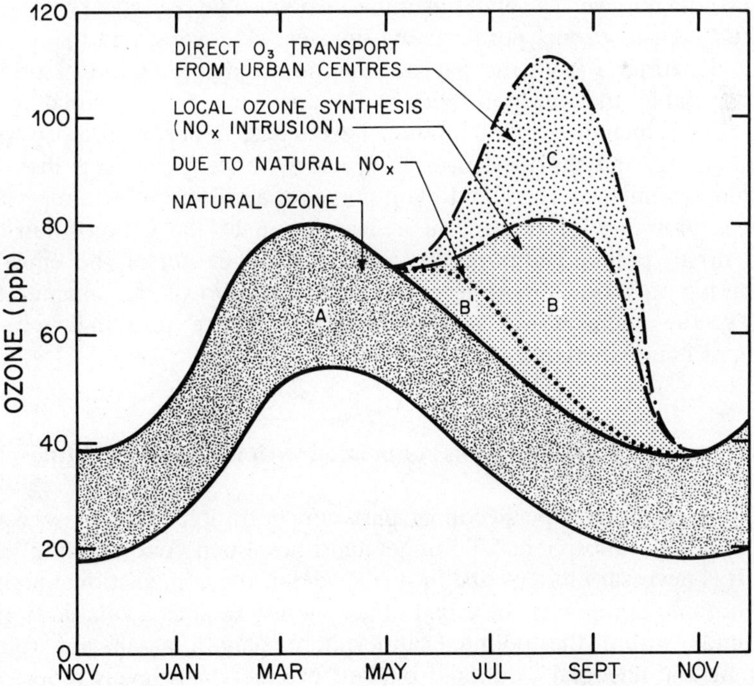

FIGURE 2. A schematic of the idealized ozone variations at remote locations. (Singh et al., 1978.)

diurnal variation, except near the earth's surface where O_3 is controlled by surface destruction and mixed layer dynamics. Idealized seasonal variations of O_3 are depicted in Figure 2, where natural O_3 effects (curve A) are expected to be at a maximum in the early spring. Perturbations in ozone levels due to localized O_3 production and transport from urban centers, both resulting from photochemical processes, are depicted by curves B and C.

3. NATURE OF THE POLLUTED ATMOSPHERE

World-wide industrialization produces rapidly growing urban areas resulting in high-density automotive traffic, enhanced power generation needs, expanding commercial and industrial activity, and as a result, the phenomenon of air pollution. The high density of emissions emanating from these geographically concentrated sources introduces a mass flux burden of such significant magnitude that the natural atmospheric ventilating and cleansing processes are not sufficient to mitigate the assault. The result is high concentrations of combustion products such as CO, NO, NO_2, SO_2, and volatile organic compounds (VOC), which, besides presenting a potential threat to human health and welfare by themselves, participate in a complex series of chemical reactions driven by sunlight leading to the phenomenon known as photochemical smog. The secondary reaction products, such as ozone, peroxyacetyl nitrates, aldehydes, and many other compounds formed from the partial oxidation and interaction of primary emitted organic and nitrogen oxide pollutants, have been correlated with health effects including eye irritation, headaches, upper respiratory irritation, coughing, and shortness of breath, and with welfare effects including restricted visibility, material and plant damage, and suppressed crop yields.

The sections to follow present a description of the composition of a typical urban polluted atmosphere and a quantification of the emissions contributing to those atmospheres. The final section of this chapter then considers the chemistry of the polluted atmosphere resulting from the impacts of human activity.

3.1 Composition and Emissions Associated with the Polluted Atmosphere

Table 1 included typical counterpart concentration levels observed in urban polluted atmospheres. To understand how such concentration levels result, it is necessary to describe in further detail the compositional makeup of the organic component of urban atmospheres. Reactive volatile organic compounds within the polluted atmosphere play a significant role in governing the nitrogen oxide and oxidant products and levels reported in Table 1.

TABLE 3. Typical Urban Hydrocarbon Ambient Concentrations (ppb C) during Early Morning Traffic Peak Hours[a]

	St. Louis, MO	Los Angeles, CA		St. Louis, MO	Los Angeles, CA		St. Louis, MO	Los Angeles, CA
Paraffins			Olefins			Aromatic		
Ethane	11	191	Ethylene	7	151	Benzene	[b]	90
Propane	12	140	Propylene	4	60	Toluene	23	259
n-Butane	16	286	Isobutylene	4	47	Ethylbenzene	5	48
Isobutane	7	65	*trans*- and *cis*-2-Butenes	4	20	*p*-Xylene	3	48
Isopentane	20	312	1,3-Butadiene	[b]	11	*m*-Xylene	3	128
n-Pentane	10	171	1-Pentene	<1	11	*o*-Xylene	6	64
Cyclopentane	9	138	2-Methyl-1-butene	1	15	Isopropylbenzene	2	27
2-Methylpentane	7	138	*trans*- and *cis*-2-Pentene	<1	32	*n*-Propylbenzene	1	18
3-Methylpentane	4	68	2-Methyl-2-butene	<1	29	*p*-Ethyltoluene	4	72
n-Hexane	4	82	Isoprene	[b]	[b]	1,3,5-Trimethylbenzene	2	27
2,4-Dimethylpentane	5	89	4-Methyl-2-pentene	<1	[b]	1,2,4-Trimethylbenzene	9	81
3,3-Dimethylpentane	<1	[b]	Hexenes	<1	15	*trans*-Butylbenzene	[b]	20
Cyclohexane	<1	16	Unknowns		9			
2-Methylhexane	1	[b]						
2,3-Dimethylpentane	1	[b]						
3-Methylhexane	2	68						
1,3-Dimethylcyclopentane	<1	[b]						
2,2,4-Trimethylpentane	1	[b]						
n-Heptane	2	40						
Methylcyclohexane	1	49						
n-Nonane	7	[b]						
n-Decane	2	[b]						
Unknowns		185						

[a] Based on data reported in Altshuller et al. (1971), Kopczynski et al. (1972), and Lonneman et al. (1978).

[b] Compound not identified/detected in analysis.

Detailed hydrocarbon analyses of urban atmospheres have been performed over the years by several groups—Lonneman et al. (1968, 1974, 1979), Stephens and Burleson (1969), Altshuller et al. (1971), and Kopczynski et al. (1972, 1975a)—utilizing gas chromatographic analysis techniques. Typical urban hydrocarbon compositions, based on ambient analyses of samples taken during early morning peak traffic hours, are reported in Table 3 for St. Louis, MO, and Los Angeles, CA. Except for differences in the sheer magnitude of levels, the compounds in the two cities are quite similar compositionally. In fact, Table 4 presents a comparison of average percentage composition of ambient hydrocarbons by organic class for scores of samples taken at several time periods and locations within three US cities. These results indicate that hydrocarbon polluted urban atmospheres across the country tend to be persistent and similar in composition. This is not surprising, given that approximately 47–50% of the VOC emission in most cities, and 41% on a national average over the 7-year period 1970–1976, is transportation related. As shown in Table 5, the overall trend in transportation VOC emissions is down. The result is attributed to the introduction of catalyst hydrocarbon controlled vehicles.

An area of significant interest and controversy is the contribution that natural organic emissions, primarily isoprene and terpene, make to the ozone photochemical production cycle. Recent biogenic emission estimates by Zimmerman (1979) (Table 6) show natural organic emissions of the order of a factor of 2 greater than anthropogenic emissions. Experimental studies show that the reactivities of isoprene and terpene are sufficiently high (Arnts and Gay, 1979) that substantial O_3 production should result in rural areas in the presence of NO_x. What is basically at issue is that there are no corroborative ambient data on isoprene or terpenes at levels that would support the estimated biogenic emission rates (Arnts and Meeks, 1980; Holdren et al., 1979).

TABLE 4. Average Percentage Composition of Ambient Hydrocarbon Classes Observed in Three U.S. Cities

Location	Year	Paraffins	Olefins	Aromatics
St. Louis, MO	1972[a]	62.0	11.3	22.7
	1977[a]	64.0	10.7	24.4
Los Angeles, CA	1968[b]	57.3	11.1	31.6
	1973[c]	66.6	7.0	26.4
Houston, TX	1978[a]	63.8	13.7	19.9

[a] From Lonneman (1979).

[b] Based on data presented in Kopczynski et al. (1972).

[c] Based on data presented in Leonard et al. (1976).

TABLE 5. Recent Nationwide NO_x and VOC[a] Emission Estimates, 1970–1976, Expressed as NO_2 and CH_4[b,c]

	1970		1971		1972		1973		1974		1975		1976	
Source Category	NO_x	VOC	NO_x	VOC	NO_x	VOC	NO_x	VOC	NO_x	VOC	NO_x	VOC	NO_x	VOC
Transportation	8.4	12.6	8.9	12.3	9.4	12.6	9.7	12.2	9.6	11.3	9.9	10.9	10.2	10.8
Highway vehicles	6.3	11.1	6.7	10.8	7.1	11.0	7.3	10.6	7.3	9.8	7.6	9.4	7.8	9.3
Nonhighway vehicles	2.1	1.5	2.2	1.5	2.3	1.6	2.4	1.6	2.3	1.5	2.3	1.5	2.3	1.5
Stationary fuel combustion	10.9	1.5	11.2	1.5	11.7	1.5	12.1	1.6	11.9	1.5	11.2	1.4	11.8	1.4
Electrical utilities	5.1	0.1	5.4	0.1	5.9	0.1	6.3	0.1	6.2	0.1	6.1	0.1	6.6	0.1
Industrial	5.1	1.3	5.1	1.3	5.1	1.3	5.1	1.4	5.0	1.3	4.5	1.2	4.5	1.2
Residential, commercial and institutional	0.7	0.1	0.7	0.1	0.7	0.1	0.7	0.1	0.7	0.1	0.6	0.1	0.7	0.1
Industrial processes	0.6	8.5	0.6	8.5	0.7	8.9	0.7	9.4	0.7	9.2	0.7	8.5	0.7	9.4
Chemicals	0.2	1.5	0.2	1.4	0.3	1.5	0.3	1.6	0.3	1.6	0.3	1.5	0.3	1.6
Petroleum refining	0.3	0.7	0.3	0.7	0.3	0.7	0.3	0.8	0.3	0.8	0.3	0.8	0.3	0.9
Metals	0	0.2	0	0.2	0	0.2	0	0.2	0	0.2	0	0.2	0	0.2
Mineral products	0.1	0	0.1	0	0.1	0	0.1	0	0.1	0	0.1	0	0.1	0
Oil and gas production and marketing	0	2.7	0	2.8	0	2.9	0	2.9	0	2.9	0	2.9	0	3.0
Industrial organic solvent use	0	2.7	0	2.7	0	2.9	0	3.1	0	2.9	0	2.4	0	2.9
Other processes	0	0.7	0	0.7	0	0.7	0	0.8	0	0.8	0	0.7	0	0.8

[a]VOC = Volatile organic carbon.

[b]Values in millions of metric tons per year. A zero in this table indicates emissions of less than 50,000 metric tons per year.

[c]From National Air Quality and Emissions Trends Report (1977).

(*Table 5 continues*)

TABLE 5. (*Continued*)

	1970		1971		1972		1973		1974		1975		1976	
Source Category	NO_x	VOC	NO_x	VOC	NO_x	VOC	NO_x	VOC	NO_x	VOC	NO_x	VOC	NO_x	VOC
Solid waste disposal	0.3	1.7	0.3	1.4	0.2	1.1	0.2	1.0	0.2	0.9	0.2	0.8	0.1	0.8
Miscellaneous	0.2	5.4	0.3	5.6	0.2	5.6	0.2	5.6	0.2	5.7	0.2	4.6	0.3	5.5
Forest wildfires and managed burning	0.1	0.7	0.2	1.0	0.1	0.7	0.1	0.6	0.1	0.6	0.1	0.4	0.2	0.8
Agricultural burning	0	0.3	0	0.3	0	0.2	0	0.1	0	0.3	0	0.1	0	0.1
Coal refuse burning	0.1	0.1	0.1	0.1	0.1	0.1	0.1	0.1	0.1	0.1	0.1	0.1	0.1	0.1
Structural fires	0	0	0	0	0	0	0	0	0	0	0	0	0	0
Miscellaneous organic solvent use	0	4.3	0	4.2	0	4.6	0	4.8	0	4.7	0	4.0	0	4.5
Totals	20.4	29.7	21.3	29.3	22.2	29.7	22.9	29.8	22.6	28.6	22.2	26.2	23.0	27.9

[a]VOC = Volatile organic carbon.

[b]Values in millions of metric tons per year. A zero in this table indicates emission of less than 50,000 metric tons per year.

[c]From National Air Quality and Emissions Trends Report (1977).

TABLE 6. Annual U.S. Biogenic Emission Inventory by Latitudinal Region in Metric Tons[a]

Latitudinal Range	Isoprene	Terpene	Total	Anthropogenic VOC[b]
45–50°	1.9×10^5	4.5×10^6	4.8×10^6	
40–45°	9.3×10^5	9.4×10^6	1.0×10^7	
35–40°	4.2×10^6	1.7×10^7	2.1×10^7	
25–35°	9.9×10^6	1.9×10^7	2.9×10^7	
Annual Total	1.5×10^7	5.0×10^7	6.5×10^7	2.8×10^7

[a] Adapted from Zimmerman (1979).
[b] VOC = Volatile organic carbon.

3.2 Chemistry of the Polluted Atmosphere

The elucidation of the elementary reactions and mechanistic pathways in photochemical smog formation has been a subject of continued and active research since the 1950s. Leighton's (1961) pioneering efforts to evaluate the alternative reaction paths which may be operative in photochemical smog have stimulated the growth of an entirely new level of sophistication in the treatment of chemical kinetic reaction mechanisms of polluted atmospheres. Modern-day computer techniques have made it quite practical to solve complex systems of ordinary differential equations resulting from detailed reaction steps thought to be operative in the mechanisms of photochemical smog formation. With this tool, kineticists and atmospheric chemists have begun to unravel the details of the complex reaction processes occurring in real urban polluted atmospheres.

Extensive discussions on the mechanism of photochemical smog and its computer simulation have been presented by Demerjian et al. (1974), Calvert and McQuigg (1975), Niki et al. (1972), Hecht et al. (1974), and Carter et al. (1979). The basic approach has been to simulate temporal concentration profiles observed under controlled laboratory conditions (smog chambers) for various volatile organic compounds with NO_x and at varying levels and ratios, using a postulated reaction mechanism. These chemical reaction mechanisms have evolved from basic kinetic studies of individual reactions and their product pathways and from theoretical estimates of rate constants and products using thermochemical kinetic techniques developed by Benson (1968). A detailed discussion of the mechanism of photochemical smog is beyond the scope of this paper; what has been considered are the salient features of the chemistry of the polluted atmosphere. The interested reader is referred to the articles by Demerjian et al. (1974), Calvert and McQuigg (1975), Niki et al. (1972), Hecht et al. (1974), and Carter et al. (1979) for more details on the subject.

Perturbations introduced as a result of human activity on the photochemical oxidation cycle within the atmosphere are predominantly due to two classes of compounds, volatile organics and nitrogen oxides. The reaction chain sequence discussed earlier for the clean troposphere has now been immensely complicated by the addition of scores of volatile organic compounds, as reported in Table 3, which participate in the chain propagating cycle. Figure 3 depicts a schematic of the polluted atmospheric photooxidation cycle. The addition of volatile organic compounds in the atmosphere introduces a variety of new peroxy radical species.

In its simple form the photochemical oxidation cycle in polluted atmospheres is governed by the following basic features. Free radical attack on atmospheric VOCs is initiated by a select group of compounds which are for the most part activated by sunlight. Formaldehyde and nitrous acid, in particular, show high potential as free radical initiators during the early morning sunrise period. After initial free radical attack, the VOCs decompose through paths resulting in the production of peroxy radical species (HO_2, RO_2, $R'O_2$, etc.) and partially oxidized products which in themselves may be photoactive radical producing compounds. The peroxy radicals react with NO converting it to NO_2 and in the process produce hydroxy and alkoxy radical species (OH, RO, R'O, etc.). Alkoxy radicals can be further oxidized, forming additional peroxy radicals and partially oxidized products, thereby completing the inner loop reaction chain illustrated in Figure 3, or they may attack, as would be the major path for hydroxyl radical, the VOC pool present in the polluted atmosphere, thereby completing the outer loop reaction chain. The resultant effects in either case are the conversion of NO

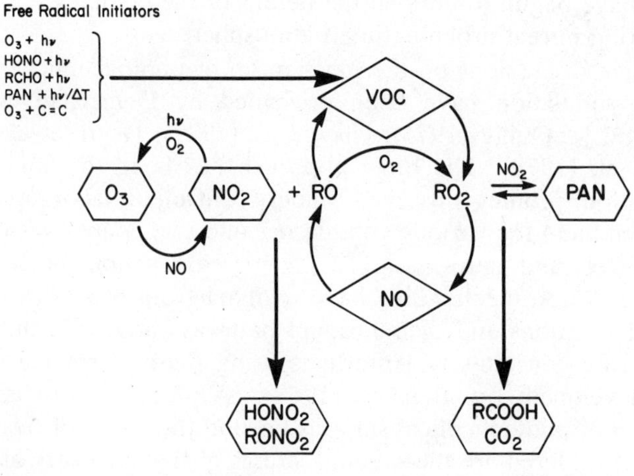

FIGURE 3. Schematic of the polluted atmospheric photooxidation cycle.

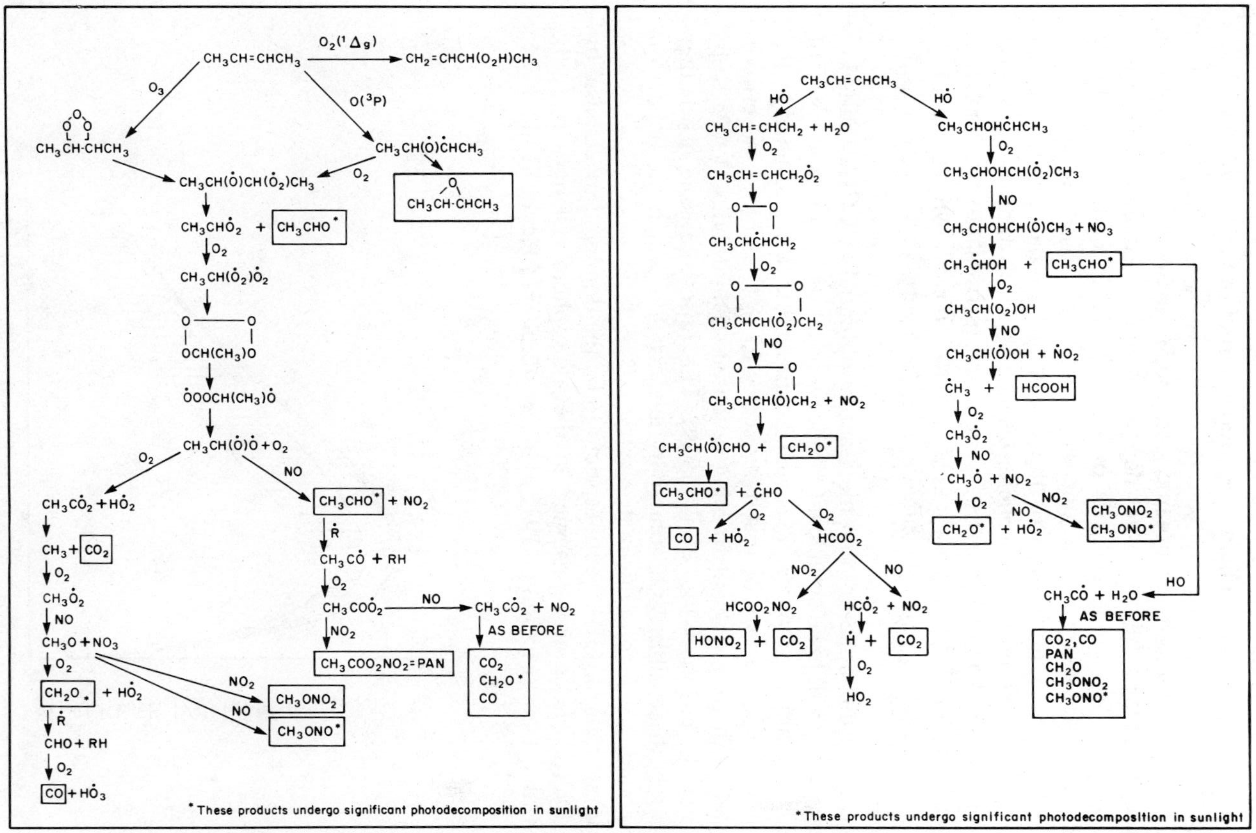

FIGURE 4. The major reaction paths for the degradation of *trans*-2-butene in an irradiated NO_x-polluted atmosphere. (Demerjian et al., 1974.)

to NO_2 with a commensurate oxidation of reactive organic carbon, shortening of the hydrocarbon chain, and production of organic particles, CO_2, and H_2O.

The complex mixture of organic compounds present in the polluted atmosphere react at different rates depending upon their molecular structure. The result is varying yields of ozone, NO_2, PAN, and other partially oxidized organic products as a function of VOC composition and VOC–NO_x levels. As an illustration of the complexity of the chemical detail operative in the polluted atmosphere, a description of the major reaction paths for the degradation of *trans*-2-butene in an irradiated NO_x-polluted atmosphere taken from Demerjian et al. (1974) is shown in Figure 4. Comparable reaction paths presumably exist for every VOC present in the atmosphere.

It becomes quite apparent that the reaction potential of VOCs emitted in the atmosphere will have significant implications on the nature and levels of secondary pollutant products. This well-established fact has resulted in the development of a variety of reactivity scales for VOCs based on criteria associated with their oxidant forming potentials (Dimitriades, 1972; Kopczynski et al., 1975b; Winer et al., 1979). This raises the issue of whether there is any discernible VOC reactivity cutoff below which the impact on oxidant buildup is inconsequential.

Hydroxyl radical (OH) reactions seem to be the dominant mechanism by which hydrocarbons are consumed in the atmosphere (Demerjian et al., 1974; Calvert and McQuigg, 1975; Niki et al., 1972; Hecht et al., 1974;

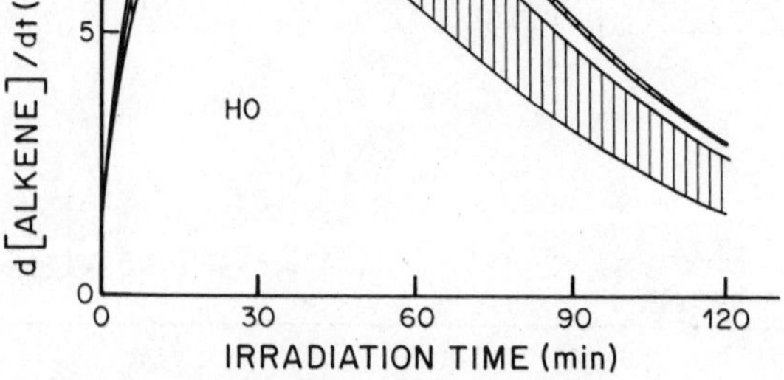

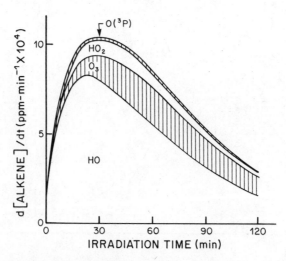

FIGURE 5. Theoretical rates of reaction of various reactive species with *trans*-2-butene in a computer-simulated, sunlight-irradiated atmosphere (solar zenith angle, 40°) containing *trans*-2-butene (0.10 ppm), NO (0.025 ppm), NO_2 (0.075 ppm), CO (10 ppm), and CH_4 (1.5 ppm) at 5% relative humidity. (Calvert and McQuigg 1975.)

Carter et al., 1979). While OH reaction rate constants with alkanes, alkenes, and aromatics are fairly reliably established (Finlayson and Pitts, 1976), much less is known about the mechanistic paths these compounds take after the initial attack by the radical. Based on the chemical reaction mechanism developed in Demerjian et al. (1974), Calvert and McQuigg (1975) estimated the theoretical rates of reaction of various reactive species with *trans*-2-butene in a computer-simulated, sunlight-irradiated atmosphere. The results shown in Figure 5 indicate the dominant role played by OH, with contributions from O_3 and to a lesser extent HO_2 becoming important in the latter part of the irradiation period when NO concentrations have diminished appreciably. It should be noted that curves for alkanes and aromatic compounds would be similar, but show very little contribution from ozone, given their very low reaction rates with ozone.

4. SCIENTIFIC ISSUES AND RECOMMENDATIONS

In the course of the discussions presented here, several unresolved scientific issues have been raised relevant to the chemistry of O_3 and NO_x in the clean and polluted troposphere. These have included:

1. The nature of the source of background ozone in the troposphere.
2. The significance of VOC reactivity to short- and long-term ozone production.
3. The relationship between biogenic emissions and ambient levels and its impact on ozone production.
4. The lifetime and fate of NO_x.
5. Carbon and nitrogen budgets.

Over the years a significant amount of research by many dedicated scientists has proceeded in an attempt to resolve these issues. Progress has been made, but many limitations in our understanding of the basic chemical and physical phenomena operative in the troposphere still exist.

In an attempt to highlight research needs and recommend approaches to the resolution of the scientific issues, three specific research areas have been identified. These include (1) the development of specialized measurement technologies, (2) implementation of specialized intensive measurement programs, and (3) the development of global and continental scale air pollution models.

To evaluate and verify chemical kinetic mechanistic theories of the atmosphere, it is necessary to establish the ambient concentration levels of species playing a dominant role in the photooxidation cycle. In this regard there is a need to develop specialized measurement technologies capable of high accuracy and sensitivity. The chemical species of major interest include aldehydes, organic acids, nitric and nitrous acid, nitrates, peroxyacetyl nitrates, hydrogen peroxide, nitrogen dioxide, and nitric oxide. In addition, measurement techniques for transient species such as hydroxyl, hydroperoxy, and alkylperoxy radicals and pernitric acid are highly desirable.

Specialized intensive measurement programs should be designed and allowed to evolve around the specific scientific issue to be addressed. Such investigations would include micrometeorological and chemical budget studies for natural organics, ozone, and nitrogen oxides; meteorological and chemical tracer studies of stratospheric–tropospheric exchange processes; and meteorological and chemical studies on the transport and fate of ozone and its precursors over regional, continental, and global scales. The programs should be phenomenologically oriented, that is, each designed to elucidate our understanding of specific processes important to the overall scientific objective.

The development of photochemical models of the troposphere has proceeded, so to speak, at two opposite ends of the air quality spectrum: the clean troposphere versus the urban polluted atmosphere. The time has come to merge these two scientific communities through the development of continental- and global-scale photochemical air quality simulation models (PAQSM). The transitional zone between the clean and urban polluted atmosphere is highly important but little understood by today's standards. It is therefore recommended that major programs in the development of continental- and global-scale PAQSM be instituted and that field studies be designed and implemented in support of the theoretical developments.

REFERENCES

Altshuller, A. P., and S. D. McPherson. Spectrophotometric analysis of aldehydes in the Los Angeles atmosphere. *J. Air Pollut. Contr. Assoc.* **13**, 109 (1963).

Altshuller, A. P., W. A. Lonneman, and S. L. Kopczynski. Non-methane hydrocarbon air quality measurements. *J. Air Pollut. Contr. Assoc.* **23**, 597 (1973).

Altshuller, A. P., W. A. Lonneman, F. D. Sutterfield, and S. L. Kopczynski. Hydrocarbon composition of the atmosphere of the Los Angeles basin—1967. *Environ. Sci. Technol.* **5**, 1009 (1971).

Arnts, R. R., and B. W. Gay, Jr. Photochemistry of some naturally emitted hydrocarbons. EPA-600/3-79-081, U.S. Environmental Protection Agency (1979).

Arnts, R. R., and S. A. Meeks. Biogenic hydrocarbon contribution to the ambient air of selected areas. EPA-600/3-80-023, U.S. Environmental Protection Agency (1980).

Benson, S. W. *Thermochemical Kinetics*. Wiley, New York (1968).

Bufalini, J. J., B. W. Gay, Jr., and K. L. Brubaker. Hydrogen peroxide formation from formaldehyde. Photooxidation and its presence in urban atmospheres. *Environ. Sci. Technol.* **6**, 816 (1972).

Calvert, J. G., and R. D. McQuigg. The computer simulation of the rates and mechanisms of photochemical smog formation. *Int. J. Chem. Kinet. Symp.* **1**, 113 (1975).

Campbell, M. J., J. C. Sheppard, and B. F. Au. Measurements of hydroxyl concentrations in boundary layer air by monitoring CO oxidation. *Geophys. Res. Lett.* **6**, 175 (1979).

Carter, W. P. L., A. C. Lloyd, J. L. Sprung, and J. N. Pitts, Jr. Computer modeling of smog chamber data: Progress in validation of a detailed mechanism for the photooxidation of propene and *n*-butane in photochemical smog. *Int. J. Chem. Kinet.* **11**, 45 (1979).

Chameides, W. L. The effects of increased CO and NO_x upon tropospheric OH, CH_4 and related species. In *Man's Impact on the Troposphere*. NASA Reference Publication 1022, Washington, D.C. (1978).

Chameides, W. L., and D. H. Stedman. Tropospheric ozone: Coupling transport and photochemistry. *J. Geophys. Res.* **82**, 1787 (1977).

Chameides, W. L., and J. C. Walker. A photochemical theory of tropospheric ozone. *J. Geophys. Res.* **78**, 8751 (1973).

Chameides, W. L., and J. C. Walker. A time-dependent photochemical model for ozone near the ground. *J. Geophys. Res.* **81**, 413 (1976).

Chan, W. H., W. M. Uselman, J. G. Calvert, and J. H. Shaw. The pressure dependence of the rate constant for the reaction: $HO + CO \rightarrow H + CO_2$. *Chem. Phys. Lett.* **45**, 240 (1977).

Cicerone, R. J., J. D. Shetter, D. H. Stedman, T. J. Kelly, and S. C. Liu. Atmospheric N_2O: Measurements to determine its sources, sinks and variations. *J. Geophys. Res.* **83**, 3042 (1978).

Cox, R. A., R. G. Derwent, and P. M. Holt. Relative rate constants for the reactions of OH radicals with H_2, CH_4, CO, NO, and HONO at atmospheric pressure and 296 K. *Chemical Society, Faraday Trans. I.* **72**, 2031 (1976).

Crutzen, P. J. Photochemical reaction initiated by and influencing ozone in unpolluted tropospheric air. *Tellus* **26**, 47 (1974).

Davis, D. D., W. Heaps, and T. McGee. Direct measurement of natural tropospheric levels of OH via aircraft borne tunable laser. *Geophys. Res. Lett.* **3**, 331 (1976).

Dawson, G. A. Atmospheric ammonia from undisturbed lands. *J. Geophys. Res.* **82**, 3125 (1977).

Demerjian, K. L., J. A. Kerr, and J. G. Calvert. The mechanism of photochemical smog formation. *Adv. Environ. Sci. Technol.* **4**, 1 (1974).

Dimitriades, B. Effects of hydrocarbon and nitrogen oxides on photochemical smog formation. *Environ. Sci. Technol.* **6**, 253 (1972).

Doyle, G. J., E. C. Tuazon, R. A. Graham, T. M. Mischke, A. M. Winerand, and J. N. Pitts, Jr. Simultaneous concentrations of ammonia and nitric acid in a polluted atmosphere and their equilibrium relationship to particle ammonium nitrate. *Environ. Sci. Technol.* **13**, 1416–1419 (1979).

Drummond, J. W. Atmospheric measurements of nitric oxide using a chemiluminescent detector. Ph.D. Thesis, University of Wyoming, Laramie (1977).

Fink, U., D. H. Rank, and T. A. Wiggins. Abundance of methane in the earth's atmosphere. *J. Opt. Soc. Am.* **54**, 472 (1964).

Finlayson, B. T., and J. N. Pitts, Jr. Photochemistry of the polluted troposphere. *Science* **192**, 111 (1976).

Fishman, J., and P. J. Crutzen. A numerical study of tropospheric photochemistry using a one-dimensional model. *J. Geophys. Res.* **82**, 5897 (1977).

Fishman, J., S. Solomon, and P. J. Crutzen. Observational and theoretical evidence in support of a significant in-situ photochemical source of tropospheric ozone. *Tellus* **31**, 432 (1979).

Hecht, T. A., J. H. Seinfeld, and M. C. Doge. Further development of generalized kinetic mechanism for photochemical smog. *Environ. Sci. Technol.* **8**, 327 (1974).

Holdren, M. W., H. H. Westberg, and P. R. Zimmerman. Analysis of monoterpene hydrocarbons in rural atmospheres. *J. Geophys. Res.* **84**, 5083 (1979).

Howard, C. J., and K. M. Evenson, Kinetics of the reaction of the hydroperoxoradical with nitric oxide. *Geophys. Res. Lett.* **4**(10), 437–40 (1977).

Hudson, R. D. (ed.) Chlorofluoromethanes and the Stratosphere. NASA Reference Publication 1010, Washington, D.C. (1977).

Huebert, B. J., and A. L. Lazrus. Global tropospheric measurements of nitric acid vapor and particulate nitrate. *Geophys. Res. Lett.* **5**, 577 (1978).

Kok, G. L., K. R. Darnall, A. M. Winer, J. N. Pitts, and B. W. Gay, Jr. Ambient air measurements of hydrogen peroxide in the California South Coast Basin. *Environ. Sci. Technol.* **12**, 1077 (1978).

Kopczynski, S. L., W. A. Lonneman, F. D. Sutterfield, and P. E. Darley. Photochemistry of atmospheric samples in Los Angeles. *Environ. Sci. Technol.* **6**, 342 (1972).

Kopczynski, S. L., R. L. Kuntz, and J. J. Bufalini. Reactions of complex hydrocarbon mixtures. *Environ. Sci. Technol.* **9**, 648 (1975a).

Kopczynski, S. L., W. A. Lonneman, T. Winfield, and R. Seila. Gaseous pollutants in St. Louis and other cities. *J. Air Pollut. Contr. Assoc.* **25**, 251 (1975b).

Leighton, P. A. *Photochemistry of Air Pollution*. Academic, New York (1961).

Leonard, M. J., E. L. Fisher, M. J. Brunelle, and J. E. Dickison. Effects of the Motor Vehicle Control Program on hydrocarbon concentrations in the Central Los Angeles atmosphere. *J. Air Pollut. Contr. Assoc.* **26**, 359 (1976).

Levy, H., II. Normal atmosphere: Large radical and formaldehyde concentrations predicted. *Science* **173**, 141 (1971).

Lonneman, W. A. Ambient hydrocarbon measurements in Houston. *Proceedings Ozone/Oxidants Interaction with the Total Environment II, Specialty Conference, APCA Houston, Texas, October* (1979).

Lonneman, W. A., T. A. Bellar, and A. P. Altshuller. Aromatic hydrocarbons in the atmosphere of the Los Angeles basin. *Environ. Sci. Technol.* **2**, 1017 (1968).

Lonneman, W. A., R. Seila, R. Kuntz, S. Meeks, and J. J. Bufalini, unpublished data report on St. Louis RAPS Field Study Results 1972–1973 and 1977. Environmental Sciences Research Laboratory, U.S. Environmental Protection Agency, Research Triangle Park, NC (1978).

Lonneman, W. A., S. L. Kopczynski, P. E. Dailey, and F. D. Sutterfield. Hydrocarbon composition of urban air pollution. *Environ. Sci. Technol.* **8**, 229 (1974).

Lonneman, W. A., J. J. Bufalini, and R. L. Seila. PAN and oxidant measurement in ambient atmospheres. *Environ. Sci. Technol.* **10**, 374 (1976).

Lowe, D. C., P. G. Guenther, and C. D. Keeling. The concentration of atmospheric carbon dioxide at Bearing Head, New Zealand, *Tellus* **31**, 58 (1979).

Nash, T. Nitric acid in the atmosphere and laboratory experiments on its photolysis. *Tellus* **26**, 175 (1974).

National Research Council, National Academy of Sciences. *Nitrogen Oxides*. Washington, D.C. (1977).

National Research Council, National Academy of Sciences, *Ozone and Other Photochemical Oxidants*. Washington, D.C. (1976a).

National Research Council, National Academy of Sciences. *Vapour—Phase Organic Pollutants—Volatile Hydrocarbons and Oxidation Products*. Washington, D.C. (1976b).

Niki, H., E. E. Daly, and B. Weinstock. *Photochemical smog and ozone reactions*, p. 16, Advances in Chemistry Series 113, American Chemical Society, Washington, D.C. (1972).

Noxon, J. F. Nitrogen dioxide in the stratosphere and troposphere measured by ground based absorption spectroscopy. *Science* **189**, 547 (1975).

Noxon, J. F. Tropospheric NO_2. *J. Geophys. Res.* **83**, 3051 (1978).

Perner, D., D. H. Ehhalt, H. W. Patz, U. Platt, E. P. Roth, and A. Volz. OH-radicals in the lower troposphere. *Geophys. Res. Lett.* **3**, 466 (1976).

Ritter, R. A., D. H. Stedman, and T. J. Kelly. Ground level measurements of NO, NO_2 and O_3 in rural air. *Proceedings ACS Symposium on Atmospheric Nitrogen Compounds, Anaheim, California, March* (1978).

Robinson, E., R. A. Rasmussen, H. H. Westburg, and M. W. Holdren. Nonurban, nonmethane, low molecular weight hydrocarbon concentrations related to air mass identification. *Geophys. Res.* **78**, 5345 (1973).

Schmidt, U. Molecular hydrogen in the atmosphere. *Tellus* **26**, 78 (1974).

Seiler, W. The cycle of atmospheric CO. *Tellus* **26**, 116 (1974).

Sie, B. K. T., R. Simonaitis, and J. Heicklen. The reaction of OH with CO. *Int. J. Chem. Kinet.* **8**, 85 (1976).

Singh, H. B. Atmospheric halocarbons: Evidence in favor of reduced average hydroxyl radical concentration in the troposphere. *Geophys. Res. Lett.* **4**, 101 (1977).

Singh, H. B., F. L. Ludwig, and W. B. Johnson. Ozone in clean remote atmospheres: Concentrations and variabilities. CRC-APRAC CAPA-15-76. Coord. Res. Council, Inc., June (1977).

Singh, H. B., F. L. Ludwig, and W. B. Johnson. Tropospheric ozone: Concentrations and variabilities in clean, remote atmospheres. *Atmos. Environ.* **12**, 2185 (1978).

Singh, H. B., W. Viezee, W. B. Johnson, and J. L. Ludwig. The impact of stratospheric ozone on tropospheric air quality. *Proceedings Ozone/Oxidants Interaction with the Total Environment II, Speciality Conference, APCA, Houston, Texas, October* (1979).

Stephens, E. R., and F. R. Berleson. Distribution of light hydrocarbons in ambient air. *J. Air Pollut. Contr. Assoc.* **19**, 929 (1969).

Stewart, R. W., S. Hameed, and J. P. Pinto. Photochemistry of tropospheric ozone. *J. Geophys. Res.* **82**, 3134 (1977).

U.S. Environmental Protection Agency. Air Quality Criteria for Ozone and Other Photochemical Oxidants. Volumes I and II. Washington, D.C. (1978).

U.S. Environmental Protection Agency. Air Quality Criteria for Oxides of Nitrogen. Washington, D.C. (1980).

U.S. Environmental Protection Agency. National Air Quality and Emissions Trends Report, 1976. EPA-450/1-77-002. Research Triangle Park, NC (1977).

Wang, C. C., L. I. Davis, Jr., C. H. Wu, S. Japar, H. Niki, and B. Weinstock. Hydroxyl radical concentrations measured in ambient air. *Science* **189**, 797 (1975).

Winer, A. M., K. R. Darnall, R. Atkinson, and J. N. Pitts, Jr. Smog chamber study of the correlation of hydroxyl radical rate constants with ozone formation. *Environ. Sci. Technol.* **13**, 822 (1979).

Wofsy, S. C., J. C. McConnell, and M. B. McElroy. Atmospheric CH_4, CO, CO_2. *J. Geophys. Res.* **77**, 4477 (1972).

Zimmerman, P. R. Testing of hydrocarbon emissions from vegetation, leaf litter and aquatic surfaces, and development of a methodology for compiling biogenic emission inventories. EPA-450/4-79-004, U.S. Environmental Protection Agency (1979).

SYNTHESIS

ATMOSPHERIC CHEMISTRY OF AIR POLLUTANTS

W. R. Pierson
Scientific Laboratory
Ford Motor Company
Dearborn, Michigan

R. B. Husar
Washington University
St. Louis, Missouri

BACKGROUND AND DISCUSSION

Effects on the terrestrial ecosystem are caused by the pollutants that arrive at the receptor, not always by what has been emitted. Some of the relevant atmospheric processes are indicated in Figure 1. In the case of SO_2 emission, much of it will arrive and be deposited as SO_2, but a fraction will be oxidized to H_2SO_4, which, in turn, may be partially or wholly neutralized in the air by naturally occurring and man-made ammonia. All three sulfur compounds (SO_2, H_2SO_4, and the NH_4^+ salts) have appreciable dry and wet deposition rates. In a time scale of a day,

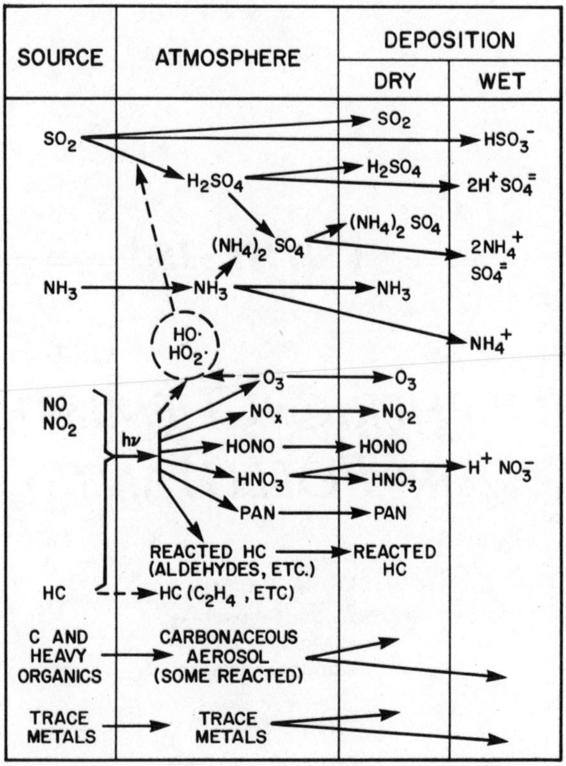

FIGURE 1. *A selection of relevant atmospheric processes. HC = hydrocarbons, especially light ones other than CH_4.*

the $NO + NO_2$ + hydrocarbon photochemical system produces O_3, nitrous acid (HONO), nitric acid (HNO_3) and peroxyacetyl nitrate (PAN), as well as partially reacted hydrocarbons (HC). The short-lived reactive radicals HO and HO_2 (and their alkoxy analogs) play key roles in the photochemistry of the $SO_x + NO_x + HC$ system, but their concentrations are low and there is as yet no evidence on their direct significance to plants.

The following points represent a consensual summary of the issues raised.

1. *The atmospheric chemists want the plant physiologists to inform them what needs to be measured in the air (and on what scales of space and time, with what time resolution, etc.), but the plant physiologists cannot do so until the atmospheric scientists tell them what is there. Amalgamation of effort is evidently needed.*

2. *O_3 and SO_2 (and NO_2 in relation to O_3 and SO_2) appear to be the species of greatest concern from the standpoint of plant effects. Aerosols, including sulfate, are of less concern. A possible exception is sulfite, which is difficult to measure correctly and which is poorly understood as to its atmospheric concentrations. Sulfur dioxide effects on photosynthesis and plant respiration have been observed in short-term (1 h) exposures at SO_2 concentrations as low as 20–35 ppb—far below the U.S. secondary 3-h standard of 500 ppb. To what extent does our present focus on these substances simply reflect their status as criteria pollutants? We know how to measure these pollutants, and some information is available regarding their atmospheric concentrations and effects on plants. There are a number of other pollutants which are potentially important, such as aldehydes, hydrogen peroxide, and nitric acid. There appears to be considerable concern by the plant physiologists about fluoride, which is relatively neglected in most atmospheric measurements. Aerosols, and particularly their heavy-metal constituents, are probably of minor influence through direct deposition onto plant surfaces; but they may have some influence through deposition on the soil and subsequent access to the plants through the roots. The understanding of these processes lags behind the understanding of the processes involving gases.*

3. *A vast amount of atmospheric chemical information exists that is directly usable by the plant physiologists. There are some deficiencies having to do with distance scale. For example, for purposes of carrying out a regional atmospheric study, it may be quite appropriate to measure O_3 at heights of 15–30 ft off the ground, as has often been done; but if O_3 concentration gradients with height can be a factor of 2 in 40 ft, then information on O_3 concentrations at the receptor—the plant—may be weak.*

4. *Joint effects of pollutants are well documented in the case of O_3 in combination with SO_2. What the atmospheric chemist can provide is the information on the joint probability of occurrence; for the most part one finds that the SO_2 concentration tends to be anticorrelated with the O_3 concentration. In any case the comprehension of such effects is inherently far more elusive and therefore more poorly understood than is the case for pollutants considered one at a time. Supposing, for example, that the plant has enough "memory" so there can be a combined effect of SO_2 and O_3 when the two are out of phase. How does this effect depend on the phase shift or the repetition rate?*

5. *There were a number of questions concerning the interaction of ozone with plants: When O_3 causes its effects, is it really the O_3*

that is doing it, or is it something else? Can anybody model the O_3 transport at the leaf surface? Apparently the reaction with the outer surface of the leaf is minimal; instead, it consists almost wholly of entrance through the stomata and reaction at the internal membranes.

6. *Free radicals at the plant interface could be important. They are part of the normal photosynthetic processes and are also associated with the functioning of membranes. If we cause an increase in free radicals at membrane surfaces, we presumably perturb the normal processes. Yet a practical method to measure free radicals at plant surfaces does not seem to exist. Moreover, there are a number of free-radical species in the atmosphere (HO, HO_2, CH_3O, and many more) which play a vital role in atmospheric chemistry but are presumed to be of no plant-physiological importance per se on account of their short lifetimes and consequent low concentrations. Given that atmospheric free-radical concentrations are low, it may not necessarily follow that they can be discounted as direct-acting pollutants to vegetation.*

7. *The question of diurnal variations of pollutant concentrations appears to be an important one with regard to the effects of pollutants on plants. Beginning with the question of whether atmospheric measurements should focus on concentrations at night or in the daytime, and with what time resolution, it was noted that the effect of O_3 depends on time of day, humidity, condition of the plant, and other factors. This being the case, the nighttime inversion breaks up in the morning and fumigates the plants just when the stomata are open to their maximum.*

8. *A corollary of the example just given is that the effect of even a fixed pulse of a pollutant depends on a number of factors. How, then, can cogent answers be expected on questions of whether integrated dose or peak exposure is the important factor?*

9. *There is a problem with the belief, held by some, that natural hydrocarbons (terpenes, isoprene) are responsible for most of the rural O_3 or oxidant: Ambient concentrations of natural hydrocarbons in the clean troposphere are several times lower than those predicted from measurements of emission from vegetation. The explanation may lie at least in part with the very fast reaction of terpenoid compounds with O_3 to form carbonaceous aerosol, although there are problems with this explanation. The question remains unsolved.*

10. *The seasonality of O_3 injection from the stratosphere into the troposphere was considered. The injection of O_3 goes through a yearly maximum. The time of this maximum depends on latitude,* *from winter in Florida to summer in Minnesota (and not necessarily simultaneous with the sulfate maximum). Plants are most vulnerable to O_3 when they are in their reproductive stages—flowering, pod formation, etc. Therefore, it is important to be aware of the seasonal injection phenomenon to make a proper evaluation of the role of man in O_3-induced effects such as crop loss.*

11. *The possibility of rural O_3 derived through transport of precursor species from polluted urban areas was also considered. The underlying idea is that less-reactive hydrocarbons, left over after all the NO_x has been depleted by reaction with the rest of the hydrocarbons, can be transported long range and presumably, therefore, cause photochemical O_3 to be produced at distances remote from the source by reaction with NO_x from rural stationary sources such as power plants. Some feel that this sequence may be the biggest contributor to rural O_3. Resolution of this question is very important.*

12. *In connection with the question of sulfur oxides emission, transformation, and transport in plumes, K. T. Whitby (University of Minnesota) presented plume data on SO_2 conversion to aerosol which were interpreted in terms of SO_2 oxidation by (probably) OH radicals and the rate of OH diffusion into the plume. D. J. Williams (C.S.I.R.O., Australia) presented data for a plume from a smelter that was situated in a remote region. These data were also interpreted in terms of SO_2 oxidation by OH, but the lateral profile was attributed not to OH diffusion but rather to OH production governed by ultraviolet light attenuation by the vertical SO_2 burden.*

PART THREE

CHARACTERIZATION AND QUANTIFICATION OF AIR POLLUTANTS

Effects research is the focus of all air pollution studies. But without adequate measurement technology to characterize and quantify the environment, the understanding of pollutant chemistry and the assessment of terrestrial ecosystem effects cannot proceed. In some cases, instrumentation developed for the routine monitoring of criteria pollutants has been adapted; in other cases, especially those with special requirements for sensitivity or for fast temporal response, new techniques must be developed. Dissemination of information on new developments in measurement technology and on the evaluation of techniques is of paramount importance in designing better experiments in effects research. Specific areas of current interest in measurement technology include artifact-free measurement of acid gases and aerosols, monitoring of organics for source identification, receptor models to identify the source of an agent by measurements at the receptor site, measurement techniques for quantifying dry deposition, and sample integrity.

This section focuses on these areas of interest. It is hoped that the mutual communication between chemists, technologists, and effects scientists will lead in the future to an enhanced capability to relate effects to environmental contaminants and eventually to proper control strategies.

Sampling, Analysis, and Interpretation of Atmospheric Particles in Rural Continental Areas

GLEN E. GORDON
Department of Chemistry
University of Maryland
College Park, Maryland

1. INTRODUCTION

Air pollution effects upon terrestrial ecosystems that are of major concern include:

1. The direct effects of pollutant gases and aerosols on vegetation.
2. Wet or dry deposition of acids or toxic species on land, water, or vegetation.
3. The modification of climate and degradation of visibility.

Discussion here will be focused on the measurement of metals and other trace substances in airborne particles. Since most of the problems noted mainly involve effects of pollutant gases or the sulfates and nitrates on particles (which are discussed by Newman in this volume), the role of trace elements in these questions may not be obvious. Except for unusual situations (e.g., near large smelters), the direct effects of the trace metals are much less intense than those of the more major species. However, as described below, the measurement of a large suite of trace substances provides detailed information that may be used to identify sources of the materials and trace their movements on a regional scale. Furthermore, since we know a great deal about the compositions of particles of different sizes, we can use the compositions of materials brought down by rain, snow, or dry deposition to infer the relative importance of various deposition mechanisms for particles of different sizes. Thus, although some effects may depend strongly upon sulfates, the measurement of sulfate concentrations alone gives little information about the origins of the sulfur species.

During the past five years, much progress has been made in the assignment of contributions of various classes of sources to urban total suspended particulate matter (TSP) based upon concentrations of large numbers of species. At the other end of the distance scale, Zoller and co-workers (Maenhaut et al., 1979; Cunningham and Zoller, 1981) have made considerable progress in determining the origins and transport of aerosols sampled at the South Pole. In this paper we are concerned with the intermediate distance scale. Can we, for example, identify particles from a specific city or large point source at distances from a few tens of kilometers up to 1000 km? What should we measure, and are methods available? How can measurements of trace elements in the atmosphere, cloud water, precipitation, and dry deposition provide information about the fundamental mechanisms involved in those processes? What additional kinds of measurements need to be developed for these applications?

2. SAMPLING AND MEASUREMENT METHODS

For the applications discussed below, it is desirable to measure concentrations of a very large number of elements or other species. To reduce the labor involved in dealing with large numbers of samples and to avoid contamination or loss of species during chemical manipulations, it is desirable to use multielement methods that are completely instrumental. Although methods such as atomic absorption spectrophotometry (AAS), spark-source mass spectrometry, and emission spectrometry can provide accurate data when performed carefully, the completely instrumental methods of X-ray fluorescence (XRF) and instrumental neutron activation analysis (INAA) are better suited to the analysis of large numbers of samples for many elements.

The elements usually measurable in particulate material from urban atmospheres by these methods are listed in Table 1. No single technique can measure all elements of interest. Because of their soft X rays, elements lighter than Al cannot be detected by XRF, and those from Al through K can be measured only with great care to correct for self-absorption effects. Since X-ray energies and production cross sections vary smoothly with atomic number, XRF does not pick out certain trace elements with unusual sensitivity, as INAA does. Thus, XRF is useful for elements with $Z > 13$ at about 0.1% or greater levels.

Instrumental neutron activation analysis can be used for many of the same elements as XRF; the major exceptions are S, P, Ni, and Pb. On the other hand, INAA measures many very trace elemental species with great sensitivity, for example, Se, As, Sb, In, I, and many rare earths. In many projects, the emphasis has been placed on measurement of major elements, since they account for much of the TSP. However, major elements usually provide little information that can be used to identify sources. Except near an unusual source, the ratios of concentrations of Si, Ti, Fe, and many other lithophiles to that of Al are within a factor of 2 of the ratio for the earth's crustal abundance pattern (Gordon et al., 1983; Rahn, 1976). By contrast, many of the chalcophiles and halogens are enriched by factors from 100 to several thousand relative to the earth's crust, for example, Zn, Sb, Se, Mo, As, Pb, Cd, Cl, Br, and I (see discussion of enrichment factors, Sec. 3). Thus, concentrations of these classes of elements provide much more variation by which to identify sources. The ability to observe many of these elements reliably is one of the major strengths of INAA.

Most of the recent detailed studies of elemental concentrations on urban aerosols have used XRF, INAA, or both. In studies of Washington, D.C. aerosols, Kowalczyk et al. (1982) measured 37 elements by INAA, augmented by AAS for Pb, Ni, and Cd. The Oregon Graduate Center used a combination of XRF and short-irradiation INAA in their studies of the Portland atmosphere (summarized by Watson, 1979). The use of XRF requires only a few minutes' irradiation of samples and the data can be analyzed immediately after irradiation. A major disadvantage of INAA is the long turn-around time for complete analysis. Samples are usually irradiated for a few minutes to analyze species with half lives up to a few hours, allowed to decay, and then reirradiated for several hours to build up large activities of long-lived species. One must wait about three weeks after

TABLE 1. Elements Measurable in the Urban Aerosol by Instrumental Methods

Method	Conditions	Quality of Measurement	Element Measurable
X-ray fluorescence (XRF)	>2.5 μm diameter	Good	Al, Si, P, S, Cl, K, Ca, Ti, Mn, Fe, Cu, Zn, Br, Rb, Sr, Pb
		Marginal	V, Cr, Ni, Se, Cd, Ba
	<2.5 μm diameter	Good	Si, P, S, Cl, K, Ca, Fe, Cu, Zn, Se, Br, Pb
		Marginal	Al, Ti, V, Cr, Mn, Ni, Rb, Sr, Cd, Sn, Sb, Ba
Instrumental neutron activation analysis (INAA)	Short irradiation	Good	Na, Mg, Al, Cl, Ca, Ti, V, Mn, Br, I
		Marginal	S, Cu, Sr, In, Ba, Dy
	Long irradiation	Good	K, Sc, Cr, Fe, Co, Zn, As, Se, Sb, Cs, La, Ce, Sm, Eu, Hf, Ta, Th
		Marginal	Ga, Mo, Ag, Cd, Tb, Yb, Lu, W

From Gordon et al. (1983).

long irradiations for short-lived activities to decay before some of the longest-lived species can be observed. The combination of XRF and "shorts" INAA used in the Portland, OR study represents a compromise that is faster and cheaper than complete INAA but obtains reliable data for more elements than XRF alone. However, leaving out the INAA longs eliminates several useful elements—especially As, Sb, and Se—that would be sensitive indicators of some sources.

The most extensive urban study was that by Loo et al. (1978), who ran 10 dichotomous samplers (mostly for 12-h periods) over a 2-yr period in the St. Louis area in connection with the Regional Air Pollution Study (RAPS). Each of the approximately 34,000 samples was analyzed automatically by XRF and the data treated and stored by computer (Dzubay and Stevens, 1975). No nuclear analyses were performed.

The only recent major urban study that did not employ XRF or INAA was that of Tucson, AZ area by Moyers et al. (1977). They used mainly AAS to analyze particles collected on Hi-Vol filters.

As shown in Table 2, concentrations of most species are typically reduced by factors of 2 to 5 in rural areas relative to urban (and even more in very remote areas); thus, some of the elements may be more difficult to measure accurately in those areas. However, one should be able to measure all elements listed in Table 1 by using filter materials with low, reproducible blank values for all species of interest, by pumping greater volumes of air through the filters, and by taking great precautions to avoid contamination during handling of the samples. By taking these precautions, Maenhaut et al. (1979) were able to measure concentrations of 36 elements in material collected at the South Pole, where atmospheric concentrations are lower by two or three orders of magnitude than in urban areas (see Table 2).

As suggested above, to make reliable measurements of trace elements, one must use filter materials that have low and reproducible blank values. The standard glass-fiber filters have enormous blank values and cannot be used. Fortunately, several types of organic-based filters (e.g., Nucleopore, Fluoropore, Whatman) with low blanks have become available in recent years. There is no particular one that is best for every application: They must be chosen on the basis of the sampler, the particulate loading of the area under study, the length of collection periods, and the analytical scheme to be used. The material selected should not only have low blank values, but also high efficiency for the sizes of particles to be collected and good flow rates over the sampling periods to be used.

The standard 8×10 in. Hi-Vol sampler is not adequate for most applications as it collects some particles of much greater size than one is interested in (giving too much emphasis to particles generated quite locally, with very short residence times), and its efficiency for large particles is strongly dependent on wind speed and direction with respect to the orientation of the shelter around the sampler (Wedding et al., 1977). If properly

TABLE 2. Typical Concentrations (ng/m^3) of Elements in Particles in Urban, Rural Continental, and Very Remote Areas

	Urban		Rural Continental				Very Remote	
Element or Species	Washington, D.C.[a]	Portland OR[b]	Prince Georges Cty, MD[c]	Walker Branch, TN[d]	Gt. Smoky Mtns.[e]	Tucson, AZ (rural sites)[f]	Northwest Territories Canada[g]	South Pole[h]
C	—	13700	—	—	4400	—	—	—
Na	300	1030	440	201	—	280	18	3.3
Al	1350	2090	910	820	205	580	66	0.82
Si	—	6200	—	—	620	3900	—	—
SO_4	8900	3800	—	—	12000	3700	—	150
Cl	530	1650	—	345	<17	—	9	2.6
K	400	470	—	223	150	530	54	0.68
Ca	860	1390	143	472	340	790	40	0.49
Sc	0.33	—	0.24	0.12	—	—	0.04	0.00016
Ti	110	170	59	66	18	100	5	0.10
V	25	24	35	3.6	<4	—	0.21	0.0013
Cr	15	47	2.8	—	—	3.1	0.59	<0.04

Mn	26	87	14	14	—	12	1.5	0.013
Fe	1000	2000	640	310	146	660	71	0.62
Ni	17	55	24	—	2	3.2	—	—
Cu	17	84	12	8	<8	110	0.9	0.029
Zn	86	130	68	23	9	110	3.8	0.033
As	3.2	5	3.8	1.4	2.2	—	0.31	0.031
Se	2.4	3	2.6	1.1	1.6	—	0.043	0.006
Br	135	340	46	23	23	—	0.54	2.6
Cd	2.5	—	—	4	—	2.0	—	<0.018
Sb	2.1	—	2.4	—	—	—	0.13	0.00084
Pb	440	1055	315	90	113	67	—	0.076

[a] Kowalczyk et al. (1982).
[b] Watson (1979) Central Air Monitoring Site.
[c] Gladney (1974).
[d] Andren (1974).
[e] Stevens et al. (1980).
[f] Moyers et al. (1977).
[g] Rahn (1976).
[h] Maenhaut et al. (1979).

mounted, 47-mm or 4-in. diameter circular filter holders can be employed if XRF or INAA are to be used. If the large volume of the Hi-Vol is needed (as might be the case for detailed analysis of the organic fraction), a new inlet is available that has a sharp upper cutoff at about 15-μm diameter.

Most particles fall into one of two modes of the size distribution curve, separated by a minimum at diameter ~2 μm (National Research Council, 1979; also see Fig. 1, Hidy and Mueller, this volume). These modes originate from different processes and the particles have quite different compositions. The coarse particles are produced by wind erosion and other abrasional processes and their compositions are similar to the crustal abundance pattern. Fine particles arise from combustion processes and formation of secondary materials from gases. They contain large amounts of sulfate, ammonium ion, carbonaceous materials, and other (fairly volatile) species. Furthermore, the fine particles will usually be transported to greater distances from their sources than the coarse particles, which are deposited at ground level more rapidly. Because of these fundamental differences, any definitive studies of particles in rural areas should employ collection devices that separate particles into two or more size groups.

Many types of cascade impactors are available, the best of which are probably those based on designs by Mitchell and Pilcher (1959). These collect particles in 6 or 7 size ranges between about 16 and 0.5 μm diameter. Many of the other designs cause severe losses of particles on the walls of the device or cause large particles to bounce off the large-particle collectors and collect on the backup filters, thus yielding distorted size distributions. While it is desirable in research studies to map out the size distribution curves for particles bearing the various species (Gladney et al., 1974), the use of cascade impactors may not be the best choice for a particular study. Even the best impactors suffer bounce-off of crustal-type particles on days of low humidity (Gordon et al., 1973). Second, the low flow rate and the division of the material into 6 or 7 fractions may reduce the amounts of many species below detection limits. Finally, the several samples for each impactor run multiplies the analytical work.

A good compromise in many applications is the dichotomous sampler (Dzubay and Stevens, 1975; Loo et al., 1976; Stevens et al., 1978). Although it collects only two size groups (usually separated at diameter of 2.5 μm), it suffers no bounce-off problems, as it uses the principle of virtual impaction, that is, it divides the incoming air into two streams bearing the different sizes of particles and passes each through a separate filter. Although the dichotomous sampler is a good collector of size-segregated particles in urban areas when coupled with sensitive analytical methods such as XRF or INAA, its low flow rate (~15 L/min) causes the amount of material collected to be inadequate for many types of analyses, for example, for organic analyses. A dichotomous sampler with a much greater flow rate is badly needed for many applications. Solomon et al. (1983) recently built and tested a prototype high volume dichotomous virtual impactor with a flow rate of 500 L/min (versus up to 50 L/min for conventional dichotomous samplers). This new sampler may be of enormous value for sampling of very clean air and/or sampling for species that require large quantities of material for analysis.

3. INTERPRETATION OF OBSERVED CONCENTRATIONS

After accumulating large sets of data in rural areas, how can one interpret the data to obtain maximum information about sources and transport of particles? Few detailed studies have been done at this distance scale in continental areas. In recent years, much progress has been made in the interpretation of urban particulate composition patterns using "receptor" models, that is, models which attempt to resolve observed concentration patterns into linear combinations of patterns characteristic of particles from various types of major sources (Gordon, 1980). The most widely used receptor model is the chemical element balance (CEB) method (Friedlander, 1973), according to which the composition pattern of the aerosol is expressed as a linear combination of the compositions of particles released by various classes of sources. The concentration of element i in a sample is given by

$$C_i = \sum_j m_j X_{ij} \tag{1}$$

where m_j is the fraction of material contributed by source j and X_{ij} is the concentration of element i in material from source j. One must determine the X_{ij} values for all of the sources that might make important contributions to the aerosol at the site of interest. The strengths of the contributing sources m_j are determined by a least-squares fit to the observed concentrations of a subset of the elements.

An example of the use of CEBs is the study by Kowalczyk (1979), who performed CEBs on a data set obtained by Gladney (1974) in southern Prince Georges County, MD, about 50 km southeast of Washington, D.C. He operated 12 air-filter stations for several weeks in August 1973. Seven 24-h filters from each station were analyzed for about 35 elements. The network was originally set up to detect emissions from a large coal-fired power plant with 122-m stacks located south of all but one of the stations. Despite the occurrence of several sampling periods with the prevailing winds from the south, plume contact with the stations was so infrequent that the samples showed almost no direct impact of the plant. Thus, the area can be taken as being fairly representative of rural areas of the Eastern United States. Largely with the use of components measured by the Maryland group, Kowalczyk (1979) resolved the observed concentrations into six components: soil, marine, motor vehicle emissions, and combustion of coal, oil, and refuse. He chose to determine the source strengths by a least-squares fit to the concentrations of eight "marker" elements carefully selected to be sensitive to the particles from the sources: Na (marine), V (oil), Zn (refuse), Pb (motor vehicles), Al and Fe (coal + soil), Mn (soil), and As (coal). Once all of the major components for a particular area have been characterized and included in the CEB, one would presumably use all of the nonvolatile elements to obtain source strengths. However, while still testing the CEB, Kowalczyk's philosophy was that of using a minimum number of elements to determine the source strengths in order to leave as many as possible "floating" elements to test the quality of the fit.

The fit to the marker elements is shown in Table 3. After eliminating several volatile elements, which are not conservative on the particles, the six-component CEBs on the average fitted the observed concentrations of 14 floating elements to within a factor of 1.7 (see Table 4). The predicted/observed ratio lies between 0.8 and 1.5 for most nonvolatile elements, but there are a few (notably Cr, Ni, Cu, Sb, La, and Ce) that are severely underpredicted, suggesting that one or more important sources were missed or that concentrations of these elements are too low in the components used. Although the fits to concentrations of the floating elements are far from perfect, they represent a considerable improvement over many earlier attempts, mainly for two reasons: (1) the components used were of high quality, and (2) there is very little industry in the Washington D.C. area.

Table 5 lists the strengths of contributions from several types of sources as indicated by CEBs performed on urban Washington and the rural Prince Georges County samples. The same components were used for both sets of CEBs except that a limestone component was needed in the Washington CEBs to fit Ca concentrations. The quality of the Prince Georges County fits was slightly better than for Washington, D.C.

The results shown in Table 5 are reasonable. The coal and oil components are comparable as expected, as most electric power plants in the area are in a ring of about 50-km radius about the city. The soil component is higher in the city because of entrainment by human activities. The urban activities of motor vehicles and refuse combustion have greater strength in the city, and the marine component is greater at the Prince Georges County sites, which are closer to Chesapeake Bay.

Chemical element balances have so far been used very little to interpret rural-area data sets. A rather good accounting for the mass concentrations in the fine and coarse fractions of particles collected in the Great Smoky Mountains was obtained in CEBs by Dzubay (1980) using five components: shale, limestone, motor-vehicle emissions, trace elements, and ammonium sulfate. In a more recent study in the Great Smoky Mountains, Stevens et al. (1980) found that 61% of the mass of the fine-particle fraction consists of sulfate and its associated cations (having an average composition equivalent to ammonium bisulfate) with only 5% of the mass contributed by elemental carbon and 10% by organic carbon. These results strongly

TABLE 3. Average Fits to the Concentrations of Eight Marker Elements in Whole-Filter Samples from Southern Prince Georges County, MD

	Contributions from Components (ng/m^3)						Total Conc. (ng/m^3)	
Element	Soil	Coal	Oil	Refuse	Motor Vehicle	Sea Salt	Predicted	Observed[a]
Na	29.0	8.0	17.0	40.0	—	345.0	440.0	440 ± 80
Al	550.0	495.0	0.6	7.0	—	—	1050.0	910 ± 150
V	0.7	1.5	32.9	0.02	—	—	35.0	35 ± 6
Mn	8.8	1.5	0.14	0.36	—	—	10.8	13.5 ± 2.5
Fe	345.0	347.0	4.0	3.0	13.6	—	710.0	640 ± 2.5
Zn	0.8	2.5	2.3	59.0	4.0	—	68.0	68 ± 12
As	0.041	3.0	0.039	0.12	—	—	3.2	3.8 ± 0.8
Pb	0.1	2.0	0.6	40.0	272.0	—	314.0	315 ± 70

Data from Gladney (1974), CEB analysis by Kowalczyk (1979).

[a]Uncertainty is the standard deviation of the mean value.

TABLE 4. Predicted and Observed Concentrations for Elements Not Used in Fitting the Prince Georges County, MD, Aerosol[a]

Element	Concentration (ng/m^3) Predicted	Observed[b]	Predicted/Observed	Major Source(s)
Mg	134.0	170±40	0.79	Soil, sea salt
Ba	15.9	12±3	1.32	Soil, coal, motor vehicle
Br	108.0	46±5	2.3[c]	Motor vehicle
I	4.1	2.9±0.8	1.4[c]	Coal, sea salt
Sc	0.27	0.24±0.8	1.12	Coal, soil
Ti	66.0	59±9	1.12	Soil, coal
Cr	1.7	2.8±1.1	0.60	Coal, soil
Co	0.62	0.41±0.11	1.5	Coal, oil, soil
Ni	7.0	24±5	0.29	Oil
Cu	4.1	12±2	0.34	Coal
Se	0.80	2.6±0.6	0.31[c]	Coal
Sb	1.17	2.4±0.4	0.49	Refuse
La	0.83	1.4±0.5	0.59	Soil
Ce	1.3	2.5±0.9	0.52	Soil, coal
Th	0.18	0.18±0.07	1.00	Coal, soil

[a]Data from Gladney (1974); interpreted by Kowalczyk (1979).

[b]Uncertainty is standard deviation of the mean value.

[c]Because of volatility perfect agreement not expected.

suggest that the whitish haze that frequently blankets the area is caused mainly by acid sulfates rather than carbonaceous particles produced by natural emissions from trees to which the visibility degradation has commonly been ascribed.

Perhaps the most spectacular remote-area CEBs are those of Cunningham and Zoller (1981), who resolved concentration patterns of particles collected at the South Pole into six components: sulfate, crust, sea salt, meteorites, other (volcanoes, anthropogenic emissions, etc.), and a special Br factor, as the Br concentration was often much greater than could be accounted for by the sea-salt component. In terms of mass, sulfate is the overwhelming contributor. Typical contributions for austral summers were (in nanograms per cubic meter): sulfate, 200–350; crust, 5–18; sea salt, 2–6; meteorites, 0.4–2; Br, 0.5–2.5; other, 0.1–0.3. Once more is known about trace element concentrations from volcanoes and other major sources, the "other" category can presumably be resolved to determine anthropogenic contributions.

TABLE 5. Contributions of Several Sources to Total Suspended Particulate Material in Washington, D.C. and Rural Prince Georges County, MD, as Indicated by Chemical Element Balances

	Contributions to TSP ($\mu g/m^3$)	
Source	Washington, D.C.[a]	Prince Georges County[b]
Soil	15.3	9.8
Coal	4.1	3.9
Limestone	2.2	—
Oil	0.4	0.5
Refuse	0.9	0.65
Motor vehicle	4.6	2.7
Marine aerosol	0.6	1.1
Total primary	28.1	18.7

[a]7 urban stations; based on data from Kowalczyk (1979).

[b]12 rural stations; based on data of Gladney (1974) as analyzed by Kowalczyk (1979).

Another important receptor model is factor analysis (Hopke et al., 1976). Unlike the CEB method, factor analysis requires no a priori assumptions about either the number or composition of the sources. Instead, one examines large data sets to determine a minimum number of common factors that will explain most of the observed variance of elemental concentrations. Since factor analysis works on variance, it has the disadvantages that (1) it does not yield compositions of the components (but only the fraction of each element's variance that is accounted for by the factor), and (2) it does not work well on data sets in which there is not a lot of variance. Regarding this point, Hopke (private communication, 1978) applied factor analysis to the data set of Tables 3 and 4, but without much success.

There is so little variance from one station to another that most treatments yielded only two (crustal and sulfate) or at most three (sea salt) components. Thus, it does not appear that factor analysis will usually be very successful when applied to rural network data. Alpert and Hopke (1980) have successfully tested a new method, target-transformation factor analysis, which combines the best features of CEB and factor analysis methods. It seems to work well in urban applications and may be useful for rural studies once it is fully developed.

Although receptor models show considerable promise for identifying the sources of aerosols and of specific elements borne on them in rural areas, one should note that they identify only the class of sources, not individual sources within a class. Perhaps if we ever acquire really definitive composition data on particles emitted by many individual sources, we shall be able to use subtle differences in concentration patterns to determine impacts of large individual sources (e.g., fossil-fueled power plants). Studies of plumes from five copper smelters in Arizona, for example, indicate that the compositions of their emitted particles are so distinctive that there are good possibilities for identifying their particles at long distances (Small et al., 1981).

Another approach for identifying particles from specific sources is that of Rheingrover and Gordon (1980), who have been working on the interpretation of the data set obtained in connection with the St. Louis RAPS project. To identify samples heavily influenced by particles from dominant sources of various elements, we have searched the data set for samples which have very high concentrations of each element (usually about 3σ greater than the average for the element at the particular station). By applying a second criterion in the selection of these samples—namely, small standard deviation of the wind direction during the sampling period—we found that for many of the elements, the mean wind directions for the sample periods selected cluster strongly about one or two directions. These directions point toward dominant sources of the element. By performing the same kind of analysis on data at all of the stations, we obtain trajectories pointing toward sources of the elements. By triangulation, using vectors from several stations, we are able to locate and identify the dominant sources of various elements.

The approaches discussed above have been developed for application in urban areas and would probably be reasonably successful on a distance scale up to about 100 km from all of the major sources influencing the area of interest. But if one goes to more remote sites, which may be influenced with about equal frequency by emissions from many distant urban areas, it may not be meaningful to sort out emissions from various sources or classes of sources within a particular urban complex. At the more remote sites it may be useful to try to identify the emissions coming from the various urban complexes. If so, one should probably use only the fine-particle fraction, as the large particles will be contaminated by locally generated dust.

Are the concentration patterns of fine particles from various cities sufficiently distinctive that there is a reasonable chance of identifying particles observed 100 km or more away with a particular city? Here we are interested not in absolute concentrations, which will be modified by dilution during transport, but some kind of relative concentration pattern. For this we can use enrichment factors, EFs, with respect to a crustal abundance pattern (Wedepohl, 1968), which are commonly defined by (Rahn, 1976):

$$\mathrm{EF} = \frac{(C_{\mathrm{X}}/C_{\mathrm{Al}})_{\mathrm{atm}}}{(C_{\mathrm{X}}/C_{\mathrm{Al}})_{\mathrm{crust}}} \tag{2}$$

where C_{X} and C_{Al} are concentrations of the element X and Al in the atmosphere and the earth's crust. If all material in the atmosphere were crustal dust, all elements would have EF values of unity. But if there are sources that release large quantities of an element relative to crustal components, the EF value for the element will be much greater than unity.

Note in Eq. (2) that there is nothing unique about the selection of Al as the normalizing element: Any nonvolatile lithophilic element that does not have special sources and which can be reliably measured would serve as well. Aluminum has often been used because many of the large data sets were obtained by use of INAA, a good technique for Al, which, unfortunately, does not allow one to measure Si at all. In general, there is considerable danger in basing the entire EF calculation of the concentration of a single element, as any errors made in that measurement will be reflected in all of the EF values. This is a special danger now that many data sets based on XRF are becoming available, as Al is the lightest element for which XRF data are normally given, and thus the values contain the largest X-ray self-absorption corrections of all elements measured. Below, I discuss EF values for fine particles only. Since much of the size data have been obtained with cascade impactors, the use of Al has the added problem that Al-bearing particles suffer severe bounce-off problems on very dry days (Gordon et al., 1973). For all these reasons, in the calculations discussed below, I have first calculated EF values by Eq. (2), but then renormalized them so that the average EF value of about five lithophile elements (usually Al, Si, K, Ca, and Ti for XRF data and Al, Mg, K, Ca, and Ti for INAA data) is set equal to unity.

Enrichment factors for fine particles collected from five cities that have been extensively studied are shown in Figure 1. First, note that the EF values of major lithophile elements Al, Si, K, Mg, Ca, Ti, and Fe are indeed close to unity. There is some variation from unity, in part because of variations of the composition of local crustal material and in part due to contributions from other sources (e.g., fly ash from coal combustion). Another effect is fractionation; the fine particles that become airborne do not necessarily have the same composition as the bulk soil or rocks. Rahn (1976) has pointed out that fine clay-mineral particles would be more predominant in airborne soil than bulk material, causing the Al/Si ratio to be greater than for bulk soil. In order to aid in the development of crustal components for CEBs of the Portland, OR aerosol, Watson (1979) measured compositions of fine and coarse particles entrained by an air stream over soils. He found the average Si/Al ratio to be 2.3 for fine particles and 4.1 for coarse particles (versus the ratio of 3.9 from Wedepohl, 1968, used in Eq. (2)). This fractionation would account for the general depletion of Si show in Figure 1. This result is in sharp contrast with reports of huge enhancements of Si on fine particles collected in the Southwest United States desert (Macias et al., 1980). Comparing the Al and Si points from the same locations in Figure 1, we see Si depletions in most areas, but a slight enhancement in Charleston.

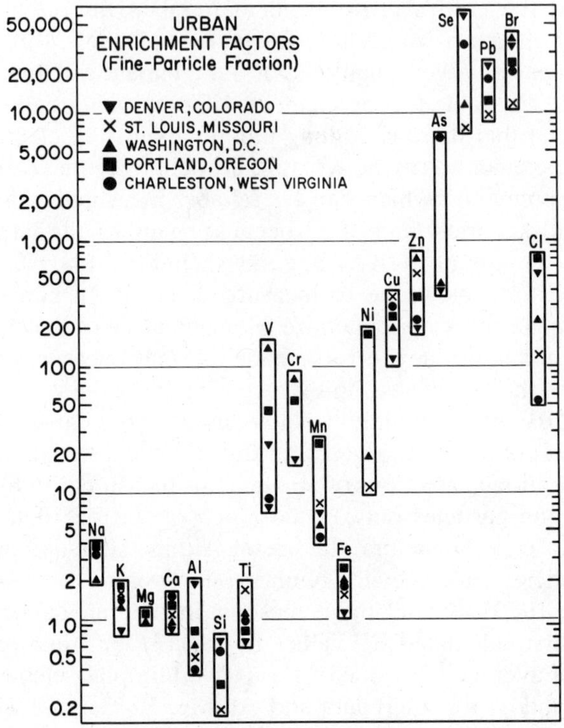

FIGURE 1. Enrichment factors for elements attached to fine particles in atmospheres of several U.S. cities. Based on data from Heisler et al. (1980) for Denver; Loo et al. (1978) for St. Louis; Kowalczyk et al. (1982) for Washington, D.C.; Watson (1979) for Portland, OR; and Lewis and Macias (1980) for Charleston, WV.

The elements to the right of Ti in Figure 1, except Fe, are strongly enriched relative to the crustal abundance pattern, usually because of large anthropogenic emissions (e.g., Pb and Br from combustion of leaded gasoline; As and Se from coal combustion; V and Ni from oil combustion; Cl from auto exhaust and marine aerosols; Mn, Cu, and Zn from a variety of sources), some of which are not well known. The EF values for many of these elements can be related to the activities known to occur in the various cities—for example, high Se and As in cities where coal is burned, and high V and Ni where large amounts of residual fuel are burned (Washington, Portland, Denver). Variations of the EF values for many of these elements span more than an order of magnitude, suggesting that it might be possible to identify the influence of particles from certain urban areas upon receptors in rural areas by use of the concentration patterns of elements borne by fine particles. Enrichment factors were calculated as described above for the fine-particle fraction collected in the Smoky Mountains (Stevens et al., 1980) with some of the values being V <40; Ni, 44; Cu, 190; Zn, 290; As,

2400; Se, 31,000; Pb and Br, 12,000. These values fall within the ranges shown for urban fine particles in Figure 1, but with both As and Se values falling rather high within those ranges, suggesting rather strong influences of coal combustion. However, the data on trace elements are not extensive enough to attempt to pin down a definite source area.

It should be noted that, since much of the size-segregated data have been taken with XRF, I have confined Figure 1 mainly to elements measured by that method. (Even so, V, Vi, Cu, As, and Se are often marginal or below detection limits.) If the samples were also analyzed by INAA, one could obtain data on about 15 additional elements (plus very accurate data for Al, V, As, and Se), several of which are highly enriched (e.g., Sb, W, I, In, Mo) and would thus provide additional identification of sources or areas of origin of the material.

The best approach for identifying the origins of particles may be the use of a combination of concentration data along with wind trajectories. As the distance increases, however, the wind direction measured at the sampling site will not be an accurate indicator of the trajectory, and synoptic scale trajectories will be required.

Concentration patterns have been used in a few instances to detect sources of material transported over continental and greater distances. For example, in the spring of 1976, Rahn et al. (1977) detected high levels of particulate matter at Barrow, Alaska, that originated from the Gobi Desert and the Loess Plateau of eastern Asia, some 12,000–15,000 km away. Rahn and McCaffrey (1978) found order-of-magnitude increases in concentrations of elements associated with anthropogenic activities (e.g., V, Pb) at Barrow in winter relative to summer. These aerosols have surely been transported from cities far to the south, but at that distance, it would be difficult to associate the aerosol with specific cities.

4. USE OF ELEMENTAL COMPOSITIONS IN DEPOSITION STUDIES

Some of our largest questions concerning the effects of air pollutants upon terrestrial ecosystems involve the deposition of acids and other substances either by dry deposition or in precipitation. In-cloud processes themselves may well be the major mechanisms for forming the acids later deposited in rain and snow (Hegg and Hobbs, 1978; Newman, this volume; Gordon et al., 1983). But despite their probable importance, we have a very poor knowledge of even the fundamental processes involving in-cloud chemistry and deposition phenomena. These processes are difficult to study using only the major species involved, such as sulfate and nitrate, because they are so ubiquitous. The use of trace elements on aerosols can provide more definitive information about the fundamentals of these processes because they provide an identification scheme by which to follow certain types of aerosols through various processes. As demonstrated by Muhlbaier (1978), one can measure the size distributions of particles bearing various elements in the air near clouds. One can contact particles with water and weak acids and analyze both the water and the insoluble residues to determine solubilities of the particles bearing the various elements to determine solubility in in-cloud processes. Likewise, one can measure the composition of cloud water droplets and ice crystals and of any precipitation that falls from the clouds to investigate those processes, as well as particles from dry deposition.

Muhlbaier's (1978) study of deposition phenomena in Prince Georges County, MD illustrates the kinds of measurements that can be done. From measurements of elements deposited by dry fallout on a horizontal surface and, simultaneously, their atmospheric concentrations, she was able to calculate deposition velocities of particles bearing various elements:

$$V_g(\text{cm/s}) = \frac{\text{Dry deposition rate } (\mu\text{g/cm}^2\text{ s})}{\text{Atm. concentration } (\mu\text{g/cm}^3)} \tag{3}$$

Although this is a crude parameter, it is often used in deposition studies for comparison with theory. Muhlbaier found that, although absolute deposition velocities are typically more than an order of magnitude greater than theoretical predictions, the shape of the curve of V_g versus particle diameter agrees with theory, being about 1 cm/s for particles bearing mostly crustal elements and with mean diameters of about 3 μm, falling to a minimum of <0.1 cm/s for 0.5-μm particles, and rising for smaller particles to about 0.3 cm/s for particles of about 0.1-μm diameter that bear elements such as As, V, Sb, and Se. The latter group includes most of the enriched elements such as As, V, Sb, and Se. The latter group includes most of the enriched elements in the atmosphere. Because of their low deposition rates, dry deposits generally have lower enrichments for these elements than the atmosphere, as shown in Table 6. (Note that Zn is anomalous because it has a bimodal size distribution curve.)

By contrast, rainfall appears to bring down elements on fine particles more efficiently than elements on large particles. Thus elements on fine particles are more enriched in rainwater than elements on suspended particles in the atmosphere (see Table 6). Muhlbaier extracted the air filters with water, which she analyzed to determine the solubilities of various elements. Solubilities ranged from 3 to 5% for many of the lithophiles to values of about 85% for many of the enriched elements such as Zn, As, and Se. A major fraction of the fine particulate matter that bears these elements consists of $(NH_4)_2SO_4$, which is also quite soluble. Thus, the high solubility of these particles enhances their interactions with atmospheric water, making them more important in cloud chemistry and precipitation than the larger, less soluble particles of more crustal composition.

These experiments are suggestive of a number of important effects involving mechanisms leading to deposition, but more work is needed if we

TABLE 6. Comparison of Average Enrichment Factors for Selected Elements at Chalk Point, MD in Dry Deposition, Suspended Particles, and Rainwater

Element	Dry Deposition	Suspended Particles	Rainwater
Ti	0.95 ± 0.35	1.5 ± 0.4	1.5 ± 0.2
Fe	1.9 ± 0.7	1.5 ± 0.2	1.8 ± 0.6
K	0.57 ± 0.07	0.73 ± 0.14	2.2 ± 1.7
Mn	0.98 ± 0.54	1.6 ± 0.8	2.6 ± 2.0
Sr	1.2 ± 0.3	1.7 ± 0.9	2.8 ± 0.8
Mg	0.87 ± 0.41	1.2 ± 0.6	3.5 ± 2.6
Ca	0.77 ± 0.61	1.2 ± 0.5	3.9 ± 3.9
Na	1.3 ± 1.7	1.2 ± 1.4	6.8 ± 8.3
In	3.4 ± 1.9	16 ± 12	10 ± 8
Ba	1.2 ± 0.7	0.93 ± 0.13	11 ± 16
V	2.5 ± 0.5	6.0 ± 4.6	13 ± 8
Co	2.6 ± 0.3	4.5 ± 2.4	45 ± 90
Zn	140 ± 160	50 ± 19	170 ± 220
As	24 ± 10	100 ± 66	800 ± 920
Sb	380 ± 430	580 ± 400	970 ± 710
Br	40 ± 29	500 ± 300	1100 ± 1100
Se	170 ± 70	1400 ± 500	4700 ± 2900

From Muhlbaier (1978).

are to gain a detailed understanding of them. However, they demonstrate that trace species can be used as an excellent tool for probing the mechanisms.

5. FUTURE RESEARCH NEEDS

Throughout this paper I have emphasized the information that can be obtained about the sources, transport, and deposition of particles by detailed study of trace species (mostly trace *element*) concentrations on airborne particles, dry deposits, precipitation, and cloud water. The experimental methods exist for doing all of these measurements, but it will be necessary to perform some rather extensive, carefully designed field studies in order to test these methods and to gain an understanding of the phenomena involved. More work is needed, however, to develop and test analytical methods for other species that could similarly be used to follow the environmental behavior of certain classes of particles. In particular, the methods described are not very effective for the study of particles from sources that release mainly carbonaceous material. Identification methods based on organic compounds and other forms of carbon are needed. There are classes of organic compounds such as large hydrocarbons or polynuclear aromatic hydrocarbons (PAHs) that have different abundance patterns for different sources (Hites et al., 1977; Fox and Staley, 1976). Preliminary attempts have been made to investigate the use of PAHs to identify sources, but much more investigation is needed (Gether and Seip, 1979). Before techniques involving compounds can become successful, one must establish that their relative abundances are not altered between release and detection, for example, because of differing chemical reactivities or volatilities. A recent study by Korfmacher et al. (1980) indicates that PAHs attached to particles are not rapidly decomposed by direct photolysis, but that they can be modified by reactions promoted on fly ash particles in the dark.

One very important new method, developed in connection with the Portland, OR study, is the use of ^{14}C/total carbon measurements to distinguish between carbonaceous aerosols from "modern" sources (recently living material) and from fossil fuel combustion (Cooper et al., 1980). With the use of this method, it was found that surprisingly large amounts of fine carbonaceous particles in the Portland atmosphere resulted from wood combustion in fireplaces and wood stoves.

Particles from many types of sources are so similar in composition to that of crustal material that it may be impossible to resolve all of them just on the basis of elemental concentrations. It may be very useful to develop microscopic methods to identify them, especially if the methods can be automated (Draftz, 1977; Graf et al., 1977).

ACKNOWLEDGMENT

This work was in part supported under EPA Grant No. R806263-01 and NSF Grant No. ENV75-02667.

REFERENCES

Alpert, D. J., and P. K. Hopke. A quantitative determination of the sources in the Boston urban aerosol. *Atmos. Environ.* **14**, 1137–1146 (1980).

Andren, A. W., B. G. Blaylock, E. A. Bondetti, C. W. Francis, S. G. Hildebrand, J. W. Húckabee, D. R. Jackson, S. E. Lindberg, F. H. Sweeton, R. I. VanHook, and A. P. Watson. Ecological research. In W. Fulkerson, W. D. Schultz, and R. I. VanHook (eds.) *Ecology and Analysis of Trace Contaminants*, pp. 61–104. Progress Report October 1973–September 1974, Oak Ridge National Laboratory Report ORNL-NSF-EATC-11 (1974).

Cooper, J. A., L. A. Currie, and G. A. Klouda. Assessment of contemporary carbon combustion source contributions to urban air particulate levels using carbon-14 measurements. *Environ. Sci. Technol.* **15**, 1045–1050 (1980).

Cunningham, W. C., and W. H. Zoller. The chemical composition of remote area aerosols. *J. Aerosol. Sci.* **12**, 367–384 (1981).

Draftz, R. G. Microscopical analysis of aerosols collected in Miami, Florida. Dept. of Physical Sciences, Florida International University, North Miami, unpublished (1977).

Dzubay, T. G. Chemical element balance method applied to dichotomous sampler data. In T. J. Kneip and P. J. Lioy (eds.) Aerosols: Anthropogenic and Natural, Sources and Transport. *Ann. NY Acad. Sci.* **338**, 126–144 (1980).

Dzubay, T. G., and R. K. Stevens. Ambient air analysis with dichotomous sampler and x-ray fluorescence spectrometer. *Environ. Sci. Technol.* **9**, 663–668 (1975).

Fox, M. A., and S. W. Staley. Determination of polycyclic aromatic hydrocarbons in atmospheric particulate matter by high pressure liquid chromatography coupled with fluorescence techniques. *Anal. Chem.* **48**, 992–998 (1976).

Friedlander, S. K. Chemical element balances and identification of air pollution sources. *Environ. Sci. Technol.* **7**, 235–240 (1973).

Gether, J., and H. M. Seip. Analysis of air pollution data by the combined use of interactive graphic presentation and a clustering technique. *Atmos. Environ.* **13**, 87–96 (1979).

Gladney, E. S. Trace element emissions of a coal-fired power plant: A study of the Chalk Point Electric Generating Station. Ph.D. thesis, Univ. of Maryland, College Park (1974).

Gladney, E. S., W. H. Zoller, A. G. Jones, and G. E. Gordon. Composition and size distribution of atmospheric particulate matter in the Boston area. *Environ. Sci. Technol.* **8**, 551–559 (1974).

Gordon, G. E. Receptor models. *Environ. Sci. Technol.* **14**, 792–800 (1980).

Gordon, G. E., E. S. Gladney, J. M. Ondov, T. Conry, and W. H. Zoller. Problems with the use of cascade impactors. In W. Fulkerson, W. D. Schultz, and R. I. VanHook (eds.) *Proceedings of the First Annual NSF Trace Contaminants Conference, Oak Ridge, Tennessee, August* 1973, pp. 138–145. CONF-730802 (1973).

Gordon, G. E., J. L. Moyers, K. A. Rahn, D. F. Gatz, T. G. Dzubay, W. H. Zoller, and M. L. Corrin. Atmospheric Trace Elements: Cycles and Measurements. Submitted (1983).

Graf, J., R. H. Snow, and R. G. Draftz. Aerosol sampling and analysis—Phoenix, Arizona. EPA-600/2-77-015 (1977).

Hegg, D. A., and P. V. Hobbs. Oxidation of sulfur dioxide in aqueous systems with particular reference to the atmosphere. *Atmos. Environ.* **12**, 241–253 (1978).

Heisler, S. L., R. C. Henry, J. G. Watson, and G. M. Hidy. The 1978 Denver winter haze study. Environmental Research and Technology, Inc. Report No. P-5417-2 to the Motor Vehicle Mfgrs. Assn. of the U.S., Inc., March (1980).

Hites, R. A., R. E. Laflamme, and J. W. Farrington. Sedimentary polycyclic aromatic hydrocarbons: The historical record. *Science* **198**, 829–831 (1977).

Hopke, P. K., E. S. Gladney, G. E. Gordon, W. H. Zoller, and A. G. Jones. The use of multivariate analysis to identify sources of selected elements in the Boston urban aerosol. *Atmos. Environ.* **10**, 1015–1025 (1976).

Korfmacher, W. A., D. F. S. Natusch, D. R. Taylor, G. Mamantov, and E. L. Wehry. Oxidative transformations of polycyclic aromatic hydrocarbons adsorbed on coal fly ash. *Science* **207**, 763–765 (1980).

Kowalczyk, G. S. Concentrations and sources of trace elements on Washington D.C. area atmospheric particles. Ph.D. Thesis, Univ. of Maryland, College Park (1979).

Kowalczyk, G. W., G. E. Gordon, and S. W. Rheingrover. Identification of atmospheric particulate sources in Washington, D.C. using chemical element balances. *Environ. Sci. Technol.* **16**, 79–90 (1982).

Lewis, C., and E. S. Macias. Composition of size-fractioned aerosol in Charleston, West Virginia. *Atmos. Environ.* **14**, 185–194 (1980).

Loo, B. W., J. M. Jaklevic, and F. S. Goulding. Dichotomous virtual impactors for large scale monitoring of airborne particulate matter. In B. Y. H. Liu (ed.) *Fine Particles*, p. 312. Academic, New York (1976).

Loo, B. W., W. R. French, R. C. Gatti, F. S. Goulding, J. M. Jaklevic, J. Llacer, and A. C. Thompson. Large-scale measurement of airborne particulate sulfur. *Atmos. Environ.* **12**, 759–771 (1978).

Macias, E. S., D. L. Blumenthal, J. A. Anderson, and B. K. Cantrell. Size and composition of visibility-reducing aerosols in southwestern plumes. In T. J. Kneip and P. J. Lioy (eds.) Aerosols: Anthropogenic and Natural, Sources and Transport. *Ann. NY Acad. Sci.* **338**, 233–257 (1980).

Maenhaut, W., W. H. Zoller, R. A. Duce, and G. L. Hoffman. Concentration and size distribution of particulate trace elements in the South Polar atmosphere. *J. Geophys. Res.* **84**, 2421–2431 (1979).

Mitchell, R. I., and J. M. Pilcher. Improved cascade impactor for measuring aerosol particle sizes in air pollutants, commercial aerosol and cigarette smoke. *Ind. Eng. Chem.* **51**, 1039–1049 (1959).

Moyers, J. L., L. E. Ranweiler, S. B. Hopf, and N. E. Korte. Evaluation of particulate trace species in southwest desert atmosphere. *Environ. Sci. Technol.* **11**, 789–795 (1977).

Muhlbaier, J. The chemistry of precipitation near the Chalk Point power plant. Ph.D. Thesis, Univ. of Maryland, College Park (1978).

National Research Council. *Airborne Particles*. University Park Press, Baltimore, MD, 1979.

Rahn, K. A. The chemical composition of the atmospheric aerosol. Technical Report, Grad. School of Oceanography, Univ. of Rhode Island, Kingston, RI, July (1976).

Rahn, K. A., and R. J. McCaffrey. Compositional differences between Arctic aerosol and snow. Paper presented at the Annual Mtg. of the Gesellschaft für Aerosolforschung, Vienna (1978).

Rahn, K. A., R. D. Borys, and G. E. Shaw. The Asian source of Arctic haze bands. *Nature* **268**, 713–715 (1977).

Rheingrover, S. W., and G. E. Gordon. Identifying locations of dominant point sources of elements in urban atmospheres from large, multi-element data sets. In J. R. Vogt (ed.) *Proc. of the Fourth Int. Conf. on Nuclear Methods in Environmental and Energy Research, Columbia, MO, April* 1980. Avail. NTIS, Springfield, VA, as CONF-800433 (1980).

Small, M., M. S. Germani, A. M. Small, W. H. Zoller, and J. L. Moyers. Airborne plume study of emissions from the processing of copper ores in Southeastern Arizona. *Environ. Sci. Technol.* **15**, 293–299 (1981).

Solomon, P. A., J. L. Moyers, and R. A. Fletcher. A high volume dichotomous virtual impactor for the fractionation and collection of particles according to aerodynamic size. *Aerosol. Sci. Technol.* **2**(4), 455–464 (1983).

Stevens, R. K., T. G. Dzubay, G. Russwurm, and D. Rickle. Sampling and analysis of atmospheric sulfates and related species. *Atmos. Environ.* **12**, 55–68 (1978).

Stevens, R. K., T. G. Dzubay, R. W. Shaw, Jr., W. A. McClenny, C. W. Lewis, and W. E. Wilson. Characterization of the aerosol in the Great Smoky Mountains. *Environ. Sci. Technol.* **14**, 1491–1498 (1980).

Watson, J. G., Jr. Chemical element balance receptor methodology for assessing the sources of fine and total suspended particulate matter in Portland, Oregon. Ph.D. Thesis, Oregon Graduate Center, Beaverton, OR (1979).

Wedding, J. B., A. R. McFarland, and J. E. Cermak. Large particle collection characteristics of ambient aerosol samplers. *Environ. Sci. Technol.* **11**, 387–394 (1977).

Wedepohl, K. H. Chemical fractionation in the sedimentary environment. In L. H. Ahrens (ed.) *Origin and Distribution of the Elements*. Pergamon Press, Oxford (1968).

Characterization and Quantification of the Oxides of Sulfur and Nitrogen and Associated Compounds in Their Gaseous, Aerosol, and Dissolved States

LEONARD NEWMAN
Environmental Chemistry Division
Department of Applied Science
Brookhaven National Laboratory
Associated Universities, Incorporated
Upton, New York

1. INTRODUCTION

The observation of a physical change frequently can be utilized as a measure of the result of the effects of air pollutants on terrestrial ecosystems. However, an understanding of the mechanisms through which pollutants bring about these changes can be accomplished only through chemical analysis of those constituents potentially causing the stress to the ecosystem. The nature and type of measurements that need to be made to come to this recognition and our ability to make them are the subjects of this discussion. The determination of the elemental composition of atmospheric particles and their role is covered in the companion discussion presented by Gordon (in this volume). It is left to this reviewer to discuss and highlight the needs and problems associated with the characterization and quantification of the oxides of sulfur and nitrogen and associated compounds in the gaseous, aerosol, and dissolved states.

The question has sometimes been raised as to why someone who wants a challenging scientific career would choose analytical chemistry. The supposition is that the field is finite, and, once you have devised a method for determining each of the chemical elements, your work is done. After all, there are only a hundred or so elements and consequently the task cannot be all that difficult. To some extent this is true: Great strides have been made in the search for a universal method, and many have been suggested. These include, among others, X-ray fluorescence, neutron activation, emission and absorption spectroscopies, and atomic absorption. The fact that there are so many approaches suggests that a true panacea has not yet been obtained. After all, once you find what you sought, you stop searching. The analytical literature is replete with discussions of the advantages and limitations of each of these methods, and they need not be dwelt upon here. However, it should be noted that, at best, these approaches generally can be used only to determine elemental composition (that aspect of the subject is covered by Gordon in this volume). No comparable methods have been developed for the pollutant *compounds* to be addressed in this review, it is the lability and transformation of these compounds that makes this subject an even greater challenge. A detailed review of methods will not be presented, as this is a subject of interest only to a very specialized audience. What will be presented is a discussion of the relevant needs and difficulties associated with the measurements, along with illustrations of typical methods that are presently employed. It will become obvious that a challenging and scientific career in analytical chemistry is still possible.

2. GENERALITIES

Let us assume you had perfect measuring tools, requiring no human attention. In principle, then, you could continuously monitor for everything, but in the end you might still not acquire a better understanding. The sheer mass of data would make it almost impossible to mount the manpower effort to look at the information, not to mention the more important question of what to look for in the data set. What this says is that one should have a clear objective in mind when setting out on a measurement program, and carefully weigh how best to deploy one's financial and scientific resources. The best of measurements are useless if there is no one to look at and interpret them.

What has developed are two types of measurement protocols: monitoring and discrete measuring campaigns. Monitoring is best suited for gathering historical records to answer such trend questions as, "Does the increase in energy production correspond to an increase in the measured average ambient sulfur dioxide concentration?" or "Does the increase in acidity of a lake correspond to the observed decrease in fish population?" However, one should bear in mind that, although cause-and-effect relationships are sometimes straightforward, a correlation by itself does not prove the existence of a cause and effect. Here it is necessary that, at the very least, a scientific basis for the conclusion be invoked to support the conclusion derived from the correlation. Searches for correlations nevertheless can be quite important, and it should be borne in mind that a lack of correlation between a pair of observations must almost invariably mean no cause-and-effect relationship. Sometimes this can be very valuable information.

Seldom is it necessary to obtain a continuous real-time record of the pollutants of interest to answer the questions posed by a monitoring protocol. However, the technology is such that some of our best and most sensitive measurement techniques result in such real-time measurements, and consequently there is a tendency to use them. Even if large capital expenditures for equipment and facilities to house them have been provided, there are many other considerations in their utilization that should not be overlooked. Manpower must be provided for servicing and calibration of equipment. Although data reduction is not trivial, seldom are sufficient funds provided to reduce the mounds of data into a form readily usable by the scientific community. Even though some automation of the measurement process is possible, it should be recognized that, as good as the technology is, the needs are such that the instruments are generally measuring at their sensitivity limitations. Many a program is doomed to die because the quality of the collected data is not worth interpreting. Ultimately, not recognizing a poor data set can set the science back many years.

In regards to monitoring programs, the value of simple and inexpensive measurement techniques should not be overlooked. Remember that, since you invariably will integrate your real-time measurements, you can let the measurement technique itself do it for you. Consequently, filters, sometimes chemically treated, are still most useful. Even passive collectors should not be overlooked. The question of specificity, in terms of both what you desire to measure and its effect are the ultimate concerns in this regard.

Campaign-type measurements are most suited to answer a mechanistic question like, "Does the rate of oxidation of sulfur dioxide increase with solar intensity, and why?" or an immediate effect question like, "At what concentration does ozone cause physical damage to the Ponderosa pine?" Here is where the utilization of real-time instrumentation is most demonstrable, especially if you need to observe changes that occur in seconds or fractions of a second. Consequently, even in campaign-type experiments the mass of data can become overwhelming, and again many an experiment has been performed for which interpretations have never been attempted. Some might consider it inexcusable that a set of results cannot be interpreted because of poorly formulated questions, but I consider it even more inexcusable that someone has overlooked the need to provide the necessary resources required to mount the interpretation effort.

The converse to a lack of interpretation sometimes occurs, namely, overinterpretation, especially when the data set is minimal. After all, any desired conclusion can be arrived at with one data point. With the vagaries of nature it takes many data points, with much redundancy, to establish a data set that restricts the number of conclusions that can be made to the point that ultimately permits the truth to be approached. Toward this end, the limitations of the measurement methods must be carefully considered. At the very least, the analytical chemist should be required to state quantitatively, with experimental justification, the measurement errors associated with the techniques employed. In this regard, what the analyst frequently underestimates is the propagation of errors (or admission) of interferents, both positive and negative. A constructive approach toward having this exercise performed successfully is to utilize an uninvolved and impartial, but knowledgeable, auditor of the complete measurement program. Finally, when interpreting the data, the basis for conclusions must include a statistical test of significance employing the stated measurement uncertainties. These are all precepts that we are well aware of but frequently choose to ignore.

3. GASES

3.1 Considerations

Pollutant gases either can have a direct impact on the terrestrial ecological system or can serve as precursors or intermediates in the formation of deleterious substances. The impact and mechanistic roles of gases can be completely understood only through measurements of their concentration in the atmosphere. A primary impact from gases can result from direct insults, such as an oxidative attack by ozone on a plant. An impact can also occur through dry deposition to the soil and a consequent alteration of the nutrient balance by participation with the chemistry therein. In addition, a mechanistic role can be important by involvement with an atmospheric chemical cycle through which secondary pollutants are produced (e.g., sulfuric acid) which themselves can have a direct impact on the ecological system.

The gases of primary interest consist of sulfur dioxide (SO_2), nitric oxide (NO), nitrogen dioxide (NO_2), nitric acid (HNO_3), ammonia (NH_3), and ozone (O_3). Concerns have been raised related to the acidic properties of SO_2, NO_2, and HNO_3 and the oxidative action of O_3. The NO is a primary pollutant which is quickly converted to NO_2 in the atmosphere and participates in the O_3 cycle. The NH_3 is generally a biogenic product and its importance derives from its ability to neutralize the acidic substances.

The gases of secondary interest include a host of biogenic sulfides including hydrogen sulfide (H_2S), carbonyl sulfide (COS), dimethyl sulfide ($(CH_3)_2S$), and carbon disulfide (CS_2). These substances generally need not be measured, since their concentrations are usually quite low (<1 ppb) and consequently have no direct ecological involvement. However, it may be necessary to measure nitrogen pentoxide (N_2O_5), hydrogen peroxide (H_2O_2), and the hydroxyl radical (OH) when one is trying to elucidate specific aspects of the atmospheric transformation processes. Such aspects include the oxidation of SO_2 to sulfate via OH or the incorporation of NO_2 in rainwater via H_2O_2. Obviously the list of gases of secondary interest can become enormous, and these would certainly include hydrocarbons resulting from biogenic sources (e.g., terpenes), pollution emissions (e.g., alkanes), or atmospheric reaction products (e.g., ketones). Again, these substances are seldom measured since they are generally of little concern as participants in a terrestrial ecological problem, even though they might be of significance in epidemiological studies. However, it must be kept in mind that these substances are sometimes involved in atmospheric processes that result in problem agents, and it might be necessary to mount measuring campaigns when the information is required. The significances of these compounds in this regard are presented in the discussions given by Hidy and Mueller and Demerjian in this volume.

3.2 Analysis

There are no particular problems in selecting methods to analyze the gases of primary interest under polluted conditions (>10 ppb). However, under relatively clean conditions, sensitivity does become a limitation, and frequently the technology still needs improvement.

The analysis of SO_2 can be performed on a sample collected with an alkaline-treated filter for sampling times as short as 15 min or as long as 24 h (Forrest and Newman, 1973). A prefilter to remove aerosol compounds is usually required (Leahy et al., 1980). The sample is dissolved and oxidized to sulfate. An end analysis is performed by any number of techniques including spectrophotometric (Lazrus et al., 1966) or ion exchange chromatography (Small et al., 1975). Continuous and real-time analysis of SO_2 can be performed with commercial instrumentation employing a measurement of the chemiluminescence arising from S_2 molecules formed through the reduction of SO_2 in a hydrogen flame (Crider, 1965). The method is sensitive to all sulfur compounds: If substances other than SO_2 are present in significant amounts, they must be separated by techniques such as gas chromatography (Stevens et al., 1971).

It is now possible to determine the oxides of nitrogen continuously and in real time, consequently, collection techniques have been almost completely abandoned. The determination of NO is accomplished almost universally by employing commercial instrumentation that measures the chemiluminescence arising from an excited state resulting from oxidation through the addition of O_3 (Fontijin et al., 1970). The determination of NO_2 can be accomplished by reducing it to NO over a molybdenum catalyst in a parallel air stream and measuring the chemiluminescence arising from the sum of $NO + NO_2$ (Thermo Electron, 1975). The concentration of NO_2 can thereby be derived from the signal difference obtained between the air streams. Joseph and Spicer (1978) attempted to measure HNO_3 acid in a similar fashion by first reducing it along with NO_2 to NO and measuring the total gaseous nitrogen by the chemiluminescent reaction. Then the signal thus derived was compared to a signal from a parallel air stream from which the HNO_3 had been removed by absorption on nylon. In both streams the aerosol nitrate had been removed previously by filtration. Another differencing approach employs selective thermal decomposition of HNO_3 to NO_2 and subsequently to NO (Kelly and Stedman, 1979). A problem associated with both of these methods is that, since HNO_3 is usually a minority species and determined by difference, the accuracy and sensitivity of the determination are still under question.

Collection techniques are still employed to determine HNO_3—for example, utilizing sodium chloride filters to retain the nitrate by displacement of the more volatile hydrochloric acid (Okita et al., 1976). Prefiltration of aerosols must be employed and this causes difficulties (see Section 4.2). The end analysis for the nitrate can be performed, after dissolution, by any conventional method, but ion exchange chromatography has quickly become a favorite (Small et al., 1975).

Collection techniques have all but disappeared for the determination of the concentration of O_3, since it can be measured so conveniently and simply in continuous and real-time fashion (Clark et al., 1974). A technique employing UV absorption in the vapor phase has been commercialized and become popular, but I prefer the commercial instruments employing a measure of the chemiluminescence arising from the reaction upon addition of ethylene to the sample air stream containing O_3. Sensitivity is no problem, but absolute calibration has been (Hodgeson et al., 1971a, b). Recently more attention has been paid to this problem (Paur and McElroy, 1979). This is particularly important since O_3 results are frequently used in a stoichiometric way to interpret mechanistic transformations in the atmosphere.

Relatively few measurements of NH_3 appear in the literature. The development of a reliable method has proved difficult because of its lability (adsorption and neutralization) and low concentration (<1 ppb). Successful continuous methods are yet in the offing. Recently Ferm (1979) proposed collecting NH_3 through diffusion denuding an air stream during its passage down a tube coated with oxalic acid. The coating is subsequently dissolved and can be analyzed by any conventional ammonium technique. Another integrative approach is adsorption on Teflon with thermal desorption for analysis (McClenny and Bennett, 1980). The question has been raised (Durham, personal communication) whether ambient NH_4NO_3 decomposes quickly enough so that a vapor pressure of NH_3 is reestablished during the residence time of the air stream in the denuder. Some of this NH_3 might then be removed by the denuder, resulting in a value higher than originally present. An alternative approach for NH_3 has been suggested that utilizes a bubbler for collection and measurement performed through *o*-phthalaldehyde fluorescence derivatization (Tanner and Lepore, 1979). Abbas and Tanner (1981) are attempting to develop this principle into a continuous real-time method. At present both of these approaches show promise, and utilization by the scientific community should be initiated.

While methods for the analysis of the gases of secondary interest are not covered in this review, they do exist to a more or less accessible extent. The reader seeking this information is advised to consult the original literature.

4. AEROSOLS

4.1 Considerations

Aerosols can have a direct impact, through dry deposition, on an ecological system in a manner similar to that of gases. Effects arise by virtue of their acidic nature or through their involvement in soil chemistry. The primary aerosols of interest for this discussion consist mainly of sulfate (SO_4^{2-}), nitrate (NO_3^-), hydrogen (H^+), and ammonium (NH_4^+). Although some SO_4^{2-} derives directly from emission sources (Dietz et al., 1978), most of the SO_4^{2-} and the NO_3^- are formed as secondary pollutants resulting from the oxidation of SO_2 and NO_2 to sulfuric and nitric acids. These acids can be subsequently neutralized by incorporation of NH_3. A calculation of the phase diagram relationships for the solid mixed salt containing SO_4^{2-}, NO_3^-, H^+, and NH_4^+ with the gaseous NH_3 and HNO_3 system, as a function of relative humidity, now appears in the literature (Tang, 1980).

Much of the aerosol mass (especially SO_4^{2-}) has been observed to exist mainly in particles significantly less than 1 μm in diameter. Partly because of this small size, it has been conjectured (and demonstrated in some experiments) that these aerosols may indeed be in phase equilibrium with NH_3 and HNO_3 (Brosset, 1980). If this supposition is true, then all aerosol particles at any moment should be uniform in composition. A test of this premise is difficult because the measurements usually require a relatively long sampling time and a changing air mass (with varying compositions of aerosol and gaseous species), which make the aerosol appear nonhomogeneous. A critical experimental test of the phase equilibrium question has been concluded with sampling and integration times as short as 15 min (Tanner, 1982). Extensive results are not yet available but it appears that, at the very least, the phase relationships are not preserved all the time.

4.2 Analysis

The analysis of aerosols is complicated by two requirements: There is a need to determine not only the concentration of each of the constituents (e.g., NH_4^+, H^+, and SO_4^{2-}) but also their molecular form (e.g., NH_4HSO_4). The techniques available for determining the chemical composition of aerosol sulfur compounds have been reviewed (Newman, 1978). Although this is a rapidly developing field, no altogether satisfactory approach exists at present. The difficulties arise mainly from the complexities of the problem associated with trying to sample small amounts of material and subsequently identify specific substances in a matrix consisting of a mixture of compounds. Further complications result either from matrix changes imposed through a sampling artifact or from changes in the composition of the air mass that is sampled. The problem of determining aerosol nitrate is especially difficult because of a vapor pressure of HNO_3 and NH_3 above the solid that varies with the acidity of the solid and the relative humidity.

Aerosols generally are sampled with a filter. Their sulfur content can be determined in situ by X-ray fluorescence (Dzubay and Stevens, 1975), or, upon dissolution (Forrest and Newman, 1973). The SO_4^{2-} can be determined by ion exchange chromatography (Small et al., 1975). An extensive intercomparison of various collection and analytical methods was conducted at Charleston, WV (Camp et al., 1978). The results showed quite good agreement, generally better than 90%, between laboratories and methods.

Although attempts have been made to determine individual compounds such as H_2SO_4 (Leahy et al., 1975) most attention has been directed toward obtaining an "average" molecular composition through ion-balance calculations derived from the measurements of SO_4^{2-}, NO_3^-, NH_4^+, and H^+ (Tanner et al., 1977). In this regard NH_4^+ can be determined by any number of convenient methods (e.g., by Indophenol; Bolleter et al., 1961) and H^+ by an approach utilizing a titration method (Gran, 1952), as employed by Askne et al. (1973).

The commercial instruments employing the flame photometric method proposed by Crider (1965) have been modified by a number of investigators for the determination of SO_4^{2-}. In one recent example SO_2 is diffusion-denuded from the air stream, upon which total SO_4^{2-} can be determined by first converting all forms to $(NH_4)_2SO_4$, through the addition of NH_3 just prior to the injection of the sample into the flame (Tanner et al., 1980). Through further developments, attempts have been made to determine specifically the H_2SO_4 concentration (D'Ottavio et al., 1981). The approach employed is to selectively thermally vaporize the H_2SO_4 as SO_3, which can then be removed along with the SO_2 by the diffusion-denuding principle. Then a determination of the H_2SO_4 can be obtained as a signal difference by comparison with the total SO_4^{2-} measurement. Since the SO_4^{2-} concentrations are quite low, frequently about 5 μg per cubic meter, the sensitivity of the flame photometric instruments is being taxed by these approaches, and much attention must be paid to the details of their operation (D'Ottavio et al., 1981). These problems notwithstanding, the recent deployment in St. Louis, MO (Husar, 1979) of various real-time SO_4^{2-} methods suggests that a high degree of success can be anticipated for their utilization in the near future.

Unfortunately, the state of the determination of aerosol nitrate does not appear to be as acceptable. The problems were highlighted by the Charleston, WV, intercomparison (Camp et al., 1978) during which investigators measured NO_3^- as a sideline interest to their determination of SO_4^{2-}. Results for NO_3^- between laboratories frequently differed by factors of 2, and differences of an order magnitude were not uncommon. The end determination, which can be performed by any number of standard methods, does not appear to have been at fault. I believe that the problems lie in the collection methods. The difficulty appears to be associated with the acidic nature of the aerosol, whereby HNO_3 can be lost from NH_4NO_3 when it comes in contact with acidic sulfates. This property has been demonstrated (Harker et al., 1977). However, one might ask, "If the aerosol had been in phase equilibrium, why should you get further release of HNO_3?" The answer might be that in sampling for many hours, as is conventionally done and was done at Charleston, WV, a changing air mass might have aerosols with varying acidity. Then upon collection the acidic and neutral aerosols could contact each other while on the filter and release HNO_3. One solution might be to sample for a shorter period of time and, with fewer particles on the filter, have a diminished probability of contact. Attempts are being mounted in this direction (Tanner, 1982), but this sets a stringent premium on the sensitivity of the end analytical method.

The problems associated with the determination of aerosol nitrate affect the determination of gaseous nitrate (HNO_3). As an example, prior to the real-time chemiluminescent determination of HNO_3, Kelly and Stedman (1979) removed NH_4NO_3 on a filter, and their results could consequently be affected by the gradual release of HNO_3. They attempted to overcome this problem by a daily change of the filter. Although the time scale seems long, it might be adequate to minimize the potential for interactions on the matrix.

An approach for measuring total nitrate (gaseous and aerosol) has been suggested by Forrest et al. (1980). The method involves the collection of aerosol nitrate on a specially prepared quartz filter followed by a NaCl-impregnated cellulose filter. The latter collects any HNO_3 that is subsequently released, along with the HNO_3 that was present in the air that was being sampled. The two filters are analyzed and the sum reported as total nitrate. During a workshop on methods for the determination of nitrate (see Stevens, 1979), a modification of this approach was suggested by Shaw et al. (1979) to obtain a measure of the HNO_3 concentration. Two identical filter packs are employed in parallel. However, in one system total nitrate is collected and in the second the HNO_3 is removed in a diffusion-denuding tube which is placed before the filter pack. Upon analyzing both filter packs, a value for the determination of HNO_3 is ascertained by difference. As a result of that workshop an experiment to intercompare methods was designed (Howes et al., 1979) and implemented at Claremont, CA, during the late summer of 1979. To measure nitric acid by the above technique, magnesium oxide was used for diffusion denuding along with Fluoropore for collecting the aerosol and nylon for collecting any released HNO_3 from it (Spicer and Schumacher, 1977). In another approach, sodium carbonate was used for diffusion denuding with the quartz NaCl filter pack described above. These methods, along with others, plus two real-time HNO_3 methods, and a long-path infrared Fourier transform spectrometer, were all deployed in the intercomparison. A preliminary look at the data indicates that we have come a long way since Charleston, WV (see Spicer et al., 1982).

5. RAIN

5.1 Considerations

Rain is the means most often indicated as the process through which pollutants are conveyed in a damaging manner to the terrestrial ecosystem. One impact that is frequently discussed is the acidification of lakes, with a resulting disappearance of fish.

Both gaseous and aerosol pollutants can be incorporated into rain drops through in-cloud processes involving condensation nuclei (rainout) and scavenged during the rain event (washout). The relative importance of gaseous or aerosol pollutants and the rainout or washout processes are poorly understood (Newman, 1979a) and certainly not documented by experimentation. By highlighting the questions, Newman (1979b) was able to conjecture that rainwater SO_4^{2-} is derived predominantly from in-cloud processes, mainly through the direct incorporation of SO_2 and not through aerosol sulfates. The conclusions were basically derived from a knowledge of the ambient concentrations of SO_2, SO_4^{2-}, and the acidity (really, relative lack of acidity) of the aerosol compared with the requirements derived from the composition of the rain. Newman was also able to conclude that NO_3^- could not be incorporated predominantly from an aerosol, since again the rainwater acidity demands that the NO_3^- not be derived solely from a neutral (NH_4NO_3) species. The question remained whether rainwater receives that portion of acidity associated with NO_3^- from NO_2 or from HNO_3. Based on recent measurements of the Henry's law constant for NO_2 (Lee and Schwartz, 1981), the suggestion can be made that NO_2 is the unlikely means. But this conclusion is made with an almost complete lack of knowledge of what the HNO_3 concentrations are at the ground, much less in the clouds. Indeed, significantly more information is required about what the concentrations are, especially at cloud levels, of all the species of gases and aerosols that are discussed in this report. If acquired then it will be possible to make firm conclusions of the relative importance of the rainout or washout processes and the relative contribution of gases or aerosols. A workshop on this subject was recently concluded, during which an attempt was made to identify research needs on the information of acid precipitation. The proceedings (Pack, 1979) should be consulted for the state-of-the-art information presented on instrumentation, atmospheric chemistry, cloud physics, precipitation formation and transport, and modeling. The research program recommendations are extensive and in a sense highlight our lack of understanding.

5.2 Analysis

Historically, sampling programs for rain involved integrated protocols such as monthly sampling with open buckets. Evaporation and dry deposition obviously compromised these values. Refinements were introduced by the employment of wet/dry collectors which were opened only during the rain events (Volchok and Graveson, 1976). Samples can now be collected and separated into single rain events, that is, initiation through completion of a given shower (Hales and Dana, 1979), and chemical differences between types of rain events can be explored (Raynor and Hayes, 1981 and 1982). An even further refinement involving collecting sequential samples within an event (Raynor and McNeil, 1979) could prove most useful for mechanistic studies.

A long analytical chemistry history has evolved for the determination of the relevant anions SO_4^{2-} and NO_3^- at trace levels in water. Hence, there should be no problems associated with their determination in rainwater. Until recently the methods of choice had been standard colorimetric procedures that could be easily automated (Lazrus et al., 1968). Now, many laboratories are adopting variations of the ion exchange chromatographic approach described by Small et al. (1975).

Little attention has been directed towards the analysis in rainwater of two related anions, sulfite (SO_3^{2-}) and nitrite (NO_2^-). Delineation of the relative importance of the rainout and washout processes could be obtained from an analysis of these species. Further efforts should be expended toward the evaluation of sulfite and nitrite. However, since these species are so labile, useful analytical results can be obtained only if provision is made to preserve them during the sampling and storage of the rain samples (Dana et al., 1976). The SO_3^{2-} can then be determined by a modified West and Gaeke approach (Scaringelli et al., 1967) and the NO_2^- by the Saltzman method (Saltzman, 1954).

Phosphate (PO_4^{3-}) and chloride (Cl^-) are two other anions that are generally measured in rainwater samples. They can both be determined with the ion exchange chromatograph. The measurement of Cl^- is particularly valuable in coastal areas as a means of determining the fraction of SO_4^{2-} derived from a sea salt aerosol. Under special circumstances it might be desired to measure two additional anions, bromide (Br^-) and fluoride (F^-), and these can also be accomplished with the ion chromatograph.

The most common and possibly important measurement made on rainwater is its pH. Unfortunately, the values obtained are not without controversy. These acidity values are determined almost universally with the glass electrode and reported either as pH or H^+ concentration. A particular difficulty has been discussed (Kadlecek and Mohnen, 1976) in which some workers claim to observe significant (tenths of a pH) change, either soon after collection or upon storage of the sample. Present wisdom mandates that the pH be measured at least within 24 h of sample collection.

The glass electrode has a number of specific limitations that are frequently ignored. As generally used, the accuracy is only ± 0.1 pH units, which in itself transcribes to an error in H^+ of approximately $\pm 20\%$. This introduces a severe limitation when the data are used to calculate ion balances. In addition, pH measurements made with the glass electrode are erratic when the solution becomes unbuffered, and this can occur at pH values greater than 5. The electrode should not be used under such circumstances to determine H^+ concentrations because the errors introduced can easily be greater than a factor of 2. A compensation, of course, is that the H^+ has become a minor species at these pH levels. Finally, the glass electrode measures only the free H^+, and if there are weak acids present they remain undetected. This is an important point because, after all, it is the total acidity that might be putting the stress on the ecosystem.

It is time that we faced up to our need and gathered more complete and useful information on rain acidity. Consequently, it is strongly recommended that in future rain-sampling programs provision be made to determine total acidity by employing the titration technique as devised by Gran (1952). Through this approach it is possible to determine both the strong and the weak acid components of the sample. The titration can be automated by employing coulometric generation of alkali (Askne et al., 1973). The accuracy of the values obtained can easily be better than 5%, and, with care, better than 1% (Biederman et al.., 1966).

The measurement of NH_4^+ is important since, if it is measured in addition to H^+, then the cation balance in rainwater is largely determined. The NH_4^+ traditionally has been measured by automatic colorimetric procedures (Lazrus et al., 1968), but again ion chromatography is taking over.

Other cations of interest include potassium (K^+) and a variety of species derived from soil, such as calcium (Ca^{2+}), magnesium (Mg^{2+}), and iron (Fe^{3+}), and from sea spray such as sodium (Na^+). Techniques like spectrophotometry, atomic absorption, and ion chromatography are useful in their analysis. Under special circumstances it might be desirable to measure any number of the elements discussed in the report of Gordon (in this volume). Finally, as a consequence of our polluted environment it will become increasingly important to pay attention to the existence of a wide range of organic molecules, including alkanes, polycyclic aromatic hydrocarbons, phthalic acid esters, and fatty acid ethylesters, as well as a variety of commonly used industrial chemicals, such as polychlorinated biphenyls (Galloway et al., 1978).

6. FUTURE NEEDS

Methods are currently available that can be deployed to study the effects of air pollutants on the terrestrial ecosystem. Some improvements, of course, are needed. Of initial concern should be the proper utilization of the currently available technology, and suggestions have been adequately presented in the discussions above. Special cases were made for the need to provide resources for data reduction and interpretation, directed utilization of real-time continuous instrumentation, and more accurate and detailed measurements of acidity, sulfite, and nitrite in rainwater.

Improvements in sensitivity and accuracy are certainly needed for the real-time determination of the gaseous oxides of sulfur and nitrogen, and advances still must be made in the determination of aerosol SO_4^{2-} and gaseous HNO_3 by these techniques. Documentation is necessary to ascertain if the proposed technology for the determination of nitrate is indeed adequate. Although collection techniques will provide useful information, new approaches for the determination of low concentrations of oxides of nitrogen are still required. The collection techniques for ammonia are just being implemented, and, although they bear watching, improvements will certainly be required. The determination of the molecular composition of aerosols is in its infancy, and new concepts must be sought.

Of course, a universal method would certainly be desirable. It would be nice to have a single approach that could measure all species of interest, determine the molecular forms, and give the results instantaneously with an infinitesimal time resolution. There is still a challenging career in analytical chemistry.

ACKNOWLEDGMENT

This work was performed under the auspices of the United States Department of Energy under Contract No. DE-AC02-76CH00016.

REFERENCES

Abbas, R., and R. L. Tanner. Continuous determination of gaseous ammonia in the ambient atmosphere using fluorescence derivatization. *Atmos. Environ.* **15**, 177–281 (1981).

Askne, C., C. Brosset, and M. Ferm. Determination of the proton-donating property of air-borne particles. Swedish Water and Air Pollution Research Laboratory (IVL) Report B157, Gothenburg, Sweden (1973).

Biederman, G., L. Newman, and H. Ohtaki. Determination of traces of protolytic impurities. Presented at Symposium on Trace Characterization—Chemical and Physical. National Bureau of Standards, Gaithersburg, MD, Oct. Proceedings NBS Monograph 100, 1969, p. 1 (1966).

Bolleter, W. T., C. J. Bushman, and P. W. Tidwell. Spectrophotometric determination of ammonia as indophenol. *Anal. Chem.* **33**, 592–594 (1961).

Brosset, C. Fate of sulfuric acid aerosol in the atmosphere. In D. S. Shriner, C. R. Richmond, and S. E. Lindberg (eds.) *Atmospheric Sulfur Deposition*. pp. 145–152. Ann Arbor Science. Ann Arbor, MI, 1980.

Camp, D. C., A. L. VanLehn, and B. W. Loo. Intercomparison of samplers used in the determination of aerosol composition. EPA-600/7-78-118 Research Triangle Park, NC (1978).

Clark, T. A., R. E. Baumgardner, R. K. Stevens, and K. J. Krost. Evaluation of new ozone monitoring instruments by measuring in nonurban atmospheres. In *Instrumentation for Monitoring Air Quality ASTM STP*555, pp. 101–111. American Society for Testing Materials, Philadelphia, PA, 1974.

Crider, W. L. Hydrogen flame emission spectrophotometry in monitoring air for sulfur dioxide and sulfuric acid aerosols. *Anal. Chem.* **37**, 1770–1773 (1965).

Dana, M. T., D. R. Drewes, D. W. Glover, and J. M. Hales. Precipitation scavenging of fossil-fuel effluents. EPA-600/4-76-031. Battelle Pacific Northwest Laboratories, Richland, WA (1976).

Dietz, R. N., R. F. Weiser, and L. Newman. Operating parameters affecting sulfate emissions from an oil fired power unit. *Proceedings Workshop on Measurement Technology and Characterization of Primary Sulfur Oxides* Volume 2. EPA-600/9-78-0206, p. 239, Research Triangle Park, NC (1978).

D'Ottavio, T., R. L. Tanner, R. Garber, and L. Newman. Detection of ambient aerosol sulfur using a continuous flame photometric detection system II. The measurement of low level sulfur concentrations under varying atmospheric conditions. *Atmos. Environ.* **15**, 197–203 (1981).

Dzubay, T. G., and R. K. Stevens. Ambient air analysis with dichotomous sampler and x-ray fluorescence spectrometer. *Environ. Sci. Technol.* **9**, 663–668 (1975).

Ferm, M. Method for the determination of atmospheric ammonia. *Atmos. Environ.* **13**, 1385–1393 (1979).

Fontijin, A., A. J. Sabadel, and R. J. Ronco. Homogenous chemiluminescent measurement of nitric oxide with ozone. *Anal. Chem.* **42**, 575–579 (1970).

Forrest, J., and L. Newman. Sampling and analysis of atmospheric sulfur compounds for isotope ratio studies. *Atmos. Environ.* **7**, 561–573 (1973).

Forrest, J., R. L. Tanner, D. Spandau, T. D'Ottavio, and L. Newman. Determination of total inorganic nitrate utilizing collection of nitric acid on NaCl-impregnated filters. *Atmos. Environ.* **14**, 137–144 (1980).

Galloway, T. N., E. B. Cowling, E. Gorham, and W. W. McFee. A national program for assessing the problem of atmospheric deposition (acid rain). Report to the Council on Environmental Quality. National Atmospheric Deposition Program NC-141, Colorado State Univ., Fort Collins, CO (1978).

Gran, G. Determination of the equivalence point in potentiometric titrations. Part II. *Analyst* **77**, 661–671 (1952).

Hales, J. M., and M. S. Dana. Regional scale deposition of sulphur dioxide by precipitation scavenging. *Atmos. Environ.* **13**, 1121–1132 (1979).

Harker, A. B., L. W. Richards, and W. E. Clark. The effect of atmospheric SO_2 photochemistry upon observed nitrate concentrations in aerosols. *Atmos. Environ.* **11**, 87–91 (1977).

Hodgeson, J. A., R. K. Stevens, and B. E. Martin. A stable ozone source applicable as a secondary standard for calibration of atmospheric monitors. Presented at Analysis Instrumentation Symposium, Instrument Society of America, Houston, TX, April (1971a).

Hodgeson, J. A., R. E. Baumgardner, B. E. Martin, and K. A. Rehme. Stoichiometry in the neutral idiometric procedures for ozone by gas-phase titration with nitric oxide. *Anal. Chem.* **43**, 1123–1126 (1971b).

Howes, J. E., C. W. Spicer, and B. P. Price. Protocol for a field study to intercompare methods for measurement of nitric acid and aerosol nitrate in ambient air. Battelle Columbus Laboratories, Columbus, OH (1979).

Husar, J. D. Intercomparison of sulfate measurement techniques for use on aircraft platforms. Private communication (1979).

Joseph, D. W., and C. W. Spicer. Chemiluminescence method for atmospheric monitoring of nitric acid and nitrogen oxides. *Anal. Chem.* **50**, 1400–1403 (1978).

Kadlecek, J. A., and V. A. Mohnen. Time dependent behavior of stored precipitation samples. Presented at the First International Symposium on acid precipitation and the forest ecosystem. Columbus, OH, May, 1975 (1976).

Kelly, T. J., and D. H. Stedman. Measurement of H_2O_2 and HNO_3 in rural air. *Geoph. Res. Let.* **6**, 375–377 (1979).

Lazrus, A. L., K. I. Hill, and J. P. Lodge. A new colorimetric micro determination of sulfate ion. In *Automation in Analytical Chemistry Technicon Symposium*, pp. 291–293. Medicad Inc. (1966).

Lazrus, A., E. Larange, and J. P. Lodge. New automated microanalysis for total inorganic fixed nitrogen and for sulfate ion in water. In *Trace Inorganics in Water.* (ACS Advances in Chemistry Series No. 73) American Chemical Society, Washington, D.C. 1968.

Leahy, D., R. Siegel, P. Klotz, and L. Newman. The separation and characterization of sulfate aerosol. *Atmos. Environ.* **9**, 219–229 (1975).

Leahy, D., M. Phillips, R. Garber, and R. L. Tanner. A filter material for sampling of ambient aerosols. *Anal. Chem.* **52**, 1779–1780 (1980).

Lee, Y-N., and S. E. Schwartz. Evaluation of the rate of uptake of nitrogen dioxide by atmospheric and surface liquid water. *J. Geophys. Res.* **86**, 11971–11893 (1981).

McClenny, W. A., and C. A. Bennett, Jr. Integrative technique for the detection of atmospheric ammonia. *Atmos. Environ.* **14**, 641–645 (1980).

Newman, L. Techniques for determining the chemical composition of aerosol sulfur compounds. *Atmos. Environ.* **12**, 113–125 (1978).

Newman, L. General considerations on how rainwater must obtain sulfate, nitrate and acid. Presented at the Am. Chem. Soc./Chem. Soc. Japan Chemical Congress Honolulu, HA, April. Preprints Div. of Environ. Chem. Paper No. 187, p. 475 (1979a).

Newman, L. Considerations on the incorporation of sulfate, nitrate and acid into rainwater. Presented at International Symposium, Sulphur Emissions and the Environment, London, England, May. Proceedings p. 223 (1979b).

Okita, T., S. Morimoto, M. Izawa, and S. Konno. Measurement of gaseous and particulate nitrate. *Atmos. Environ.* **10**, 1089–1099 (1976).

Pack. D. H. (ed.) Proceedings: Advisory Workshop to Identify Research Needs on the Formation of Acid Precipitation, Alta, UT, August 1978. EA-1074, Special Study Project

WS-78-98. Prepared by Sigma Research, Richland, WA, for Electric Power Research Institute, Palo Alto, CA (1979).

Paur, R. J., and F. F. McElroy. Technical assistance document for the calibration of ambient ozone monitors. EPA-600/4-79-057. Research Triangle Park, NC (1979).

Raynor, G. S., and J. V. Hayes. Acidity and conductivity of precipitation on central Long Island, New York in relation to meteorological variables. *Water Air Soil Pollut.* **15**, 229–245 (1981).

Raynor, G. S., and J. V. Hayes. Concentrations of some ionic species in central Long Island, New York precipitation in relation to meteorological variables. *Water Air Soil Pollut.* **17**, 309–335 (1982).

Raynor, G. S., and J. P. McNeil. An automatic sequential precipitation sampler. *Atmos. Environ.* **13**, 149–155 (1979).

Saltzman, B. E. Colorimetric microdetermination of nitrogen dioxide in the atmosphere. *Anal. Chem.* **26**, 1949–1955 (1954).

Scaringelli, F. P., B. E. Saltzman, and S. P. Frey. Spectrophotometric determination of atmospheric sulfur dioxide. *Anal. Chem.* **39**, 1709–1719 (1967).

Shaw, R. W., T. G. Dzubay, and R. K. Stevens. The denuder difference experiment. In R. K. Stevens (ed.) *Current Methods to Measure Atmospheric Nitric Acid and Nitrate Artifacts*, pp. 79–84. Reports EPA-600-2-79-051, U.S. Environmental Protection Agency. Research Triangle Park, NC. (1979).

Small, H., T. Stevens, and W. Bauman. Novel ion exchange chromatographic method using conductimetric detection. *Anal. Chem.* **47**, 1801–1809 (1975).

Spicer, C. W., and P. M. Schumacher. Studies of the effects of environmental variables on the collection of atmospheric nitrate and the development of a sampling and analytical method. Interim Report on Phase I to U.S. EPA from Batelle Columbus Laboratories. Columbus, OH (1977).

Spicer, C. W., J. E. Howes, Jr., T. A. Bishop, and L. H. Arnold. Nitric acid measurement methods: an intercomparison. *Atmos. Environ.* **16**, 1487–1500 (1982).

Stevens, R. K. (ed.) Current Methods to Measure Atmospheric Nitric Acid and Nitrate Artifacts. Proceedings of Workshop, Southern Pines, NC, Oct. 1978. EPA-600/2-79-051 Research Triangle Park, NC (1979).

Stevens, R. K., J. Mulik, A. E. O'Keefe, and J. Krost. Gas chromatography of reactive sulfur gases in air at the parts-per-billion level. *Anal. Chem.* **43**, 827–830 (1971).

Tang, I. N. On the equilibrium partial pressures of nitric acid and ammonia in the atmosphere. *Atmos. Environ.* **14**, 819–828 (1980).

Tanner, R. L. An ambient experimental study of phase equilibria in the atmospheric systems: Aerosol H^+, NH_4^+, SO_4^{2-}, NO_3^-–NH_3(g), HNO_3(g). *Atmos. Environ.* **16**, 2935–2942 (1982).

Tanner, R. L., and J. Lepore. Continuous determination of ammonia at ambient atmospheric concentrations using fluorescence derivatization. Presented at the 177th National Meeting. Am. Chem. Soc. Honolulu, HA, April (1979).

Tanner, R. L., T. D'Ottavio, R. Garber, and L. Newman. Determination of ambient aerosol sulfur using a continuous flame photometric detection system I. Sampling system for aerosol sulfate and sulfuric acid. *Atmos. Environ.* **14**, 121–127 (1980).

Tanner, R. L., R. Cederwall, R. Garber, D. Leahy, W. Marlow, R. Meyers, M. Phillips, and L. Newman. Separation and analysis of aerosol sulfate species at ambient concentrations. *Atmos. Environ.* **11**, 955–966 (1977).

Thermo Electron. Manual for Model 14D Dual Chamber Chemiluminescent NO–NO_2–NO_x Analyzer. Thermo Electron Corp., Waltham, MA. (1975).

Volchok, H. L., and R. T. Graveson. Wet/dry fallout collection. *Proceedings Second Federal Conf. On the Great Lakes*, pp. 259–264. Great Lakes Basin Commission (1976).

SYNTHESIS

CHARACTERIZATION AND QUANTIFICATION OF AIR POLLUTANTS

W. A. McClenny
U.S. Environmental Protection Agency
Research Triangle Park,
North Carolina

T. H. Risby
The Johns Hopkins University
Baltimore, Maryland

BACKGROUND

All levels of the ecosystem are exposed to combinations of physical, chemical, and biological agents. These agents are produced by various natural processes and by human activities. It is the task of the effects scientists to assess the potential of these agents to modify the ecosystem. This requires characterization and quantification. The most practical way of expressing effect is in terms of a dose–response curve in which there is a threshold below which no measurable response is detected. The *dose–response curve and the threshold are then used by regulatory agencies to develop legislation aimed at protecting the ecosystem by controlling the emission of these agents. Implicit to this process is the fact that, apart from biological processes, the only source of pollution that can be controlled by regulation is the result of human activity. This has often meant that control legislation has become "technology forcing."*

It should be recognized that dose–response curves require an accurate quantification of dose, and that dosage is the integrated exposure as a function of time. For an agent, or agents, to interact with the receptor, that agent must reach the receptor in a form that is compatible with the receptor; this means that in terms of chemical agents, speciation is important. This requirement can provide the analytical chemist with an almost impossible task since, in the case of inorganic particulate matter, chemical speciation is a difficult task with existing instrumentation. Also, the quantification of the agent interacting with the receptor as a function of time (effective dose) is often almost impossible to measure. Indeed, the only way it can be estimated is to quantify the response of the receptor, which results in a vicious circle of cause and effects. In addition, it is only in rare cases that the receptor is insulted by a single agent, for it normally interacts with an agent contained in a complex matrix and, as a result, the response is modified by enhancement or depletion. Therefore, it is essential to characterize and quantify those agents and matrices which are reaching the receptors in the hope that the effects scientists may model response by representative exposure studies.

If it is established that an agent can cause a response in the ecosystem that is deemed to be deleterious, then its origin must be determined before its control can be implemented. The origin of the agent can only be established as a result of a complete knowledge of any reactions and transport which occur between source and receptor. The two processes also require characterization and quantification of the agents and their matrices.

This background has shown the need for quantification and characterization of agents and their matrices but has not discussed the need for temporal resolution of these measurements. The concentration of agents that leave the source and reach the receptors are dependent upon the operation of the source, the factors involved in transport, deposition, and the condition of the receptor. These features seldom result in a constant exposure of the receptor to the agent. Indeed, this is complicated further, since the matrix in which the agent is contained is dynamic and therefore has a temporal component. It should, therefore, be obvious that all the scientific disciplines which are concerned with emission, fate, transport, and the effect of agents should pool their knowledge to produce an understanding of the effects of agents on the ecosystem. The common thread through this discourse is the need for characterization and quantification of the agents that are present in the terrestrial ecosystem.

DISCUSSION

A. Ambient Air Monitoring Techniques—Adaptation to Effects Research

During the 1970s, concerted efforts were made to develop instrumental techniques to adequately characterize and quantify certain of the ambient air pollutants. Today a number of real-time monitors as well as their corresponding calibration techniques are available. In the United States the ambient air monitoring procedures for the so-called criteria pollutants are standardized with respect to accuracy and precision. This standardization simplifies comparison of data bases and interpretation of results. Many of the new monitoring techniques are based on the physical properties of the analyzed species and involve the measurement of the emission or absorption of radiation. Examples of those monitors are as follows: chemiluminescence monitors for NO, NO_2, SO_2, HNO_3, NH_3, and O_3; fluorescence monitors for SO_2 and NO_2; infrared monitors for CO; and ultraviolet absorption monitors for O_3. These instruments can, in some cases, be modified to meet special requirements for effects research.

In addition to these established monitors, new and improved techniques are being developed. For the gases, there is current emphasis on monitors with increased sensitivities for the detection of NH_3, HNO_3, H_2O_2, HCHO, nonmethane hydrocarbons (NMHC), and various organic vapors. With respect to aerosols, new techniques are under development for measurement of elemental and total carbon and for monitoring individual organics, especially those that originate from diesel exhaust. Improved instrumentation is available for particulate size separation using the principles of virtual impaction as employed in the so-called dichotomous sampler to collect both fine ($<2.5\ \mu m$) and coarse ($>2.5\ \mu m$) particles. Efforts are nearly completed on a new design for a particle collector inlet with an upper cutoff that is reasonably independent of wind speed while maintaining sharp cutoff features. Similarly, measurements of H_2SO_4 by several techniques are being improved.

The availability of these instruments allows two important activities to occur:

1. *The monitoring of simulated or real levels of air pollutants so their effects may be delineated.*
2. *The documentation of pollutant variability.*

The latter activity is of great importance since effects are known to be related to excess concentration above a certain threshold level and not just to the product of concentration and time. Thus, realistic simulation must be programmed to account for variability.

B. Summary

The following subheadings (1–6) correspond to the main topic areas covered in the two preceding papers and the period of discussion following the two presentations. The text associated with each subheading is our subjective summary of the more important material. The reader will find the detailed discussion of topics in the two Part Three papers.

1. ACID AEROSOL

A subject of great current interest in atmospheric chemistry is the characterization of aerosol sulfur. The aerosol sulfur of concern occurs in the fine particle size range originating primarily from the atmospheric reactions of SO_2 released from man-made sources. The resulting aerosol is in dynamic equilibrium and involves the partition of components between the gas and aerosol phase, as, for example, NH_3, SO_2, H_2O and NH_4^+, SO_4^{2-}, H^+. The prediction of the equilibrium conditions toward which the atmospheric components tend under constant ambient conditions can be attempted by interpretation of appropriate phase diagrams of the type being developed by C. Brossett of Sweden.

The availability of gaseous NH_3 is often thought to result in neutralization of acidic aerosols, implying an irreversible reaction. In reality the system is in a dynamic equilibrium in which NH_3 can be released as well as captured by the aerosol phase. An experiment, monitored by C. Brossett, demonstrates this fact. In the experiment two specially prepared diffusion tubes were used. The interior wall of one tube was coated with $(NH_4)_2SO_4$ and of the second tube with $(NH_4)HSO_4$. After passing ambient air through the two tubes for 24 h at a flow rate of 3 L/min, the coatings were extracted and examined; they revealed a common ratio of NH_4^+/SO_4^{2-} of 0.8—that is, NH_3 was exchanged between tube coatings and the ambient air until equilibrium was established. While this example demonstrates the release of ammonia to achieve a more acid coating, a second example was given by B. Ottar of Norway, in which the uptake of NH_3 resulted in a less acid aerosol. Air masses arriving in Scandinavia, after remaining for some days over the North Sea, contained extremely acid sulfate particulates, with pH values as low as 2.4. However, once these air masses encountered land, the aerosol reacted with NH_3 originating from biogenic sources. The aerosol was partially neutralized after passing over Denmark and, by the time the air mass was carried to Stockholm, the aerosol was fully neutralized.

Agreement is not observed when the ionic components of rainwater and the gaseous species at ground level are compared with equilibrium predictions. A difference between the conditions at the site of droplet formation and droplet collection may be the reason. This means that values of NH_3 and relative humidity at ground level and aerosol composition, as inferred from collected rainwater, may not constitute a system subject to equilibrium. Observed NH_3 is often a factor of 10^2 greater than predicted. The question of the origin of the acidity observed in rainwater still appears to be unresolved. This question hinges on whether the scavenging of aerosol sulfur by rainwater could account for the high acidity observed, or whether SO_2 is assimilated through in-cloud processes as mentioned by L. Newman (in this volume). Other nonequilibrium conditions can occur. For example, when different air masses mix, the release of H_2O and/or NH_3 from aerosols may require a long equilibration time.

Research on the nature of equilibrium in mixed-phase systems is continuing. The overall objective of this research is to provide insight

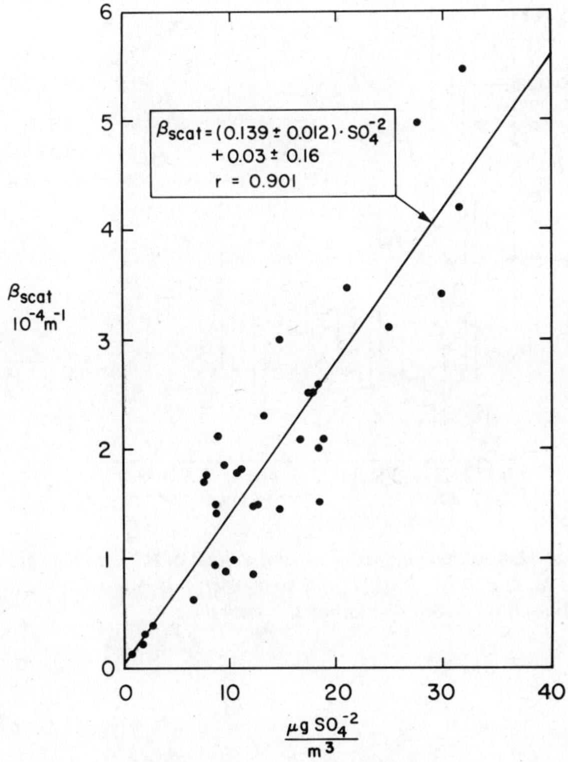

FIGURE 1. *Plot of β_{scat} versus aerosol SO_4^{2-} (12-hour averages). (From Pierson et al., 1980.)*

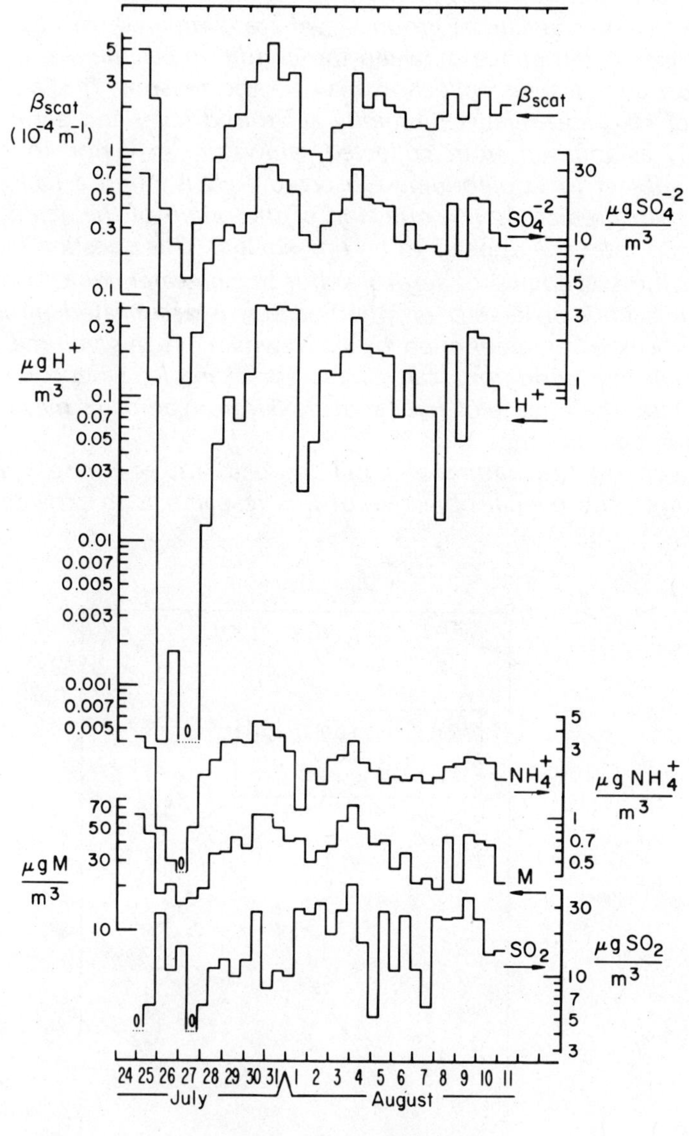

FIGURE 2. *Plot of 12-hour average (run-by-run) values of β_{scat}, SO_4^{2-}, aerosol H^+, NH_4^+, total aerosol mass M, and SO_2; 2000 EDT July to 0800 EDT 11 August 1977. M was evaluated at constant 39% humidity. (From Pierson et al., 1980.)*

concerning conditions in more complex mixed phase systems. Such knowledge will be a guide for excluding unrealistic atmospheric mixes as well as a starting point for a consideration of the time dependent processes which establish equilibrium. This research will be aided by new real-time aerosol monitors such as the sulfur aerosol monitor mentioned by K. Whitby of the University of Minnesota.

One manifestation of acidic sulfate in the ambient air is the particle growth due to the hydrophilic nature of H_2SO_4. This leads to increased light scattering as measured by the coefficient of extinction due to scattering (β_{scat}). Pierson documented this as early as the summer of 1977 in measurements made on the Allegheny Mountains in Pennsylvania. Three effects are of particular note in these experiments; the correlation between sulfate concentration and β_{scat} as shown in Figure 1, the correlation of the concentration of H^+ with SO_4^{2-} and β_{scat} as shown in Figure 2, and a persistent SO_4^{2-} aerosol consisting mostly of H_2SO_4.

2. GASEOUS MEASUREMENT WITH DIFFUSION TUBES

A number of air quality measurements are now being made by selective preconcentration of diffusion tubes with subsequent release and analysis. The diffusion tubes function as separators of gases and particulate matter. The gases diffuse to the walls of the hollow, coated diffusion tubes as ambient air is pulled through them. Particulate matter is carried through the tube in laminar flow.

The measurement of either ammonia or nitric acid by use of diffusion tubes inherently assumes that no aerosol-to-gas conversion is occurring during the residence time of the particulate matter in the diffusion tube. This may not be true for a volatile compound such as NH_4NO_3, since NH_3 and HNO_3 could be eliminated from the sample stream to create a nonequilibrium condition. No consensus was evident during the discussion.

3. ORGANIC TRACERS

The use of receptor models to identify sources may be applicable to organics as well as to the inorganic elements. Specific organic compounds could be used as markers for source identification. The marker compounds would be extracted from collected particulate matter along with other organics and identified by gas chromatography–mass spectrometry procedures. Supplemental data on wind direction and persistence would be used. However, two questions arise: (1) Do organic species retain their integrity between the release point and the receptor point? (2) Do the organics remain stable between the time of collection and time of analysis? Because of the reactions between organics and other species, the most reactive organics could change, depending on transit time and ambient conditions (e.g., night versus day). Examples also exist of extensive degradation of polyaromatic hydrocarbons on glass fiber filters which have been stored in the dark at room temperature for a few days. This degradation can probably be minimized by reducing the storage temperature. To make organic source identification possible using marker compounds, it will be necessary that the various sources release classes of organic compounds in which the relative concentrations of the members have different patterns. For simplicity of interpretation, it would be desirable, but not necessary, that a specific compound be uniquely associated with each source. Attempts to identify source-specific compounds associated with major sources such as diesel and gasoline engines have been made, but so far the observed differences are in degree and not in kind.

4. RECEPTOR MODELS

With the use of elemental composition of aerosols as determined by X-ray fluorescence, neutron activation, and emission spectroscopic analysis, chemical balance or factor analysis procedures can be applied to deduce the relative contribution of a variety of pollution sources to a given receptor site. G. Gordon of the University of Maryland explained that the combination of elemental analysis and wind direction information can be used to identify dominant sources of a particular element. This process involves selection of samples that have high concentrations of the element and that were acquired during periods of reasonably constant wind direction. The process is highly selective, using only about 1% of the available filters when applied to source identification. H. R. Krouse of The University of Calgary suggested a way to make more efficient use of samples by locating several of them at a given site and triggering each one to respond to a different wind direction. Siting of sampling stations is important to optimize the chance that the samplers are in a downwind direction from likely sources.

Fractionation (or preferential loss) of particles may occur in transit from source to receptor. There are several causes of fractionation. Sedimentation is obviously a main source of fractionation if fine and coarse particles are collected together, but this can be minimized effectively by collecting fine and coarse particles separately, as with a dichotomous sampler. Fractionation caused by solubility of the particle to which elements are attached is documented (Muhlbaier, 1978), as are differences in deposition rate depending on the size distribution of the particles bearing particular elements.

Separation of contributions from different sources can be a problem, as mentioned by A. Chamberlain (in this volume) in the case of soil contribution and coal contribution (fly ash) where elemental composition is similar. Secondary methods of analysis, such as identifying characteristic shapes (i.e., fly ash particles are round, glassy spheres, whereas soil particles are platey aggregates), may be used to resolve such similarities. The elemental composition of soil is currently determined by collecting and sieving soil samples to obtain a fine particulate fraction. This composition is assumed to correspond to that entrained from the soil. D. Williams of CSIRO, Australia, suggested that soil contribution may depend on sampler height above ground, although the soil contribution seems to vary from 10 to 50% of the total even for sampler locations at least 10 ft above the ground. Soil contribution is surprisingly high in urban areas—for example, a factor of 5 higher in Tucson, AZ than at nearby rural sites.

Automatic scanning electron microscopy coupled with energy dispersive X-ray fluorescence detectors and X-ray diffraction analyzers are currently being used to determine the sources of aerosols. These direct methods of determining the sources of aerosols, if perfected, will serve as a means of validating the source apportionment models.

5. MONITORING GUIDELINES

One recurring theme concerned the design of intensive field studies in which the objective is to identify atmospheric mechanisms related to transport and transformation. A distinction was made between this type of field study and monitoring to establish compliance with standards or to identify the existence of trends. There is an important need to integrate measurements of air quality with meteorological parameters in intensive field studies to make possible interpretation of air quality data, especially with regard to secondary pollutants.

Sample integrity was mentioned by S. Krupa of the University of Minnesota with regard to the collection of rain samples. A rain sampler is currently available to sample specific rain events or segments thereof into sealed containers. Since no gas exchange is allowed between the rain inside the sampler and the gas phase outside, sample integrity or, more specifically rain pH values, are maintained. In general the mere process of sampling, that is, of transporting the sample from the ambient to the measurement device, alters the sample and hence should be avoided or minimized where possible. In situ optical absorption measurements are probably the best way to ensure sample integrity, although these systems are not now practical for monitoring applications.

6. *SPECIFIC NEEDS IN EFFECTS RESEARCH*

W. Heck of North Carolina State University mentioned that fast response instruments are needed for time-sharing arrangements in which one instrument is used for several exposure chambers. For this application the sum of times necessary to flush lines leading to the instruments from the chambers and the response time of the instruments are the important parameters. T. C. Weidensaul of the Ohio Agricultural Research and Development Center mentioned the monitoring need to separate effects due to short duration and high concentrations, and long durations and low concentrations, since the effective dose in these two cases can be quite different for the same integrated product of concentration and exposure time. There is also a monitoring need to separate effects due to short duration and low concentrations. B. Hicks of the Atmospheric Turbulence and Diffusion Laboratory reiterated a continuing need for two types of instruments in studies to determine the velocity of deposition of various gases. For the gradient technique a 1% accuracy for time-averaged numbers is needed, while for the eddy correlation technique a fast response (<1 s) is required.

C. Needs for New Instrumentation

The following developments appear to be indicated to satisfy effects research needs:

1. *A general continuation of development of real-time or near real-time monitors to characterize the elemental, chemical, and temporal nature of agents in the ambient air.*
2. *The development of specific monitors to use in determining velocity of deposition by the gradient technique in which 1% accuracy for time averaged numbers is needed and by the eddy correlation technique in which fast response (typically less than 1 s) is required.*
3. *Real-time or semi-real-time monitors for certain gases (e.g., HNO_3, NH_3, HCHO) as well as for the ionic groups in collected particulate matter (such as NH_4^+, NO_3^-, SO_4^{2-}).*
4. *Total strong and weak acid measurements.*
5. *New data handling systems which simplify the storage and retrieval of data bases and provide fast turnaround analysis of data.*
6. *Ways to characterize pollutant variability using manipulation of data bases.*
7. *New techniques to develop receptor models for organics.*

8. *Improved optical (including SEM) and X-ray diffraction techniques to quantify the impact of industrial and nonindustrial emissions to ambient aerosols.*
9. *Development of real-time tracer techniques for use in effects and modeling studies.*

REFERENCES

Muhlbaier, J. The chemistry of precipitation near the Chalk Point power plant. Ph.D. Thesis, Univ. of Maryland, College Park (1978).

Pierson, W. R., W. W. Brachaczek, T. J. Truex, J. W. Butler, and T. J. Korniski. Ambient sulfate measurements on Allegheny Mountain and the question of atmospheric sulfate in the Northeastern United States. In T. J. Kneip and P. J. Lioy (eds.) Aerosols: Anthropogenic and Natural, Sources and Transport. Ann. N.Y. Acad. Sci. **388**, 145–173 (1980).

Part Four

Transport and Deposition of Air Pollutants on Terrestrial Vegetation and Soils

Consistent with the science of epidemiology, the transport and deposition of air pollutants are critical in terrestrial ecosystem effects research. Long-distance transport of pollutants is becoming a global concern. Specific problems in this area are still being defined and described. Meteorology for each area is specific and complex. Long-distance transport of ozone, nitrogen compounds, and sulfur compounds is of extreme importance at this time. However, we do not sufficiently understand the transport of other pollutants. Currently, acidic deposition is of major concern.

Deposition of air pollutants relates to effects. Several scientists have evaluated the deposition rates and velocities of O_3 and SO_2, but similar examples with NO_x and fine particulates provide little more than estimates. Plant canopies and soil act as pollutant sinks. Little duplicative information is available on pollutant transport through plant canopies.

Refined techniques and derivations are still in the development stage.

This section addresses these as well as other scientific areas of concern and helps provide a point of reference for the remaining sections.

Deposition of Gases and Particles on Vegetation and Soils

A. C. CHAMBERLAIN
Atomic Energy Research Establishment
Harwell, England

1. AERODYNAMIC AND CANOPY RESISTANCES

The processes of transport and deposition of gases and particles can be considered in four stages:

1. Transport by wind in the free air.
2. Transport by eddy diffusion and by sedimentation across the boundary layer to the canopy.
3. Transport by eddy and molecular diffusion, sedimentation and impaction to the surfaces of vegetation and soil.
4. Transport by molecular diffusion and other processes within stomata of plants and within pores of soil, and adsorption by solution, ion exchange, or physical adherence to surfaces.

Processes (1) and (2) are almost purely meteorological in character, and the only influence of the type of vegetation is on the eddy diffusivity of the air in the boundary later. Processes (3) and (4) are determined equally by the physics and chemistry of the aerosol or vapor and of the vegetation, and the physiological state of the vegetation usually has to be taken into account.

The state of knowledge of processes (1) and (2), although by no means complete, is probably sufficient for most practical purposes, bearing in mind that the purpose of any model of aerial transport is either to explain what has been observed in the past or to forecast what will happen in the future. Either way, the accuracy is likely to be determined by limited knowledge of the actual wind and eddy structure of the atmosphere between the source of emissions and the location of deposition. One area of uncertainty is the so-called oasis effect, whereby transport of gases and particles to and from isolated areas of crop, forest, natural vegetation, or bodies of water differs from that of extended areas. For example, a recent study of Oke (1979) has shown that evapotranspiration per unit area from an irrigated lawn of area 106 m^2 in Vancouver, British Columbia, was about twice that from the surrounding district as a whole. The oasis effect must be expected to increase the uptake of SO_2 or O_3 by an isolated canopy because the boundary layers of wind speed and of SO_2 concentration are not fully developed over a small area. The deposition of particles by impaction on vegetation is probably even more susceptible to oasis effects than the deposition of gases. The reason is that the deposition coefficient (ratio of velocity of deposition to wind speed) increases with increasing windspeed when impaction is the main mechanism.

In the case usually considered, where the fetch over a reasonably uniform canopy is sufficient to allow the development of the boundary layer to a height of several meters, deposition can be considered as the downward flux of matter brought about, for gases, by eddy diffusion down a concentration gradient, and for particles, by a combination of eddy diffusion and sedimentation. With the assumption of constant vertical flux within the fully developed boundary layer, the velocity of deposition v_g is defined by

$$v_g = F/X$$

where F is the flux and X the concentration at some reference height.

The reciprocal of v_g, which is the resistance to transfer, denoted by r, is considered as the sum of three resistances:

r_a Aerodynamic resistance. This depends on the turbulence of the air above the canopy.

r_b Additional boundary layer resistance for transfer of matter compared with transfer of momentum can be considered as resistance to transport between z_0, roughness length of canopy, and z_v, the analogous length for vapor transport. This depends on the molecular diffusivity of the gas as well as on the nature of the roughness of the vegetation. For most crops it is small compared with r_a but becomes appreciable for surfaces with bluff roughness elements, such as soils, and also for gases with low molecular diffusivity.

r_c Canopy or surface resistance. This is the residual resistance depending on the affinity of the surface for the gas. For surfaces which are a perfect sink, r_c is zero. It appears probable that the moist surfaces of mesophyll cells within leaves are perfect sinks for reactive gases such as SO_2. The rate of uptake to them is controlled by stomatal diffusion. For an individual leaf, the resistance to stomatal diffusion is denoted by r_s. For a canopy, the stomatal resistances of different leaves are in parallel with each other and also with the resistance for cuticular uptake (direct deposition on the leaf surface) and the resistance for diffusion through the crop to the underlying soil (Fowler and Unsworth, 1979).

The estimation of r_a and r_b for various crops is described, for example, in the book *Vegetation and the Atmosphere* (Monteith, 1975), and, although there are many uncertainties, these are mostly small in comparison with the variation in r_c. In comparing experiments in which v_g or r has been measured, it is best to attempt to estimate r_a (and if possible r_b) in the conditions of the experiment and subtract from r to derive r_c. If values of v_g or r are compared, there is the danger that variation in the aerodynamic conditions of the experiment will be confused with variation due to physiological or morphological parameters in the crop.

For deposition of particles in the 1–30 μm size range it may be assumed that

$$v_g = v_s + v_T$$

where v_s is the sedimentation velocity and v_T is the velocity of deposition due to impaction. To a first approximation v_s and v_T may be considered as independent. The resistance to deposition by impaction (v_T^{-1}) may be considered as aerodynamic and canopy resistances in series. In theory, the deposition of particles to vegetation with leaves of simple geometry—for example, pine forests (Belot, 1975; Belot et al., 1976)—can be calculated by considering the impaction efficiency of individual leaves. In practice, as will be considered later, the accuracy of the calculation depends on whether or not all impacts of particles on leaves are equivalent to captures.

2. TRANSPORT OF GASES TO VEGETATION

2.1 Oxides of Sulfur

There is now extensive information on the rate of deposition of SO_2 to grass, crops, forest, soil, and building surfaces, which has been reviewed by Garland (1978, 1979), McMahon and Denison (1979), and Chamberlain (1980). Most of the uptake of SO_2, like the rate of loss of water vapor from the leaf, is controlled by the rate of stomatal diffusion. Since this is proportional to the molecular diffusivity of the diffusing gas, the resistances are inversely proportional to the diffusivities.

Belot (1975) enclosed shoots of pine (*Pinus sylvestris*) in small chambers, fumigated them with SO_2, and simultaneously measured the transpiration of water from the leaves. To enable the gas phase uptake by leaves to be distinguished from root uptake, Belot used SO_2 enriched with the stable isotope ^{34}S. Garland and Branson (1977) did similar experiments with radioactive $^{35}SO_2$. In both series of experiments, the rates of uptake of SO_2 and transpiration of water varied together, depending on the degree of stomatal opening, and the conductances (i.e., velocities of deposition) were proportional to the molecular diffusivities (Fig. 1). There was no evidence of cuticular uptake of SO_2 by pine needles.

Fowler and Unsworth (1979) measured the uptake of SO_2 to a wheat crop by the gradient method. They concluded that stomatal resistance determined the canopy resistance r_c during the daytime, provided the crop was dry, and their results in these conditions were consistent with the hypothesis that uptake of SO_2 and transpiration of water are controlled by the same mechanism. Fowler and Unsworth, however, found evidence of cuticular deposition of SO_2, and by night this was the main route of deposition. When dew formed on the crop, deposition of SO_2 increased and was limited at first only by aerodynamic resistance. As deposition continued, there was evidence that r_c increased as acidity increased to a level (about pH 3) at which further uptake became inhibited. Uptake of SO_2 by wet vegetation is likely to be of considerable importance, particularly on northern and upland forests, and there is need for more work in this field.

Fowler and Unsworth found that nighttime values of the canopy resistance of wheat were lower than the probable resistances for water vapor,

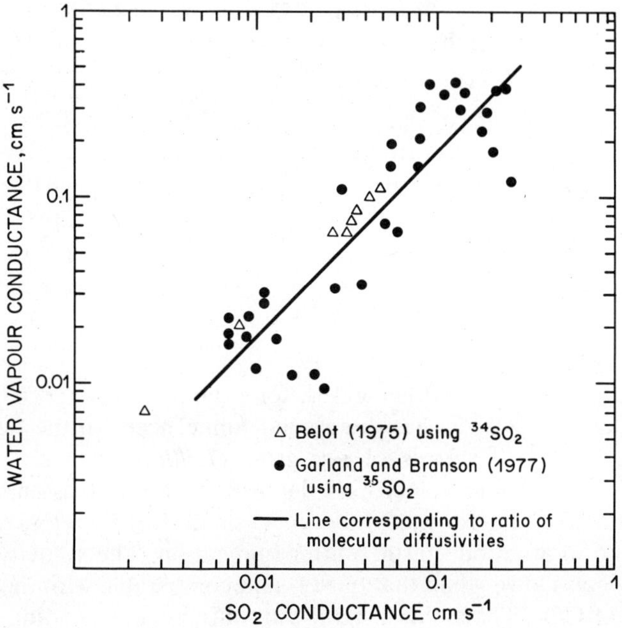

FIGURE 1. Conductances for water vapor and SO_2 of Scots pine shoots.

and it appeared that cuticular uptake is appreciable. Damage to the cuticles of grasses and cereals can be caused by grazing and trampling by animals and by friction between leaves in windy weather. When the cuticle is damaged, the uptake of SO_2 may well be increased.

From these data, Fowler and Unsworth deduced that the total gaseous sulfur uptake in a wheat crop exposed to 50 $\mu g/m^3$ of SO_2 (25 $\mu g/m^3$ of SO_2–S) during a 90-day growing period as 1.05 g S/m^2 of crop. This corresponds to an average velocity of deposition of 0.54 cm/s. Sprugel and Miller (1979) found deposition velocities ranging from 0.5 to 0.9 cm/s by determining the sulfur balance of soybean crops exposed in the field to SO_2 fumigation. For comparison, Denmead (1976) estimated the average daytime uptake of CO_2 by a wheat crop as 33 g/m^2 day (about half of this was lost during nighttime respiration). Assuming an atmospheric CO_2 concentration of 300 ppm, the corresponding daytime velocity of deposition is about 0.1 cm/s. Thus the uptake of SO_2 is greater in proportion to the atmospheric concentration than the uptake of CO_2. The reason for this is that there is a "biochemical" resistance to the uptake of CO_2, determined by the rate of conversion of bicarbonate to carbonate in the plant cells. The biochemical resistance to uptake of SO_2 is an order of magnitude, at least, less than the minimal stomatal resistance, and may be zero (Black and Unsworth, 1979).

Although most of the sulfur in the gas phase in the atmosphere is present

TABLE 1. Deposition of Labeled H_2S and COS to Rye Grass in a Wind Tunnel[a]

Gas	Illumination	Crop Density (mg/cm^2)	v_g (cm/s)
H_2S	On	92	0.64
COS	On	72	0.080
COS	Off	71	0.031

[a] From Chadwick (1977).

as SO_2, trace amounts of other gases, for example, H_2S and COS are also present. Chadwick (1977) has done wind tunnel tests on the deposition of $H_2{}^{35}S$ and $CO^{35}S$ to swards of rye grass (*Lolium perenne*), with results shown in Table 1. The velocity of deposition of H_2S was similar to that found with SO_2. There was only very small (0.13%) loss of H_2S by volatilization in a period of 100 h after fumigation. The velocity of deposition of COS was lower than that of SO_2 and comparable with the velocity of deposition of CO_2. There was a significant difference depending on whether the grass was illuminated during the fumigation. As it was thought that extra CO_2 in the air might inhibit the uptake of COS, either by closing the stomata or by a biochemical effect, tests were done in which 6% CO_2 was added to the air in the tunnel during the fumigation with the labeled H_2 and COS, but this was found to make no difference to the rates of deposition.

There is clearly room for more work to elucidate mechanisms of uptake of these gases.

2.2 Oxides of Nitrogen

Hill and Chamberlain (1976) reported deposition velocities of 1.9 cm/s for NO_2 and 0.1 cm/s for NO to an alfalfa canopy in a wind tunnel. Omasa et al. (1980) found that the concentrations of both NO_2 and O_3 at the gas–liquid interface in the leaf were effectively zero, and that the sorption rates were controlled by boundary layer and stomatal resistances. However, Garland (1979) found no gradient of NO or NO_2 above grassland, which suggested deposition velocities less than 0.1 cm/s. Judeikis and Wren (1978) measured deposition of NO and NO_2 onto soil and cement surfaces. On freshly prepared surfaces, v_g was about 0.3–0.8 cm/s for NO_2 and 0.1–0.3 cm/s for NO. Although irreversible, deposition decreased with time, indicating a finite capacity of the surfaces to absorb the gases. Surfaces could, however, be reactivated by washing, and in the case of NO_2 but not that of NO, by interaction with atmospheric ammonia. Judeikis and

Wren (1978) concluded that dry deposition of NO was probably of small importance compared with that of NO_2 in urban areas. In country areas, where concentrations of NO_2 are small anyway, dry deposition of NO_x was not thought to be of significance to the atmospheric budget. The more reactive oxides of nitrogen are likely to be removed more rapidly than NO or NO_2 at the earth's surface. In wind tunnel experiments, Garland and Penkett (1976) found that v_g for deposition of peroxyacetyl nitrate (PAN) to both grass and soil was about 0.25 cm/s (which is about a factor of 3 lower than the value of Hill and Chamberlain (1976)). Under inversion conditions, a noticeable diurnal variation of concentrations of both PAN and ozone was found, due to deposition at the ground depleting the surface layers of air more rapidly than it could be replenished by downwards diffusion from above. Kelly et al. (1979) report gaseous HNO_3 in rural air at concentrations up to 5 ppb. The ratio of HNO_3 to NO_x was about 1:7. It is likely, but unproven, that the rate of dry deposition of HNO_3 would be similar to that of PAN, or possibly higher. In this case, the dry deposition of HNO_3 may be greater than the deposition of NO_x.

Brice et al. (1977) found a small diurnal variation in the concentration of N_2O in the atmosphere near the ground, with lower concentrations by night, which they attributed to deposition of N_2O at the earth's surface. However, Pierotti et al. (1978) found no diurnal variation, and Matthias et al. (1979) found a diurnal variation in the opposite sense, with higher concentrations in the early morning, which they attributed to emissions from the soil.

2.3 Ozone

A comprehensive paper on deposition of ozone, including new measurements and a review of the literature, has been published recently by Galbally and Roy (1980). The daytime surface resistance for transfer to soil and grassland is about 1 cm/s. The uptake by plants is controlled by the opening of the stomata, and surface resistance increases several-fold when they close. The resistance for water vapor transpiration and ozone uptake are found to be identical. However, unlike SO_2, ozone is not very soluble in water, and surface resistance to uptake by soils increases with water content. It is conjectured that ozone is catalytically destroyed at active sites on soil particles or reacts by oxidizing the organic fraction. There is still uncertainty about the mode of uptake by vegetation and the possible interaction with gases such as SO_2 at plant surfaces.

The destruction of ozone at the ground causes a marked diurnal cycle of concentrations with higher values by day, which can be distinguished from the effects of daytime photochemical ozone production (Garland and Derwent, 1979).

3. TRANSPORT OF PARTICLES TO VEGETATION

3.1 Submicron Particles

Particles in the submicron size range conform almost perfectly to the streamlines of flow in the atmosphere, even following small eddies. The relaxation time of a 1-μm-diameter unit density particle is 4 μs. This means that if it is given a velocity v relative to air, this will be reduced to v/e, where e is Euler's constant (2.7) in a time of 4 μs. In this respect, submicron particles behave almost like gases, but there is a big difference in the rate at which they migrate to surfaces, since this requires transfer across the laminar sublayer. The diffusivity of a particle is independent of its density and depends only on its size and shape. For 0.1-μm and 0.01-μm-diameter particles, the Brownian diffusivity $D = 7 \times 10^{-6}$ and 5×10^{-4} cm^2/s, respectively, whereas for gases D is of order 0.1 cm^2/s. Chemical engineers have found that transfer to surface of gases and particles with varying diffusivity can be correlated if the product ShSc$^{1/3}$ is plotted against the Reynolds number of the flow over the surface. The Sherwood number Sh is a deposition velocity made nondimensional by multiplying by L/D, where L is the characteristic length of the surface. The Schmidt number Sc is ν/D, where ν is the kinematic viscosity of air (equal to 0.15 cm^2/s at 15°C).

Figure 2 shows the correlation of ShSc$^{-1/3}$ with Re (equal to uL/ν, where u is the windspeed) for the transport of gases and particles to leaves, petioles, and stems of various plant species. The results of Grace and Wilson (1976) were obtained by measuring the rate of evaporation of water ($D = 0.20$ cm^2/s) from a flat filter paper replica of a leaf of poplar (*Populus euramericana*). Those of Chamberlain (1974) refer to transport of ^{212}Pb vapor ($D = 0.054$ cm^2/s) to leaves of bean (*Vicia faba*). The bean leaves were in their normal posture in a canopy in the wind tunnel, whereas the model leaves of Grace and Wilson (1976) were glued to aluminum templates and placed parallel to the mean wind flow. Grace and Wilson give references to several other correlations of heat or water vapor transport to model leaves which show similar results. The lines drawn in Figure 1 are theoretical correlations of mass transfer to smooth surfaces in laminar (A) or turbulent (B) flow.

Also shown in Figure 2 are the results of experiments by Belot (1975) on transport of 0.17-μm particles ($D = 3.2 \times 10^{-6}$ cm^2/s, Sc $= 4.7 \times 10^4$) to leaves of oak (*Quercus sessiliflora*) and pine (*Pinus sylvestris*), and experiments by Little and Wiffen (1977) with 0.03-μm particles ($D = 7 \times 10^{-5}$ cm^2/s, Sc $= 2 \times 10^3$) to leaves of nettle (*Urtica dioica*), beech (*Fagus sylvatica*), and white poplar (*Populus alba*). The values of ShSc$^{-1/3}$ for both particles and gases lie above the theoretical line. This is attributed to the effects of surface roughness, leaf aspect, turbulence, and aspect ratio (limited cross-wind dimension of leaf surface). Considering that the range of diffusivity in the experiments is five orders of magnitude, the correlation is

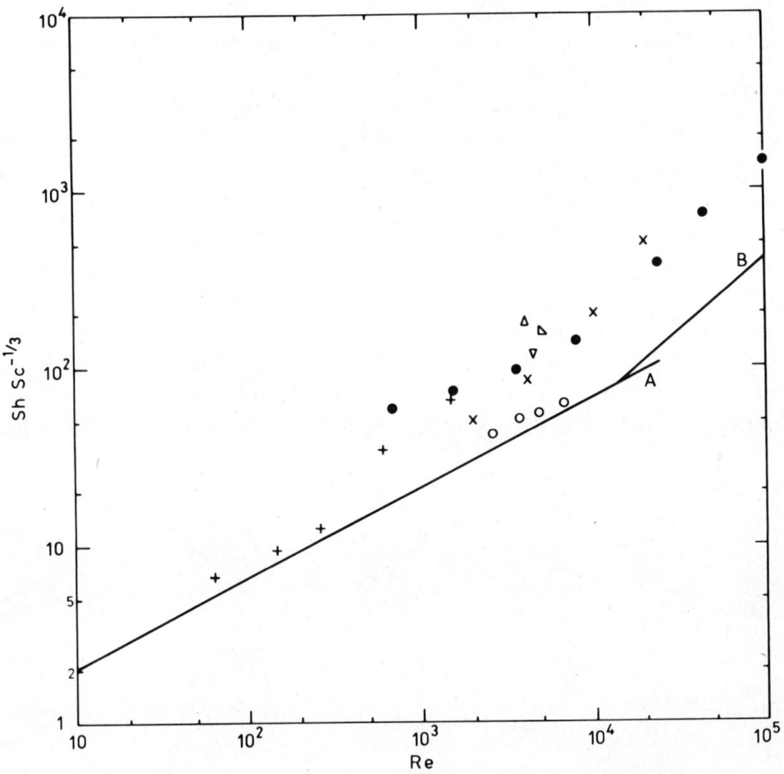

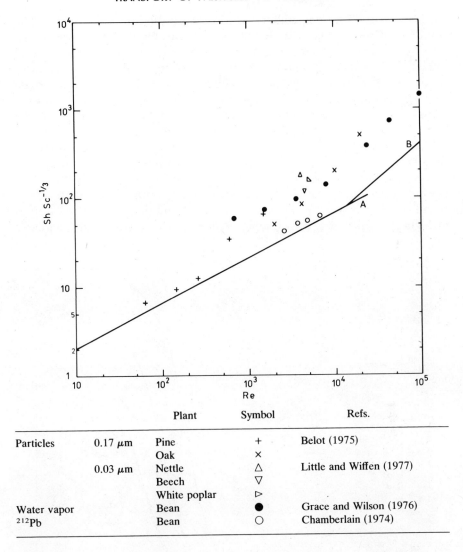

		Plant	Symbol	Refs.
Particles	0.17 μm	Pine	+	Belot (1975)
		Oak	×	
	0.03 μm	Nettle	△	Little and Wiffen (1977)
		Beech	▽	
		White poplar	▷	
Water vapor		Bean	●	Grace and Wilson (1976)
^{212}Pb		Bean	○	Chamberlain (1974)

FIGURE 2. Transport of particles and gases to and from leaves by Brownian diffusion. Line A: theory, laminar flow; $Sh = 0.66 Re^{0.5} Sc^{0.33}$. Line B: theory, turbulent flow; $Sh = 0.037 Re^{0.8} Sc^{0.33}$.

good, and this confirms that Brownian diffusivity across the sublayer is the limiting factor. It follows from Figure 2 that the velocity of deposition by Brownian motion varies with wind speed raised to a power less than unity, so the deposition coefficient v_g/u (termed C_p by analogy with the drag coefficient C_D) decreases with increasing wind speed.

Belot (1975) used his measurements of deposition and submicron particles to pine shoots, and also measurements of the deposition of larger particles to develop a computer model of deposition to a pine forest. The results are shown in Figure 3, together with the results of earlier experiments by Chamberlain (1966) on the deposition of particles to grass in a wind tunnel. The wind speeds shown are those at the top of the canopy, denoted $u(h)$. The corresponding friction velocity at canopy height can be derived from the relations $u(h)/u_* = 3.6$ for the pine forest and $u(h)/u_* = 4$ for the grass where u_* is the friction velocity. The characteristic shape of the curves in Figure 3 is familiar from many experimental and theoretical estimates of deposition to a variety of surfaces, and it is almost always found that there is a minimum of deposition velocity in the particle size range 0.1 to 1 μm, where particles are too large for Brownian motion and too small for impaction to be effective. However, the depth of the trough in the curve is still a matter of dispute. The apparent agreement in the two sets of curves in Figure 3, indicating, for example, that v_g for 0.5-μm particles increases

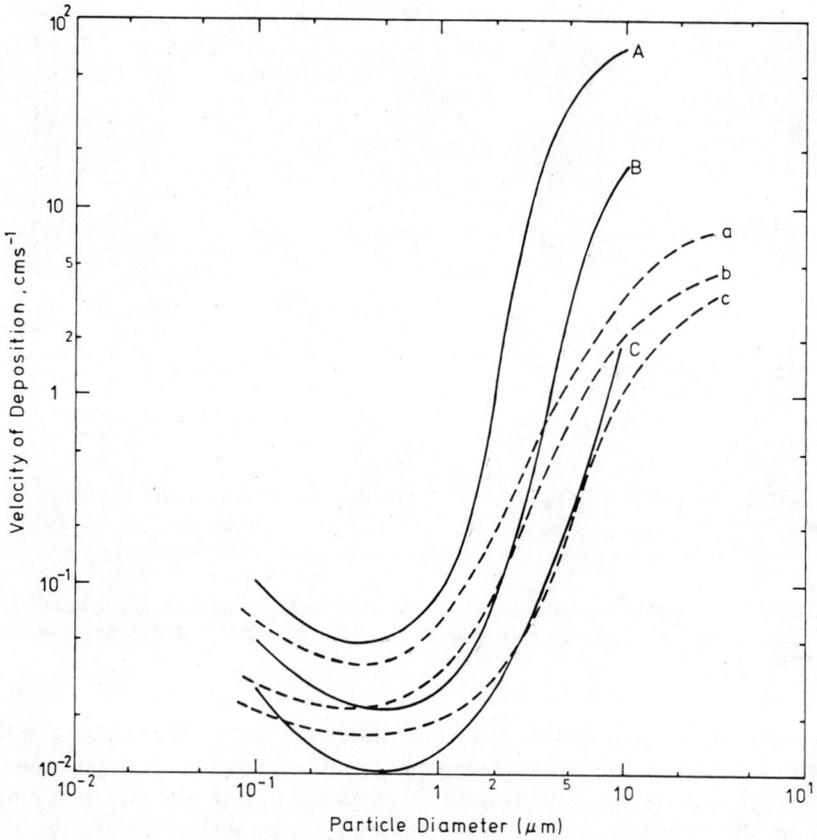

FIGURE 3. Velocity of deposition of particles to canopies. Solid lines: pine forest, theoretical (Belot, 1975), (A) $u(h) = 5$ m/s, (B) 2 m/s, (C) 1 m/s; dashed lines: grass in wind tunnel (Chamberlain, 1966), (a) $u(h) = 4$ m/s, (b) 2 m/s, (c) 1 m/s.

from about 1.5×10^{-2} cm/s at $u(h) = 1$ m/s to about 5×10^{-2} cm/s at $u(h) =$ 5 m/s, irrespective of whether the canopy is a grass sward 6 cm high or a pine forest 15 m high, does not reflect the range of values reported elsewhere in the literature. Measurements of deposition of submicron aerosols in the field appear in many cases to imply higher v_g values than are derived theoretically or by wind tunnel studies. For example, Cawse (1974) measured the dry deposition of a number of trace elements in the atmosphere to filter paper pads (Whatman 541 paper: 25×20 cm) mounted horizontally and sheltered from rain by a sheet of Perspex (1 m^2) placed 12 cm above. Many elements (V, Cr, Zn, As, Se, Sb, Cs, Ce, Pb) showed deposition velocities in the range of 0.1 to 1 cm/s. The soil-derived elements (which may have been present in resuspended soil particles or in fly ash) showed deposition velocities mostly exceeding 1 cm/s.

Clough (1973) attempted to reconcile Cawse's results with his own wind tunnel measurements of deposition to filter paper pads, and could only do so by assuming that the atmospheric aerosols in fact had a wide size distribution of the Junge type. In this case, the small fraction of particles in the 1–10 μm size range contribute most of the deposition.

This explanation, whether correct or not, would not apply to the report by Wesely et al. (1977) that the deposition velocity of particles in the size range 0.05 to 0.1 μm to a moderately rough surface, determined by the eddy correlation method, varies from 0.1 to 1 cm/s in light winds. By contrast, Garland and Cose (1982) failed to find any gradient of concentration of sulfate particles (typically about 0.5-μm diameter) with height above grassland. A recent review of this subject by Slinn (1978) shows that there is clearly still a need for further work in this area, which might include the following:

1. Wind tunnel and field experiments of deposition of radioactively labeled submicron particles.
2. Gradient and eddy correlation measurements.
3. Die-away experiments in closed-cycle wind tunnels or vigorously stirred growth chambers.

3.2 Particles in the 1 to 5 μm Diameter Size Range

In this size range the detailed structure of the surface roughness is all important, particularly the presence or absence of hairs on leaves and stems. Wells and Chamberlain (1967) found that the deposition of 1 and 2 μm diameter tricresylphosphate droplets to extended vertical surfaces was increased 20–100-fold when a filter paper covered surface was used instead of a smooth brass surface. This was despite the fact that the filter paper surface was aerodynamically smooth so that no change in velocity gradient near the surface nor in fraction velocity was induced. The reason for the dependence on microstructure is probably that interception by roughness elements is an important mode of deposition.

Chamberlain (1966) determined the influence of leaf surface texture on deposition of particles by cutting strips from rye grass (*Lolium perenne*), clover (*Trifolium pratense*), and plantain (*Plantago lanceolata*) and fixing them to PVC spills. Particles of various diameters were tagged with radioactivity and introduced into the airstream which flowed over the spills at a velocity sufficient to give a friction velocity of 70 cm/s. The results were expressed as the deposition on the various segments relative to that on the smooth PVC, which had been made sticky by treating with a suitable solution. With 5-μm (polystyrene) and 1-μm (tricresylphosphate) particles, the leaf surfaces and also the filter paper collected more particles than the sticky smooth PVC surface. With clover, the factor was 5.5 for 1-μm particles and 3.25 for 5-μm particles. With ragweed pollen (19-μm diameter) and *Lycopodium* spores (32-μm diameter), it was the other way around; the sticky PVC collected 5 times as many particles as the clover because the particles bounced off the leaf surface.

Leaves and stems vary greatly in their surface roughness. Little (1977) found a 10-fold difference in the deposition of 5-μm polystyrene particles to leaf discs of various tree species, which correlated well with the hairiness of the leaf surface. The leaf discs were of standard size fixed onto an aluminum plate so as to exclude differences due to the leaf size, shape, and orientation.

As the dependence on the microscale surface roughness is so great, it is unlikely that it will ever be possible to predict the deposition of particles in this size range with any accuracy.

3.3 Particles in the 5 to 30 μm Diameter Size Range

For particles in this size range, deposition is by impaction and sedimentation. The efficiency E of impaction of a particle which approaches an obstacle in a wind velocity u is defined as the number of impacts on the obstacle divided by the number which would have passed through the space occupied by it if it had not been there. It is readily seen that E is equal to v_g/u, if v_g is defined relative to the projected area of the obstacle perpendicular to the wind direction. Thus, E is analogous to a deposition coefficient and may be alternatively designated C_p. E is found to depend on the Stokes number, which is equal to S/L, where S is the stopping distance of the particle at the wind speed u and L is the characteristic dimension of the obstacle (i.e., diameter of cylinder). Since S is equal, to a first approximation, to uv_s/g, where v_s is the terminal velocity of the particle and g is the acceleration of gravity, S/L increases with u, and so does E. Thus v_g, which is proportional to uE, is a rapidly increasing function of u.

Deposition by impaction is also a strongly increasing function of the ratio of the size of the particle to the size of the obstacle, since it depends on the particle being projected from the streamlines of flow where these bend around an obstacle. The combined effects of wind speed, size of particle, and size of obstacle are illustrated in Belot's model calculations in Figure 3. For 10-μm particles and $u(h) = 5$ m/s, Belot's calculation gives v_g to a pine forest as 70 cm/s, corresponding to a resistance of 1.4 s/m. The corresponding friction velocity is 1.4 m/s (see Belot et al., 1976, Fig. 5) and the aerodynamic resistance $u(h)/u_*^2$, is 2.5 s/m, so that in fact such a high rate of deposition could not be supported by eddy diffusion from above the canopy. In other words, the canopy would be a perfect sink for the deposition of 10-μm particles.

Chamberlain (1966) had previously concluded that his artificial grass canopy of PVC spills, when made sticky, was a perfect sink for deposition of *Lycopodium* spores in the wind tunnel. This was also the case with a canopy of wetted cereal stems and heads in the wind tunnel (Chamberlain and Chadwick, 1972).

The difficulty is that impacts of particles on obstacles at high wind speed are not always followed by capture. This is so even with obstacles artificially made sticky. Gregory (1951) measured the deposition of *Lycopodium* spores to sticky cylinders in a wind tunnel and found good agreement with the theoretical values expected from the Stokes number except at the highest wind speeds and with the smallest cylinders. When the Stokes number exceeded about 10, the efficiency of impaction decreased with increasing Stokes number instead of increasing as it should theoretically, and this was almost certainly due to bounce-off.

This effect is well known to investigators of the filtration efficiency of fibrous filters. Walkenhorst (1974) showed that capture of particles as small as 3 μm by 0.22-mm fibers was ineffective at air speeds of more than about 2 m/s, and, even when adhesive was used, capture efficiency declined when air speeds exceeded 10 m/s. For larger particles, the critical velocities are smaller. Esmen et al. (1978) impacted particles of uranine on brass and aluminum surfaces coated with a very thin layer of carbon black. After an experiment, particles which had impacted but had bounced off could be detected under the microscope by the "footprint" of the particle in the carbon black. With 15.5-μm particles, there was 50% bounce-off from a brass surface when the incident velocity was 50 m/s. Bounce-off depends, among other things, on the degree to which either the particle or the surface is deformed by the collision, since this is a means whereby the energy of the incident particle may be absorbed.

Loffler and Umhauer (1971) studied the bounce-off of particles from fibers by high-speed spark photography, and this method has been applied by Kyaw Tha Paw U and Reifsnyder (1979) to study the bounce-off of *Lycopodium* spores from leaves of American elm (*Ulmus americana*). When bounce-off occurred, the rebound coefficient (ratio of rebound to incident velocities) was 0.28, and it was deduced that a spore impinging at 1.5 m/s on a leaf at a 45° angle would strike the leaf again after rebound less than

4 mm from the initial point of contact. Thus, for relatively large obstacles, there may still be capture even though rebound has taken place. For small obstacles, for example, conifer needles, for which the impaction efficiencies are theoretically high, a particle which rebounds is unlikely to strike the surface again. Also, for a given wind speed, the velocity of impact and probability of rebound are greater on a small than on a large obstacle as the boundary layer of stagnant air around it is smaller.

Before the development of methods for visualizing rebound, the evidence for it was indirect, namely the additional catch on plant surfaces made wet or sticky compared with untreated surfaces. Chamberlain and Chadwick (1972), in field experiments with radioactively tagged *Lycopodium* spores dispersed over a wheat crop, found about 10 times as many spores deposited on vaselined as on untreated plants. The crop was approaching senescence and the leaves and stems were dry. Also, *Lycopodium* spores are themselves free of any particular tendency to adhere to each other or to surfaces, which is why they are so readily dispersed. The experiments of Chamberlain and Chadwick were therefore possibly rather an extreme case.

Little (Chamberlain and Little, 1980) has recently measured the deposition of monodisperse polystyrene particles to the needles and stems of pine shoots (*Pinus sylvestris*) in the wind tunnel. The methods were the same as those of Little (1977). The velocity of deposition to the untreated shoots and the ratios of deposition to vaselined versus untreated needles and stems, are shown in Table 2. The ratio becomes large for 8.5-μm particles at 2.5 m/s and for 5-μm particles at 5 m/s. A wind speed of 2.5 m/s within a pine canopy would correspond to about 10 m/s at a height of 10 m above the top of the canopy (Oliver, 1971) and this is a very strong wind.

Thus, the conclusion from Little's experiments might be that particles of less than 10-μm diameter will not bounce off vegetation except when the

TABLE 2. Velocity of Deposition (cm/s) of Polystyrene Spheres to Pine Shoots and Ratio of Catches on Vaselined to Untreated Needles and Stems

Wind speed (m/s)	1.5	2.5	2.5	2.5	2.5	5	5
Particle size (μm)	5	2.75	5.0	8.5	2.75	5.0	8.5
Velocity of deposition (cm/s)							
Needles	1.6	1.0	2.6	9.6	5.6	14.8	11.7
Stems	0.6	0.6	1.0	6.2	3.2	16.4	15.5
Whole shoot	1.5	0.9	2.5	9.4	5.1	14.9	12.0
Ratio vaselined/untreated							
Needles	1.3	1.3	1.4	3.2	1.2	4.7	9.5
Stems	0.8	0.6	0.8	0.9	—	1.6	1.9

wind speed is high and the roughness elements very small. For larger particles, the threshold velocity for bounce-off will be lower. Legg and Powell (1979) dispersed *Lycopodium* spores over a crop of ripe barley and compared the concentration profiles downwind with the results of a mathematical theory which took into account diffusion and impaction of the crop. The theoretical trapping efficiency of the awns of the barley was high, but the observed profiles indicated that awns were ineffective in reducing the concentration of spores at the appropriate height, probably because bounce-off occurred.

It is possible that bounce-off of pollen grains from leaves and stems is advantageous since it increases the likelihood of the grain reaching the stigma of the plant. Pathogen spores, which transfer disease to leaf tissue, are often moist or sticky. Carter (1965) placed apricot wood infected with the fungus *Eutypa armeniacae* in a wind tunnel and measured the impaction of the ascospore octads on fresh shoots downwind. The efficiency of impaction on twigs corresponded to theoretical predictions, indicating that bounce-off was unimportant. The presence of hairs on leaves and stems is likely to reduce the incidence of bounce-off by cushioning the impact. Tauber (1967) found 40,000 pollen grains on the hairy twig ($\sim$15 cm long by 2 mm in diameter) of a willow shoot but only 200 on a similar but smooth birch twig from a forest. Some of this difference may have been due to more rapid wash-off from the smooth twig. White and Turner (1970) found a contrasting effect when they sprayed shoots of various tree species with droplets of salt solution. Hazel, with hairs on both leaf surfaces, was least efficient in collecting the salt. Ash and birch, with no hairs, were most efficient. This apparent anomaly needs further investigation. Possibly hairs reduce the impaction efficiency of droplets, by increasing the boundary layer thickness, but increase the collection efficiency for dry particles by reducing bounce-off.

4. REMOVAL OF DEPOSITED MATERIAL FROM PLANTS

4.1 Mechanism of Removal

If deposition of a pollutant continues at a constant rate, as measured in units of mass (or radioactivity) per unit ground area and per unit time, the concentration in plant tissues will generally increase initially pro rata and then level off when a quasi-equilibrium is reached between deposition and removal from the plant tissues. In some systems, equilibrium may not be reached in the lifetime of the plant.

Removal from the above-ground parts of the plant may occur through:

1. Translocation to roots.
2. Revolatization and emission in gas phase.

3. Washing away as a solution or suspension by rain.
4. Loss of plant parts, particularly cuticle, by various mechanisms.
5. Die-back of leaves.

Some translocation of ^{34}S, absorbed from the atmosphere as SO_2, from the leaves to the roots of soybean (*Glycine max*. var. Bilosa) has been observed (Garsed and Read, 1974), but it seems that this route of loss is generally small. Aneja et al. (1979) have observed the release of CS_2 and COS from areas of marshland with *Spartina alterniflora*, and have experiments underway to test for CS_2 emissions from land vegetation.

The concentrations of the reduced gaseous sulfur species over land are small compared with the concentration of SO_2, and this suggests that reemission is small. The deposition velocities which can be deduced from chamber or field fumigation experiments lasting many weeks are similar to those found in short-term measurements using ^{35}S or the gradient method (Chamberlain, 1980; Sprugel and Miller, 1979). It appears that the rate of loss of sulfur from vegetation may be determined partly by leaching and partly by die-back of foliage. It would be feasible to determine the field loss for SO_2 (rate of loss of sulfur from a crop by all routes) by sequential analyses following fumigation with $^{35}SO_2$, but measurements do not seem to have been reported. Chadwick (1977), by sampling air in a wind tunnel downwind from a crop of rye grass which had been fumigated with $H_2{}^{35}S$, showed a field loss half-life of 20 days. This was similar to the rate of field loss of ^{89}Sr and ^{51}Cr from a grass at the same location (Chadwick and Chamberlain, 1970) and is ascribed to weathering and die-back of leaves.

The loss of radioactive isotopes from foliage following aerial contamination has been studied extensively (references in Scott, 1965, 1966, and Chamberlain, 1970). Most of the fission products of interest are bivalent and multivalent cations which can move upwards in shoots in the transpiration stream but, having entered leaves, are redistributed little if at all. In experiments in which carrier-free solutions were sprayed onto the foliage of young cabbage plants, the subsequent rate of loss of Sr, Zr, Ru, I, Cs, and Ce were found to be similar. This suggested that the rate of loss was determined by the life-cycle of the plant rather than by the chemistry of the nuclide. Moorby and Squire (1963) showed that loss of radioactivity from the leaves of plants occurred in the absence of rain, and that the rate of loss appeared to be greatest when the plants were growing rapidly. They thought this might be related to the loss of waxy cuticle which is continuously produced and removed during leaf growth.

Chadwick and Chamberlain (1970) found that the rate of field loss of ^{89}Sr from a canopy of fixed grasses and weeds in the field was greater in summer than in winter, even though the results were calculated relative to unit ground area and not to unit weight of crop so that dilution by new growth would not affect the results. Chamberlain (1970), in a review of other literature, also found evidence of longer retention times in winter than in summer and in slow growing as against rapidly growing crops. This lends support to Moorby and Squire's suggestion that the shedding of cuticle during growth is an important mechanism of removal of foliar contaminants. On the microscale, this might also help to explain the very nonuniform distribution of particulate deposit on leaves which has been observed. However, Crump and Barlow (1982) consider that die-back of leaves is more important than shedding of cuticle.

4.2 Normalized Specific Concentration

It often happens that the rate of fallout of a pollutant, whether by wet or dry deposition, is monitored by use of some form of deposit gauge, giving deposition rates in micrograms (or microcuries) per square meter per day. Also periodically vegetation exposed to the fallout is collected and the concentration of the pollutant in it is measured in micrograms (or microcuries) per kilogram. The ratio of these quantities was termed by Chamberlain (1970) the normalized specific concentration (NSC) (or activity), so that

$$(\text{NSC}) = \frac{\text{Amount per kg dry matter in foliage}}{\text{Amount deposited per m}^2 \text{ of ground per day}}$$

The use of the NSC presupposes that foliar uptake is the dominant mode of contamination.

Chamberlain (1974) derived values of the NSC by making use of published data on the proportion of the deposited activity initially retained by the foliage and the field loss coefficient (units per day). Miller (1979) has shown that the parameters derived for assessing the proportion initially deposited are supported by more recent published data for grasses, although not necessarily for vegetation of other types. It seems that the proportion depends on the herbage density and character but not very strongly on the physical and chemical nature of the fallout material.

Table 3 shows the NSC values derived by Chamberlain from the data on interception and rates of field loss. As the length of the growth period increases, the NSC tends towards a constant value as field loss balances continued deposition. Chamberlain (1970) reviewed the available measurements of the NSC for both radioactive and stable nuclides. For grassland in good growing conditions, measurements of the NSC for several pollutants were found to lie in the range 30–60 m^2 days/kg. In poor growing conditions in winter considerably higher NSC values were found. Measurements of fission products in arctic vegetation (birch, willow, and cowberry) gave NSC values in the range of 100–700 m^2 days/kg.

It is suggested that the expression of results in terms of the NSC provides a useful comparison which may help in elucidating the mechanisms of retention and field loss of pollutants. Sometimes the reason for particularly

TABLE 3. Normalized Specific Concentration (NSC) of Herbage Exposed to Constant Daily Fallout[a]

Period of growth (days)	30	50	80
NSC (m^2 days/kg)			
$\lambda = 0.054$/day	27	35	41
$\lambda = 0.037$/day	32	44	55

[a]The values of λ (field loss coefficient) are those found in the field experiments of Milbourn and Taylor (1956) and Chadwick and Chamberlain (1970). The value of μ has been taken as 3 m^2/kg, where μ is the proportion p of deposited material retained initially by the herbage and is related to the herbage density w (kilograms dry matter per square meter) by $1 - p = \exp(-\mu w)$.

high or low NSC values is obvious; at other times it is not. This can be illustrated by referring to NSC values for the accumulation of lead fallout by grass. The results of Roberts et al. (1974) were obtained by transferring turf grown in clean conditions to an area where there was a fallout of lead from a refinery at a rate of 8.3 mg/m day. The lead content of the grass increased and leveled off at about 270 mg/kg dry weight, giving a NSC value of 32 m^2 days/kg. Tjell et al. (1979) grew Italian rye grass (*Lolium multiflorum* Lam.) in plots and measured the concentration of lead emitted from distant vehicular traffic and deposited on the grass. The rate of fallout of the lead was 0.03 mg/m day, much less than in Roberts' experiment. The lead content of the grass increased to 7 ppm in September at the end of the growing season and to 14 ppm in the spring of the following year. The corresponding NSC values are 230 and 470 m^2 days/kg. It can only be supposed that the vehicle lead was retained more strongly on account of its smaller particle size. Tjell et al. showed, by incorporating ^{210}Pb in the soil of the plots, that root uptake of lead by the grass was not important.

Retention of fallout materials by nonvascular plants, such as lichens and bryophytes, is particularly strong. These plants grow very slowly and they have no cuticle to shed. Radionuclides such as ^{137}Cs accumulate in lichens in the Arctic, and this gives high concentrations in the flesh of grazing animals, particularly reindeer. Andersen et al. (1978) have reported concentrations of trace metals (Cu, Pb, Zn, V, Fe) in lichens and bryophytes from the Copenhagen district of Denmark and have also given the rates of fallout in deposit gauges. The NSC values for lichens range from 300 to 1200 and for bryophytes from 400 to 6000 m^2 days/kg.

5. SUMMARY OF PRINCIPAL UNKNOWN FACTORS

The following are among the main areas of uncertainty in the deposition of gases and particles to vegetation:

1. Nature of the reaction between O_3 and other oxidant gases and vegetation. Possible synergistic effects on deposition of concurrent levels of SO_2.
2. Extent to which translocation and reemission as gaseous reduced sulfur species affects accumulation of sulfur in plants.
3. Effect of damage to cuticles on uptake of SO_2 and O_3.
4. Effect of leaf hairs and other microroughness elements on the capture and retention of particles by leaves and stems.
5. Importance of bounce-off and factors influencing it, such as presence of moisture, surface characteristics, and resilience of particles and plant surfaces.
6. Mechanisms of removal of particles from leaves by rain and wind and by loss of cuticle. Relation between deposition rates and equilibrium foliage concentrations.

REFERENCES

Andersen, A., M. F. Hovman, and I. Johnson. Atmospheric heavy metal deposition in the Copenhagen area. *Environ. Pollut.* **17**, 133–151 (1978).

Aneja, V. P., J. H. Overton, C. T. Cupitt, J. L. Durham, and W. E. Wilson. Carbon disulphide and carbonyl sulphide from biogenic sources and their contributions to the global sulphur cycle. *Nature* **282**, 493–496 (1979).

Belot, Y. Etude de la captation des polluants atmospheriques par les vegetaux. C.E.A. Fontenay aux Roses, France (1975).

Belot, Y., A. Baille, and J-L. Delmas. Modele numerique de dispersion des polluants atmospheriques en presence de couverts vegetaux. *Atmos. Environ.* **10**, 89–98 (1976).

Black, V. J., and M. H. Unsworth. Resistance analysis of sulphur dioxide fluxes to *Vica faba*. *Nature* **282**, 68–69 (1979).

Brice, K. A., A. E. J. Eggleton, and S. A. Penkett. An important ground surface sink for atmospheric nitrous oxide. *Nature* **268**, 127–129 (1977).

Carter, M. V. Ascospore deposition in *Eutypa armeniacae*. *Austr. J. Agric. Res.* **16**, 825–836 (1965).

Cawse, P. A. A survey of atmospheric trace elements in the UK (1972–3). Atomic Energy Research Establishment report AERE R 7669 (HMSO) (1974).

Chadwick, R. C. Uptake of H_2S and COS by vegetation. Atomic Energy Research Establishment report AERE M 2898 (HMSO) (1977).

Chadwick, R. C., and A. C. Chamberlain. Field loss of radionuclides from grass. *Atmos. Environ.* **4**, 51–56 (1970).

Chamberlain, A. C. Transport of *Lycopodium* spores and other small particles to rough surfaces. *Proc. Roy. Soc. A.* **296**, 45–70 (1966).

Chamberlain, A. C. Interception and retention of radioactive aerosols by vegetation. *Atmos. Environ.* **4**, 57–78 (1970).

Chamberlain, A. C. Mass transfer to bean leaves. *Boundary Layer Met.* **6**, 477–486 (1974).

Chamberlain, A. C. Dry deposition of sulfur dioxide. In D. S. Schriner, C. R. Richmond, and S. E. Lindberg (eds.) *Atmospheric Sulfur Depositions: Environmental Impact and Health Effects*, pp. 185–197, Ann Arbor Science, Ann Arbor, MI (1980).

Chamberlain, A. C., and R. C. Chadwick. Deposition of spores and other particles on vegetation and soil. *Ann. Appl. Bio.* **71**, 141–158 (1972).

Chamberlain, A. C., and P. Little. Transport and capture of particles by vegetation. In J. Grace, E. D. Ford and P. G. Jarvis (eds.) *Plants and Their Atmospheric Environment.* Blackwell, Oxford (1980).

Clough, W. S. Transport of particles to surfaces. *J. Aerosol Sci.* **4**, 227–234 (1973).

Crump, D. R., and P. J. Barlow. Factors controlling the lead content of a pasture grass. *Environ. Pollut.* **3**, 181–192 (1982).

Denmead, O. T. Temperature cereals. In J. L. Monteith (ed.) *Vegetation and the Atmosphere.* Academic, New York (1976).

Esmen, N. A., P. Zeigler, and R. Whitfield. The adhesion of particles upon impaction. *J. Aerosol Sci.* **9**, 547–556 (1978).

Fowler, D., and M. H. Unsworth. Turbulent transfer of sulphur dioxide to a wheat crop. *Quart. J. Roy. Met. Soc.* **105**, 767–783 (1979).

Galbally, I. E., and C. R. Roy. Destruction of ozone at the earth's surface. *Quart. J. Roy. Met. Soc.* **106**, 599–620 (1980).

Garland, J. A. Dry and wet removal of sulphur from the atmosphere. *Atmos. Environ.* **12**, 349–362 (1978).

Garland, J. A. Dry deposition of gaseous pollutants. WMO Symposium No. 538, pp. 95–103, WMO, Geneva, Switzerland (1979).

Garland, J. A., and J. R. Branson. The deposition of sulphur dioxide to pine forest assessed by a radioactive tracer method. *Tellus* **29**, 445–454 (1977).

Garland, J. A., and L. C. Cose. Deposition of small particles to grass. *Atmos. Environ.* **16**, 2699–2702 (1982).

Garland, J. A., and R. G. Derwent. Destruction at the ground and the diurnal cycle of concentration of ozone and other gases. *Quart. J. Roy. Met. Soc.* **105**, 169–183 (1979).

Garland, J. A., and S. A. Penkett. Absorption of peroxyacetylnitrate and ozone by natural surfaces. *Atmos. Environ.* **10**, 1127–1131 (1976).

Garsed, S. G., and D. J. Read. The uptake and translocation of $^{35}SO_2$ in the soybean *Glycine max* var. Bilosa. *New Phytol.* **73**, 229–307 (1974).

Grace, J., and J. Wilson. The boundary layer over a *Populus* leaf. *J. Exptl. Bot.* **27**, 231–241 (1976).

Gregory, P. H. Deposition of airborne *Lycopodium* spores on cylinders. *Ann. Appl. Biol.* **38**, 357–376 (1951).

Hill, A. C., and E. M. Chamberlain. Removal of water soluble gases by vegetation in atmosphere–surface exchange of particulate and gaseous pollutants. ERDA Symposium Series 38, CONF 740921, NTIS, Springfield, VA (1976).

Judeikis, H. S., and A. G. Wren. Laboratory measurements of NO and NO_2 deposition onto soil and cement surfaces. *Atmos. Environ.* **12**, 2315–2319 (1978).

Kelly, T. J., D. S. Stedman, and G. L. Kok. Measurements of H_2O_2 and HNO_3 in rural air. *Geophys. Res. Letts.* **6**, 375–378 (1979).

Kyaw Tha Paw U, and W. E. Reifsnyder. The physics of spore and pollen rebound from vegetation surfaces. Am. Met. Soc., 14 Conf. Agricultural and Forest Met., Minneapolis, MN, pp. 244–246 (1979).

Legg, B. J., and F. A. Powell. Spore dispersal in a barley crop: a mathematical model. *Agricultural Met.* **20**, 47–67 (1979).

Little, P. Deposition of 2.75, 5.0 and 8.5 μm particles to plant and soil surfaces. *Environ. Pollut.* **12**, 293–305 (1977).

Little, P., and R. D. Wiffen. Emission and deposition of petrol engine exhaust. I. Deposition of exhaust Pb to plants and soil surfaces. *Atmos. Environ.* **11**, 437–447 (1977).

Loffler, F., and H. Umhauer. An optical method for the determination of particle separation on filter fibres. *Staub* (in English) **31**, 9–14 (1971).

McMahon, T. A., and P. J. Denison. Empirical atmospheric deposition parameters—A survey. *Atmos. Environ.* **13**, 571–585 (1979).

Matthais, A. D., A. M. Blackmer, and J. M. Bremner. Diurnal variability in the concentration of nitrous oxide in surface air. *Geophys. Res. Letts.* **6**, 441–443 (1979).

Milbourn, G. M., and R. Taylor. The contamination of grassland by radioactive strontium. (I) Initial retention and loss. *Rad. Bot.* **5**, 337–347 (1956).

Miller, C. W. Validation of a model to predict aerosol interception by vegetation. IAEA-SM-237/53 (1979).

Monteith, J. L. (ed.). *Vegetation and the Atmosphere*, Vols. I and II. Academic, New York (1975).

Moorby, J., and H. M. Squire. The loss of radioactive isotopes from the leaves of plants in dry condition. *Rad. Biol.* **3**, 163–167 (1963).

Oke, T. R. Advectively-assisted evapotranspiration from irrigated urban vegetation. *Boundary Layer Met.* **17**, 167–173 (1979).

Oliver, H. R. Wind profiles in and above a forest canopy. *Quart. J. Roy. Met. Soc.* **97**, 548–553 (1971).

Omasa, K., F. Abo, T. Natouri, and T. Totsuka. Analysis of air pollution sorption by plants. (3) Sorption under fumigation with NO_2, O_3 or NO_2+O_3; studies on the effects of air pollutants on plants and mechanisms of phytotoxicity. Res. Rep. Natl. Inst. Environ. Stud. No. 11, Ibaraki, Japan (1980).

Pierotti, D., R. A. Rasmussen, and R. Chatfield. Continuous measurements of nitrous oxide in the troposphere. *Nature* **274**, 574–576 (1978).

Roberts, T. M., J. J. Paciga, R. E. Jervis, J. C. Van Loon, T. C. Hutchinson, A. Chattopadhyay, and F. Hahn. Lead contamination around two secondary smelters in downtown Toronto. Institute for Environmental Studies, University of Toronto (1974).

Scott, R. R. Interception and retention of airborne material on plants. *Health Physics* **11**, 1305–1315 (1965).

Scott, R. R. (ed.) *Radioactivity and the Human Diet*. Pergamon, New York (1966).

Slinn, W. G. N. Parameterizations for resuspension and for wet and dry deposition of particles and gases for use in radiation dose calculations. *Nuclear Safety* **19**, 205–219 (1978).

Sprugel, D. G., and J. E. Miller. A field estimate of SO_2 deposition velocities to rapidly growing soybeans. *Water Air Soil Pollut.* **12**, 233–236 (1979).

Tauber, H. Investigations of the mode of pollen transfer in forested areas. *Rev. Palaeobotan. Palyon.* **3**, 277–286 (1967).

Tjell, J. C., M. F. Hormand, and H. Mosbaek. Atmospheric lead pollution of grass grown in a background area of Denmark. *Nature* **280**, 425–426 (1979).

Walkenhorst, W. Investigations on the degree of adhesion of dust particles. *Staub* **34**, 149–153 (1974).

Wells, A. C., and A. C. Chamberlain. Transport of small particles to vertical surfaces. *Br. J. Appl. Phys.* **18**, 1793–1799 (1967).

Wesely, M. L., B. B. Hicks, W. P. Dunnerick, S. Frissela, and R. R. Husar. An eddy correlation measurement of particulate deposition in the atmosphere. *Atmos. Environ.* **11**, 561–563 (1977).

White, E. J., and F. Turner. A method of estimating income of nutrients in catch of airborne particles by a woodland canopy. *J. Appl. Ecol.* **7**, 441–461 (1970).

Precipitation Chemistry: Its Behavior and Its Calculation*

JEREMY M. HALES
Battelle Pacific Northwest Laboratories, Richland, Washington

*Precipitation-chemistry measurements and modeling techniques have progressed markedly since this paper was presented at the Banff Conference in 1980. For more current information the reader is referred to a forthcoming publication entitled,The Mathematical Characterization of Precipitation Scavenging and Precipitation Chemistry, to be published in Volume II of the Handbook of Environmental Chemistry, Springer.

1. INTRODUCTION

As our understanding of the atmospheric sciences has evolved it has been marked increasingly by the compelling need to develop generalized and simple, yet reliable methods for assessing the impacts of man-made change. Development of such procedures has always been characterized by a tradeoff between simplicity on the one hand and reliability on the other; and although limited accuracy and overextended application have continued to pose problems, some rather remarkable successes have been achieved. One has only to consider the extended application of the Gaussian plume model, as presented in Turner's *Workbook of Dispersion Estimates* (Turner, 1973), to illustrate this point.

Similar successes in the field of precipitation chemistry have been comparatively limited, owing to the complexity of the scavenging process. Some notably elegant inroads have been established (e.g., Chamberlain, 1953), but these have focused on limited *subsets* of the overall scavenging problem, and cannot be extended for generalized, reliable usage. In aggregate, however, these assorted techniques compose a useful means of attacking the extended scavenging problem; and while it is probably unreasonable to ever expect a straight-forward Turner's *Workbook* type of document to emerge for scavenging calculations, one can at least look to a composite set of techniques which is generally useful on an applied basis.

The purposes of this paper are two-fold. The first of these is to present a rational basis for examining this aggregate set of scavenging-calculation techniques, and for guiding the reader in a course toward choosing the most appropriate technique for a particular application. The second purpose of this paper is to present a somewhat brief survey of our current understanding of scavenging and precipitation chemistry. Both objectives will be implemented by a flowchart approach, which attempts to draw the various facets of scavenging calculations together and present a generalized approach to the problem in total.

The mathematical level of this paper is restricted to the presentation of the equations necessary to provide the reader with a basic appreciation of the fundamental concepts involved. References to more detailed mathematical treatments are cited at appropriate juncture points, for the reader interested in more detailed pursuit. (The chapter by Slinn (1984) in the DOE Publication *Atmospheric Science and Power Production* is recommended as a key reference in this regard.) Within this format it is hoped that this article will find extensive usage as a first reference, and will allow the user to scope particular problems in a valid manner, which will direct the user rapidly to the most expedient solution technique.

2. MATERIAL BALANCES: SOURCES OF SPATIAL AND TEMPORAL VARIABILITY

Since the preponderance of scavenging calculations are based on one sort of material balance or another, it is appropriate at this point to examine briefly some qualitative aspects of the general material balance of pollution in the atmosphere. This is shown schematically in Figure 1, which depicts a given pollutant as it is emitted from a source and ultimately delivered to a receptor, via the atmosphere. Important points to note from this diagram are the competing effects of wet and dry deposition, and the potential for reversible *cycling* of pollutants through various combinations of steps before ultimate delivery to the surface. It should be noted also that material balances can be formulated around various individual steps, substeps, and combinations of steps in Figure 1; and in assessing a particular type of scavenging calculation it is important to ascertain just what portion of this scheme has been covered.

Mathematical characterization of the processes in Figure 1 can be accomplished by defining some chosen volume of atmosphere, and then formally summing the effects of all of these processes over this space.

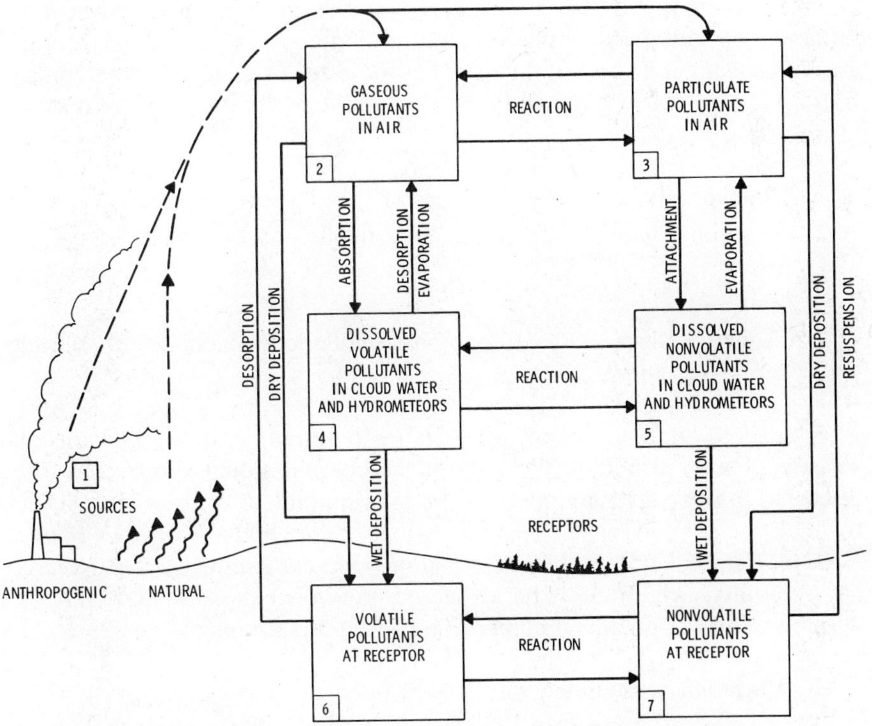

FIGURE 1. Schematic description of material balance for pollution in the atmosphere.

Depending on the volume element chosen for this summation, the resulting characterization can be either *integral* or *differential* in form. Differential material balances are normally based on small volume increments and yield differential equations, which must be integrated subsequently to produce the desired computations of concentrations and removal rates. Integral balances typically are performed over much larger regions, and result either in integral equations or else algebraic forms derived from some sort of implied integration process. Quite often material balances are mixed in nature, and yield correspondingly mixed mathematical forms.

Examples of *integral* approaches are simple box models and storm-scale material balances. One particularly important result of the *differential* material balance can be expressed by the forms (see Bird et al., 1960; Hales, 1972; or Slinn, 1984, for a more detailed discussion)

$$\frac{\partial c_{Ay}}{\partial t} = -\nabla \cdot c_{Ay}\tilde{v}_{Ay} - w_A + r_{Ay} \quad \text{(gaseous phase)} \tag{1}$$

and

$$\frac{\partial c_{Ax}}{\partial t} = -\nabla \cdot c_{Ax}\tilde{v}_{Ax} + w_A + r_{Ax} \quad \text{(aqueous phase)} \tag{2}$$

which describe the net input of some arbitrary pollutant A over a small volume increment of the atmosphere, as it is interchanged between the precipitation and gaseous-phase medium (denoted here by the subscripts x and y respectively). In Eqs. (1) and (2) the rates of change in the concentrations of gaseous-phase and aqueous-phase pollutant are expressed in terms of:

1. Transport across the boundaries of the element (divergence terms).
2. Transport between gaseous and aqueous phases within the element (w_A).

   and
3. Aqueous-phase and gaseous-phase chemical reaction with the element (r_{Ax} and r_{Ay}).

$\tilde{v}_{Ax}$ and $\tilde{v}_{Ay}$ denote velocity vectors for pollutant A in the aqueous and gaseous phases, respectively. Many of the computational approaches to be discussed in this paper are based on various simplified forms of Eqs. (1) and (2).

From Figure 1 and the above equations one can identify several sources of variability, which may be expected to induce spatial and temporal differences in the chemical composition of precipitation:

1. Variability associated with source fluctuation and configuration.
2. Variability associated with normal atmospheric transport and mixing processes.

3. Variability induced by storm dynamics.
4. Variability caused by atmospheric transformation processes prior to the precipitation event.
5. Variability associated with microphysical cloud processes; physical attachment and aqueous-phase transformation.
6. Variability caused by pollutant depletion via wet- and dry-removal processes.

These features are difficult to isolate, and their relative effects will vary, depending on the averaging times associated with the precipitation-chemistry measurements at hand. In performing and assessing scavenging calculations, however, it is important that one keep these factors in mind, and attempt to define the spatial and temporal averaging times appropriate to one's own particular requirements.

So little is known presently with regard to spatial and temporal variability in precipitation chemistry that it is difficult to draw any really meaningful or helpful conclusions regarding its behavior. Some limited insight can be obtained, however, by considering some typical case examples. Figure 2, for instance, shows the results of a *sequential* sampling of rain from a particular precipitation event measured at the Brookhaven National Laboratory (Raynor and Hayes, 1978). This is a relatively remote site located on upper Long Island; and while it reflects the presence of the east-coast megalopolis, it is considered to be a reasonably valid representation of *regional* precipitation chemistry. Key features to note from this figure are the pronounced variability of concentration during storm passage, and the obvious continua of the time–concentration curves.

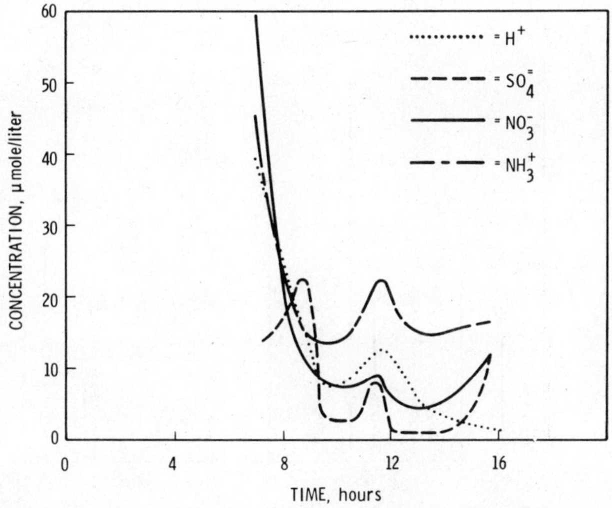

FIGURE 2. Concentrations of selected chemical species in rain during a rain event at the Brookhaven MAP3S site.

Figure 3 (data from MAP3S sampling site at State College, Pennsylvania, MAP3S, 1980) is a typical result of averaging precipitation-borne pollutant concentrations over entire precipitation periods, and plotting several events in sequence. Here discrete plotting is necessary, owing to the episodic nature of precipitation. The fact that large fluctuations exist in spite of the longer averaging times should not be surprising, in view of the introduction of additional sources of variability from the candidates itemized above.

Figure 4 pertains to an expanded data set that originated from this same sampling site, but now has been averaged over one-month periods (computed as

$$\sum_{1}^{N} C_e R_e \Big/ \sum_{1}^{N} R_e$$

where C_e and R_e are the concentrations and rainfall amounts associated with a particular event, and N is the number of events occurring during a particular month). At this point the averaging process appears to have smoothed the concentration excursions somewhat, and suggests a seasonal cycling of species such as SO_4^{2-}, H^+, and NH_4^+. This apparent smoothing should be observed with some caution, however, in view of pronounced excursions typically observed from longer data sets. This is illustrated by some of the long-term European Air Chemistry Network data, as presented in Figure 5. From this it can be recognized that one must exercise appropriate caution in interpreting limited data sets such as given in Figure 4, *especially* for trend analysis.

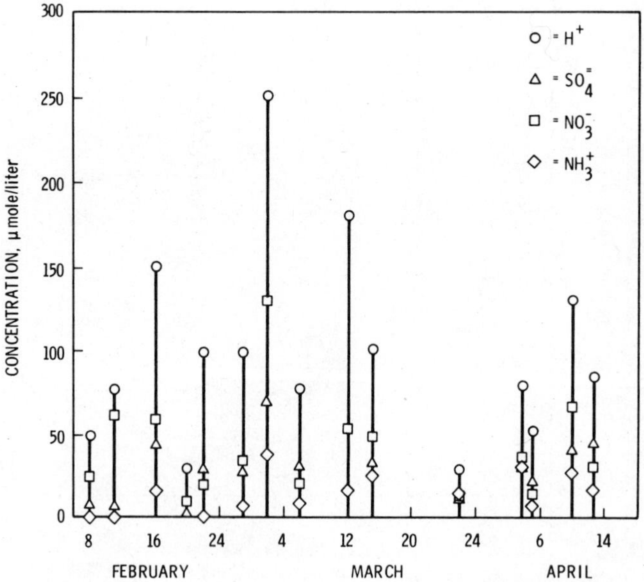

FIGURE 3. Concentrations of selected chemical species in precipitation for several precipitation events. Event-averaged concentrations from Penn State MAP3S site.

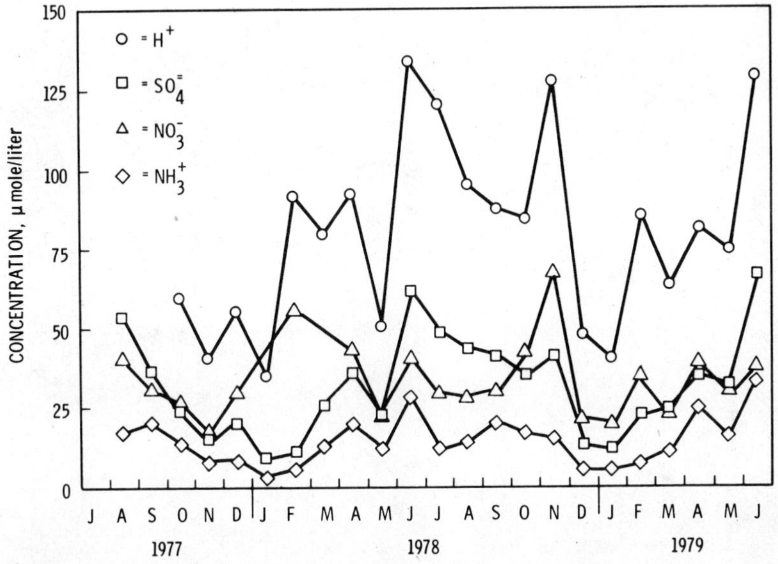

FIGURE 4. Concentrations of selected chemical species in precipitation. Monthly averaged concentrations from Penn State MAP3S site.

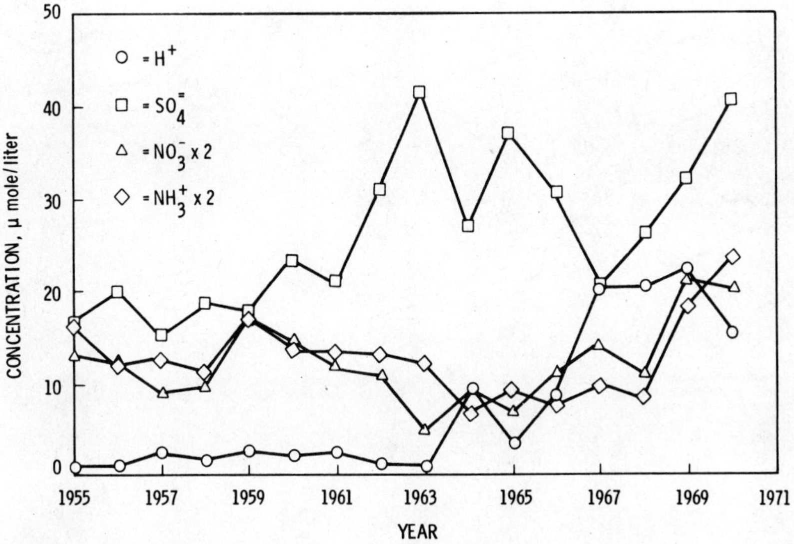

FIGURE 5. Annually averaged precipitation chemistry data: EACN Arjeplög site, north-central Sweden.

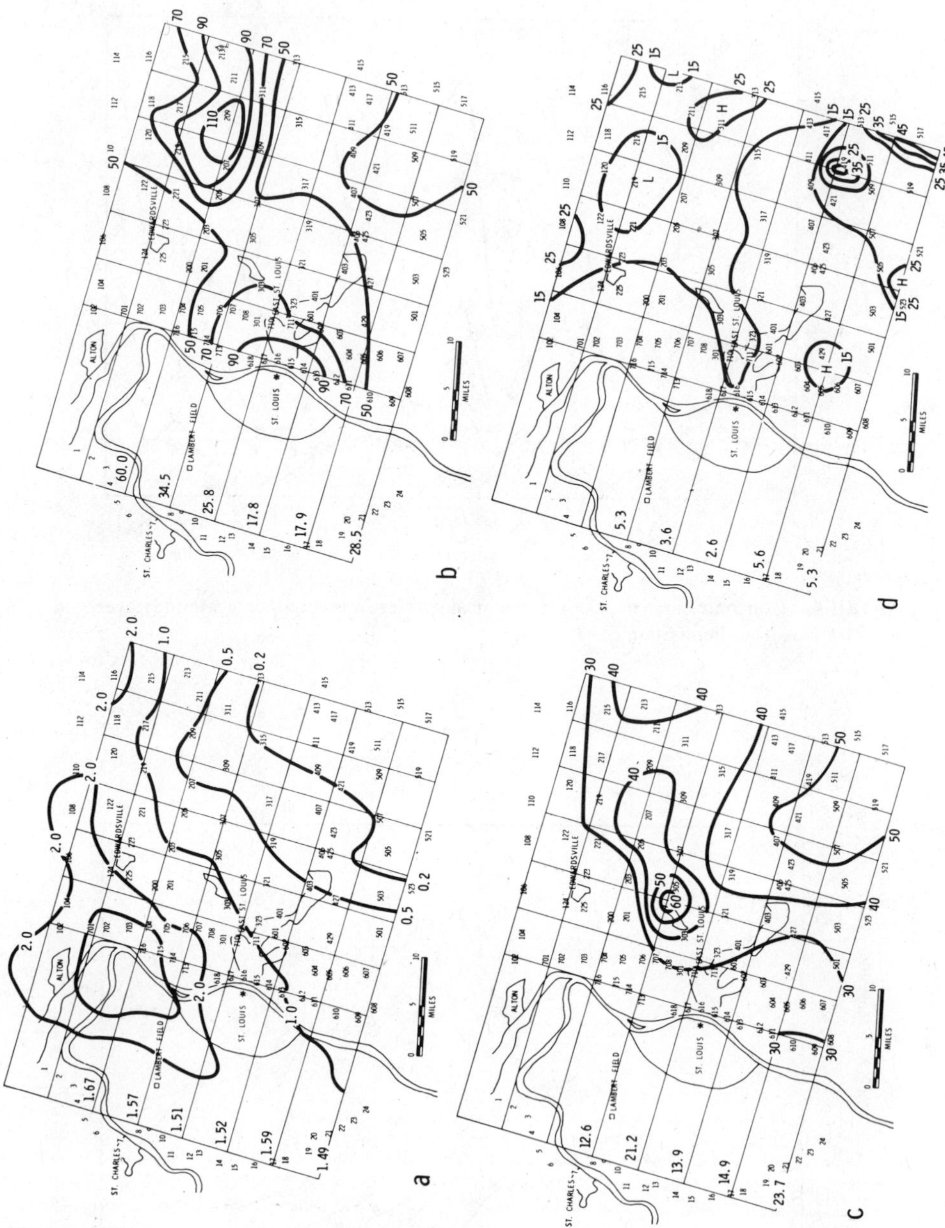
a
b
c
d
ALTON
ST. CHARLES
LAMBERT FIELD
EDWARDSVILLE
ST. LOUIS
EAST ST. LOUIS
MILES

In addressing *spatial* variability, it should be noted that point-to-point differences in rainborne pollutant concentrations will be strongly related to *temporal* variability in most cases. Although spatial variability has been considered carefully by Granat and his co-workers in siting studies (Granat, 1978), and several statistical interpretations of variability over regional networks have been presented (Pack and Pack, 1980; Munn and Rodhe, 1971), this whole question remains at a highly unresolved state. Figure 6, which is a concentration and rainfall map for a convective event in the vicinity of St. Louis, Missouri, indicates the type of complexity that can be observed in spatial structure.

3. GENERAL SCAVENGING CALCULATIONS: FLOWCHART APPROACH

The diversity of methods that have been applied in precipitation scavenging calculations presents a composite set of alternate pathways that can be rather bewildering, even to those who are relatively familiar with the field. A serious problem associated with this situation is that it is not difficult at all to choose a particular technique of calculation within this set, which appears superficially to be a reasonable approach but in reality is totally inappropriate. Errors of several orders of magnitude (and even in sign) can be (and have been) experienced because of such pitfalls.

One useful approach to minimizing these dangers and to analyzing the composite of possible scavenging calculations is to prepare a decision tree, which, by presenting series of questions about the specific problem at hand, allows one to proceed in a logical fashion to determine the most expedient computational approach. Such a decision tree is presented in Figure 7. The remainder of this paper is addressed to an examination of various branches of this tree, in a manner designed to guide the reader rapidly to appropriate modeling techniques and extended literature sources.

Several features of Figure 7 should be noted. First, it should be emphasized that this flow diagram is certainly not the *only* one that could be presented for this purpose. Its form depends to some extent on the relationships existing in the atmospheric material balance shown in Figure 1, but is highly dependent on the existing state of our scientific understanding as well. Figure 7 discriminates between scavenging processes that take place in the condensing region of a cloud and those that occur in precipitation falling through clear air. This is somewhat artificial in a scientific sense, because common physical mechanisms are operative in both types of

←

FIGURE 6. Measured rain concentrations and amounts in the St. Louis region for convective storm of 23 July 1973: (*a*) rain amount (cm); (*b*) SO_4^{2-} concentration (μmole/L); (*c*) NO_3^- concentration (μmole/L); (*d*) NH_4^+ concentration (μmole/L). Data from numbered network stations interpolated to indicated grid centers for isopleth analysis. (From Hales and Dana, 1979.)

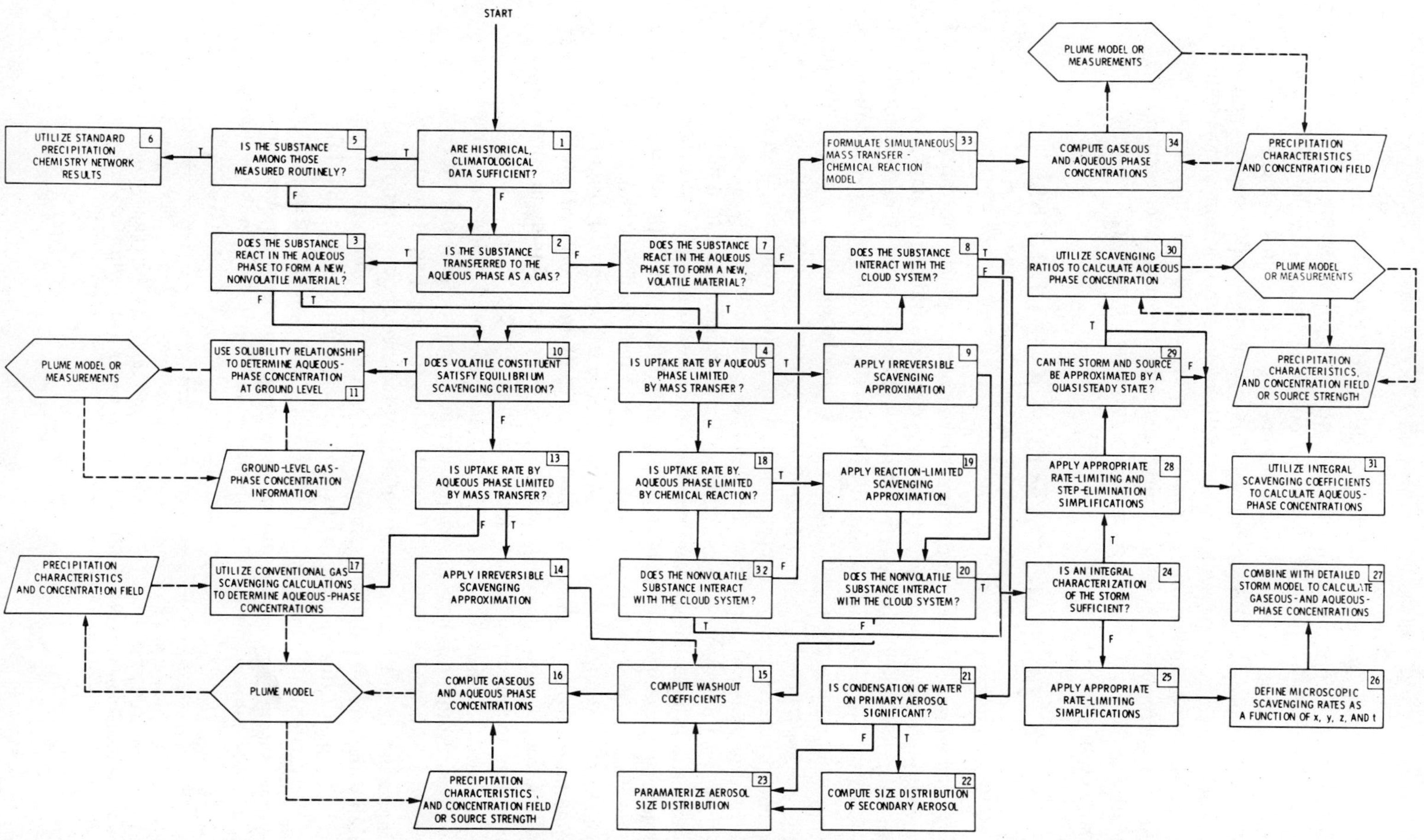

FIGURE 7. Flow chart for scavenging calculations.

systems; but it presents a rather convenient format in view of most traditional approaches to scavenging assessment. Finally it should be noted that, while Figure 7 is addressed primarily to mathematical material-balance applications, it provides a route for climatologically based predictions as well. Although this latter class of predictions must be utilized with careful consideration of the variability described in Section 2, it provides an expedient route for many types of evaluative applications, and should be considered seriously as an alternative candidate to the more modeling-oriented approaches.

3.1 Pathway 1–5–6: Use of Climatological Precipitation-Chemistry Data

There are many circumstances where one is interested in obtaining reasonable estimates for actual values of wet deposition or concentration

TABLE 1. Selected Sources of Regional Precipitation-Chemistry Data

Network	Location	Sample Period	Contact
EPA/NOAA/WMO	United States	Monthly	John Miller NOAA/ARL 8060 13th Street Silver Spring, MD 20910
MAP3S	Eastern United States	Event	Terry Dana Battelle-Northwest P.O. Box 999 Richland, WA 99352
NADP	United States	Weekly	James Gibson Natural Resources Ecology Lab Colorado State University Fort Collins, CO 80523
CANSAP	Canada	Monthly	Douglas Whelpdale Atmospheric Environment Service 4905 Dufferin Street Downsview, Ontario M3H 5T4 Canada
European Air Chemistry Network	Western Europe	Monthly	Lennart Granat Meteorological Institute Stockholm University Stockholm, Sweden
LRTAP[a]	Western Europe	Event	Director of Information, OECD 2, rue Andre-Pascal 75775 Paris Cedex 16 France

[a]Operational 1972 through 1975.

but is not at all concerned about long-term trends or the impacts of new, localized sources. Under such conditions it is often appropriate to disregard any potential model application, and base precipitation-chemistry estimates solely on climatological data. In the absence of any better information one could, for example, estimate that the average rainborne sulfate concentration at State College, Pennsylvania for the month of July 1981 will be roughly equal to that shown in Figure 4 for July 1978. Obviously one must beware of the potential pitfalls involved in making such a prediction; but given the present uncertainties in regional modeling procedures, such an application of climatological persistence is often the most logical and productive approach.

Data sources for this purpose are somewhat difficult to access; and although there are current plans to implement a centralized precipitation-chemistry data repository within the United States (this repository is currently intended to become a component of the EPA SAROD system), one must currently obtain data directly from the individual sources in most cases. Table 1 itemizes some of the major sources of such data for North America and western Europe; a more detailed listing of North American networks is provided in the recent report by Niemann and his co-workers (1979).

3.2 Pathway 2–7–8–21–23–15–16: Below-Cloud Scavenging of Inert Aerosols

The scavenging of inert aerosol by falling raindrops is a comparatively straightforward problem, and thus is a logical starting point for this overview of modeling techniques. The major problem envisioned here is the determination of the local rate of uptake of aerosol by the raindrops (particles per unit volume per unit time), as characterized by the term w_A in Eqs. (1) and (2). The terms r_{Ax} and r_{Ay} are zero (inert aerosol), and *we shall assume for the time being that other features of these equations are sufficiently well known to permit final computation*, once the nature of w_A is established. Some simple examples of such computations are presented later in this section.

Despite the relative simplicity indicated above, some rather troublesome features emerge during the application of aerosol-scavenging computations. These stem primarily from particle-size modifications during the scavenging process and from the size distributed nature of the raindrop and aerosol size spectra. Because of these features, the following discussion will be presented sequentially, starting with the relatively simple system involving a homogeneous aerosol.

3.2.1 Homogeneous Nonnucleating Aerosol

The simple case of a homogeneous aerosol collection by raindrops can be analyzed most conveniently by visualizing a volume element of air as

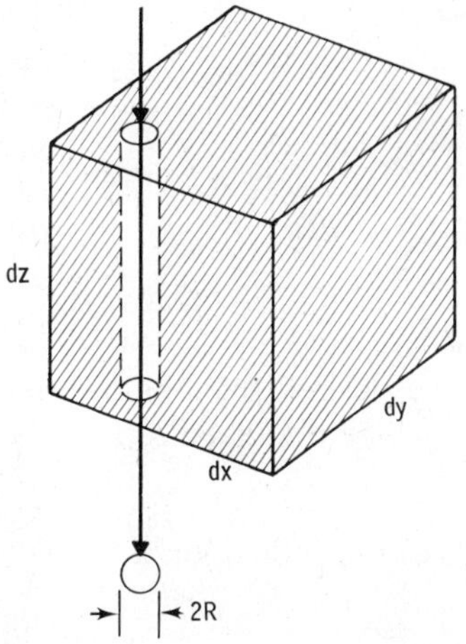

FIGURE 8. Schematic of raindrop penetrating unit volume element of atmosphere.

shown in Figure 8. If a single raindrop falls through this element, one can define a *collection efficiency* in terms of the total aerosol existing in the pathway of the drop and the amount actually collected during the raindrop's passage through the element; that is,

$$E(R, a) = \frac{\text{Mass of particles collected during drop's passage through element}}{\pi R^2 m\, \Delta z} \tag{3}$$

where R is the raindrop's projected radius, a is the (effective) aerosol particle radius, and m is the mass of particles per unit volume existing within the element prior to the drop's passage.

From Eq. (3) and Figure 8, it is obvious that the accumulation rate of particle mass by the falling drop should be

$$-\pi R^2 v_z m E(R, a)$$

where v_z is the vertical velocity of the raindrop (negative downward). Now for an ensemble of raindrops falling through the element, whose size distribution is characterized by the probability-density function $f_R(R)$, the total rate of pickup can be derived by integration over the total range of raindrop sizes. This total pickup rate is by definition equal to minus the interphase transport rate [w_A in Eqs. (1) and (2)], thus:

$$w_A = -\pi N_T m \int_0^\infty R^2 v_z(R) E(R, a) f_R(R)\, dR \tag{4}$$

where N_T is the total number of raindrops resident in the unit volume element. This relationship also can be expressed in terms of a *washout coefficient* Λ, defined as

$$w_A = \Lambda m \tag{5}$$

From Eq. (4),

$$\Lambda = -\pi N_T \int_0^\infty R^2 v_z(R) E(R, a) f_R(R)\, dR \tag{6}$$

relating the washout coefficient to the efficiency.

From Eq. (4) one can in principle compute the desired scavenging rate w_A if the entities E, v_z, and f_R are known; these will be discussed separately in the following paragraphs.

3.2.1.1 $E(R, a)$**.** The efficiency term $E(R, a)$ depends upon a host of possible collection mechanisms. These include:

impaction of aerosol particles on the raindrop,
interception of particles by the raindrop,
Brownian motion of particles to the raindrop,
nucleation of a water drop by the particle,
electrical attraction,
thermal attraction (thermophoresis), and
diffusionphoresis.

These mechanisms have been discussed at length by numerous authors (e.g., Dingle and Lee, 1973; Hidy, 1973). The last three of these mechanisms are of secondary importance in the case of below-cloud scavenging, except for rather special circumstances (Wang et al., 1978). The nucleation mechanism, while potentially significant in many applications, is disregarded in the present context on the presumption that the aerosol in question is hydrophobic, and thus will maintain its fixed particle size a. Slinn (1977) has analyzed the first three of these mechanisms and has suggested the following three formulae for computing the corresponding component efficiencies:

$$e_{\text{impaction}} = [(S - S_*)/(S + C)]^{3/2} \tag{7}$$

$$e_{\text{impaction}} = 4a / R \tag{8}$$

$$e_{\text{diffusion}} = 4\text{Sh}/(\text{Re Sc}) \tag{9}$$

Here the Sherwood number can be calculated from the Froessling equation

$$\text{Sh} = \frac{2k_v R}{Dc} = 2 + 0.6\text{Re}^{1/2}\text{Sc}^{1/3} \tag{10}$$

where

$$S \text{ (Stokes number)} = -2a^2\rho_p v_z/9R\rho_a\nu,$$
$$S_* \text{ (critical Stokes no.)} = (1.2 + L/12)/(L+1),$$
$$\text{Sc (Schmidt number)} = \nu/D,$$
$$\text{Re (Reynolds number)} = -2Rv_z/\nu,$$
$$C = 2/3 - S_*,$$
$$L = \ln(1 + \text{Re}/2),$$

c = molar concentration of air molecules,
D = molecular (Brownian) diffusivity,
k_y = mass-transfer coefficient,
ν = kinematic viscosity of air,
ρ_a = density of air, and
ρ_p = density of the aerosol particle.

More refined and involved estimates of these component efficiencies are available in the more recent literature (Slinn, 1984).

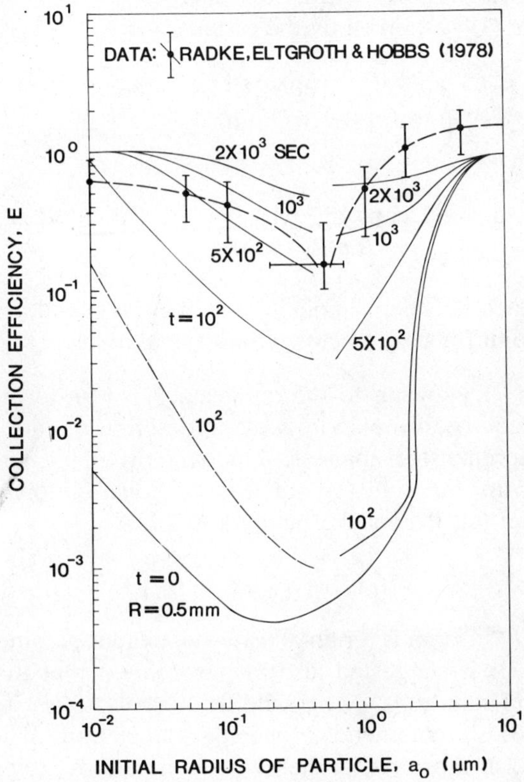

FIGURE 9. Theoretical washout efficiencies for 0.5-mm-radius raindrops, for nucleating aerosols. (Calculations of Slinn, 1984.)

The corresponding numerical values of $E(R, a)$ obtained by summing Eqs. (7), (8), and (9) exhibit the well-known tendency to become large for both very large and very small particle sizes, and to become low at intermediate sizes in the range of 0.1 μm (cf. Fig. 9). Since contributions of secondary mechanisms are neglected in this approach, $E(R, a)$ values computed in this manner can be considered to be conservatively *low* estimates of actual behavior. One can, of course, establish a corresponding *upper-limit* estimate of $E(R, a)$ by simply setting it to unity. Since this practice can lead in some cases to efficiency-values three orders of magnitude higher than those obtained from Eqs. (7)–(9), it is somewhat limited in value—at least in the present context where nucleation is assumed unimportant. Because of this, Eqs. (7)–(9) are recommended for practical use under these conditions.

3.2.1.2 v_z**.** Estimation of the fall velocity of raindrops is complicated by the presence of temperature and pressure gradients, and internal circulations and deformations within the drop (cf. Pruppacher and Klett, 1978). For practical application, however, empirical fits to measured data provide the most practical means for characterization. The equations of Dingle and Lee (1972), given by the forms

$$v_z = 27.2692 - 1206.2884R + 348.0768R^2 \quad (0.05 \leq R \leq 0.7 \text{ mm}) \tag{11}$$

and

$$v_z = -155.6745 - 613.4914R + 123.3392R^2 \quad (0.7 \leq R \leq 2.9 \text{ mm}) \tag{12}$$

provide a balance between simplicity and accuracy, and are recommended as a starting point for use in below-cloud calculations.

3.2.1.3 $f_R(R)$**.** Owing to the complexity of rain-formation processes, no really satisfactory formulation exists to describe raindrop size spectra in a totally comprehensive manner. Undoubtedly the most-often applied probability-density function for raindrop size distributions is that of Marshall and Palmer (cf. Pruppacher and Klett, 1978):

$$f_R(R) = C_2 \exp(-C_2 R) \tag{13}$$

Here $C_2 = 8.2J^{-0.21}$/mm is a rainfall rate–dependent parameter (J = rainfall rate in mm/h). It is suggested also in this context that the total number-concentration of raindrops, N_T, should lie close to $980J^{0.21}$ drops/m^3.

Equation (13) is recommended for initial calculations in conjunction with Eqs. (4) and (6). If more comprehensive computations are desired, one may choose to utilize other types of spectral equations, or employ field measurements of the actual rain spectra for the specific case at hand.

3.2.2 Size-Distributed, Nonnucleating Aerosol

When (as is the usual case) *both* the raindrop and aerosol spectra are size-distributed, an extension of Eq. (4) is required. If the aerosol *mass* concentration is described as the probability density function $f_m(a)$, then

$$w_A = -\pi N_T m \int_0^\infty \int_0^\infty R^2 v_z(R) E(R, a) f_R(R) f_m(a)\, dR\, da \qquad (14)$$

$$\Lambda = -\pi N_T \int_0^\infty \int_0^\infty R^2 v_z(R) E(R, a) f_R(R) f_m(a)\, dR\, da \qquad (15)$$

It should be noted that in Eq. (15) Λ is defined as a *mass* washout coefficient. If one were interested in actual *numbers* of particles washed out, one could define a *number* washout coefficient simply by inserting a number-density function in place of $f_m(a)$ in Eq. (15). Several examples of washout coefficient curves for various rain and aerosol spectra are given by Dana and Hales (1976).

Although size-distributed aerosol systems do not cause any great computational difficulty in principle, they do tend to pose extreme complications in practice. Calculations using Eq. (15) demonstrate that aerosol scavenging rates are in general strongly dependent on both particle size and spread of the particle-size distribution. Thus, there is a definite tendency for $f_m(a)$ (and thus Λ) to change radically during the course of a rain event, simply by action of the washout process. This combined with the fact that aerosol size distributions are seldom known with any acceptable degree of certainty, even before washout commences, imposes rather large limits of uncertainty in associated washout computations in a majority of practical applications.

3.2.3 Condensational Growth of Aerosols and Its Influence on Below-Cloud Scavenging

The discussion of below-cloud scavenging in Sections 3.2.1 and 3.2.2, which is based on the presumption that aerosol particles do not change their sizes during the scavenging process, is somewhat unrealistic. Most common aerosol particles do indeed act as nuclei for water condensation at high humidities (Junge, 1963), and appreciable changes in their sizes can be expected to occur as a result. This combined with the rather radical changes in E with particle size predicted from Eqs. (7)–(9) (compare lower curve of Fig. 9), suggests that considerable modifications of below-cloud scavenging rates can occur via the condensation process.

Size and growth rates for nucleated droplets depend on the nature of the nucleating particle and the water-vapor content of its surrounding environment. At high humidities, growth can be very rapid for small droplet sizes; as the droplets become larger, however, the process slows significantly. Given a supersaturation of 1% for example, a 1-μm droplet will double its size via condensation within a few seconds; for a 10-μm particle the corresponding doubling time is of the order of several minutes (Mason, 1971).

Our theoretical capability to deal with the prospect of nucleation and condensational growth in below-cloud scavenging is presently at an unsatisfactory state. Slinn (1984) has taken the rather straightforward approach of:

1. Selecting an aerosol particle of dry size a_0.
2. Calculating the size of the particle as it grows by condensation, assuming specific growth conditions.
3. Calculating revised values of E as a function of time, corresponding to the increasing size of the droplet.
4. Repeating the procedure over a range of dry particle sizes, to obtain the revised efficiency curves shown in Figure 9. Here the bottom curve corresponds to a dry aerosol and is essentially that which would be computed from Eqs. (7)–(9). The higher curves pertain to washout efficiencies of a growing aerosol, after the indicated growth times.

The efficiency curves on the right-hand side of Figure 9 are relatively simple, owing to the fact that particles in this size range do not interact significantly with each other via the Brownian diffusion process. They do, however, interact strongly with smaller aerosol particles, and thus the collection efficiencies of the latter are altered appreciably. Slinn has attempted to account for this in preparing the left-hand curves in Figure 9; his efforts have been limited, however, by the assumptions needed with regard to the characteristics of the large-particle end of the droplet spectrum. This has led to the variety of curves and the discontinuities that appear on the figure.

All of the above uncertainties, plus the generally unknown time–humidity history of an air parcel in a below-cloud scavenging environment add up to the fact that we have very little competence in prediction of below-cloud scavenging rates of aerosols under conditions where nucleation occurs. This effect undoubtedly serves to push scavenging efficiencies in the direction of the upper "asymptote" ($E = 1$) conditions mentioned in Section 3.2.1.1. Just how effective this process is, however, is understood very poorly. Rather comprehensive analyses of aerosol growth with condensation are available (Mason, 1971; Fitzgerald, 1974; Johnson, 1979), and some fragmentary field measurements of plume scavenging exist (Radke et al., 1978), but much remains to be accomplished before a really satisfactory understanding of this phenomenon is attained.

3.2.4 Below-Cloud Scavenging of Aerosols by Snow

The irregular and varied geometries of snow particles lead to difficulties in assessment of their size distributions, fall velocities, and scavenging efficiencies; thus the computation of below-cloud scavenging by snow emerges as a problem fraught by even more difficulty than that described previously for rain. The usual mathematical approach to this problem is to define some sort of "efficiency," which is comparable to that defined in Eq. (3), and is based on an equivalent diameter of one type or another. Slinn (1984) suggests:

D_e = diameter of the sphere circumscribing the snow particle, and proceed to express a corresponding washout coefficient by the form

$$\Lambda = -\frac{\pi N_T}{4} \int_0^\infty D_e^z v_z(D_e) E(D_e, a) f_{D_e}\, dD_e \tag{16a}$$

(cf. Eq. (6)). Combining this with an expression describing equivalent precipitation rate J in terms of D_e, he proceeds to the simplified form

$$\Lambda = \gamma J E(D_e, a) / D_m \tag{16b}$$

where γ is a constant of the order of unity, and D_m is a characteristic length scale whose numerical values are summarized in Table 2.

Slinn also provides a semiempirical equation for E (not given here), which is similar in form to Eqs. (7)–(9). An upper asymptote for the system, of course, is simply $E = 1$.

An alternative approach to snow-scavenging calculations, which is based on a more empirical framework, is that outlined by Knutson and Stockham (1977). These authors give explicit expressions for Λ which are functions of J, a, and temperature, and are based upon direct experimental observations.

In comparing the above results as well as the computations and measurements by additional investigators, it becomes readily apparent that several orders of magnitude uncertainty exist in typical applications of snow-scavenging calculations. Much more research needs to be accomplished, especially in the area of physical measurements of E, before a satisfactory computational capability will exist in this area.

TABLE 2. Characteristic Lengths of Ice Crystals for Use in Eq. (16b)[a]

Crystal Type	D_m (cm)
Graupel	0.014
Rimed plates and stellar dendrites	0.0027
Powder snow and spatial dendrites	0.001
Plane dendrites	0.00038
Needles	0.0019

[a]From Slinn (1984).

3.2.5 *Example Integrations of Continuity Equations for Aerosol Scavenging*

Thus far this text has been addressed to evaluation of the microphysical processes leading to the *microscopic* features of the pollution material-balance equations. Usually, however, the desired products of a scavenging calculation are *macroscopic* features, such as delivery fluxes and concentrations. These features are typically calculated via solution of the material-balance equations, and it is appropriate at this point to illustrate this procedure using some rather simplified, yet practical, examples.

EXAMPLE 1: SCAVENGING THROUGH A GAUSSIAN PLUME

If below-cloud scavenging occurs through an aerosol plume which is distributed in a Gaussian manner, and furthermore the scavenging interaction is characterized by the constant coefficient Λ, then

$$w_{\mathrm{A}} = \Lambda c_{\mathrm{Ay}}$$

and the solution to Eq. (1), subject to appropriate restrictions and boundary conditions, is (Slade, 1968)

$$c_{\mathrm{Ay}} = \frac{Q}{2\pi\sigma_y\sigma_z u}\exp\left(\frac{-y^2}{2\sigma_{y^2}}\right)\left\{\exp\left[-\frac{(z-h)^2}{2\sigma_{z^2}}\right] + \exp\left[-\frac{(z+h)^2}{2\sigma_{z^2}}\right]\right\}\exp\left(-\frac{\Lambda x}{u}\right) \tag{17}$$

where Q and h are the plume's source strength and release height, u is the wind velocity, and σ_y and σ_z are the plume spread parameters.

If one assumes constant, vertical rainfall with homogeneous drop size, then the corresponding reduced form of Eq. (2) is

$$-v_{\mathrm{Axz}}\frac{dc_{\mathrm{Ax}}}{dz} + \Lambda c_{\mathrm{Ay}} = 0 \tag{2a}$$

The average rainborne polluted concentration c_{Ax} can be calculated at any point x, y, z simply by inserting Eq. (17) into Eq. (2a) and integrating. In particular, c_{Ax} at ground level is

$$c_{\mathrm{Ax}} = \frac{-\Lambda Q}{\sigma_y v_{\mathrm{Axz}} u\sqrt{2\pi}}\exp\left(\frac{-y^2}{2\sigma_{y^2}}\right)\exp\left(\frac{-\Lambda x}{u}\right) \tag{18}$$

One should be careful to note here that c_{Ax} is the concentration of rainborne pollutant in terms of *total space* occupied by both the gaseous and aqueous phases. The relationship between c_{Ax} and the concentration of pollutant in collected rain, $\hat{C}_{\mathrm{A}}$, can be derived by considering once again the ensemble of raindrops in the volume element of Figure 8. If $\hat{c}_{\mathrm{A}}(R)$ is the aqueous-phase concentration of pollutant (pollutant per unit volume of water) associated with size-R hydrometeors, then

$$c_{Ax} = N_T \int_0^\infty V(R)\hat{c}_A(R) f_R(R)\, dR \tag{19}$$

Here $V(R)$ is the volume associated with size-R raindrops. Now $\hat{C}_A$ can be expressed simply as the vertical flux of rainborne pollutant, i.e.,

$$c_{Ax} v_{Ax}$$

divided by the vertical flux of rain:

$$\hat{C}_A = \frac{N_T \int_0^\infty V(R)\hat{c}_A(R) v_z(R) f_R(R)\, dR}{N_T \int_0^\infty V(R) v_z(R) f_R(R)\, dR} \tag{20}$$

Under the present special conditions of uniform raindrop size, the vertical velocity of rainborne pollutant is equal to the rainfall velocity, i.e.,

$$v_{Axz} = v_z$$

and Eq. (20) reduces to the form

$$\hat{C}_A = -\frac{c_{Ax} v_z}{J}$$

where J is the rainfall rate in arbitrary length per unit time units, giving

$$\hat{C}_A = \frac{\Lambda Q}{J \sigma_y u \sqrt{2\pi}} \exp\left(\frac{-y^2}{2\sigma_{y^2}}\right) \exp\left(\frac{-\Lambda x}{u}\right) \tag{21}$$

upon application to Eq. (12).

By studying this simple example, one can note that direct solution of Eq. (2) can become extremely cumbersome if the hydrometeor system is size-distributed in nature. Under such conditions it is often much more expedient to approximate a partial solution to Eq. (2) by abandoning the use of the washout coefficient, and utilizing instead a material balance over a single droplet. If one defines a particle-mean efficiency $\bar{E}(R)$, then from Section 3.2.1 this balance becomes

$$\frac{d\hat{c}_A(R)}{dz} = -\frac{3 c_{Ay} \bar{E}(R)}{4R} \tag{22}$$

By repeated integrations of Eq. (22) in conjunction with a descriptor of c_{Ay}, such as Eq. (17), and with subsequent distribution according to Eq. (20), one can compute a corresponding concentration in collected rain. This type of solution technique will be discussed further in conjunction with the discussion of gas scavenging in Section 3.3.

EXAMPLE 2: SCAVENGING THROUGH A UNIFORM AIR MASS

Perhaps the simplest example of a scavenging process is that where there are no gradients in the gas-phase pollutant concentration and no chemical reaction, thus reducing Eq. (1) to the form

$$\frac{dc_{Ay}}{dt} = -w_A = -\Lambda c_{Ay} \tag{23}$$

This can be integrated immediately to obtain the form

$$c_{Ay} = c_{Ay}|_{t=0} \exp(-\Lambda t) \tag{24}$$

Corresponding solutions of Eqs. (2) can be obtained as well, if desired. For example, integration of Eq. (2a) for the situation of (homogeneous distributed) rain falling a distance z_0 into a uniformly distributed plume of concentration c_{Ay} gives

$$c_{Ax} = -\frac{\Lambda z_0}{v_z} c_{Ay} \tag{25}$$

$$\hat{C}_A = \frac{\Lambda z_0}{J} c_{Ay} \tag{26}$$

Although often applied for atmospheric modeling purposes, Eq. (23) is usually too restrictive to be a truly useful or accurate descriptor. In general the divergence terms in Eq. (1) *are* important, and their truncation in this manner is not usually justified. One should note as well that Eq. (23) is *not* a *definition* of Λ; rather, it is a mathematical description of a highly specialized set of circumstances. Confusion of this point has led to erroneous applications in some past efforts.

3.3 Scavenging of Nonreactive Gases

3.3.1 Pathway 2–3–10–13–17: General Conditions

In the preceding discussion of aerosol scavenging it was assumed tacitly that interphase transport of pollutant between the atmosphere and a falling drop was *irreversible*; that is, once collected the aerosol could not escape back to the air from the aqueous phase. This feature is reflected in Eqs. (4) and (5), which imply that w_A is always positive, that is, interphase transport should always be *from* the gas phase *to* the drop.

In the case of gases, which can both *absorb in* and *desorb from* water, the irreversibility assumption is generally invalid; and under such conditions it is usually necessary to reformulate expressions for w_A which take reversibility into account. This is done most conveniently by discontinuing use of the efficiency concept [as expressed in Eq. (3)] and employing instead a corresponding expression for *flux* of pollutant from the falling hydrometeor:

$$F = -\frac{K_y}{c}(c_{Ay} - h'\hat{c}_A) \tag{27}$$

Here K_y is an overall mass-transfer coefficient, and h' accounts for the solubility of the gas. One should note that both absorption and desorption are predicted by Eq. (27), depending on the relative magnitudes of c_{Ay} and $h'\hat{c}_A$. One should observe also that, because of the small molecular masses and relatively high diffusivities of gaseous pollutants, diffusion predominates as an interphase transport mechanism; and thus all mechanisms in Section 3.2.1 other than diffusion become insignificant.

Diffusive transport in *both* the gaseous and aqueous phases is important in determining gas scavenging rates, and it is usually convenient to consider these effects individually in terms of gas- and liquid-phase coefficients k_y and k_x, such that

$$\frac{1}{K_y} = \frac{1}{k_y} + \frac{h'}{k_x} \tag{28}$$

(Bird et al., 1960). k_y can be estimated from Eq. (10). Evaluation of k_x is somewhat more difficult, although for many gases of high or moderate solubility (small h') its relative effect in Eq. (28) is small and it can be neglected (Barrie, 1978; Hales, 1972).

On the assumption of spherical raindrops, Eq. (27) can be integrated to provide a general expression for the interphase transport, which is a gas-scavenging counterpart to Eq. (4):

$$w_A = \frac{4\pi N_T}{c} \int_0^\infty R^2 f_R(R) K_y(R)[c_{Ay} - h'\hat{c}_A(R)]\, dR \tag{29}$$

3.3.2 Example Integrations of Continuity Equations for Gas Scavenging

Equation (29) can be incorporated with Eqs. (1) and (2) and utilized to calculate spatial concentration fields and delivery fluxes in a manner similar to that described previously for aerosol washout. The increased complexity of the coupling term w_A requires that additional attention be focused on the interactive nature of the rain and the gas-phase plume and one often is forced to make further simplifying assumptions, or else increase the complexity of the calculation appreciably. This plus the size-distributed nature of the rain spectrum often discourages direct solution of Eq. (2), in favor of an approximation in terms of individual hydrometeors, similar to that described in Section 2.3.5.1.

From Eq. (27) the single-drop material balance [cf. Eq. (22)] is

$$\frac{d\hat{c}_A(R)}{dz} = \frac{3K_y}{v_z R c}(c_{Ay} - h'\hat{c}_A) \tag{30}$$

Specific applications of this equation are presented in the following paragraphs.

EXAMPLE 1: SCAVENGING THROUGH A GAUSSIAN PLUME

In the event that scavenging does not deplete the plume appreciably, the conventional Gaussian plume equation [Eq. (17)] (with $\Lambda = 0$) may be incorporated with Eq. (30), and the results integrated to obtain the following expression for pollutant concentration in raindrops at ground level:

$$\hat{c}_A(R) = \frac{-Q\xi}{2\sqrt{(2\pi)}\sigma_y u} \exp\left(-\frac{y^2}{2\sigma_{y^2}} + \frac{\sigma_z^2 \zeta^2}{2}\right)$$
$$\times \left\{ \exp(\zeta h)\left[1 - \operatorname{erf}\left(\frac{-\sigma_z^2\zeta - h}{\sigma_z\sqrt{(2)}}\right)\right] + \exp(-\zeta h)\left[1 - \operatorname{erf}\left(\frac{-\sigma_z^2\zeta + h}{\sigma_z\sqrt{(2)}}\right)\right]\right\} \tag{31}$$

where

$$\zeta = \frac{3K_y h'}{c v_z R} \quad \text{and} \quad \xi = \frac{3K_y}{v_z R}$$

This may be considered to be a gas-scavenging counterpart of Eq. (21) although it is somewhat more restrictive because the coupling term w_A was removed from the gas-phase equation (Hales et al., 1973).

A more comprehensive model of this type, which allows numerical computations to be performed for general plume types, nonlinear solubility behavior, and nonvertical rainfall has been presented by Drewes and Hales (1980). An elegant analytical solution of Eqs. (1) and (2) which does *not* decouple the equations and thus accounts for plume distortion via the absorption-desorption process has been given by Slinn (1974a).

EXAMPLE 2: SCAVENGING THROUGH A UNIFORM AIR MASS

A gas-scavenging counterpart of Eq. (26) can be derived via simple integration of Eq. (30) for an initially clean drop as it is allowed to fall through a uniformly distributed pollutant gas of concentration c_{Ay} from height z_0. The result is

$$\hat{c}_A(R) = \frac{c_{Ay}}{h'}[1 - \exp(\zeta z_0)] \tag{32}$$

It is of some interest to observe the differences between Eq. (32) and its irreversible counterpart, Eq. (26). Visual inspection of Eq. (32) shows that a raindrop falling through a uniformly distributed polluted gas will approach a limiting concentration c_{Ay}/h'. Equation (25), on the other hand, suggests that the raindrop should scavenge pollutant indefinitely as long as it is able to fall. This difference can result in orders-of-magnitude deviations in computed values. Accordingly, one must exercise proper care in performing such calculations to ensure that the formulations employed are appropriate to the specific pollutant of interest.

3.3.3 Pathway 2–3–10–11: Equilibrium Scavenging

Under the special conditions where the raindrops are known to be at a state of solubility equilibrium with regard to the pollutant, scavenging computations become especially simple. If c_{Ay} is the ground-level gas-phase concentration, then

$$\hat{C}_A = c_{Ay}/h' \tag{33}$$

can be employed immediately for scavenging calculations. Situations where Eq. (33) is known to hold are referred to as *equilibrium scavenging conditions*. These conditions, promoted by short relaxation times for the absorption-desorption process, and slowly varying gas-phase concentration fields in the vicinity of the falling drops, are observed to occur whenever the dimensionless group

$$\Gamma_{eq} = \frac{-3K_y h' c_{Ay}}{cv_z R[dc_{Ay}/dz]} \tag{34}$$

becomes greater than about 10 (Hales, 1972). Here the term $[dc_{Ay}/dz]$ should be interpreted conservatively as the maximum gas-phase concentration gradient experienced by the raindrop throughout its fall.

3.3.4 Pathway 1–2–3–4–9: Mass-Transfer-Limited Gas Scavenging

One situation where the assumption of irreversible capture may be valid, even in the case of gases, is that where the pollutant is highly soluble or reactive. Under these conditions gas-phase mass transfer is the sole limiting factor, and Eqs. (4)–(6) (with $E = e_{\text{diffusion}}$) still apply. HCl is a prime example of a gas having a sufficiently high solubility to provide mass-transfer limited conditions under a large variety of circumstances (Pellet, 1977).

3.4 Scavenging of Reactive Gases

The prospect of chemical reaction of a dissolved gas in rainwater introduces the possibility of several alternate types of behavior, which are itemized below:

1. If the chemical reaction is *rapid* and *reversible* with a nonvolatile product, i.e.,

$$A \leftrightharpoons B$$

then the scavenging interactions usually can be treated as a *pseudo-physical absorption process* (Sherwood and Pigford, 1952). With this treatment pathways 2–3–10–11 or 2–3–10–13–17 can be utilized directly for calculation, as long as an appropriate means for describing solubility is available. Dimensionless criteria describing conditions acceptable for pseudophysical absorption calculations are available (Hales, 1972).

2. If the chemical reaction is rapid, irreversible, and leads to a nonvolatile product, then mass transfer to the raindrop's surface usually can be considered as the rate-limiting step in the scavenging process. Under such conditions pathway 2–3–4–9–20–15 can be employed for direct calculations. Dimensionless criteria (Hales, 1972) for fast-reaction, mass-transfer-limited conditions are available.
3. With the relatively slow chemical reactions (e.g., $SO_2 \rightarrow SO_4^{2-}$), two modes of below-cloud scavenging may be isolated. These correspond to the nonreactive pickup of gas by physical (or pseudophysical) absorption and the reactive depletion of gas within the drop. If relaxation times for the absorption step are short compared to those for reaction, then the first of these modes may be treated via steps 3–10–11 or 3–10–13–17 in a quasi-independent manner. Likewise, the reactive mode can be simplified under some circumstances to allow a relatively straightforward calculation to be performed. Again, dimensionless criteria may be derived (Hales, 1972) to describe conditions where such assumptions are allowable.
4. In the more general case, involving the possibility of multiple reactions, competitive effects, or volatile reaction products, one usually has little choice other than formulating a detailed mathematical description of the mass-transfer-chemical-reaction process (pathway 3–4–18–32–33–34). A generalized numerical framework for scavenging calculations of this class has been reported by Drewes and Hales (1980). Specific computations for the SO_2–SO_4^{2-} system in well-mixed environments have been presented by Overton et al. (1979) and Hill and Adamowicz (1977).

3.5 In-Cloud Scavenging of Gases and Aerosols

As indicated previously, common mechanisms contribute to the scavenging process regardless of whether or not it occurs within a visible cloud system; and thus the distinction between in- and below-cloud scavenging is somewhat artificial. There is, however, a definite shift in the relative importance of these mechanisms. Readdressing the collection pathways itemized in Section 3.2.1.1, it seems obvious that, for cloud environments where condensation is occurring, *nucleation* should play a much more dominant role in contacting the pollutant with condensed water. Also, because of the importance of evaporation–condensation cycles in typical cloud systems, *electrical*, *thermal*, and *diffusiophoretic* forces should be expected to become relatively important (Dingle and Lee, 1973). Interception and impaction, on the other hand, can be expected to become insignificant for the attachment of primary pollutant particles to cloud droplets, although they definitely remain important as mechanisms of accretion of pollutant-laden droplets to falling hydrometeors.

In-cloud scavenging computations tend to become highly involved, owing to the complex flows that typically occur in condensing and evaporating systems. The previous discussion of below-cloud scavenging was based on the rather casual assumption that flow-fields were defined, or at least could be estimated with adequate precision. This assumption is usually invalid for in-cloud systems, however, and quite often one is faced with the additional need to derive these flow fields via modeling of storm-dynamics processes. Such modeling entails solving the appropriate equations of conservation for energy, momentum, and mass of the storm system, and lends significant increases to the complexity of the problem.

Because of the involved nature of this subject, it is convenient to subdivide the in-cloud scavenging process into a number of sequential steps, which can be treated individually to isolate key aspects of the process. These are portrayed in the simplified visualization shown in Figure 10, and can be itemized as:

1. Transport of the pollutant to the cloud system from its source.
2. Transport of pollutant within the cloud system.
3. Interphase transport of airborne pollutant to the aqueous phase.
4. Removal of the pollutant-laden cloud water as precipitation.

Typically the *first* of these events is treated in terms of a transport model or else it is ignored, assuming that the pollutant is already in the region of the storm. Numerous models of this type exist, ranging from rather

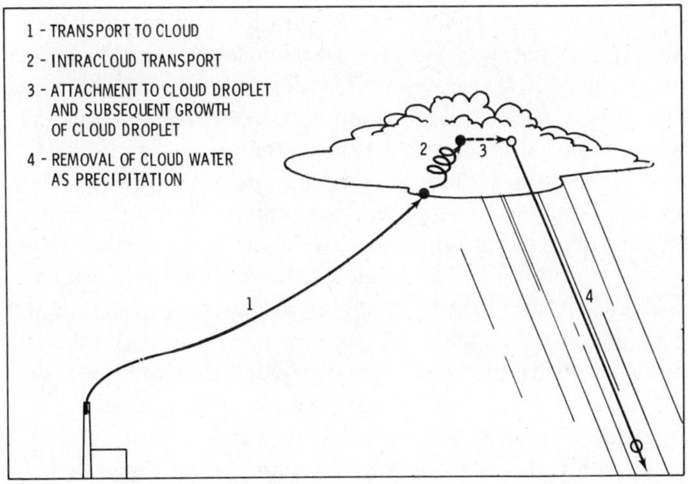

FIGURE 10. Steps in the overall in-cloud scavenging process.

straightforward trajectory calculations (Wendell et al., 1976; Samson, 1980; Hefter, 1980; Bolin and Persson, 1975), to detailed numerical solutions of variants of Eq. (1) and its momentum- and energy-conservation counterparts (Kreitzburg and Leach, 1978). While such trajectory modeling efforts are being performed rather routinely at the present time, it is important to note that, owing to the complexities associated with air motions near precipitation regions, these calculations must be conducted with due care to produce meaningful results.

Other approaches to the analysis of source-cloud transport have been statistical in nature, and are potentially useful if general climatological analysis, rather than specific source–receptor information, is required. The reader is referred especially to the paper of Rodhe and Grandell (1972) for an example of this type of analysis.

The *second* event of the above sequence often can be simplified, since pollution is generally well mixed in the atmosphere and usually enters the cloud system in the same manner as the water vapor from which the cloud is formed. An example of where this might *not* be the case is that of a plume that mixes into an already existing cloud system by diffusion from below. Since the attachment process can occur through the cloud volume, and also since the plume particles will themselves affect cloud microphysics and thus scavenging, it is not unreasonable to expect that their removal will depend upon their intracloud mixing to some degree. Such cases are of relatively minor importance, however, and the remainder of this discussion shall be based upon the assumption that pollutant and cloud-makeup water are introduced to the storm system in the same manner and are thus intimately mixed throughout.

The *third* event, that of the microphysical attachment of aerosol particles to cloud droplets and hydrometeors, has been the subject of extensive debate over the past three decades. Despite the noted profusion of mechanisms, however, there seems to be a rather general consensus that the primary attachment pathway for aerosol particles in the 0.1–1 μm range is nucleation. Junge (1963), for example estimated that anywhere from 50 to 80% of the mass of a general continental aerosol will be active as condensation-nucleus material in a typical storm situation.

As noted in Section 3.2.3, the times required for growth of nucleated droplets via the condensation process are sufficiently long to make this step a potential rate-influencing feature of the overall scavenging process. This combined with additional known complexities of cloud processes results in a rather involved picture of the in-cloud attachment phenomenon. Quite obviously mechanisms such as coalescence serve as additional factors to modify the size distribution of cloud particles. Moreover, the natural fluctuations in supersaturation within typical clouds tend to complicate matters considerably (Junge, 1963); and it seems obvious that, even with the simplistic notion that all attachment occurs via nucleation, the problem of rigorous mathematical characterization becomes overwhelming.

An important feature to note from this discussion, however, is that many of the mechanisms for extraction of pollutant in storm systems (e.g., nucleation, coagulation, accretion) are related intimately to those for removal of water. This is a feature that can be used to advantage in many practical scavenging calculations, and will be considered in more detail in the following discussion.

The *fourth* event in the in-cloud scavenging sequence—that of removal of pollutant-laden cloud water as precipitation—quite obviously involves a close relationship between scavenging and water removal as well. Those cloud particles that have grown sufficiently by condensation and coalescence to achieve significant fall velocity drop through the cloud, accreting other droplets and finally emerging as precipitation, carrying their associated pollutant burden to the ground. Owing to the size-distributed nature of the cloud and precipitation elements the mathematical description of this process can become rather complex, although simplified parameterizations have been formulated (Mason, 1971). It is sufficient for this discussion to note, however, that here again is a process that takes sufficient amounts of time to be a significant rate-influencing step in the overall scavenging process.

The calculation of in-cloud scavenging rates can be simplified appreciably if one or more of the events in the above sequence can be disregarded. As indicated previously, one way to accomplish this is simply to begin the modeling process at a late point in the sequence, thus assuming that the consequences of all previous steps are already known, or else have been predicted by other models. Quite obviously this approach demands specific information regarding either concentrations at the beginning of the modeled sequence, or else the rates of a concurrent phenomenon, such as rain production.

A second possible way to disregard steps in the sequence is to establish events that occur slowly compared to others and thus can be considered as rate limiting. If such steps are indeed shown to exist, the remaining ones can be ignored, thus simplifying the modeling problem. This is a procedure identical to that employed in chemical reaction-rate modeling, and has been examined previously in the context of events 3 and 4 by Slinn (1974b). In view of the above discussion it appears unlikely that either event 3 or event 4 will become rapid enough in a sufficiently large number of cases to permit their general neglect as rate-influencing steps. In formulating working models, therefore, one must either begin at event 3 (or earlier) and model the process through to completion, or else utilize additional information to permit a beginning at event 4.

Regardless of this starting point, most practical assessments of this situation can be categorized into a manageable number of classes, depending on whether

1. the material balance used for calculation is integral or differential in nature; and

2. the derivation requires explicit solution of momentum and/or energy equations to derive thermodynamic and/or flow features.

The following discussion of calculation methods will be subdivided according to these classes.

3.5.1 Pathways 20–24–28–29–30 and 20–24–28–29–31: Integral Material Balances

Perhaps the most straightforward example of an integral material-balance approach to in-cloud scavenging analysis is the derivation of washout ratios. This is a particularly appealing approach, because it allows most of the essential features to be lumped into a small number of parameters; and although these are difficult to estimate from first principles, they can be force-fit to experimental observations in a rather convenient manner.

The washout ratio is defined as

$$\xi = \hat{C}_{A}/c_{Ay} \tag{35}$$

Its basic features can be derived (Engelmann, 1971) by assuming that the storm can be characterized as a quasi-steady-state phenomenon, and then by performing a material balance over a total precipitating cloud system for both the pollutant and water. If one denotes the overall extraction efficiencies for the storm as ϵ_w (water vapor) and ϵ_p (pollutant), then it can be shown (Hales and Dana, 1979) from such a balance that

$$\xi = \epsilon_p \rho_w / \epsilon_w H \tag{36}$$

where ρ_w is the density of water, and H is the mass concentration of water vapor entering the cloud with the pollutant.

Considerable effort has been placed on the elucidation of ϵ_p and ϵ_w. Based on the concept of common mechanisms for water and pollutant removal (e.g. nucleation, accretion) it has been suggested that ϵ_p and ϵ_w should be roughly equal to one another, giving

$$\xi = \rho_w / H \tag{37}$$

which typically assumes numerical values in the range of 10^5.

Field measurements have shown that washout ratios of this order of magnitude often occur (Gatz, 1972; Engelmann, 1971). The associated variability is rather large, however, and this combined with more detailed examination of the scavenging mechanisms suggests that Eq. (37) is applicable only as a general rule of thumb, and then only for particulate pollutants and rather specialized storm types.

Scott (1978) has recently extended scavenging-ratio theory by providing a more sophisticated model of mechanisms operating within the cloud environment. In this model a pollutant aerosol is attached to cloud droplets via a nucleation step, with subsequent incorporation into snow or rainwater via the processes of coagulation and accretion. The relative values of ϵ_p and ϵ_w depend on storm type and intensity, and Scott has subdivided his derived scavenging ratios into three storm types:

warm rain storms,

cold storms, where the Bergeron process is important in defining the character of the precipitation, and

convective storms.

Scott's initial calculations are summarized in Figure 11, which indicates that significant differences in ξ should occur as a function of storm type. While these curves are extremely convenient for applied calculations, they are based strongly on assumed nucleating capabilities of the pollutant aerosol. They therefore should be applied with some caution, especially if the pollutant tends to be hydrophobic or distributed as a very fine aerosol with a correspondingly low nucleating capability. It also should be applied with caution in circumstances where competitive mechanisms may contribute to the presence of pollutants in precipitation. Examples of the occurrence of competitive mechanisms include the processes leading to the presence of sulfate and nitrate in precipitation, which include both aerosol scavenging and the uptake of reactive gases.

In a subsequent analysis Scott (1979) has extended the above treatment to nonreactive gas scavenging via snowstorms. (In-cloud scavenging of

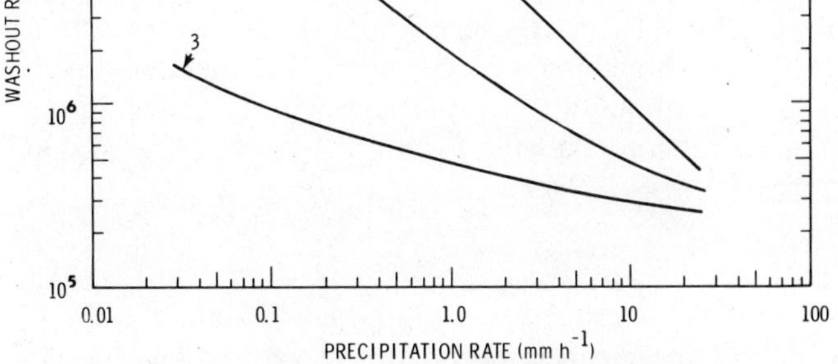

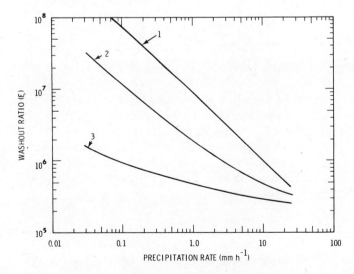

FIGURE 11. Washout ratios for aerosols for different storm types: (1) convective storms; (2) warm-rain process storms; (3) Bergeron-process storms. (Adapted from Scott, 1978.)

nonreactive gases by *rainstorms* is usually a nonessential problem, owing to the reversible nature of the scavenging process, which focuses primary emphasis on processes that occur close to the ground.) Here the primary gathering mechanism is expected to be dissolution of the gas in supercooled droplets near the cloud base. From the equilibrium scavenging criterion given by Eq. (34), the extent of this dissolution should be dictated simply by a solubility relationship. Subsequent delivery of SO_2 to the surface occurs by accretion of the supercooled droplets via a riming process.

Since the amount of supercooled cloudwater in a snowstorm is inversely related to temperature, Scott's model suggests that gas scavenging should be comparatively limited for storms occurring in very low temperature environments; this type of behavior has been observed experimentally in the case of SO_2 scavenging (Hales et al., 1971).

Other types of integral material balances have adopted the concept of a storm-averaged scavenging coefficient. If, for example, one expands the size of the volume element used for deriving the microscopic Eqs. (1) and (2) so that it encompasses a total cloud system, then the interphase transport rate can be expressed in terms of an averaged scavenging coefficient and an average concentration:

$$\bar{w} = \overline{\Lambda c_{\mathrm{Ay}}}$$

By defining corresponding macroscopic terms for the divergence and (if necessary) the reaction terms, one can proceed directly to formulate an expression for the total system, which may be applied directly for a practical use. (A simple example of such an expression is that for a spatially homogeneous storm system, stationary in space, and involving a constant wind speed $\bar{u}$. Under such conditions the governing equation is

$$\frac{dc_{\mathrm{Ay}}}{dx} = -\frac{\bar{\Lambda}}{\bar{u}}\,\bar{c}_{\mathrm{Ay}}) \tag{38}$$

Before this is possible, however, one must determine appropriate values for $\bar{\Lambda}$.

A few examples of modeling efforts leading to storm-averaged scavenging coefficients are available. Slinn (1977) for example, begins with a rather general form of Eq. (1), inserts terms appropriate to describe the above-noted microphysical attachment mechanisms, and then averages over space and particle size to obtain a space-particle average scavenging coefficient. The resulting expression takes the form

$$\bar{\Lambda} = \frac{J\bar{\epsilon}}{2\bar{R}_m} \tag{39}$$

where $\bar{\Lambda}$ is a spatially averaged scavenging coefficient, J is the rainfall rate, and $\bar{R}_m$ is the volume-mean raindrop size at ground level. $\bar{\epsilon}$ is given by a rather complicated expression reflecting particle attachment and droplet growth behavior, and is predicted to vary with time. Although very few data exist to test Slinn's expression, tracer-release tests have been shown to fit Eq. (39) reasonably well with an $\bar{\epsilon}$ value of $\frac{1}{3}$.

A second expression for a storm-average scavenging coefficient has been derived by Klett (1977) in his analysis of wet removal of nuclear debris. Basically this author has assumed complete attachment of pollutant to cloud particles of a given size distribution, and has integrated expressions for droplet capture by accretion to obtain formula for the scavenging coefficient and the rain rate. Combining these Klett arrives at the expression

$$\bar{\Lambda} = 4.2 \times 10^{-4} \bar{E} J^{0.79} \quad (\mathrm{s}^{-1}) \tag{40}$$

where $\bar{E}$ is an average scavenging efficiency and the rainfall rate is expressed in units of millimeters per hour. Comparing with numerical solutions of the accretion equation for assumed cloud droplet spectra Klett suggests an $\bar{E}$ value of 0.83.

Integral material-balance approaches using storm-averaged scavenging coefficients offer some advantage over washout-ratio applications, because the former do not depend intrinsically on the assumption of a steady state. Moreover, the scavenging-coefficient approach is somewhat more satisfactory when vertical gradients of c_{Ay} are pronounced. Both approaches have had virtually no serious application for cases where reactive scavenging of gaseous materials occurs, although there have been some attempts to guess at values of ξ and $\bar{\Lambda}$, where the chemical conversion rate r_{Ax} has been rather crudely lumped with the physical removal terms of the governing equations. Much remains to be accomplished in this important research area.

3.5.2 Differential Material Balances

Given flow-field and thermodynamic properties, one can integrate Eqs. (1) and (2), subject to appropriate initial and boundary conditions, and compute scavenging features that are more detailed than those made possible via the integral approach. Several examples of such computations exist (Lange and Knox, 1977; Watson et al., 1977). These tend to be mathematically involved and require machine computation; and up to the present time they have been limited by uncertainties regarding the local interphase transport term w_{A}. These uncertainties stem from the multitude of possible attachment mechanisms, and the usual procedure is simply to choose some constant value for the scavenging coefficient and incorporate it into the overall formulation. The errors associated with this process are usually of such a magnitude that gross scavenging rates computed by this method are not significantly superior to those obtained from the simpler, integral formulations. Increased resolution of spatial and temporal variability provided by this modeling approach can be useful in specific situations, however, especially in the case of diagnosing physical behavior in field experiments. As with the integral approaches, very little progress has been made to date in the field of reactive-scavenging analysis.

3.5.3 Scavenging Models Involving Storm-Dynamics Computations

The material-balance approaches described in Sections 3.5.1 and 3.5.2 were based largely on the presumptions that velocity and/or temperature features of the storm systems were known in sufficient detail to allow essential computations to be performed. There are in existence, however, a number of rather detailed storm models which generate these features explicitly, and thus can be utilized potentially for more detailed scavenging analysis. Owing to relatively large computation requirements, the applications of such models to date has not been extensive. Molenkamp (1977) has performed some limited scavenging calculations for aerosol using a one-dimensional convective-storm model, and Hane (1978) has utilized a two-dimensional model for this same purpose. Kreitzburg and Leach (1978) have performed more extensive scavenging computations using a detailed mesoscale model, which provides significant new insights regarding the characteristics of cyclonic storm systems.

As with the differential material balances described in Section 3.5.2, these models have been constrained largely to computations based on rather gross assumptions regarding interphase transport rates. Future applications with more detailed treatment of microphysical extraction mechanisms can be expected to provide some noteworthy advances to our understanding during future years.

3.6 Composite Regional Models

Thus far, this discussion has centered mainly on phenomena occurring in the vicinity of precipitating systems. This focus has allowed a much more detailed examination of the individual mechanisms of the scavenging process; and while the discussion has been conducted at a somewhat superficial level, some idea of the complexities of these mechanisms has emerged.

Against this backdrop, it is of interest to consider the evolving set of *regional* models of pollution behavior. Because of their large time and distance scales, these models cannot afford the luxury of concentrating on one particular atmospheric pathway, such as wet removal; indeed, *all* pathways, including long-range transport, chemical reaction, and dry deposition, must be considered simultaneously.

The necessarily composite nature of such models introduces several new areas of uncertainty, and these combined with the mathematical complexity of solving equations such as (1) and (2) over extended distances has generally forced the characterization of most processes in highly parameterized form. In particular, wet removal has been treated principally in terms of simplified expressions for washout coefficients or washout ratios.

In principle, there is nothing particularly limiting about the expression of wet removal in terms of Λ and ξ in such models. These parameters can vary with time and space as computation proceeds, and as long as one stipulates the *correct* values of these parameters at each computation point, valid results can be obtained. The challenge, of course, is in the selection of these values—a task which can be guided to some extent by the considerations in the preceding text.

Regional-scale modeling is currently developing at an extremely rapid rate, and several examples exist where wet removal is treated in terms of different expressions for either ξ or Λ, as functions of storm type, precipitation rate, chemical species, and so forth. Many of these models are summarized in the recent review by Drake and his co-workers (1979).

The question regarding whether to select washout *coefficients* or *ratios* as a parameterization basis depends to some extent on the particular model being used, and also on the pollutant species in question. Regional scavenging of nonreactive (or pseudononreactive) gases such as SO_2 is usually treated most appropriately in terms of washout ratios, because of the natural relationship between ξ and the equilibrium-scavenging expression, Eq. (33). Nonreactive aerosol scavenging, at the present state of understanding, can be treated equally well using either approach, unless serious vertical stratification of the gaseous-phase pollutant exists. Under such conditions the washout coefficient, which allows specific vertical integrations to be performed, is usually more appropriate. When reactive scavenging is involved, both techniques of parameterization are on rather uncertain grounds. Certainly reactive removal can be treated more specifically in terms of a nonintegrated parameter such as Λ; but thus far there has been little progress in this area, other than the practice of fitting values empirically to observed precipitation-chemistry data. Much remains to be accomplished on this important research topic.

4. CONCLUSIONS

This paper has provided an overview of scavenging calculation techniques, and has summarized our present state of knowledge in this area of the atmospheric sciences. Obviously, there is much that is needed to be learned before we can attain a totally satisfactory capability in this regard; and in concluding this presentation it is worthwhile to reconsider the data shown in Figures 2–6, with the following question:

Given the techniques for wet-removal calculation that currently exist, how well can we explain and/or predict the observed concentrations and their variability?

The answers to this question are somewhat mixed. Certainly the techniques discussed in Section 3, if used appropriately, are capable of order-of-magnitude determinations in many circumstances; and under restricted conditions they can even generate predictions having factor-of-two accuracy or better. Moreover, there is ample explanation in existing theories of wet removal to account easily for the spatial and temporal variability exemplified in Figures 2–6.

These capabilities, however, cannot be considered to be very satisfactory in the context of current needs. The noted ability to *explain* spatial and temporal variability on a semiquantitative basis has not resulted in any real competence in *predicting* such variability in specific instances. Moreover, we possess very little competence in identifying specific sources responsible for wet deposition at a given receptor site. Finally, the order-of-magnitude predictive capability noted above hardly can be judged satisfactory for most assessment purposes. In reviewing the discussion of Section 3 against the backdrop of these deficits, several research needs become apparent. The most important of these are itemized in the following paragraphs.

1. Much more definitive information is needed with regard to the scavenging efficiencies of submicron aerosols, for both rain and snow. Especially important in this regard is the effect of condensational growth of such aerosols in below-cloud environments.
2. We need to know much more about aqueous-phase conversion processes, which are potentially important as alternate mechanisms resulting in the presence of species such as sulfate and nitrate in precipitation. Since virtually nothing is known presently regarding the chemical formation of such species in clouds and precipitation, there is a tendency to lump these effects with *physical* removal processes in most modeling efforts, expressing them in terms of pseudo scavenging coefficients or collection efficiencies. Such phenomena must be resolved in finer mechanistic detail than this before a satisfactory treatment is possible, and this requires a knowledge of chemical transformation process that is much more advanced than existing at the present time.
3. Much more extensive understanding of the competitive nucleation capability of aerosols in in-cloud environments is needed, especially for those substances that do not compete particularly well in the nucleation process. The influence of aerosol-particle composition—especially for "internally mixed" aerosols (those containing individual particles composed of mixture of chemical species)—is particularly important in this regard.
4. The identification of specific sources responsible for chemical deposition at a given receptor location requires that we possess a much more accomplished capability to describe long-range pollution transport. Progress in this area during recent years has been encouraging, but much more remains to be achieved before we have a proficiency that is really satisfactory for reliable source–receptor analysis.

5. We will need to enhance our understanding of the detailed microphysical and dynamic processes that occur in storm systems. Besides providing required knowledge of basic physical phenomena, such research is important in providing valid parameterizations of wet removal for subsequent use in composite regional models.

As a final note, it is useful to reflect on the fact that scavenging modeling research—as treated in the context of this report, at least—has been in a rather continuous state of development over the past 30 years (Fuquay, 1970). While progress has been indeed significant during this period, a number of important and unsolved problems still exist. Accordingly, one must be cognizant of this perspective in judging our rate of advancement during future years. Reasonable progress in resolving the above items can be expected over the next decade; but the complexity of these problems demands that a serious and sustained effort be applied to this purpose.

5. NOMENCLATURE

a	Aerosol particle radius (l)
A	Component identifier
c	Molar concentration of air (moles/l^3)
c_{Ax}	Concentration of pollutant A associated with the aqueous phase (moles/unit volume of total space)
c_{Ay}	Concentration of pollutant A associated with the gaseous phase (moles/unit volume of total space)
$\hat{c}_{\mathrm{A}}$	Concentration of pollutant A in a single raindrop (moles/unit volume of water)
$\hat{C}_{\mathrm{A}}$	Concentration of pollutant A in collected rainwater (moles/unit volume of water)
D	Molecular (or Brownian) diffusivity (l^2/t)
D_{e}	Diameter of sphere circumscribing a snow particle (l)
E	Collection efficiency
f	Probability-density functions ($1/l$)
F	Pollutant flux from a drop surface (moles/$l^2\ t$)
h	Emission height (l)
h'	Solubility parameter
H	Mass concentration of water vapor (m/l^3)
J	Precipitation rate (l/t) or flux ($l^3/l^2\ t$)
k_{x}	Aqueous-phase mass-transfer coefficient (moles/$l^2\ t$)
k_{y}	Gaseous-phase mass-transfer coefficient (moles/$l^2\ t$)
K_{y}	Overall mass-transfer coefficient (moles/$l^2\ t$)

m	Mass concentration of aerosol particles (m/l^3)
N_T	Number concentrations of raindrops (drops/l^3)
Q	Pollutant source strength (moles/t)
r_{Ax}, r_{Ay}	Aqueous- and gaseous-phase chemical reaction rates (moles/l^3 t)
R	Raindrop radius (l)
Re	Reynolds number
S	Stokes number
S_*	Critical Stokes number
Sc	Schmidt number
Sh	Sherwood number
t	Time (t)
u	Wind velocity (l/t)
v	Velocity (l/t)
$\tilde{v}_{Ax}$	Velocity vector of aqueous-phase pollutant (l/t)
$\tilde{v}_{Ay}$	Velocity vector of gaseous-phase pollutant (l/t)
v_{Axz}	Vertical component of aqueous-phase pollutant velocity vector (l/t)
v_z	Hydrometeor fall velocity (negative downward) (l/t)
w_A	Interphase transport rate of pollutant *to* the aqueous phase *from* the gaseous phase (moles/l^3 t)
x	Coordinate
x	Aqueous phase designation
y	Coordinate
y	Gaseous phase designation
z	Coordinate (l)
ϵ_w, ϵ_p	Cloud extraction efficiencies for water and pollutant
Γ_{eq}	Dimensionless group defining equilibrium scavenging conditions
ζ	Parameter used in Eqs. (31) and (32), (1/l)
ξ	Washout ratio; also parameter used in Eq. (31) (moles/l^4)
Λ	Washout coefficient (1/t)
σ_y, σ_z	Plume-spread parameters (l)
ρ	Density (m/l^3)
∇	Vector divergence operator (1/l)

ACKNOWLEDGMENTS

Preparation of this paper was supported by the U.S. Environmental Protection Agency under the MAP3S Program, and I would like to express my sincere appreciation to this Agency for this assistance. I would also like to thank Dr. George Slinn for his helpful comments during preparation of this text, and Dr. Alvin Vanderpol for his kind assistance in supplying some of the reported precipitation-chemistry data.

REFERENCES

Barrie, L. A. An improved model of reversible SO_2 washout by rain. *Atmos. Envir.* **12**, 407–412 (1978).

Bird, R. B., W. E. Stewart, and E. N. Lightfoot. *Transport Phenomena*. Wiley, New York, (1960).

Bolin, B., and C. Persson. Regional dispersion and deposition of atmospheric pollutants with particular application to sulfur pollution over Western Europe. *Tellus* **XXVII**, 281–208 (1975).

Chamberlain, A. C. Aspects of travel and deposition of aerosols and vapour clouds. AERE Harwell Report R1261, HMSO, London (1953).

Dana, M. T., and J. M. Hales. Statistical aspects of the washout of polydisperse aerosols. *Atmos. Envir.* **10**, 45–50 (1976).

Dingle, A. N., and Y. Lee. An analysis of in-cloud scavenging. *J. Appl. Met.* **12**, 1295–1302 (1973).

Dingle, A. N., and Y. Lee. Terminal fallspeeds of raindrops. *J. Appl. Met.* **11**, 877–879 (1972).

Drake, R. L., D. J. McNaughton, and C. Huang. Mathematical models for atmospheric pollutants. Appendix D: available air quality models. Battelle-Northwest Report to EPRI. EA 1131 (1979).

Drewes, D. R., and J. M. Hales, SMICK—A Scavenging Model Incorporating Chemical Kinetics. Final Report to EPRI (1980).

Engelmann, R. J. Scavenging prediction using ratios of concentrations in air and precipitation. *J. Appl. Met.* **10**, 493–497 (1971).

Fitzgerald, J. W. Effect of aerosol composition on cloud-droplet size distribution: A numerical study. *J. Atm. Sci.* **31**, 1358–1367 (1974).

Fuquay, J. J. Scavenging in perspective. In R. J. Engelmann and W. G. N. Slinn (eds.) *Precipitation Scavenging 1970. USAEC Symposium Series 22* (1970).

Gatz, D. E. Washout ratios in urban and non-urban areas. *Proc. AMS Conf. on Urban Environment. Philadelphia* (1972).

Granat, L. Sulfate in precipitation as observed by the European Air Chemistry Network. *Atmos. Envir.* **13**, 413–424 (1978).

Hales, J. M. Fundamentals of the theory of gas scavenging by rain. *Atmos. Envir.* **6**, 635–659 (1972).

Hales, J. M., and M. T. Dana. Precipitation scavenging of urban pollutants by convective storm systems. *J. Appl. Met.* **18**, 294–316 (1979).

Hales, J. M., J. M. Thorp, and M. A. Wolf. Field investigation of sulfur dioxide washout from the plume of a large coal-fired power plant by natural precipitation. Battelle-Northwest Report to EPA NTIS PB 203–129 (1971).

Hales, J. M., M. A. Wolf, and M. T. Dana. A linear model for predicting the washout of pollutant gases from industrial plumes. *AICHE Journal* **19**, 292–297 (1973).

Hane, C. E. Scavenging of urban pollutants by thunderstorm rainfall: Numerical experimentation. *J. Appl. Met.* **17**, 699–710 (1978).

Hefter, J. L. Air Resources Laboratories atmospheric transport and dispersion model (ARL-ATAD). NOAA Tech. Memo ERL-81 (1980).

Hidy, G. Removal of gaseous and particulate pollutants. In S. I. Rasoul (ed.) *Chemistry of the Lower Atmosphere*. Plenum, New York (1973).

Hill, F. B., and R. F. Adamowicz. A model for the reversible washout of sulfur dioxide, ammonia, and carbon dioxide from a polluted atmosphere, and the production of sulfates in raindrops. *Atmos. Envir.* **11**, 917–927 (1977).

Johnson, D. B. The Role of Coalescence Nuclei in Warm Rain Initiation. Ph.D. Thesis, University of Chicago (1979).

Junge, C. E. *Air Chemistry and Radioactivity*. Academic, New York (1963).

Klett, J. Precipitation Scavenging. LASL Report to ERDA LA-6763 (1977).

Knutson, E. O., and J. D. Stockham. Aerosol scavenging by snow: Comparison of single-flake and entire snowflake results. In R. G. Semonin and R. W. Beadle. *Precipitation Scavenging (1974)*. ERDA Symposium Series CONF 741003 (1977).

Kreitzburg, C. W., and M. J. Leach. Numerical prediction of precipitation for computing atmospheric cleansing. *Proc. 85th National AICHE Meeting, Philadelphia* (1978).

Lange, R., and J. B. Knox. Adaptation of a three-dimensional atmospheric transport-diffusion model to rainout assessments. In R. G. Semonin and R. W. Beadle. *Precipitation Scavenging (1974)*. ERDA Symposium Series CONF 741003 (1977).

MAP3S. The MAP3S Precipitation Chemistry Network: Third Periodic Summary Report. Battelle, Pacific Northwest Laboratory (1980).

Mason, B. J. *The Physics of Clouds*. Oxford, London (1971).

Molenkamp, C. R. Numerical modeling of preciptation scavenging of convective clouds. In R. G. Semonin and R. W. Beadle. *Precipitation Scavenging (1974)*. ERDA Symposium Series CONF 741003 (1977).

Munn, R. E., and H. Rodhe. On the meteorological interpretation of the chemical composition of monthly precipitation samples. *Tellus* **XXIII**, 1–12 (1971).

Niemann, B. L., J. Root, N. VanZwalenburg, and A. L. Mahan. An integrated monitoring network for acid deposition: A proposed strategy interim report to EPA, R-023-EPA-79 (1979).

Overton, J. H., V. P. Aneja, and J. L. Durham. Production of sulfate in rain and raindrops in polluted atmospheres. *Atmos. Environ.* **13**, 355–367 (1979).

Pack, D. H., and D. W. Pack. Seasonal and annual behavior of different ions in acidic precipitation. WMO Special Environmental Report No. 14, WMO No. 549, 303–313, Geneva, Switzerland (1980).

Pellett, G. L. Washout of HCl and application to solid rocket exhaust clouds. In R. G. Semonin and R. W. Beadle. *Precipitation Scavenging (1974)*. ERDA Symposium Series Conf. 741003 (1977).

Pruppacher, H. R., and J. D. Klett. *Microphysics of Clouds and Precipitation*, Reidel, Boston (1978).

Radke, L. F., M. W. Elgroth, and P. V. Hobbs. Precipitation scavenging of aerosol particles. *Proc. Cloud Physics and Atmospheric Electricity*. Am. Met. Soc., Boston (1978).

Raynor, G. S., and J. V. Hayes. Experimental data from analysis of sequential precipitation samples at Brookhaven National Laboratory. BNL 50826 (1978).

Rodhe, H., and J. Grandell. On the removal time of aerosol particles from the atmosphere by precipitation scavenging. *Tellus* **XXIV**, 442–454 (1972).

Samson, P. J. Ensemble trajectory analysis of summertime sulfate concentrations in the Northeastern United States. *J. Appl. Met.* **19** (1980).

Scott, B. C. Parameterization of sulfate removal by precipitation. *J. Appl. Met.* **17**(7), 1375–1389 (1978).

Scott, B. C. Modeling of atmospheric wet deposition. ACS Symposium, Washington, D.C. September (1979).

Semonin, R. G., and R. W. Beadle. *Precipitation Scavenging (1974)*. ERDA Symposium Series CONF 741003 (1977).

Sherwood, T. K., and R. L. Pigford. *Absorption and Extraction*. McGraw-Hill, New York, (1952).

Slade, D. H. Meteorology and Atomic Energy, USAEC TID-24190 (1968).

Slinn, W. G. N. The redistribution of a gas plume caused by reversible washout. *Atmos. Envir.* **8**, 233–239 (1974a).

Slinn, W. G. N. Rate-limiting aspects of in-cloud scavenging. *J. Atm. Sci.* **31**, 1172–1173 (1974b).

Slinn, W. G. N. Some approximations for the wet and dry removal of particles and gases from the atmosphere. *Water, Air and Soil Poll.* **7**, 513–543 (1977).

Slinn, W. G. N. Precipitation scavenging. In D. Randerson (ed.) *Atmospheric Science and Power Production*. U.S. Department of Energy DOE/TIC-27601. Available NTIS DE 84005177 Springfield, VA, pp. 466–532 (1984).

Turner, D. B. Workbook of Dispersion Estimates. USEPA AP-26 (1973).

Wang, P. K., S. N. Grover, and H. R. Pruppacher. On the effect of electric charges on the scavenging of aerosol particles by clouds and small raindrops. *J. Atm. Sci.* **35**, 1735–1743 (1978).

Watson, C. W., S. Barr, and R. E. Allenson. Rainout assessment: The ACRA system and summaries of simulation results. LASL Report to ERDA LA-6763 (1977).

Wendell, L. L., D. C. Powell, and R. L. Drake. A regional-scale model for computing deposition and ground-level air concentrations of SO_2 from an elevated source. Battelle Pacific Northwest Laboratory Annual Report to ERA (1976).

SYNTHESIS

TRANSPORT AND DEPOSITION OF AIR POLLUTANTS ON TERRESTRIAL VEGETATION AND SOILS

B. B. Hicks
National Oceanic and Atmospheric Administration
Atmospheric Turbulence and Diffusion Laboratory
Oak Ridge, Tennessee

W. B. Johnson
Atmospheric Science Center
SRI International
Menlo Park, California

BACKGROUND

The rates of delivery of pollutants to vegetation will sometimes be controlled by physical, chemical, and biological factors associated with the surface itself, and sometimes by properties associated with the nature of the pollutant or by meteorological factors. In the case of wet deposition, concentrations of particular species in rainfall will be strongly influenced by the amount of that species (or its precursors) in the air in which the precipitation is originating and in the layer of air through which the precipitation is falling. With dry deposition, the efficiency of turbulent transfer through the layer of air immediately above the surface will be a limiting factor in some cases, such as for soluble gases over water, and molecular diffusion or particle inertia will be important in others. The rate of dry deposition, then, will be determined by factors such as these and the quantity of material available in the airborne source. The common factor in all such considerations is the concentration in air. If pollutants are not transported to the vicinity of receptors at risk or are somehow made unavailable to the corresponding delivery mechanisms, then pollutant impacts cease to be a problem.

An obvious complication is that the deposition of pollutants to the surface is the natural sink mechanism that limits atmospheric concentrations. Without deposition (and in the absence of other decay or transformation mechanisms), atmospheric concentrations would increase without bound, as determined by the appropriate emission rates. Thus we cannot divorce considerations of deposition from those of concentration.

The subject of the transport and deposition of pollutants is clearly too broad for detailed scrutiny here. The two major mechanisms by which pollutants are delivered to vegetation are dry and wet processes. Even these subjects are sufficiently broad that detailed coverage is beyond the scope here. The complexities of these problems, and the uncertainties involved, become apparent as one reads the information presented in other sections of this volume and compares the viewpoints expressed by the various authors.

It is of interest to ask where in the atmosphere a pollutant should reside in order to be available to these mechanisms of dry and wet deposition. Clearly, dry deposition occurs in response to pollutant concentration in the immediate vicinity of the surface. In the case of wet deposition, in-cloud (rainout) processes draw air from well above the surface, and so it is important to know concentrations throughout the mixed layer and possibly above it. Subcloud (washout) mechanisms scavenge pollutants from the entire layer of air through which the precipitation falls. A critical question is how we can predict or account for concentration profiles in the lower atmosphere. This, then, is the question of transport.

There is a tendency for nonspecialists to be unaware of recent advances in meteorological and air pollution modeling. The references that are often used represent modeling capabilities about 10 years ago. Recent developments have been largely in response to the need to relate sources and effects over relatively large distances, and to provide a defensible basis for assessing the environmental impacts of alternative emission control strategies or regulatory actions. A number of improved numerical models are now available, ranging in complexity from relatively simple modified Gaussian plume models (suitable for use over flat, uniform terrain up to distances of 50–100 km), to trajectory and statistical trajectory models that include descriptions of the diurnal cycle and permit use over distances that correspond to several days (the so-called regional scale, e.g., 500–2000 km). More sophisticated models are presently in use as research tools, both to address specific nonuniform situations such as complex terrain or coastlines and to provide guidance for the continued development of simpler models.

The large number of available models is an illustration of the complexity of the problems that are being addressed in contemporary research. Most of the models are designed to address specific problems and are constructed to emphasize corresponding critical factors. For example, a model intended for use in studies of fluorocarbons is unlikely to have the dry and wet deposition emphasis that would be required of a model intended to address the dispersion of sulfur oxides. The selection of an appropriate model for use in some particular assessment exercise is thus a matter that requires familiarity with both current modeling and modelers.

Even after careful selection of an appropriate model, it must be remembered that the answers obtained will be limited by the inadequacies of the meteorological data on which the model is based. Both spatial and temporal resolution are influenced by the corresponding resolution of radiosonde data which form the basis for most calculations and provide the checks necessary in dynamic models. Nevertheless it is clear that modern simulations provide evaluations of airborne concentrations that are indeed adequate for many purposes.

The importance of atmospheric transport processes has been emphasized in the earlier parts of this book (refer to Demerjian), and many current models and a brief outline of significant features of each kind are listed. It can be concluded that there is a need for accurate transport and dispersion models. If we consider some particular pollutant, and if all we know is an emission inventory, then models are required to produce the spatial and temporal concentration distributions from which surface fluxes and hence ecological doses can be evaluated. However, if we have access to accurate monitoring data, then the models are required to interpolate between stations and to infer undetected short-term variability. In any case, atmospheric transport and transformation models are required to link observed effects to suspected causes and hence play a critical role in any consideration of possible emission control strategies or regulatory actions. A further role, which has been anticipated but which has yet to be fully explored, concerns their use as a predictive tool, both to enable more accurate warning of pollution episodes and to permit some corresponding remedial emission changes to be made (such as by fuel switching).

DISCUSSION

A. Dry Deposition

The processes associated with the transfer of soluble gases to vegetation appear to be relatively well understood. Discussion is directed to the biological factors and three gaseous pollutants: sulfur dioxide, ozone, and hydrogen fluoride. With the first two gases, daytime stomatal control is well accepted; the results of all recent field experiments support the earlier work in chambers in this regard. For highly soluble species such as SO_2 we have confidence in extending the stomatal and mesophyll uptake models developed in agrometeorology primarily for studies of evapotranspiration (see the description given in the paper by Hosker in this book), since the most likely site of SO_2 absorption is the moist mesophyll tissue inside stomates. It is only when stomates are closed that cuticular resistance appears to be an important factor. In the case of ozone, the experimental finding of stomatal control presents something of an enigma since we know that ozone is far less soluble than gases like SO_2, therefore we must propose some currently unknown chemical mechanism for the rapid destruction of ozone upon contact with mesophyll tissue. It is well known that ozone is indeed rapidly destroyed upon contact with some surfaces; but in this particular instance we must propose that epidermal tissue does not provide a site for efficient ozone reaction, nor is cuticular uptake significant when stomates are open. Exchange of hydrogen fluoride is strongly influenced by concentrations within leaves; in some circumstances (such as when affected by acidic rainfall), previously absorbed hydrogen fluoride can be released from foliage. The complexity of this behavior might not be unrepresentative. Even the uptake rates of SO_2 and O_3 are strongly affected by factors such as the presence of dew, which serves to promote SO_2 deposition (at least until limited by the pH of the liquid) but inhibits that of O_3. A further complexity is introduced by the sensitivity of some leaf tissue to particular chemicals, with the consequence that exposure to some pollutants (probably including SO_2 and O_3, for example) will result in a reaction which might modify or even help regulate the rate of uptake.

The transfer of particles to vegetation remains a contentious issue. Large particles fall from the air at a velocity that depends on their size, density, and shape. The capacity of leaf surfaces to retain large particles that are deposited upon them is also complex. Particles may bounce from surfaces, even leaf surfaces, unless some sort of sticky substance is available to retain them. Pollen particles, for example, are more efficiently deposited to sticky stigmatic surfaces than to less sticky leaves.

The deposition of very small particles is limited by their low Brownian diffusivity. However, the exceedingly low deposition velocities predicted by many diffusive models might be considerable underestimates because of the role played by wax on leaves, leaf hairs, moisture films, etc. Most atmospheric sulfate is associated with particles about 0.8 μm in diameter, corresponding to the minimum deposition velocity predicted by some of the diffusive models. The model results appear to be confirmed by wind-tunnel studies. Yet there is evidence from other modeling studies and from some field experiments that these predicted deposition velocities might be too low. In this area of research, there is considerable variability between the results of different kinds of field experiments. Gradient investigations have largely failed to detect the vertical differences in sulfate concentration that would be expected on the basis of some recent eddy correlation and mass balance experiments. At this time it appears that the major processes controlling the deposition of small particles remain to be identified.

The dry deposition of gases and particles is influenced by meteorological and biological processes that both display strong diurnal cycles. In consequence, we must expect to find great variability of the deposition flux of any atmospheric trace quantity. Measurements of the flux of atmospheric sulfur to pine plantations, measured by micrometeorological methods in independent field experiments conducted in England and the United States during 1977, reflect the influence of both meteorological and biological factors. Diurnal cycles in the deposition flux are similar to those in the more familiar meteorological quantities, such as sensible heat and momentum. Spatial variability is also likely to be high, especially beneath the top of a canopy or in areas of nonuniform surface.

B. Wet Deposition

Whereas dry deposition provides a fairly regular rate of delivery of pollutants to a surface, precipitation concentrates air contaminants and delivers them in highly irregular, intense doses. There is, in fact, a spectrum of related events. Dew-fall can enhance the deposition of some materials and inhibit that of others. Fog provides a mechanism for scavenging aerosol from low-level air through droplet nucleation, condensation and coagulation, and subsequent interception. Sulfate deposition rates to a moist forest canopy can be more than three times the values reported for a dry canopy.

Precipitation characteristics and distributions vary considerably with both time and location. Different areas of the world experience substantially different kinds of precipitation—for example, the long, slow, synoptic rains of the United Kingdom; the orographic rains of western North America; and the violent thunderstorms of the American midwest. These three kinds of systems are chosen to illustrate the major differences in the part of the atmosphere from which pollutants are scavenged. The drizzle that is commonly associated with English winters originates well above the polluted mixed layer; falling raindrops scavenge airborne particles by impaction and coagulation, a subcloud process that is often referred to as washout. Orographic clouds are often stationary, deriving their water and much of their pollution from air flowing through them. The in-cloud processes by which pollutants are incorporated in precipitation are often known as rainout mechanisms. In the case of the highly convective storms that typify spring and summer in the midwestern United States and Canada, air is drawn from the entire mixed layer, and both subcloud and in-cloud mechanisms must be taken into account. If we then consider the obvious differences between scavenging by snow, hail, rain, and mixes between them, it becomes clear that the case-to-case and place-to-place variability of precipitation chemistry could be more than that of precipitation alone. It is informative, in this regard, to recognize that efforts to identify trends in any meteorological quantity have been contentious, especially in the case of precipitation. Excursions from the long-term mean lasting for a few years are sometimes erroneously interpreted as evidence of trends in precipitation, and are accepted as true long-term trends rather than local aberrations.

The generally accepted trend in rainfall acidity appears to be real, even though its origins and potential causes remain obscure. Historical data provide an inadequate reference against which to compare contemporary results; many of these data were obtained using methods and instrumentation that are not now acceptable. Interpretation of these older results is correspondingly difficult, and many of the conclusions drawn from them remain contentious. Moreover, it is often unclear whether detected trends in pH are due to changes in precipitation, in emissions of materials such as SO_2 (as is frequently assumed) or in the emission rates of potentially neutralizing fugitive dust. In any particular circumstance, the correct answer might be any one of these factors, or some combination of them. To understand the causes of rainfall acidity and to enable appropriate regulatory or control steps to be designed, the mechanisms of scavenging of air pollutants by precipitation must be understood.

Although single observations of precipitation chemistry can vary widely, averages derived from the U.S. Multistate Atmospheric Power Production Pollution Study's precipitation-chemistry network in the northeast seem well behaved, and it appears that the concentrations of species like sulfate, nitrate, and various trace metals can be extended into Tennessee without much change.

Models are available for relating concentrations of material in rain to concentrations in air and to other important meteorological and pollutant characteristics. These models offer the ability to derive concentrations of specific species in precipitation, and hence to estimate acidity (and pH) from an ion balance. However, it is known that some storms cause in-cloud oxidation of SO_2 and hence give rise to higher sulfate concentrations than would be expected on the basis of scavenging of sulfate aerosol alone. This result appears to be fairly well founded for North American convective storms; it is the result of studies of both ion balances and scavenging ratios (ratios of concentrations in precipitation to simultaneous concentrations of the same material in the mixed layer).

The detection of considerable quantities of sulfite in rain in the eastern United States indicates a previously unsuspected seasonal dependence. Sulfite concentrations can be predicted with greater accuracy than many other parameters, since sulfur dioxide solubility and its pH dependence are well known.

Much of the detected rainfall acidity can be attributed to sulfate and nitrate ions, but the apportionment between these species is seasonally dependent. In winter in the eastern half of North America, there is usually more nitrate than sulfate; in summer the opposite is true. This difference is not so evident in the European data. Ammonium and trace metal concentrations vary widely from place to place, as well as with season. Furthermore, weak acids appear to contribute significantly to the overall ion balance. It is suspected that a number of these weak acids might be organic. The matter of precipitation scavenging of airborne organic materials remains to be addressed, although there are some indications of relatively low scavenging ratios of hydrophobic organics.

Not all clouds result in precipitation. Most evaporate without raining, and sometimes falling precipitation evaporates before reaching the ground. In this regard, clouds act as an agent for redistributing material and exposing it to a saturated atmosphere in which many chemical reactions may be promoted or accelerated.

With the present knowledge, we can predict rainfall concentrations of most pollutants from air concentrations and meteorological properties within an order of magnitude and sometimes within a factor of two. We can explain the observed variability quite well, but as yet we cannot predict it. Moreover, existing precipitation scavenging and transport models impose a smoothing so that it is often not possible to relate single observations with suspected causes or emissions. To improve the present capabilities, we need to know more about the scavenging of submicron particles and about aqueous-phase chemical conversion. We need to investigate the behavior of mixes of different kinds of particles and gases, and we must generate an improved understanding of cloud microphysics.

C. Transport and Dispersion

Several relatively sophisticated numerical models have been constructed in response to the need to relate observations to specific sources at distances of up to 1000 km to assess the present impact of selected sources and to study the probable effects of alternative control strategies. These models are capable of estimating the transport of pollutants across international (or other) boundaries. Presently available models include descriptions of diurnal cycles of meteorological properties like the depth of the surface mixed layer. They are driven by observed meteorological variables, usually derived from routine radiosonde ascents. In some of these models, dry deposition is permitted to proceed at a rate which is controlled by the meteorology and by surface characteristics such as vegetation type and roughness. Wet deposition is likewise introduced in varying complexities, but application of a simple scavenging ratio (which is sometimes rainfall rate dependent) is common. Atmospheric chemistry is usually simplified, but major features of reactions such as the oxidation of sulfur dioxide are generally included. Transport is simulated according to observed winds, sometimes along isentropic trajectories and sometimes with some allowance for major terrain features underneath.

Papers by Demerjian and Hosker, presented elsewhere in this book, consider many of the available models and modeling techniques. Hidy and Mueller (in this book) present examples of the data now available to test the results of different numerical models, such as the European Regional Model of Air Pollution (EURMAP) and Eastern North American Model of Air Pollution (ENAMAP) developed by SRI International; the Advanced Statistical Trajectory Regional Air Pollution (ASTRAP) model of Argonne National Laboratory; and the Regional Air Pollutant Transport (RAPT) model of the Pacific Northwest Laboratory. Models such as these provide a large range of capabilities. Some models (e.g., EURMAP-2 and RAPT) permit consideration of short-term events and can be used to investigate the origins and causes of particular episodes. Other models (e.g., ASTRAP, EURMAP-1, ENAMAP-1) are used for calculating longer-term averages of concentrations and deposition. The latter models are computationally inexpensive and therefore are attractive for use in scenario studies that require the consideration of a large number of different control or regulation options.

D. Conclusions

The considerations of the available transport models leads to the conclusion that the link between specific sources and sites of interest can be described with some confidence. Naturally, the accuracy of the association and the confidence in a prediction or assessment will increase with the averaging time involved. Although existing models concentrate on sulfur oxides, extension to other pollutants and properties of ecological interest presents no large problems, provided sufficiently simple yet accurate equations can be used to describe the appropriate chemical reactions. More sophisticated models are currently in use as research tools and are being developed for routine use at this time.

Present understanding of dry deposition is most complete for soluble gases such as sulfur dioxide and worst for submicron particles. Wet deposition of hydrophilic particles and soluble gases by subcloud processes is also relatively well understood, but in-cloud processes can only be poorly approximated at this time.

PART FIVE

TERRESTRIAL VEGETATION–AIR POLLUTANT INTERACTIONS: GASEOUS AIR POLLUTANTS

On a global scale, terrestrial vegetation is intermittently exposed to gaseous air pollutants. Under specific conditions the effects are beneficial, but in general the effects are negative. The effects can be acute, chronic or subtle. Further, the effects can be direct through canopy absorption and/or through changes in soil physics, chemistry, and biology. Complicating these features are the effects of mixtures of pollutants and loss in growth and/or productivity without visible injury symptoms. Possible interactions between air pollutants and biotic and other abiotic environmental factors require better understanding. The difficulty in translating laboratory data to field situations is of concern, as is the small amount of information on the response of plants to varying pollutant concentrations simulating ambient conditions. Further, additional efforts should be made to interpret air quality (pollutant measurement) data and related technology for use with effects research.

Selection and measurement of the response parameter itself must be carefully determined. Translation of data to economic and/or aesthetic consequences is another question that must be addressed. Biologists need to communicate with physical scientists to better understand the physics and chemistry of air pollution and how they relate to effects.

These and other considerations are the focus of this part.

Photochemical Oxidants

V. C. RUNECKLES
Department of Plant Science
University of British Columbia
Vancouver, British Columbia, Canada

1. PREAMBLE

The photochemical oxidant air pollutants comprise the gases ozone, peroxyacetyl and other nitrates, nitric oxide, and nitrogen dioxide. With the possible exception of the peroxyacyl nitrates, the major components of the atmospheric oxidants are naturally occurring (Valley, 1965), and hence are only considered to be pollutants when and where higher than normal atmospheric loadings occur, usually as a result of human activities.

The origins of photochemical oxidants, their atmospheric chemistry, their effects on human health, on plants and animals, and on materials and their regulation, have been the subject of numerous reviews in the past. Many aspects are dealt with at length in other chapters in this volume. Two reviews, however, deserve special mention, because of their comprehensive nature, namely that of the National Research Council of Canada's Associate Committee on Scientific Criteria for Environmental Quality (NRCC, 1975), and that of the National Academy of Sciences' Committee on Medical and Biologic Effects of Environmental Pollutants (NAS, 1977). Each of these provides an extensive overview of the effects of photochemical oxidants on vegetation, while the latter provides the additional benefit of an examination of the effects of oxidant-pollutant stress on ecosystems. More recently, various aspects of the effects of oxidants on vegetation have been reviewed in the proceedings of the 32nd School in Agricultural Science held at the University of Nottingham in 1981 (Unsworth and Ormrod, 1982), in the proceedings of the international symposium on the biochemical effects of gaseous air pollutants on plants (Koziol and Whatley, 1984), and in Treshow (1984).

Taking these and other reviews as the baseline of information about the effects of oxidants on terrestrial vegetation, this chapter discusses the shortcomings of the current data base and offers suggestions as to where deficiencies can be overcome by the application of future research efforts. Wherever possible, an attempt has been made to develop proposals on the basis of a critical review of the available data, by asking three questions: What could be done? What should be done? What must be done?

The nature of the subject matter is such that no ideal subdivision exists. There is inevitable overlap between topics; some aspects have received greater emphasis than others. However, the chapter is organized to focus discussion on the major issues that should receive attention in the future. Furthermore, I have been guided by a personal philosophy that emphasizes the need for experimentation and observation that is relevant to the pollution problems of the real world and not merely dictated by the necessity or simple desire to publish data. The problems of oxidants in the real world are too pressing to condone the diversion of research effort into busy work that leads to little advance either in understanding the problems or in developing solutions.

2. UPTAKE OF OXIDANTS

2.1 Sorption and Deposition

McCune (1973) has summarized the action of pollutants on vegetation in terms of the four basic elements: receptor, pollutant, event, and environment, and has stressed that the definition of receptor has three parts: it is an object or a population of objects which receives a pollutant, and it has a locus along the scale of biological organization.

At the risk of stating the obvious, oxidant pollutants in the ambient air may be harmless to vegetation, regardless of concentration, since effects on vegetation (events) only occur when the pollutants come in contact with receptor sites, whether on the surface of the plants or within them.

Hence, all attempts at reaching an understanding of the effects of oxidants on vegetation require an appreciation of the ways in which pollutants reach receptors, the factors which influence the concentration of pollutant that reaches the receptor site, and its duration.

The leaf and its cells and tissues are usually the primary organs in which the effects of oxidants are demonstrated, because they are the organs with the greatest surface/volume ratio exposed to the atmosphere and are primarily responsible for the exchange of gases. However, little attention has been paid to the effects of sorption of oxidant pollutants onto the surfaces of leaves, although Bystrom et al. (1968) showed that the morphology of the cuticular waxes of the leaves of *Beta vulgaris* was changed by oxidants.

In terms of dynamics, Hill (1971) developed the concept of vegetation as a sink for air pollutants, but concluded that sorption through open stomata was much greater than surface deposition. Thorne and Hanson (1972) reached similar conclusions for the absorption of ozone by several plant species, although their data also revealed significant deposition velocities for nontranspiring leaves and paper surfaces. Several workers have shown that the soil surface may provide an effective sink for ozone (Macdowall, 1974; Turner et al., 1974), and there is no a priori reason to suppose that the surfaces of leaves may not also act as significant sinks, without penetration, nor that considerable interspecific variation may occur because of the diversity of cuticular deposits and epidermal structures, and the high chemical reactivity of ozone and other oxidant species. Nevertheless, the overwhelming bulk of evidence clearly demonstrates a close relationship between ozone uptake and transpiration (Rich et al., 1970; Thorne and Hanson, 1972). (The recent review by Galbally and Roy (1980) has brought together much of the relevant data.) As a consequence, the passage of the pollutant through the stomata and into the intercellular spaces of the leaf and the regulation of this process have been the focus of attention of several plant physiologists, and have led to the development of several models of pollutant uptake which will be discussed in Section 2.2.

The micrometeorological approach to deposition which has been extensively investigated with respect to other pollutants, especially SO_2 (Fowler and Unsworth, 1974; Fowler, 1978; Garland, 1978; Schwela, 1979) has also been applied to ozone uptake (see Galbally and Roy, 1980). An early example is the work of Mukammal (1965), which incorporated an experimentally determined coefficient of evapotranspiration as a measure of mass transfer in establishing a relationship between tobacco injury and oxidant dosage under field conditions.

2.2 Models of Gas Exchange

Several workers have developed models of pollutant uptake along the lines of those established for photosynthetic and transpiratory gas exchange by leaves. The model proposed by Bennett et al. (1973) was improved upon by O'Dell et al. (1977). Both used the electrical analogue simulation approach to gas diffusion, and the latter authors were able to demonstrate a close agreement between observed and predicted uptakes and uptake rates for several gases, although none of those reported on were in the oxidant group.

While models of pollutant uptake have intrinsic value, they are of additional importance as a tool in determining the effectiveness of experimental exposure facilities. Bennett (1979) in his review of foliar gas exchange, has stressed that

> *plant assimilation studies conducted in controlled environments should select conditions that approximate plant microenvironments in the field. Mass and energy budgets . . . need to be monitored . . . for use in characterizing the integrity of the system . . . This requires information . . . ranges of [uptake] rates to be expected . . . Regression data and multivariate analyses stressing functional dependencies are to be preferred over "point in time and space" data.*

The work of Leuning et al. (1979) compared the micrometeorological and physiological approaches to estimating ozone flux to field tobacco and concluded that the physiological approach involving measurements of leaf diffusive resistances provided a better basis for the prediction of biological response, and for estimating the magnitude of the role played by vegetation as an oxidant sink.

Both approaches require an appreciation of the extent of turbulence in the ambient air. In its simplest terms, turbulence is related to wind speed. Ashenden and Mansfield (1977) clearly demonstrated the particular importance of using realistic wind speeds in studies of response conducted in experimental chambers, to avoid the existence of unrealistically large boundary layers with their concomitant increased aerodynamic or boundary layer resistances, leading to exaggerated values of concentration required to elicit a given response.

2.3 Leaves and Canopies

Little recognition has been made of the inherent differences in gas exchange (and hence pollutant flux) between plants treated at wide and close spacings. In most experimental work involving exposure chambers, groups of individual plants are exposed simultaneously, but the conditions of air circulation around the leaves of such plants are probably very different from those which occur in most field situations, where, even in row crops, a canopy is present. Bennett et al. (1973) stressed the importance of studies using plant canopies to better understand both their effects on pollutant penetration (and hence response) and their role in removing pollutants from air. Their studies with ozone, NO_2, and other gases clearly showed the relationships between the solubility characteristics of the individual pollutant gases studied and the shape of the concentration gradients established vertically through alfalfa canopies. In the case of ozone, ready deposition occurred onto or into leaves close to the canopy surface, but the deeper leaves had relatively less effect upon the concentration profile.

Some of the discrepancies between the reports of different workers regarding dose–response relationships may thus be attributable to canopy effects, for dosages received by individual plants at close spacings will be less than those received by isolated plants exposed to the same general ambient concentration of pollutant. However, there appear to have been no specific studies reported which have focussed on this problem of relating dose or concentration of oxidants to response, although such studies have been undertaken with respect to SO_2.

2.4 Effective Dose

The purpose of the foregoing has been to point out that knowledge of the ambient oxidant concentration is only one of several factors which are essential prerequisites to establishing meaningful dose–response models. Deferring for the moment such matters as interactions among pollutants, and the effects of environmental factors, both abiotic and biotic, there has been too little emphasis placed in the past upon the definition of the experimental conditions used for many studies of oxidant effects, particularly with regard to the meaning of the pollutant concentrations quoted in many publications. With few exceptions, such concentrations are usually described as "being within the chamber," with little or no information to help the reader to determine whether this represents the concentration supplied, or the concentration close to the leaf surface. Such concerns led

Runeckles (1974) to stress the need for an appreciation of effective dose, rather than a mere statement of ambient concentration, and further led to the development of the continuous stirred tank reactor (CSTR) system for exposing plants to air pollutants (Rogers et al., 1977). Black and Unsworth (1979) used the effective dose concept in their examination of SO_2 fluxes to bean plants. Fowler and Cape (1982) have extended this idea further by the introduction of the concept of pollutant adsorbed dose, which incorporates concentration, time, and stomatal (or canopy) conductance.

The lack of radioactive isotopic forms of the elements of the major oxidants (ozone, oxides of nitrogen) has prevented the use of the approach used to determine the uptake and ultimate fate of pollutants such as SO_2. The potential for using stable isotopes such as ^{18}O and ^{15}N exists, but only recently have studies of this type been reported. Rogers et al. (1979) used $^{15}NO_2$ at concentrations of 0.1–0.32 ppm to demonstrate the ready uptake and metabolism of the gas by bean plants.

3. DOSE–RESPONSE RELATIONSHIPS

3.1 Acute, Chronic, and Subtle Effects

The bulk of the studies on dose–response relationships reported in the literature related to acute injury, in which oxidant (usually ozone) dosages have been sufficient to result in visible necrosis of leaf tissues within 24 to 48 h. While there has been a growing number of studies carried out over the past few years on the chronic and the subtle effects of oxidants, only recently have studies begun to be reported on dose–response models which involve responses such as growth and yield following long-term exposures. On the other hand, several theoretical or empirical relationships have been formulated which relate damage (the economic consequence of injury; Guderian et al., 1960) to air pollution, regardless of type of pollutant or type of injury. For example, Liu and Yu (1977) developed a crop-loss model incorporating both oxidant and SO_2 responses, while Oshima et al. (1976) utilized the range of ambient oxidant levels which occur within the South Coast Air Basin of California to determine ozone–crop-loss functions for alfalfa based upon defoliation (and hence crop quality) and yield. Such models do not attempt to define the nature and type of injury response, but their ability to generate approximately linear relationships between yield and accumulated dose (above a selected threshold) suggests that acute and chronic responses may form a continuum, even though the nature and mechanisms of the responses are different. On the other hand, Bucher and Keller (1978) have suggested a reversed sigmoid relationship between concentration and duration for the threshold response to SO_2 exposures covering the range of durations from less than one hour to one year, with a marked discontinuity occurring at the transition between acute and chronic responses. However, no comparable proposals have yet been made with regard to oxidants. What is clear is that effects on young plants cannot be used to predict long-term effects (Runeckles, 1976; Heagle et al., 1979c). Currently, the National Crop Loss Assessment Network (NCLAN) program in the United States is developing both dose–response and economic loss models for major agricultural crops, and has shown that the form of the dose–response function is crop-specific, and may be linear, plateau-linear, or curvilinear overall. The responses appear to be best described by Weibull functions (Heck et al., 1983, 1984a, b).

3.2 Acute Dose–Response Models

Various models of acute injury to vegetation have been developed. Those of O'Gara (1922), Guderian et al. (1960), and Zahn (1963), which were developed for SO_2, are essentially hyperbolic and relate concentration and duration of exposure for a given level of response (threshold or greater). Heck et al. (1966) investigated the interrelationships of ozone concentration, duration of exposure and injury (response) on tobacco and pinto bean, and showed that although both injury–concentration (at constant duration) and injury–duration (at constant concentration) relationships were sigmoidal, the relationship between ozone dose and injury was nonlinear, in confirmation of the findings of Guderian et al. (1960) and Zahn (1963) with SO_2. Nevertheless, there are innumerable differences between exposure chamber studies and field exposures to ambient oxidants, which may account for nonlinearity on the one hand and apparent linearity on the other (cf. Mukammal, 1965 and Macdowall et al., 1964, 1966).

The first, albeit simplistic, model of acute oxidant injury, was that of Heck and Tingey (1971):

$$C = A_0 + A_1 I + A_2 / T$$

where C is ozone concentration, I is the response measure (percent injury or percent reduction from control), T is time, and A_0, A_1, and A_2 are constants which relate to inherent and external factors affecting sensitivity.

A more elaborate model was developed by Larsen and Heck (1976) on the basis of analysis of the foliar response of 14 species to ozone. They recognized that, for a given exposure duration, response and concentration are log-normally related, and that the concentration required to achieve a given response is an inverse function of duration. Their final equation,

$$c = m_{\mathrm{ghr}} s_{\mathrm{g}}^{z} t^{p}$$

incorporated concentration (c), the geometric mean concentration for a 1-h exposure to produce 50% injury (m_{ghr}), the standard geometric mean deviation of concentration (s_{g}), the number of standard deviations of injury from median injury (z), exposure duration (t), and the slope of the log–log plot of concentration (ordinate) and duration (abscissa) for a given level of injury (p). By rearrangement, the model can be used to predict injury (in terms of standard deviation) for those species for which the various parameters m_{ghr}, s_{g}, and p have been determined.

More recently, Nouchi and Aoki (1979) have presented a somewhat simpler model for oxidant injury to *Pharbitis nil*, in which injury is related to oxidant dosage, which takes the form of $C^{2.2} \cdot t$—that is, they chose to raise concentration by an exponent, rather than modifying duration by means of a negative exponent (Larsen and Heck, 1976). Their model gives a slightly better fit to experimental data than does that of Larsen and Heck but is at present limited to the single species studied.

A somewhat different approach has been used by Naveh et al. (1978) in employing the injury response of tobacco Bel-W_3 to estimate average ambient field concentrations. They established a response curve of the type

$$P = P_k(1 - e^{-kt})$$

where P is the percentage of injured leaves at time t, P_k is the equilibrium percentage (i.e., the maximum percentage of leaves susceptible to injury), and k is a factor which is linearly related to oxidant concentration. This study is of particular interest because, although the response used was acute injury, the study extended over several weeks in ambient air in which the daily oxidant concentrations followed a typical diurnal fluctuation with a peak shortly after midday. Since accelerated senescence and leaf drop occurred, the study may have some relevance to dose–chronic-injury relationships.

A further development is that of Bicak (1978), who proposed acute injury relationships for bean and radish which encompassed constant dosages and constant durations in which concentration was varied over the range ±100% of the average, steady-state concentration. He was able to show that a linear relationship existed between observed visible injury and that predicted by

$$I = k\frac{\text{dose} \cdot C_{\text{max}}}{t_{\text{max}}}$$

where I is injury (percent of leaf area), C_{max} is the maximum concentration occurring at time t_{max} within the exposure period, and k is a constant related to species. Dose is defined as $\sum(c \cdot 1/r_s)$, where c is the average 1-h concentration, and r_s is the concurrent mean stomatal diffusive resistance, and is thus an approximation of flux, effective dose, or pollutant absorbed dose (Section 2.4). This model, therefore, has some similarity to that of Mukammal (1965) in that it incorporates an estimate of pollutant flux, rather than being solely dependent upon ambient pollution concentration (see also Section 3.3).

3.3 Dosage Forms

Most dose–response models have been developed on the basis of steady-state concentrations maintained for various durations. However, several workers have pointed out that dosages, defined simply as the product of concentration and time, do not elicit a constant response. Variations in concentration within a given period may range from the extreme, in which case the overall dosage essentially consists of a series of short, intermittent exposures, to the gradual, in which the concentration fluctuates around an average level. In extreme cases, the duration of the individual exposures and the length of the pollutant-free recovery time have a pronounced effect upon injury (Zahn, 1963; Heck and Dunning, 1967). However, such observations relate to the more general matter of predisposition and sensitivity and are discussed more fully in Section 4.2.1. The matter of the effect of fluctuating concentrations within a constant dosage or within a constant exposure duration has recently been addressed for SO_2 (Jones et al., 1979; McLaughlin et al., 1979; Male et al., 1983), but the work of Bicak (1978) appears to be the only study on ozone in which ambient dose and duration were kept constant. Several other researchers have allowed dose, concentration, and duration to vary (Heck and Dunning, 1967; Menser and Hodges, 1968). Bicak (1978) observed that, for 7-h exposures to ozone with constant dosage in which the steady-state condition was compared with treatments in which the peak concentration occurred in hours 1, 4, or 7, injury in relation to the steady-state conditions was increased by 69, 21, and 13%, respectively, for the three times of maximum concentration in bean, and by 26, 22, and 4% in radish. The significance of these data lies in their relevance to the types of fluctuation in pollutant concentration which occur in field situations, and add a further dimension to the problem of establishing reliable dose–response models.

The model of Nouchi and Aoki (1979) is unique in that it attempts to provide for accumulative doses resulting from exposures over several days. By incorporating a term based upon the sum of the log doses accumulated over the three days prior to a given exposure, they were able to increase the correlation coefficient between observed and predicted injury from 0.76 to 0.84.

One may question whether the search for the perfect model is worthwhile. Certainly, the wide range of genetic and environmental factors which influence plant response make it unlikely that a workable model will be developed to cover all species (or individual plants) in all situations. And in all attempts to express biological responses to treatment in mathematical terms there is the risk of excessive curve-fitting now made possible by numerous sophisticated computer programs, but which exceeds the reliability of the primary data. Nevertheless, there is still room to improve upon the existing models, or validate them for a wider range of species, to increase the accuracy of prediction of injury as a prerequisite to corrective action. Such an improvement is the recent introduction of the concept of effective mean concentration of ozone (Larsen and Heck, 1984).

4. INTERACTIONS OF OXIDANT POLLUTANTS

4.1 Simultaneous Reactions

While oxidants may occur in isolation in a given air mass, there is a substantial and growing body of evidence which indicates clearly that the concurrence of oxidants with other air pollutants is widespread, and that simultaneous exposures to pollutant mixtures give rise to various interactive responses which may differ both qualitatively and quantitatively from those caused by the individual pollutants. Recently, Lefohn and Tingey (1984) have analyzed the frequency of co-occurrences and sequences of various gaseous pollutants in several locations in the United States. Following the initial report of Menser and Heggestad (1966) that ozone–SO_2 combinations in ambient air acted synergistically in injuring tobacco leaves, the effects of interactions of ozone and SO_2 have been extensively studied. Macdowall and Cole (1971) showed that the synergism demonstrated on tobacco did not occur below the ozone response threshold, and Jacobson and Colavito (1976) determined that, for both bean and tobacco, the direction of the interaction was dependent upon species and pollutant concentration. Heagle and Johnston (1979) have demonstrated that the nature of the response (synergistic, antagonistic, or additive) of two soybean cultivars is furthermore dependent upon exposure duration and the severity of injury caused by each gas singly. Numerous other investigations have led to a growing body of information on additive, synergistic and less than additive responses of a wide range of species to ozone–SO_2 mixtures, but little work has been done on other pollutant combinations involving oxidants. Within the oxidant family itself, the only other investigations appear to be those of Matsushima (1971) and Fujiwara (1973) on ozone–NO_2 mixtures and of Kohut et al. (1976) on ozone–PAN mixtures, although such mixtures occur regularly during the buildup of photochemical oxidants in many locations.

Several studies have investigated the interactive effects of oxides of nitrogen and SO_2, since these are frequently produced simultaneously in the combustion of fuels. NO_2–SO_2 interactions have received the greatest attention. However, many of the reports are essentially empirical in their approach, and the results of studies such as those reported by Tingey et al. (1971) indicate a wide range of responses, varying from no effect to greater than additive, with a decrease in the injury response threshold in most species studied. Of particular interest are the reports of synergistic effects on the suppression of growth of several grasses by low level combinations of

NO_2 and SO_2 maintained for up to 20 weeks (Ashenden and Mansfield, 1978; Ashenden, 1979), and on the suppression of soybean yield following several weeks of intermittent exposure to the two pollutants (Irving et al., 1982).

These various studies serve to make one thing clear: Our knowledge of such interactions is still only fragmentary. Considerably more investigation is needed before we get beyond the stage of mere observation of the effects, as confirmed by the recent review by Ormrod (1982).

4.2 Sequential Interactions

4.2.1 Predisposition and Protection

While our knowledge of the interactive effects of simultaneous exposures to mixtures of gases is still limited, the same is true of our understanding of sequential exposures. The special case of sequential exposure to the same pollutant (e.g., ozone), has been mentioned already in Section 3.3, in the context of dose–response modeling. A further aspect of this special case situation is that in which there is an interaction between low levels of a pollutant (below the acute injury threshold) and subsequent acutely injurious exposures, that is, the lower limit of the range of types of sequential exposures. Macdowall (1965), Heagle and Heck (1974), and Runeckles and Rosen (1974, 1977) have demonstrated that several species may be predisposed to subsequent acute ozone injury by being subjected to previous dosages of ozone which may be either noninjurious or only mildly so. Where a sequence of severely injurious doses occur, the later doses result in relatively less injury than might be expected (Macdowall, 1965). In contrast, when the early doses are low, the magnitude of the injury response to a subsequent high dose is frequently enhanced. However, for bean leaves, Runeckles and Rosen (1977) have shown that the nature of the interaction between doses is dependent upon leaf ontogeny, in that the leaves pass through three distinct stages during their lifetimes; an early stage in which pretreatments which are noninjurious per se cause predisposition, followed by a stage in which a marked decline in susceptibility occurs, and terminated by a stage in which predisposition again occurs. The significance of these observations lies in their implications as to the relevance of dose–injury responses obtained by exposing plants which may have been either sensitized or desensitized by the prevailing ambient air conditions. This concern is thus related to that expressed by Bennett et al. (1974), who suggested that greater attention be paid to the avoidance of the use of truly pollutant-free (i.e., filtered) air as a control for dose–response studies, since the complete absence of low levels of a pollutant such as ozone from ambient air is unnatural. Other types of pollutant sequence involve oxidants and other pollutants. Although Hofstra and Beckerson (1981) showed that

SO_2 pretreatment could significantly change the response of four species to ozone or ozone–SO_2 mixtures, we have very little information about the effects of such real-world sequences.

4.2.2 Sequences of Oxidant Pollutants

Surprisingly, little attention has been directed towards the interactive effects of the sequence of gases which are typical of the process of photochemical oxidant accumulation (discussed in NAS, 1977) (i.e., oxides of nitrogen, ozone, PAN, etc.). Kress (1972) reported that sequential exposure of hybrid poplar to ozone–PAN or PAN–ozone resulted in synergistic injury responses in most experiments, but less than additive (protective) responses also occurred, particularly when PAN preceded ozone, or when the ozone–PAN sequence was immediately followed by darkness. NO_2–ozone sequences have also been shown to result in different interactions with different species (Runeckles et al., 1978). Thus, in radish, low concentration NO_2–ozone sequences administered daily exhibited synergistic suppression of growth; in wheat, a synergistic suppression over the first 40 days of growth gradually disappeared, leading eventually to no significant differences between NO_2–ozone and ozone alone, while in bean, NO_2–pretreatment reduced the magnitude of the response to ozone.

Such observations combined with the variability of the results obtained with various mixtures of pollutants indicate clearly the lack of a true understanding which still prevails with regard to the effects of pollutant combinations involving oxidants. A notable exception is the NO_2–SO_2 interaction, in which Wellburn (1982) has shown that synergistic growth reductions may be the result of SO_2-induced inhibition of nitrite reductase.

5. BIOCHEMICAL EFFECTS OF OXIDANTS

In seeking subcellular explanations for the observable responses of plants to oxidants, a wide variety of biochemical investigations has been undertaken. NAS (1977) and Koziol and Whatley (1984) present recent summaries, covering observations on such metabolites as sugars, amino acids, ascorbic acid, phenolic compounds, pigments, and alkaloids such as nicotine, and on many metabolic functions and processes. The problem, however, is to decide which changes are mechanistic, that is, related to cause, and which are consequences of cell injury. At both the metabolite and macromolecular levels, there is frequent conflict between the results of different workers, for example, in the effects of ozone on the soluble protein content of leaves. In part, the difficulties in interpreting the observations of biochemical change are a function of the specificity of the approach of the investigators. Plant cells have a well-defined (though not necessarily well understood) organizational structure, and investigations which tend to regard the cell merely as a bag containing a mixture of chemical constituents will rarely be able to progress beyond the empirical. Such investigations may yield knowledge, but rarely understanding. Thus, the investigations which have perhaps the greatest significance are those which have recognized that, in the case of an air pollutant penetrating the tissues of a leaf, there is a spatial and temporal sequence in the catalogue of likely events. The first biological "contact" with the cell will be at the plasmalemma, and only subsequently will the pollutant reach internal receptors (whether they be organelles or the constituents of the cytoplasm or vacuole), either directly or via the products of its chemical reactions. Since the cell, its organelles, and the vacuole are bounded by membranes, investigations of the effects of oxidants on membrane structures and function have been particularly useful in attempting to differentiate between cause and effect. Mudd (1982) and Mudd et al. (1982) summarized the results of such studies with ozone and noted that, while the chemical reactions of ozone with saturated lipids and with the sulphydryl groups and other substituents of protein amino acids may be well understood, we have no clear picture of the nature of the changes undergone by the plasma membrane as a result of ozone treatment. Indeed, the picture is further obscured by the observation of deep-seated effects, for example, the degradation of chloroplast polysomes (Chang, 1971), which may occur before any measurable effect on membrane structure is apparent, and which suggests that ozone can pass through the plasma membrane. The generation of free superoxide radicals from ozone, leading to the formation of H_2O_2, catalyzed by superoxide dismutase, can result in the formation of both singlet oxygen and hydroxyl radicals, which can seriously disrupt membranes lipids (Mead, 1976). Since several studies of animal membranes have implicated both sulphydryl groups and unsaturated fatty acids as precursors of ozone-induced free radicals, it appears possible that membrane integrity could be impaired as a consequence of both ozone-induced free radical formation itself, and the reactions of the powerful oxidant radicals formed. The increased activity of peroxidase following exposure to ozone (Dass and Weaver, 1972; Curtis and Howell, 1971) may represent a detoxification mechanism, as has also been postulated for the increased superoxide dismutase activity detected in lung tissue (Mustafa et al., 1975) and reported in bean leaves (Lee and Bennett, 1982) after ozone exposure. The implications of these free radical oxidants to ozone-induced plant cell injury must remain speculative for the moment, although lipid peroxidation and superoxide dismutase have been suggested as components of the mechanism for the maintenance of membrane function in drought resistance (Bewley, 1979). Tanaka and Sugahara (1980) have postulated a role of superoxide dismutase in the tolerance of poplar and spinach leaves to SO_2, and Lee and Bennett (1982) have similarly suggested that tolerance of bean leaves to ozone is directly related to superoxide dismutase activity. However, recent studies by Chanway and Runeckles (1984) and McKersie et al. (1982) tend to refute this relationship. Nevertheless, Tingey and Taylor (1982) have suggested that the inability of the cell to dissipate excess reluctants, allowing free radical formation, may provide the explanation for the observation that ozone injury is frequently seen first in the chloroplast. It is worth noting that many free radical reactions occur in normal cellular processes, including photosynthesis, and may only become pathologic when exogenously stimulated. Indeed, it has been suggested, on energetic grounds, that ozone itself could be derived from the superoxide radical; since the latter is formed during several cellular processes, ozone might thus be a normal transient intermediate in vivo (Shoaf et al., 1974). Such a mechanism might account for the apparent stimulation of plant growth afforded by low ozone levels in relation to ozone-free air (see Section 7.3).

In spite of the limited general understanding afforded by most biochemical studies, however, there are several approaches which seem to offer considerable promise. Firstly, advances in techniques for the fractionation of proteins have led to the observation that changes in the pattern of the isoenzymes of several enzymes may be a sensitive indication of injury (Curtis and Weinstein, 1979). The earlier reports of changes in the levels of several leaf enzymes, for example, peroxidase (Curtis and Howell, 1971; Dass and Weaver, 1972), glucose-6-phosphate dehydrogenase (Tingey, 1974), and nitrate reductase (Leffler and Cherry, 1974; Tingey et al., 1973) certainly warrant more detailed investigation as to isoenzyme changes. While the foregoing has focused upon the effects of ozone, several studies have been made upon the biochemical effects of PAN (see NAS, 1977). However, many of the effects of PAN are difficult to interpret because of differences between in vitro and in vivo experimentation. Nevertheless, studies with both oxidants have implicated reactivity with sulphydryl groups; hence isoenzyme changes are likely to occur as a result of PAN treatment.

Ozone, PAN, and the oxides of nitrogen have been shown to reduce the rate of photosynthesis, both in terms of CO_2-uptake and O_2-evolution; at low dosages, the reductions are usually reversible (reviewed in NAS, 1977; Mudd and Kozlowski, 1975). Similarly, inhibitions of O_2 uptake by plant mitochondria have been reported. Many of these observations have been associated with changes in the activities of various enzymes associated with specific metabolic pathways; many of the changes obviously contribute to the disruption of normal cell function and hence may contribute to permanent injury. However, because of the key role of the photosynthetic process, it seems reasonable to expect that interference with this process is likely to be of particular importance to the ongoing well-being of leaf cells, a view supported by the work of Pell and Brennan (1973). Furthermore, although the assimilation of CO_2 is essential for sustained growth, the light-dependent processes which are responsible for the generation of reducing power and the capture of energy are more fundamental and warrant more detailed examination as to the effects of oxidants. One technique which holds considerable promise in this regard is the exploitation of the so-called Kautsky effect, as suggested by Arndt (1972). The availability of instrumentation for measuring chlorophyll fluorescence in situ in leaves (Schreiber et al., 1975) led to an investigation of the effects of O_3 on bean leaves, which permitted Schreiber et al. (1978) to predict the severity of acute injury 20 h before visible signs of necrosis occurred. They interpreted the observed changes in chlorophyll fluorescence as indicating an early effect on the water-splitting enzyme systems associated with Photosystem II, followed by inhibition of electron transport between the photosystems. Of further interest is the observation that, while the visible injury response of the bean leaves investigated showed typical dose–response characteristics, in which, at constant dosage, high ozone concentrations of short duration were more injurious than longer exposures to lower concentrations, low concentrations over extended periods resulted in a greater inhibition of photosynthesis, but this inhibition was reversible, and recovery occurred on removal of the ozone. Since the durations of the highest concentrations were 1 h or less, the explanation probably lies in the rate of uptake into the leaves, cells, and chloroplasts, since fluorescence measurements must start with leaves in which the photosystems are relaxed, a condition induced by dark adaptation of the plants for 1 h during which time stomatal closure would have occurred to some extent. Such an explanation is in agreement with Tingey and Taylor (1982), who stressed the importance of recognizing that leaf injury is the resultant of two regulatory processes, one controlling penetration (leaf conductance), and the other related to capacity to repair or compensate for altered metabolic states (homeostasis).

Of particular importance to the long-term effects of oxidants are effects on endogenous growth regulators. Several reports have confirmed the effect of ozone on stimulating the release of ethylene by leaves (Craker, 1971; Abeles and Abeles, 1972; Tingey et al., 1976), and there have been scattered reports of effects on other growth regulators. However, because of the importance of such regulators in the processes of growth and development, the effects of oxidants on their synthesis and distribution warrant considerably more attention, particularly with respect to long-term, chronic responses.

6. PHYSIOLOGIC EFFECTS OF OXIDANTS

As in the case of biochemical effects, the physiologic effects of oxidants have been quite widely investigated and have been the subject of several reviews (Heck, 1968; Dugger and Ting, 1970; NRCC, 1975; NAS, 1977). Physiologic investigations have tended to fall into one of three groups:

1. Studies of the physiologic processes in organs or tissues (e.g. respiration, photosynthesis, stomatal function).

2. Studies of the effects on growth and differentiation.
3. Studies of the interactions between oxidants and other environmental factors on physiological processes.

In this section, I shall deal only with the first category; others are addressed in Sections 7 and 8, respectively.

Stomatal function has been shown by numerous investigators to be affected by exposure to oxidants (NAS, 1977). Stomatal closure has been shown to be under genetic control in onion, in which the stomata of a sensitive cultivar did not close when exposed to ozone (Engle and Gabelman, 1966). Since stomatal function also depends upon a complexity of interacting factors, including water relations, ion movement, and light, it is not surprising that ozone and PAN have been shown to interact with these factors in affecting stomatal opening and closing. The general observation that stomatal closure follows exposure to oxidants is an oversimplification; in many instances such exposures merely reduce the degree of diurnal stomatal opening controlled by light intensity. This effect has been discussed above in Section 3, with regard to its implications for uptake and effective dose. However, stomatal aperture is not the only factor which controls plant response, as shown by Ting and Dugger (1968) and others, and as discussed in NAS (1977) and by Tingey and Taylor (1982).

Net photosynthesis, the resultant of the concurrent processes of photosynthesis and respiration, has also been shown by several workers to be inhibited by oxidants; changes in net photosynthetic rate can be attributed, in part, to differential responses of the photosynthetic and respiratory systems (Pell and Brennan, 1973). Such processes and many others which are based on biochemical activities, including the development of visible necrosis, show marked variability in response as a function of developmental age. For example, in tobacco, Macdowall (1965) showed that maximum sensitivity to oxidant (ozone) occurred at the completion of leaf expansion, whereas in cotton, Ting and Dugger (1968) found that maximum sensitivity occurred at the time of the greatest rate of expansion. Such changes in sensitivity must of necessity be the resultant of a complex mixture of biochemical and physiological developments, and reference has been made in Section 4.2.1 to the ways in which sensitivity can be further affected by previous exposures to oxidants.

A technique that does not appear to have been employed in oxidant (or other) air pollution investigations and which might offer some assistance in overcoming the difficulties resulting from changes in susceptibility of tissues with age is the use of the plastochron index as a measure of developmental age (Erickson and Michelini, 1957).

Several investigations which have focused upon total plant growth have shown that the yields of leaves, stems, and roots may be differentially affected by ozone, as discussed in Section 7. Adverse effects of ozone on root growth, which lead to increased shoot/root ratios, appear to be widespread, provided that leaf injury is not so severe as to result in defoliation; Bennett (1975) argued that this was to be expected on the basis of general stress theory, which suggests that the plant organ biologically farthest removed from a given stress will exhibit the greatest response (Levitt, 1972; Mitchell, 1970). However, while such effects have been attributed to an impairment of translocation (Tingey et al., 1971), there appear to have been no studies directed specifically to effects on translocation per se.

7. EFFECTS OF OXIDANTS ON GROWTH AND MORPHOGENESIS

7.1 Effects on Yield and Biomass

There has been an exponential increase in the number of investigations reporting effects of oxidants on yield and biomass over the last decade, including a growing number of publications in which effects on the growth of different parts of the plant are described. NAS (1977) reviews the earlier reports, but there have been several recent significant developments in our understanding of effects on growth. In particular, greater emphasis is now being placed on long-term effects, in which visible injury, if it occurs at all, is of the chronic rather than the acute type. Reports of effects on yield have particular relevance in the context of agricultural crops and the establishment of standards of air quality. Many studies have used polluted ambient air, with charcoal-filtered air as a control, in greenhouse or field-exposure chamber studies, and have reported yield reductions in a wide range of crops. For example, Bennett and Oshima (1976) reported that season-long treatment of carrot plants with intermittent exposures to 0.19 or 0.25 ppm ozone resulted in drastic reductions in root growth, which were related to the weight of injured foliage. Thompson et al. (1976) found that the growth of two corn cultivars was significantly decreased by ambient oxidants. Heagle et al. (1979a) further showed that, while the yield of corn was reduced by exposure to chronic ozone doses, the threshold for yield reduction was higher than that for foliar injury. Both bean and tomato yields were shown to be reduced significantly by long-term exposures to ambient oxidants (MacLean and Schneider, 1976). Oshima et al. (1977) obtained similar results on a range of tomato cultivars and showed that foliar injury was a poor indicator of yield response. Studies on spinach (Heagle et al., 1979b), parsley (Oshima et al., 1978), cotton (Oshima et al., 1979), rice (Nakamura and Ota, 1977), and winter wheat (Heagle et al., 1979c) similarly have demonstrated yield reductions due to oxidants, especially ozone. The current NCLAN studies are expanding our knowledge of yield reductions attributable to ozone (Heck et al., 1984b).

Adverse effects of ozone on growth have also been reported for various nonagricultural species, including native grasses and trees; summaries of these effects are given in NAS (1977). Overall, there is a clear body of evidence indicating that yield and many other measures of growth are adversely affected by ozone over a wide range of exposure conditions. There is, however, much less information concerning such effects of PAN and the oxides of nitrogen. In the case of PAN, there appear to have been no long-term definitive studies carried out; one might question the need for such studies because of the usual co-occurrence of PAN with other oxidants, especially ozone. Indeed several studies which have been undertaken using ambient oxidants probably included PAN. With regard to oxides of nitrogen, numerous studies have reported effects of NO_2 on growth and development. Most of the early studies used extremely high concentrations ($\geqslant 1$ ppm), which caused visible injury. However, the studies of Thompson et al. (1970) on orange trees, and of Taylor and Eaton (1966) and Spierings (1971) on tomato, using low concentrations for extended periods, demonstrated that the effects of NO_2 were time dependent: Early apparent stimulation of elongation growth eventually led to increased leaf fall and yield reduction, even though NO_2 uptake did not result in exceptionally high N levels within the plants. Extended exposures of pasture grass species to low levels of NO_2 significantly inhibited leaf growth at 56 and 84 days (Ashenden, 1979), but the trend lines indicated recovery and possible stimulation after 140 days, again illustrating the time-dependence of such effects.

One aspect of growth which has received considerable attention concerns nodulation of legumes. Although no thorough survey of the Fabaceae has been undertaken, there are consistent reports of reduction or suppression of nodulation by ozone in species such as soybean (Blum and Tingey, 1977), pinto bean (Engle and Gabelman, 1967), and ladino clover (Letchworth and Blum, 1977). The mechanism involved is unclear, but Blum and Tingey (1977) showed conclusively that the effect was mediated by the leaves and was not a direct effect on the root system; hence, translocation or changes in the composition of the translocated assimilates appear to be involved.

7.2 Growth Dynamics

Few investigations have been made of the effects of ozone on the dynamics of plant growth. The use of the techniques of plant growth analysis (Evans, 1972; Hunt, 1978) and the application of curve-fitting procedures (Hunt and Parsons, 1974, 1977; Hunt, 1982) provide ready means of determining the effects of oxidants on the process of growth, and not merely the end results. With regard to ozone, Bennett and Runeckles (1977), working with ryegrass–clover mixtures, showed that long-term chronic exposures resulted in significantly decreased leaf area ratios, but there was no concomitant significant effect on either net assimilation rate (unit leaf rate) or relative growth rate. Oshima et al. (1978) partitioned the root and leaf relative growth rates of parsley grown with intermittent exposures to ozone, and showed that the marked reduction in root weight observed was brought about by an early reduction in root relative growth rate, with a concurrent increase in leaf relative growth rate, in spite of the fact that it was the leaves which were the primary receptor organs. Such studies and those on cotton (Oshima et al., 1979) clearly indicate that, in addition to inhibiting overall growth, ozone must also exercise a subtle effect on translocation to account for the reduced root growth. In fact, Bennett and Oshima (1976) have also shown that in carrot the negative allometric relationship between chlorotic leaf weight and root weight suggested that for every gram of chlorotic leaf tissue produced there was a 1.5-g loss of root weight. The techniques of growth analysis are thus capable of providing considerable insight into the processes of growth over extended periods, and provide information on the rates at which growth processes occur, and not merely the end results of such processes at harvest. Indeed, it should be stressed that the use of terms such as growth rate should be reserved for such studies, since useful rate data are not generated by single harvest studies. Since one of the consequences of long-term exposure to oxidants is accelerated senescence and death of leaves, the recently developed relative death rate concept (Runeckles, 1982) will have applications in oxidant studies.

Effects on growth are likely to be influenced by adaptation of the plant to oxidant stress. Such adaptation has been suggested (Section 7.1) in connection with Ashenden's (1979) work on the effects of NO_2 on grasses. A dramatic demonstration was provided by the effects of continuous exposure of radish plants to 0.17 ppm ozone, as a result of which, although hypocotyl growth was reduced, assimilates were partitioned so as to accelerate the development of new leaves (Walmsley et al., 1980).

7.3 Growth Stimulations

Bennett et al. (1974) noted the frequent observation of apparent stimulations of various aspects of growth following exposures to low concentrations of ozone and other pollutants, in comparison with controls grown in pollutantfree, filtered air, and suggested that the explanation lay in adaptation by many species to ambient levels of pollutant. In few cases do such stimulations appear to be of long-term benefit to the plant, and they may merely represent various types of accommodation to changes in overall assimilation and translocation. However, Maas et al. (1973) reported increased storage root–shoot ratios for table beets exposed to ozone, although total plant weight was reduced.

A more subtle type of growth stimulation is related to long-term accommodation. For example, Oshima et al. (1978) showed that, in spite of the early reductions of parsley relative growth rates for leaves and roots, recoveries occurred, leading to relative growth rates in excess of those of control plants. The maintenance of such stimulated growth rates may ultimately lead to a minimization of the overall effect on final yield or total biomass. This was shown by Runeckles et al. (1978) for wheat, exposed daily to 5–6 h of 0.08 ppm ozone for 178 days, in which biomass 40 days after germination was reduced 23% yet showed no significant reduction at final harvest. Heagle et al. (1979c), on the other hand, showed significant yield reductions of winter wheat exposed to a somewhat higher daily ozone concentration (0.13 ppm, 7 h/day) for 54 days during the latter part of the growing period, but no reduction with 0.10-ppm daily doses in spite of the fact that there was considerable foliar injury.

Hence the potential for growth stimulations requires further investigation before its true significance is established. At present, such stimulations tend to be relegated to the sidelines and labeled as interesting but unexplained observations.

7.4 Effects on Reproduction

Effects of oxidants on reproductive processes have not been widely studied. At the chromosome level, ozone has been reported to be a weak mutagen on the basis of studies of somatic mutations in *Tradescantia* (Sparrow and Schairer, 1974). Pollen germination has been shown to be inhibited by ozone in tobacco (Feder, 1968), *Lycopersicon* spp. (Gentile et al., 1971), corn (Mumford et al., 1972), and petunia (Harrison and Feder, 1974).

Ozone has also been shown to inhibit inflorescence growth (Craker and Feder, 1972) and flower bud formation (Feder, 1970), and hence may give rise to decreased seed yields for a variety of reasons.

Arabidopsis thalliana, which has a 35-day life cycle, has been used in one study of the long-term effects of ozone (Bruton, 1974). Although reductions in seed production and biomass were reported within generations, no demonstrable mutations could be detected over seven generations. On the other hand, preliminary studies in my laboratory have provided some evidence of selection in rapeseed grown through two generations in the presence of low levels of ozone (Runeckles, 1976). While the evidence of genetic effects transmitted via sexual reproduction is far from compelling, nevertheless, the topic warrants further investigation. The implications are far reaching and are discussed further in Section 10.2.

With regard to vegetative reproduction, low-level ozone has been reported to decrease lateral branching in carnation and petunia (Feder, 1970), although it stimulates tillering of annual ryegrass (Bennett, 1975) and wheat (Runeckles et al., 1978). Although definitive studies are lacking, it appears that such responses are the indirect results of effects on assimilation and the partitioning of assimilates, although a role for ethylene cannot be ruled out.

8. ENVIRONMENTAL INFLUENCES

Plant response to oxidants is modified by a wide range of factors. The physical environment plays an important role in susceptibility, as does the chemical environment (for example, with regard to nutrition, and interactions with other pollutants, in the air or in the soil). The biological environment is also important, and embraces interactions with plant pathogens, with insects and other pests, or with other individuals of the same species.

8.1 Physical Factors

The components of the physical environment have been studied extensively with regard to their influence on plant response to ozone. NAS (1977) provides a useful summary of the effects of light (quality, duration, and intensity), temperature, relative humidity, and water stress on responses to both ozone and PAN. Dunning and Heck (1977) have shown that significant interactions may occur among the effects of light intensity, temperature, and relative humidity, and that these interactions differ for soybean and tobacco. They further showed that there were interactions involving each factor independently between the conditions under which the plants were grown as compared with the conditions under which they were exposed to ozone. Their data clearly demonstrate the complex interrelationships between the different environmental factors under study and make clear the necessity for further investigation before we shall be in a position to predict injury. Their work also summarizes the dilemma posed by the abundance of conflicting data in the literature. For example, while most plants appear to be more susceptible to ozone when grown at low light intensities, there are reports of the reverse being true. The same situation exists with regard to temperature, with some species being reported sensitized by cool temperatures and others by warmer temperatures. The problem with such data is that, as Dunning and Heck (1977) have shown, significant interactions occur between such factors; unfortunately, many of the reports in the literature ignore the possibility of such interactions and do not provide sufficient information about the totality of the environmental conditions used to permit valid interpretations. Hence, many of these data must be treated with caution.

On the other hand, the evidence regarding the importance of water relations is much more consistent (NAS, 1977) and shows that increased water stress almost invariably results in decreased susceptibility.

8.2 Chemical Factors

Soil nutrients are the most important chemical factors that influence plant response, but our knowledge of their importance is both fragmentary and confused. Various studies have been conducted on the effects of the major nutrients (N, P, K) of different plant species in response to ozone, and again numerous contradictory results have been reported. The problem frequently appears to be related to lack of inclusion of a treatment at the optimal level of fertility appropriate to the other environmental conditions in use. As a result, no clear interrelationship between nutrient level and sensitivity has emerged. Even when optimal levels for growth are reported to have been used, conflict still exists with regard to sensitivity; for example, Macdowall (1965) found that tobacco grown at optimal nitrogen levels was least susceptible to ozone, whereas Leone et al. (1966) found the reverse. Other workers have found nitrogen levels to have little effect on susceptibility. Undoubtedly, other factors, both genetic and environmental, have influenced the outcomes of these studies.

Phosphorus and potassium have been shown to interact by several workers (NAS, 1977); a change from low to high P results in a change in K response, from a direct relationship with injury to an inverse relationship. Independent effects of phosphorus (Leone and Brennan, 1970) and potassium (Leone, 1976) on tomato show direct relationships, with increased levels (in soil and plants) resulting in increased susceptibility to ozone. In the case of potassium (Leone, 1976), it was shown that the effect was related to stomatal function. Heagle and Johnston (1979) have shown that maximum sensitivity of soybean occurs at a suboptimal level of general fertility, and they stress the need for nutrient levels to be controlled and reported upon in all studies of plant response to ozone.

Given that our knowledge of the interactions between oxidants and major soil nutrients is fragmentary and confused, we know even less about the ways in which other soil-borne chemicals interact with oxidants in terms of plant response. There is a dearth of information on interactions with soil pollutants, such as the heavy metals, and with essential trace elements. The few reports which have been published have shown that Zn increases ozone-sensitivity of cress and lettuce (Czuba and Ormrod, 1974) and soybean (McIlveen et al., 1975). Heavy metals such as Cd and Ni have been shown by Ormrod (1977) to increase ozone sensitivity of peas, at levels up to those at which the elements become toxic per se, above which ozone sensitivity is reduced. Recently, Harkov et al. (1979) have shown that increased ozone sensitivity of tomato induced by Cd is only demonstrable under conditions in which slight-to-moderate ozone injury occurs.

Several studies have been undertaken on interactions between ozone and levels of soil salinity, following the early observations of Oertli (1959) that salinity reduced the susceptibility of sunflower. Studies of pinto bean showed that salinity suppressed plant growth but that it also reduced sensitivity at higher ozone concentrations or dosages (Hoffman et al., 1973; Maas et al., 1973). Further studies by Hoffman et al. (1975) showed that salinity reduced ozone sensitivity of alfalfa, to the extent that at intermediate salinities, at which there was no yield reduction due to salinity itself, there was significant protection against ozone. In part, these observations can be accounted for by the effects of salinity on the plants' water status, but salinity undoubtedly also induces interactive effects in its own right.

Finally, there have been several reports of interactions between ozone and pesticides. Sherwood and Rolph (1970) showed that ozone and 2,4-dichlorophenoxyacetic acid (2,4-D) behaved antagonistically in terms of the response of tomato, zinnia, and elm. Similarly, Ordin et al. (1972) found that 2,4-D reduced the inhibitory effects of PAN on oat coleoptile growth. Greater than additive interactions have also been reported (Carney et al., 1973). The most extensive studies have been on ozone–diphenamid interactions, which have shown that ozone modifies the time course of diphenamid metabolism (Hodgson et al., 1974) and the nature of the diphenamid conjugates formed (Hodgson and Hoffer, 1977).

Several crop protection agents have been found to exercise a protective effect against oxidants; the subject has been extensively reviewed in NAS (1977). The most widely used materials have been fungicides and compounds with known or postulated antioxidant activity. Benomyl (methyl 1-butylcarbamoyl-2-benzimidazolecarbamate) has been widely studied, but, as with many other compounds or formulations, results have been mixed. Although protection against ambient oxidant or ozone has been reported for many species, on several it has been shown to have no effect. Since it does not induce stomatal closure (Taylor and Rich, 1974), an effect on internal physiologic processes is probable in those cases where it reduces ozone injury, since benzimidazole itself has been shown to provide protection, presumably because of an effect at the membrane level (Spotts et al., 1975). More recently, the antioxidant *N*-[2-(oxo-1-imidazolidinyl)ethyl]-*N*⁻-phenylurea (ethylene diurea, EDU) has been shown to protect beans from ozone (Carnahan et al., 1978; Hofstra et al., 1978). The linear relationship between EDU dose and the ozone dosage for 50% foliar injury to pinto bean suggests a defined antagonistic interaction within the plant. Protection against foliar injury has also been reported for potato (Clarke et al., 1978) and seedlings of black cherry and white ash (McClenahen, 1979). Although it is too early to be certain, it appears that EDU may have wide applicability as a protectant, but much still has to be done to determine its range of effectiveness, its cost-effectiveness, and its negative environmental qualities, if any.

8.3 Biological Factors

Oxidant air pollutants have been reported to affect the suceptibility of plants to plant pathogens, as well as the pathogens themselves. The earlier studies have been extensively reviewed by Heagle (1973), and are summarized in NAS (1977). Established fungal, bacterial, and viral infections have been frequently shown to confer localized resistance to foliar ozone injury, and in the case of viruses, the infection may be latent and symptomless (Davis and Smith, 1974). No recent studies have been reported on these phenomena, and the mechanisms involved remain speculative. A converse relationship between ozone and pathogenesis has also been reported, in which previous ozone injury increases susceptibility to subsequent infection, presumably by facilitating entry of the pathogen. Such relationships have been demonstrated for herbaceous crops (Manning et al., 1969; Wukasch and Hofstra, 1979) and forest trees (Costonis and Sinclair, 1972) and appear to be indicative of a general situation in which oxidant injury causes host species to become more susceptible to fungal pathogens and may also render them susceptible to weak parasitism by fungi that are normally saprophytic. Several studies have investigated the simultaneous growth of host and pathogen, and have shown that, in general, ozone reduces pathogenesis and the severity of infection by reducing spore germination and hyphal growth. There are also several reports in which the progress of infection and the development of ozone injury appear to be independent. Effects of ozone in reducing pathogen development are in keeping with the well-known toxicity of relatively high concentrations of ozone to microorganisms, which has been used in a variety of commercial applications for the control of fungal and bacterial growth.

Related to these observations on oxidant–host–pathogen relationships are observations of the inhibitory effects of ozone on nodule formation on the roots of legumes (Section 7.1), and on the reduction of numbers of mycorrhizal rootlets of pines (Parmeter et al., 1962), although this type of effect is undoubtedly an indirect one mediated through effects on assimilate partitioning.

Interaction studies have been extended to plant-parasitic nematodes. Weber et al. (1979) showed that the development of several nematode species was impaired by the exposure of infected soybean or begonia plants to ozone. In addition, ozone increased the suppression of host growth caused by infection. The reduction in the development of certain nematodes appeared to be related to the reduction in suitable feeding sites caused by ozone, whether because of foliar injury itself, or because of reduced root growth.

In spite of these studies, it would be misleading to imply that our knowledge of disease–oxidant interactions is anything more than superficial, in spite of the magnitude of the general problem of crop losses caused by disease, and the uncertainties as to whether oxidant pollutants may affect these losses, for better or worse.

Although many insects are herbivores and are vectors of plant pathogens, there appears to have been only one study conducted on a direct plant–insect–oxidant interaction. Rosen and Runeckles (1976) demonstrated a synergistic chlorotic injury to bean leaves resulting from simultaneous infestation with the greenhouse whitefly and daily exposure to extremely low ozone concentrations (0.02 ppm).

As in the case of some plant pathogens whose incidence rises dramatically following the onset of oxidant injury to the host species, so there are examples of increased insect attack following oxidant injury. In particular, it has been clearly shown that it is subsequent increased infection by bark beetles that delivers the coup de grace to ponderosa pines in Southern California weakened by oxidant injury (Stark and Cobb, 1969).

9. EFFECTS ON PLANT COMMUNITIES

Reference has already been made in Section 2.3 of the need to recognize the differences between individual plants and groups of plants which form a canopy of foliage, in the context of pollutant uptake. The effects of pollutants on groups of plants, whether of the same or of different species, in field situations, also constitute a further type of potential biological interaction, with both agricultural and ecological ramifications. In terms of crop plants, a growing number of reports is appearing in which field stands exposed to oxidants or ozone have been studied in large chambers of various types. Much of our understanding of the chronic effects on growth and yield has come from such studies, in which the planting densities used are those typical of local agronomic practice. However, there have been no direct experimental investigations of the effects of oxidants on intraspecific competition per se and few studies of effects on interspecific competition. In the latter context, most of the investigations which have been reported have related to the analysis of ecological change in situations where elevated oxidant levels occur.

9.1 Plant Competition

Harper (1964) wrote:

> *The essential qualities which determine the ecology of a species may only be detected by studying the reaction of its individuals to their neighbours and ... the behaviour of the species in isolation may be largely irrelevant to understanding their behaviour in the community. Differences between species which are crucial in determining their success or failure when grown together may only be exposed and demonstrated when the species are grown together.*

These remarks are equally appropriate if we consider a single species and the individuals which constitute a community or crop, since such individuals will interact with each other and their growth will be the outcome of the availability of resources to each individual plant. How do oxidants affect the individual's ability to utilize these resources? Many of the physiological investigations already referred to provide some answers. How do oxidants affect the community's or crop's ability to utilize these resources? In most instances we know very little, except in terms of final yield. But if final yield is reduced, this provides no information as to the mechanism involved within the crop as a whole.

Similarly, with interactions between species, which may be agricultural (e.g., forage and pasture mixtures) or of natural occurrence, there is a dearth of information on the effects of oxidants on plant competition or interference. Kochhar (1974) reported that the root exudates from ozone-treated fescue plants inhibited growth and nodulation of clover and hence demonstrated a potentially important effect of ozone on competition. Bennett and Runeckles (1977) studied the effects of ozone on competition between crimson clover and annual ryegrass using the replacement series model of deWit (1960). They found that the clover was more adversely affected by ozone than the ryegrass, but that the 1 : 1 mixture of the species performed somewhat better than would have been expected on the basis of their growth in monoculture. In a study of four species (barley, flax, crimson clover, and radish) grown from seed as mixed plantings with the same initial density, Runeckles (1976) found that daily exposures to 0.1 ppm ozone over a period of 160 days led to several significant differences in growth, relative to that in filtered air. For example, crimson clover shoot weight was reduced while the aerial biomass of flax and radish was increased, per unit area.

9.2 Ecological Analysis

While several workers have noted the ecological implications of oxidant injury to individual species, there have been relatively few studies of the direct ecological effects of oxidants. Outstanding are the investigations of the impact of oxidants on the forest ecosystem of southern California, reviewed in NAS (1977) and by Miller and McBride (1975). Both reviews also relate the observations made in the eastern United States and Canada with regard to the relationship between oxidant (or oxidant–SO_2 mixtures) and various types of dieback and injury to both softwood and hardwood species.

The tragedy of such studies is that they are only undertaken when the ambient air quality has deteriorated to a sufficient degree that obviously injurious effects begin to become visible on individual plants or species. When this happens, other more subtle ecological effects may be measured. Although the obvious visible effects may be the result of chronic or acute injury, the time frame within which observations have to be maintained is sufficiently long as to be a deterrent to all but the most dedicated and "wealthy" investigators. Yet to be reported upon are studies with oxidants along the lines of the investigations of the effects of SO_2 on prairie communities undertaken in connection with the Colstrip coal-fired power plant study (Preston and Gullett, 1979), which are aimed at developing an understanding of the ecological effects of deliberately increased ambient concentrations of an air pollutant for which baseline data have been obtained. But the ecological consequences of widespread oxidant pollution may be of equal or greater importance than those of other pollutants, such as SO_2, particularly within a given air shed. For example, there is evidence to suggest the involvement of oxidants together with acid rain in the widespread forest decline syndrome currently occurring in parts of Europe and eastern North America (Lefohn and Brocksen, 1984).

10. GENETIC IMPLICATIONS

10.1 Genetic Variability

Although plant response to oxidants may be modified by a host of environmental variables, there is also a genetic basis for variation in response. At the genus level, there is an obvious range in susceptibilities to ozone, PAN, and other oxidants, with different genera being typed along the spectrum of general response from highly sensitive (or susceptible) to highly tolerant (or resistant). The same principle is demonstrated frequently over a somewhat narrower range, by different species within a genus, or even between different varieties or the cultivars of cultivated plant species. Numerous tabulations of the sensitivities of individual species to different oxidants have been published (NRCC, 1975; NAS, 1977). Whether at the genus, species, or variety level, genetically based susceptibility or tolerance has obvious implications with regard to changes in ecological community structure, and with regard to such tabulations. Since the printed word may at times assume oracular proportions, it must always be remembered that tables of sensitivity are frequently composites of data culled from many sources, many investigators, and many different experimental facilities; in some cases, environmental factors may have been responsible for greater (or lesser) sensitivity than normal.

Nevertheless, there is ample evidence that gene action is sometimes directly related to sensitivity; for example, the work of Engle and Gabelman (1966) on ozone sensitivity of onions, and of Povilaitis (1967) on ozone tolerance of tobacco.

It is also worth remembering that, for every susceptible species in a given community or location, there are likely to be a much greater number of tolerant species, so that, while changes in the structure of natural communities may result from continued exposure to oxidants, the total denudation of vegetation such as has resulted from industrial emissions of other pollutants such as SO_2 or fluorides is unlikely, and has not to my knowledge been reported. However, such concepts are irrelevant in the context of agricultural production, with its predilection for monocultural systems; recourse must be made to the selection of tolerant cultivars or to the use of protectant chemicals in locations where high oxidant levels are prevalent.

10.2 Selection and Plant Breeding

Since oxidants as pollutants are relatively new to the earth's atmosphere, except as extremely localized events related to violent atmospheric disturbances, the processes of natural selection have yet to be demonstrated, except in the grossest sense of the elimination or reduction in densities of sensitive species from highly polluted regions (Miller and McBride, 1975). However, it has been argued (Bennett et al., 1974) that, because low levels of ozone (0.02 ppm) are normal in unpolluted air, widespread selection for such levels has occurred over time, and that this may account for the apparent stimulations of plant growth which have been observed in low concentrations of ozone in relation to growth in filtered air.

On the other hand, artificial selection for tolerance to oxidants has undoubtedly occurred. Ryder (1973) has summarized the requirements for breeding programs aimed at selecting tolerant cultivars, and stresses the necessity for identifying tolerant individuals, strains, and cultivars of a given species. He also emphasizes the uncertainties of outcome of breeding programs and their inevitable long duration. While the initial screening of hybrids and selections within a breeding program may be undertaken in greenhouse or growth chamber conditions, the ultimate test is tolerance under field conditions. In turn, this eventually requires evaluation in locations in which the pollutant levels are predictably likely to be injurious.

It is worth noting that cultivars which have been selected for specific desirable characteristics other than tolerance to oxidants may, nevertheless, perhaps unwittingly, have been selected for tolerance, if the breeding program has been undertaken in an area with measurable oxidant pollution. Indeed, some of the tolerant varieties which have previously been identified may have had such origins, but this aspect does not appear to have been directly studied.

11. MONITORING

Sensitive plant species and cultivars have been used as monitors of air quality by several investigators (NAS, 1977). Unlike chemical procedures and instrumental methods, which are capital intensive and subject to malfunction and breakdown, plants themselves offer the potential for integrating the effects of pollutants over time. While the early studies tended to stress visible symptoms of acute injury, the approach of Oshima et al. (1976) which led to the establishment of crop-loss functions of alfalfa grown under conditions of oxidant pollution in the field, is really a quantitative development of the monitoring technique, which is equally applicable to chronic or acute injury situations. While the intent of these studies was to permit the prediction of crop loss, the relationships could equally well be used to predict average air quality from measured plant performance.

The sensitive tobacco cultivar Bel-W_3 has been widely used as a monitor of oxidants and ozone in many locations throughout the world. In spite of the caution (NAS, 1977) that its use only indicates oxidant phytotoxicity and not oxidant concentration, monitoring programs based on Bel-W_3 have provided useful data on regional oxidant distributions; recently Naveh et al. (1978) reported on the use of Bel-W_3 as an indicator of oxidants in Israel, and established a relationship that enabled them to predict average oxidant concentrations on the basis of acute injury. Similar studies have been conducted in the Netherlands (Posthumus, 1982) and in the United Kingdom (Ashmore et al., 1980). (see Section 3.2)

A differential monitoring approach has been used to evaluate the yield responses of bean in southern Ontario (G. Hofstra, personal communication). The effective protection against oxidant injury afforded by the antioxidant EDU (Hofstra et al., 1978) has provided a means whereby crop damage could be assessed throughout the area, by comparing EDU-treated and untreated field plots, with regard to both leaf injury (bronzing) and bean yield. As with all field surveys, a variety of factors other than pollutant concentration influenced the quantitative response, but the method yielded data which showed a good correlation between injury and reduction in yield when other environmental factors (such as rainfall or the lack of it) were taken into account.

Native vegetation has also been used in ground surveys and surveys using the techniques of remote sensing. As these techniques become further developed and refined, there is no doubt that they will permit assessments of injury to be made for a wide range of crops and plant communities. The early studies of Larsh et al. (1970) clearly demonstrated that severity of injury to ponderosa pine assessed from aerial photographs was correlated with ground truth. The major problem is that resulting from situations in which several stress factors may be affecting plant growth simultaneously. With forest trees, Murtha (1972) has developed a dichotomous key for identifying various types of injury and die-back based on air photo interpretation, but while several types of injury caused by disease, insect, or other causes present reasonably distinct characteristics when using densitometric techniques and tree color and false color images, there are limitations, and ground checks are very necessary.

12. EPILOGUE

The preceding sections have dealt with a wide range of topics related to the effects of oxidants on plants. Inevitably, I have selected some topics for review at the expense of others. The reader has my apologies if he or she feels that a favorite topic has been glossed over or neglected completely. However, constraints of space have required that only those aspects which appear to me to need emphasis should be included. My guide throughout has been a consideration of what I believe must be done in terms of future research, bearing in mind that understanding and not merely the acquisition of knowledge should be our ultimate goal.

Note added in proof. Considerable development has taken place in effects research of oxidants on plants since this manuscript was originally prepared in 1980 and revised in 1983. At the galley stage, a few additional changes were made to permit reference to significant new reviews and developments such as the NCLAN program.

REFERENCES

Abeles, A. L., and F. B. Abeles. Biochemical pathway of stress-induced ethylene. *Plant Physiol.* **50**, 496–498 (1972).

Arndt, U. The Kautsky effect as a sensitive proof for air pollution effects on plants. *Chemosphere* **5**, 187–190 (1972).

Ashenden, T. W. The effects of long-term exposures to SO_2 and NO_2 pollution on the growth of *Dactylis glomerata* L. and *Poa pratensis* L. *Environ. Pollut.* **18**, 249–258 (1979).

Ashenden, T. W., and T. A. Mansfield. Influence of wind speed on the sensitivity of ryegrass to SO_2. *J. Exper. Bot.* **28**, 729–735 (1977).

Ashenden, T. W., and T. A. Mansfield. Extreme pollution sensitivity of grasses when SO_2 and NO_2 are present in the atmosphere together. *Nature* **273**, 142–143 (1978).

Ashmore, M. R., J. N. B. Bell, and C. L. Reilly. The distribution of phytotoxic ozone in the British Isles. *Environ. Pollut.* (*Ser. B*) **1**, 195–216 (1980).

Bennett, J. H. Foliar exchange of gases. In W. W. Heck, S. V. Krupa and S. N. Linzon (eds.) *Handbook of Methodology for the Assessment of Air Pollution Effects on Vegetation.* Chapter 10. Air Pollut. Control Assoc., Pittsburgh, Pennsylvania, 1979.

Bennett, J. H., A. C. Hill, and D. M. Gates. A model for gaseous pollutant sorption by leaves. *J. Air Pollut. Control Assoc.* **23**, 957–962 (1973).

Bennett, J. P. Effects of low levels of ozone on plant populations. Ph.D. Thesis. University of British Columbia, Vancouver, B.C. (1975).

Bennett, J. P., and R. J. Oshima. Carrot injury and yield response to ozone. *J. Am. Soc. Hort. Sci.* **101**, 638–639 (1976).

Bennett, J. P., and V. C. Runeckles. Effects of low levels of ozone on plant competition. *J. Appl. Ecol.* **14**, 877–880 (1977).

Bennett, J. P., H. M. Resh, and V. C. Runeckles. Apparent stimulations of plant growth by air pollutants. *Can. J. Bot.* **52**, 35–42 (1974).

Bewley, J. D. Physiological aspects of desiccation tolerance. *Ann. Rev. Plant Physiol.* **30**, 195–238 (1979).

Bicak, C. J. Plant response to variable ozone regimes of constant dosage. M.Sc. Thesis. University of British Columbia, Vancouver, B.C. (1978).

Black, V. J., and M. H. Unsworth. Resistance analysis of sulphur dioxide fluxes to *Vicia faba*. *Nature* **282**, 68–69 (1979).

Blum, U., and D. T. Tingey. A study of the potential ways in which ozone could reduce root growth and nodulation in soybean. *Atmos. Environ.* **11**, 737–739 (1977).

Bruton, V. C. Environmental influence on the growth of *Arabidopsis thalliana*. Ph.D. Thesis. North Carolina State University, Raleigh, North Carolina (1974).

Bucher, J. B., and T. Keller, Einwirkungen niedriger SO_2-konzentrationen im mehrwochigen

Begasungsversuch auf Waldbäume. *Proc. Seminar on Oxidized Sulphur Compounds*, Verein Deutscher ingenieure (VDI), Augsberg (1978).

Bystrom, B. C., R. B. Glater, R. M. Scott, and E. S. C. Bowler. Leaf surface of *Beta vulgaris*—electron microscope study. *Bot Gaz*. **129**, 133–138 (1968).

Carnahan, J. E., E. L. Jenner, and E. K. W. Wat. Prevention of ozone injury to plants by a new protectant chemical. *Phytopathology* **68**, 1225–1229 (1978).

Carney, A. W., G. R. Stephenson, D. P. Ormrod, and G. C. Ashton. Ozone–herbicide interactions in crop plants. *Weed Sci*. **21**, 508–511 (1973).

Chang, C. W. Effect of ozone on ribosomes in pinto bean leaves. *Phytochemistry* **10**, 2863–2868 (1971).

Chanway, C. P., and V. C. Runeckles. The role of superoxide dismutase in the susceptibility of bean leaves to ozone injury. *Can. J. Bot*. **62**, 236–240 (1984).

Clarke, B., M. Henninger, and E. Brennan. The effect of two anti-oxidants on foliar injury and tuber production in "Norchip" potato plants exposed to ambient oxidants. *Plant Dis. Reptr*. **62**, 715–717 (1978).

Costonis, A. C., and W. A. Sinclair. Susceptibility of healthy and ozone-injured needles of *Pinus strobus* to invasion by *Lophodermium pinastri* and *Aureobasidium pollulans*. *Eur. J. For. Pathol*. **2**, 65–73 (1972).

Craker, L. E. Ethylene production from ozone injured plants. *Environ. Pollut*. **1**, 299–304 (1971).

Craker, L. E., and W. A. Feder. Development of the inflorescence in petunia, geranium, and poinsettia under ozone stress. *Hort. Science* **7**, 59–60 (1972).

Curtis, C. R., and L. H. Weinstein. Special Techniques. A. Electrophoresis. IN W. W. Heck, S. V. Krupa and S. N. Linzon (eds.) *Handbook for the Assessment of Air Pollution Effects on Vegetation*. Chapter 16. Air Pollut. Control Assoc., Pittsburgh, Pennsylvania, 1979.

Curtis, C. R., and R. K. Howell. Increases in peroxidase isoenzyme activity in bean-leaves exposed to low doses of ozone. *Phytopath*. **61**, 1306–1307 (1971).

Czuba, M., and D. P. Ormrod. Effects of cadmium and zinc on ozone-induced phytotoxicity in cress and lettuce. *Can. J. Bot*. **52**, 645–649 (1974).

Dass, H. C., and G. M. Weaver, Enzymatic changes in intact leaves of *Phaseolus vulgaris* following ozone fumigation. *Atmos. Environ*. **6**, 759–763 (1972).

Davis, D. D., and S. H. Smith. Reduction of ozone-sensitivity of pinto bean by bean common mosaic virus. *Phytopathology* **64**, 383–385 (1974).

deWit, C. T. *On Competition*. Agricultural Research Report 668. Center for Agricultural Publishing and Documentation, Wageningen (1960).

Dugger, W. M., and I. P. Ting. Air pollutant oxidants—their effects on metabolic processes in plants. *Ann. Rev. Plant Physiol*. **21**, 215–234 (1970).

Dunning, J. A., and W. W. Heck. Response of bean and tobacco to ozone: effect of light intensity, temperature and relative humidity. *J. Air Pollut. Control Assoc*. **27**, 882–886 (1977).

Engle, R. L., and W. H. Gabelman. Inheritance and mechanism for resistance to ozone damage in onion, *Allium cepa* L. *Proc. Am. Soc. Hort. Sci*. **89**, 423–430 (1966).

Engle, R. L., and W. H. Gabelman. The effects of low levels of ozone on pinto beans, *Phaseolus vulgaris* L. *Proc. Am. Soc. Hort. Sci*. **91**, 304–309 (1967).

Erickson, R. O., and F. J. Michelini. The plastochron index. *Am. J. Bot*. **44**, 297–305 (1957).

Evans, G. C. *The Quantitative Analysis of Plant Growth*. Blackwell, Oxford and Univ. Calif. Press, Berkeley, 1972, p. 734.

Feder, W. A. Reduction in tobacco pollen germination and tube elongation, induced by low levels of ozone. *Science* **160**, 1122 (1968).

Feder, W. A. Plant response to chronic exposure of low levels of oxidant type pollutant. *Environ. Pollut.* **1**, 73–79 (1970).

Fowler, D. Dry deposition of SO_2 on agricultural crops. *Atmos. Environ.* **12**, 369–373 (1978).

Fowler, D., and J. N. Cape. Air pollutants in agriculture and horticulture. In M. H. Unsworth and D. P. Ormrod (eds.) *Effects of Gaseous Air Pollution in Agriculture and Horticulture.* Chapter 1. Butterworth Scientific, London, 1982.

Fowler, D., and M. H. Unsworth. Dry deposition of sulphur dioxide on wheat. *Nature* **249**, 389–390 (1974).

Fujiwara, T. Damage to plants by combined air pollution. *Shokobutsu Boeki* **27**, 233–236 (in Japanese) (1973).

Galbally, I. E., and C. R. Roy. Destruction of ozone at the earth's surface. *Quart. J. Roy. Met. Soc.* **106**, 599–620 (1980).

Garland, J. A. Dry and wet removal of sulphur from the atmosphere. *Atmos. Environ.* **12**, 349–362 (1978).

Gentile, A. G., W. A. Feder, R. E. Young, and Z. Santner. Susceptibility of *Lycopersicon* spp. to ozone injury. *J. Am. Soc. Hort. Sci.* **96**, 94–96 (1971).

Guderian, R., H. Van Haut, and H. Stratmann. Probleme der Erfassung und Beurteilung von Wirkungen gasförmiger Luftverunreinigungen auf die Vegetation. *Z. Pflanzenk. Pflanzensch.* **67**, 257–265 (1960).

Harkov, R., B. Clarke, and E. Brennan. Cadmium contamination may modify response of tomato to atmospheric ozone. *J. Air Pollut. Control Assoc.* **29**, 1247–1249 (1979).

Harper, J. L. The individual in the population. *J. Ecol.* **52** (Suppl.), 149–159 (1964).

Harrison, B. H., and W. A. Feder. Ultrastructural changes in pollen exposed to ozone. *Phytopathology* **64**, 257–258 (1974).

Harward, M., and M. Treshow. Impact of ozone on the growth and reproduction of understory plants in the aspen zone of western U.S.A. *Environ. Conserv.* **2**, 17–23 (1975).

Heagle, A. S. Interactions between air pollutants and plant parasites. *Ann. Rev. Phytopathology* **11**, 365–388 (1973).

Heagle, A. S., and W. W. Heck. Predisposition of tobacco to oxidant air pollution injury by previous exposure to oxidants. *Environ. Pollut.* **7**, 247–252 (1974).

Heagle, A. S., and J. W. Johnston. Variable responses of soybeans to mixtures of ozone and sulfur dioxide. *J. Air Pollut. Control Assoc.* **29**, 729–732 (1979).

Heagle, A. S., R. B. Philbeck, and W. M. Knott. Thresholds for injury, growth and yield loss caused by ozone on field corn hybrids. *Phytopathology* **69**, 21–26 (1979a).

Heagle, A. S., R. B. Philbeck, and M. B. Letchworth. Injury and yield responses of spinach cultivars to chronic doses of ozone in open-top field chambers. *J. Environ. Qual.* **8**, 368–373 (1979b).

Heagle, A. S., S. Spencer, and M. B. Letchworth. Yield response of winter wheat to chronic doses of ozone. *Can. J. Bot.* **57**, 1999–2005 (1979c).

Heck, W. W. Factors influencing expression of oxidant damage to plants. *Ann. Rev. Phytopathology* **6**, 165–188 (1968).

Heck, W. W., and J. A. Dunning. The effects of ozone on tobacco and pinto bean as conditioned by several ecological factors. *J. Air Pollut. Control Assoc.* **17**, 112–114 (1967).

Heck, W. W., and D. L. Tingey. Ozone. Time-concentration model to predict acute foliar injury. In H. M. Englund and W. T. Beery (eds.) *Proceedings of the Second International Clean Air Congress, Washington, D.C.* Academic, New York, 1971.

Heck, W. W., J. A. Dunning, and I. J. Hindawi. Ozone: nonlinear relation of dose and injury in plants. *Science* **151**, 577–578 (1966).

Heck, W. W., R. M. Adams, W. W. Cure, A. S. Heagle, H. E. Heggestad, R. J. Kohut, L. W. Kress, J. O. Rawlings, and O. C. Taylor. A reassessment of crop loss from O_3. *Environ. Sci. Technol.* **17**, 573A–581A (1983).

Heck, W. W., W. W. Cure, J. O. Rawlings, L. J. Zaragoza, A. S. Heagle, H. E. Heggestad, R. J. Kohut, L. W. Kress, and P. J. Temple. Assessing impacts of ozone on agricultural crops: I. Overview. *J. Air Pollut. Control Assoc.* **34**, 729–735 (1984a).

Heck, W. W., W. W. Cure, J. O. Rawlings, L. J. Zaragoza, A. S. Heagle, H. E. Heggestad, R. J. Kohut, L. W. Kress, and P. J. Temple. Assessing impacts of ozone on agricultural crops: II. Crop yield functions and alternative exposure statistics. *J. Air Pollut. Control Assoc.* **34**, 810–817 (1984b).

Hill, A. C. Vegetation: a sink for atmospheric pollutants. *J. Air Pollut. Control Assoc.* **21**, 341–346 (1971).

Hodgson, R. H., K. E. Dusbabek, and F. L. Hoffer. Diphenamid metabolism in tomato: time course of an ozone fumigation effect. *Weed Sci.* **22**, 204–210 (1974).

Hodgson, R. H., and B. L. Hoffer. Diphenamid metabolism in pepper and an ozone effect. II. *Herbicide metabolite characterization. Weed Sci.* **25**, 331–337 (1977).

Hoffman, G. J., E. V. Maas, and S. L. Rawlins. Salinity–ozone interactive effects on yield and water relations of pinto bean. *J. Environ. Qual.* **2**, 148–152 (1973).

Hoffman, G. J., E. V. Maas, and S. L. Rawlins. Salinity–ozone interactive effects on alfalfa yield and water relations. *J. Environ. Qual.* **4**, 326–331 (1975).

Hofstra, G., and D. W. Beckerson. Foliar responses of five plant species to ozone and a sulphur dioxide/ozone mixture after a sulphur dioxide pre-exposure. *Atmos. Environ.* **15**, 383–389 (1981).

Hofstra, G., D. A. Littlejohns, and R. T. Wukasch. The efficacy of the antioxidant ethylene-diurea (EDU) compared to carboxin and benomyl in reducing yield losses from ozone in navy bean. *Plant Dis. Reptr.* **62**, 350–352 (1978).

Hughes, A. P., and P. R. Freeman. Growth analysis using frequent small harvests. *J. Appl. Ecol.* **4**, 553–560 (1967).

Hunt, R. *Plant Growth Analysis: Studies in Biology No. 96*, Edward Arnold, London (1978).

Hunt, R. *Plant Growth Curves: Functional Approach to Plant Growth Analysis*. Edward Arnold, London, 1982.

Hunt, R., and I. T. Parsons. A computer program for deriving growth-functions in plant growth-analysis. *J. Appl. Ecol.* **11**, 297–307 (1974).

Hunt, R., and I. T. Parsons. Plant growth analysis: further applications of a recent curve-fitting program. *J. Appl. Ecol.* **14**, 965–968 (1977).

Irving, P. M., P. B. Xerikos, and J. E. Miller. The combined effect of sulfur dioxide and nitrogen dioxide gases on the growth and productivity of soybeans. *Radiological and Environmental Research Division Annual Report*, ANL-81-85, Argonne Natl. Laboratory, Argonne, Illinois (1982).

Jacobson, J. S., and L. J. Colavito. The combined effect of sulfur dioxide and ozone on bean and tobacco plants. *Environ. Exper. Bot.* **16**, 277–275 (1976).

Jones, H. C., F. P. Weatherford, J. C. Noggle, N. T. Lee, and J. R. Cunningham. Power plant siting: assessing risks of sulfur dioxide effects on vegetation. *Proc. 72nd Ann. Meeting, Air Pollut. Control Assoc.*, Cincinnati, Ohio, Paper #79-13.5 (1979).

Kochhar, M. Phytotoxic and competitive effects of tall fescue on ladino clover as modified by ozone and/or *Rhizoctonia solani*. Ph.D. Thesis. North Carolina State Univ., Raleigh, North Carolina (1974).

Kohut, J., D. D. Davis, and W. Merrill. Response of hybrid poplar to simultaneous exposure to ozone and PAN. *Plant Dis. Reptr.* **60**, 777–780 (1976).

Koziol, M. J., and F. R. Whatley. *Gaseous Air Pollutants and Plant Metabolism*. Butterworths, London, 1984.

Kress, L. W. Response of hybrid poplar to sequential exposures of ozone and PAN, M.Sc. Thesis, Pennsylvania State Univ.; Center for Air Environment Studies Publ. 259-72, Pennsylvania State Univ., University Park, Pennsylvania, 39 pp. (1972).

Larsh, R. N., P. R. Miller, and S. L. Wert. Aerial photography to detect and evaluate air pollution damaged ponderosa pine. *J. Air Pollut. Control Assoc.* **20**, 289–292 (1970).

Larsen, R. I., and W. W. Heck. An air quality data analysis system for interrelating effects, standards and needed source reductions. 3. Vegetation injury. *J.. Air Pollut. Control Assoc.* **26**, 325–333 (1976).

Larsen, R. I., and W. W. Heck. An air quality data analysis system for interrelating effects, standards, and needed source reductions: Part 8. An effective mean O_3 crop reduction mathematical model. *J. Air Pollut. Control Assoc.* **34**, 1023–1034 (1984).

Lee, E. H., and J. H. Bennett. Superoxide dismutase. A possible protective enzyme against ozone injury in snap beans (*Phaseolus vulgaris*). *Plant Physiol.* **69**, 1444–1449 (1982).

Leffler, H. R., and J. H. Cherry. Destruction of enzymatic activities of corn and soybean leaves exposed to ozone. *Can. J. Bot.* **52**, 1233–1238 (1974).

Lefohn, A. S., and R. W. Brocksen. Acid rain effects research—a status report. *J. Air Pollut. Control Assoc.* **34**, 1005–1013 (1984).

Lefohn, A. S., and D. T. Tingey. The cooccurrence of potentially phytotoxic concentrations of various gaseous air pollutants. *Atmos. Environ.* **18**, 2521–2526 (1984).

Leone, I. A. Response of potassium-deficient tomato plants to atmospheric ozone. *Phytopathology* **66**, 734–736 (1976).

Leone, I. A., and E. Brennan. Ozone toxicity in tomato as modified by phosphorus nutrition. *Phytopathology* **60**, 1521–1524 (1970).

Leone, I. A., E. Brennan, and R. H. Davies. Effect of nitrogen nutrition on the response of tobacco to ozone in the atmosphere. *J. Air Pollut. Control Assoc.* **15**, 191–196 (1966).

Letchworth, M. B., and U. Blum. Effects of acute ozone exposure on growth, nodulation and nitrogen content of ladino clover. *Environ. Pollut.* **14**, 303–312 (1977).

Leuning, R., M. H. Unsworth, H. H. Neumann, and K. M. King. Ozone fluxes to tobacco and soil under field conditions. *Atmos. Environ.* **13**, 1155–1163 (1979).

Levitt, J. *Responses of Plants to Environmental Stresses*. Academic, New York, 1972.

Liu, B. C., and E. S. H. Yu. Air pollution and vegetation damage functions. *Environ. Planning A*. **9**, 643–652 (1977).

Maas, E. V., G. J. Hoffman, S. I. Rawlins, and G. Ogata. Salinity–ozone interactions on pinto bean: integrated response to ozone concentration and duration. *J. Environ. Qual.* **2**, 400–404 (1973).

Macdowall, F. D. H. Predisposition of tobacco to ozone damage. *Can. J. Plant Sci.* **45**, 1–12 (1965).

Macdowall, F. D. H. Importance of soil in absorption of ozone by a crop. *Can. J. Soil Sci.* **54**, 239–240 (1974).

Macdowall, F. D. H., and A. F. W. Cole. Threshold and synergistic damage to tobacco by ozone and sulfur dioxide. *Atmos. Environ.* **5**, 553–559 (1971).

Macdowall, F. D. H., E. I. Mukammal, and A. F. W. Cole. Direct correlation of air-polluting ozone and tobacco weather-fleck. *Can. J. Plant Sci.* **44**, 410–417 (1964).

Macdowall, F. D. H., E. I. Mukammal and A. F. W. Cole. Ozone and plant injury. *Science* **153**, 1552 (1966).

MacLean, D. C., and R. E. Schneider. Photochemical oxidants in Yonkers, New York: effects on yield of bean and tomato. *J. Environ. Qual.* **5**, 75–78 (1976).

Male, L., E. Preston, and G. Neely. Yield response curves of crops exposed to SO_2 time series. *Atmos. Environ.* **17**, 1589–1593 (1983).

Manning, W. J., W. A. Feder, I. Perkins, and M. Glickman. Ozone injury and infection of potato leaves by *Botrytis cinerea*. *Plant Dis. Reptr.* **53**, 691–693 (1969).

Matsushima, J. On composite damage to plants by sulfurous acid gas and oxidant. *Sangyo Kogai* **7**, 218–224 (in Japanese) (1971).

McClenahen, J. R. Effects of ethylene diurea and ozone on the growth of tree seedlings. *Plant Dis. Reptr.* **63**, 320–323 (1979).

McCune, D. C. Summary and synthesis of plant toxicology. In J. A. Naegele (ed.) *Air Pollution Damage to Vegetation*. Chapter 5. American Chemical Society, Washington, D.C., 1973.

McIlveen, W. D., R. A. Spotts, and D. D. Davis. The influence of soil zinc on nodulation mycorrhizae and ozone-sensitivity of pinto bean. *Phytopathology* **65**, 645–647 (1975).

McKersie, B. D., W. D. Beversdorf, and P. Hucl. The relationship between ozone insensitivity, lipid soluble antioxidants and superoxide dismutase in *Phaseolus vulgaris*. *Can. J. Bot.* **60**, 2686–2691 (1982).

McLaughlin, S. B., D. S. Shriner, R. K. McConathy, and L. K. Mann. The effects of SO_2 dosage kinetics and exposure frequency on photosynthesis and transpiration of kidney beans (*Phaseolus vulgaris* L.). *Environ. Exptl. Bot.* **19**, 179–191 (1979).

Mead, J. F. Free radical mechanisms of lipid damage and consequences for cellular membranes. In W. A. Pryor (ed.) *Free Radicals in Biology*, pp. 51–68. Academic, New York, 1976.

Menser, H. A., and H. E. Heggestad. Ozone and sulfur dioxide synergism: injury to tobacco plants. *Science* **153**, 424–425 (1966).

Menser, H. A., and G. H. Hodges. Varietal tolerance of tobacco to ozone dose rate. *Agron. J.* **60**, 349–352 (1968).

Miller, P. R., and J. R. McBride. Effects of air pollutants on forests. In J. B. Mudd and T. T. Kozlowski (eds.) *Responses of Plants to Air Pollution*. Chapter 10. Academic, New York, 1975.

Mitchell, R. L. *Crop Growth and Culture*. Iowa State Univ. Press, Ames, Iowa, 1970.

Mudd, J. B. Effects of oxidants on metabolic function. In M. H. Unsworth and D. P. Ormrod (eds.) *Effects of Gaseous Air Pollution in Agriculture and Horticulture*. Chapter 9. Butterworth Scientific, London, 1982.

Mudd, J. B., S. K. Banerjee, M. M. Dooley, and K. L. Knight. Pollutants and plant cells. In M. J. Koziol and F. R. Whatley (eds.) *Gaseous Air Pollutants and Plant Metabolism*. Chapter 8. Butterworths, London, 1982.

Mudd, J. B., and T. T. Kozlowski. *Response of Plants to Air Pollution*. Academic, New York, 1975.

Mukammal, E. I. Ozone as a cause of tobacco injury. *Agr. Meteorol.* **2**, 145–165 (1965).

Mumford, R. A., H. Kipke, D. A. Laufer, and W. A. Feder. Ozone-induced changes in corn pollen. *Environ. Sci. Technol.* **6**, 427–430 (1972).

Murtha, P. A. *A Guide to Air Photo Interpretation of Forest Damage in Canada*. Publ. 1296. Canadian Forestry Service, Ottawa (1972).

Mustafa, M. G., S. M. Macres, B. K. Tarkinton, C. K. Chow, and M. Z. Huzzein. Lung superoxide dismutase (SOD): stimulation by low-level ozone exposure. *Clin. Res.* **23**, 138A (Abstract) (1975).

Nakamura, H., and Y. Ota. Investigations on injury to rice plants from photochemical oxidants in Japan. In *Proc. Fourth Int'l. Clean Air Congr.*, pp. 103–105 (1979).

NAS. *Ozone and Other Photochemical Oxidants*. Committee on Medical and Biologic Effects of Environmental Pollutants. National Academy of Science, Washington, D.C. (1977).

Naveh, Z., S. Chaim, and E. H. Steinberger. Atmospheric oxidant concentrations in Israel as manifested by foliar injury in Bel-W_3 tobacco plants. *Environ. Pollut.* **16**, 249–262 (1978).

Nouchi, I., and K. Aoki. Morning glory as a photochemical oxidant monitor. *Environ. Pollut.* **18**, 289–303 (1979).

NRCC. *Photochemical Air Pollution: Formation Transport and Effects*. Associate Committee on Scientific Criteria for Environmental Quality. Report No. 12: N.R.C.C. 14096. National Research Council of Canada, Ottawa (1975).

O'Dell, R. A., M. Taheri, and R. L. Kabel. A model for uptake of pollutants by vegetation. *J. Air Pollut. Control. Assoc.* **27**, 1104–1109 (1977).

Oertli, J. J. Effect of salinity on susceptibility of sunflower plants to smog. *Soil Sci.* **87**, 249–251 (1959).

O'Gara, P. J. Sulfur dioxide and fume problems and their solutions. *J. Ind. Eng. Chem.* **14**, 744 (1922).

Ordin, L., M. J. Garber, and J. I. Kindinger, Effect of 2,4-dichlorophenoxyacetic acid on growth and on β-glucan synthetases of peroxyacetyl nitrate pretreated *Avena* coleoptile sections. *Physiol. Plant* **26**, 17–23 (1972).

Ormrod, D. P. Cadmium and nickel effects on growth and ozone sensitivity of pea. *Wat. Air Soil Pollut.* **8**, 263–270 (1977).

Ormrod, D. P. Air pollutant interactions in mixtures. In M. H. Unsworth and D. P. Ormrod (eds.) *Effects of Gaseous Air Pollution in Agriculture and Horticulture*. Chapter 15. Butterworth Scientific, London, 1982.

Oshima, R. J., M. P. Poe, P. K. Braegelmann, D. W. Baldwin, and V. Van Way. Ozone dosage crop loss function for alfalfa: a standardized method for assessing crop losses from air pollutants. *J. Air Pollut. Control Assoc.* **26**, 861–865 (1976).

Oshima, R. J., P. K. Braegelmann, D. W. Baldwin, V. Van Way, and O. C. Taylor. Responses of five cultivars of fresh market tomato to ozone: a contrast of cultivar screening with foliar injury and yield. *J. Am. Soc. Hort. Sci.* **102**, 286–289 (1977).

Oshima, R. J., J. B. Bennett, and P. K. Braegelmann. Effect of ozone on growth and assimilate partitioning in parsley. *J. Am. Soc. Hort. Sci.* **103**, 348–350 (1978).

Oshima, R. J., P. K. Braegelmann, R. B. Flagler, and R. R. Teso. The effects of ozone on the growth yield and partitioning of dry matter in cotton. *J. Environ. Qual.* **8**, 474–479 (1979).

Parmeter, J. R., R. V. Bega, and T. Neff. A chlorotic decline of ponderosa pine in Southern California. *Plant Dis. Reptr.* **46**, 269–273 (1962).

Pell, E. J., and E. Brennan. Changes in respiration, photosynthesis, adenosine-5′-triphosphate, and total adenylate content of ozonated pinto bean foliage as they relate to symptom expression. *Plant Physiol.* **51**, 378–381 (1973).

Posthumus, A. C. Biological indicators of pollution. In M. H. Unsworth and D. P. Ormrod (eds.) *Effects of Gaseous Air Pollution in Agriculture and Horticulture*. Chapter 2. Butterworth Scientific, London, 1982.

Povilaitis, B. Gene effects for tolerance to weather fleck in tobacco. *Can. J. Genet. Cytol.* **9**, 327–336 (1967).

Preston, E. M., and T. L. Gullett. The Bioenvironmental Impact of a Coal-Fired Power Plant. Fourth Interim Report. U.S. Environmental Protection Agency, Corvallis, Oregon (1979).

Rich, S., P. E. Waggoner, and H. Tomlinson. Ozone uptake by bean leaves. *Science* **169**, 79–80 (1970).

Rogers, H. H., Jeffries, E. P. Stahel, W. W. Heck, L. A. Ripperton, and A. M. Witherspoon. Measuring air pollutant uptake by plants: a direct kinetic technique. *J. Air Pollut. Control. Assoc.* **27**, 1192–1197 (1977).

Rogers, H. H., J. C. Campbell, and R. J. Volk. Nitrogen-15 dioxide uptake and incorporation by *Phaseolus vulgaris*. *Science* **206**, 333–335 (1979).

Rosen, P. M., and V. C. Runeckles. Interaction of ozone and greenhouse whitefly in plant injury. *Environ. Conserv.* **3**, 70–71 (1976).

Runeckles, V. C. Dosage of air pollutants and damage of vegetation. *Environ. Conserv.* **1**, 305–308 (1974).

Runeckles, V. C. Assessment of Long-Term Effects of Air Pollutants on Vegetation and Simple Ecosystems. *Report to the Associate Committees Secretariat*. National Research Council of Canada, Ottawa (1976).

Runeckles, V. C. Relative death rate: a dynamic parameter describing plant response to stress. *J. Appl. Ecol.* **19**, 295–303 (1982).

Runeckles, V. C., and P. M. Rosen. Effects of pretreatment with low ozone concentrations on ozone injury to bean and mint. *Can. J. Bot.* **52**, 2607–2610 (1974).

Runeckles, V. C., and P. M. Rosen. Effects of ambient ozone pretreatment on transpiration and susceptibility to ozone injury. *Can. J. Bot.* **55**, 193–197 (1977).

Runeckles, V. C., K. Palmer, and K. Giles. Effects of sequential exposures to NO_2 and O_3 on plants. *3rd Int. Cong. Plant Pathology, Munich* (Abstract) (1978).

Ryder, E. J. Selecting and breeding plants for increased resistance to air pollutants. In J. E. Naegele (ed.) *Air Pollution Damage to Vegetation*. Chapter 7. American Chemical Society, Washington, D.C., 1973.

Schreiber, U., L. Groberman, and W. Vidaver. Portable, solid-state fluorometer for the measurement of chlorophyll fluorescence induction in plants. *Rev. Sci. Instrum.* **46**, 538–542 (1975).

Schrieber, U., W. Vidaver, V. C. Runeckles, and P. M. Rosen. Chlorophyll fluorescence assay for ozone injury in intact plants. *Plant Physiol.* **61**, 80–84 (1978).

Schwela, D. H. An estimate of deposition velocities of several air pollutants on grass. *Ecotoxicol. Environ. Safety* **3**, 174–189 (1979).

Sherwood, C. H., and G. D. Rolph. Ozone protects plants from air pollution with 2,4-D. *Hort. Science* **5**, 190 (Abstract) (1970).

Shoaf, A. R., R. C. Allen, and R. H. Steele. Electronic excitation state generation in mammalian systems: mechanism–role–pathology. *2nd Ann. Meeting Am. Soc. Photobiology*, Univ. of British Columbia, Vancouver (Abstract) (1974).

Smith, W. H. Air pollution-effects on the structure and function of the temperate forest ecosystem. *Environ. Pollut.* **6**, 111–129 (1974).

Sparrow, A. H., and L. A. Schairer. Mutagenic response of *Tradescantia* to treatment with X-rays, EMS, DBE, ozone, SO_2, N_2O and several insecticides. *Mutat. Res.* **26**, 445 (Abstract) (1974).

Spierings, F. H. F. G. Influence of fumigations with NO_2 on growth and yield of tomato plants. *Neth. J. Plant Pathol.* **77**, 194–200 (1971).

Spotts, R. A., F. L. Lukezic, and N. L. Lacasse. The effect of benzimidazole, cholesterol and a steroid inhibitor on leaf sterols and ozone resistance of bean. *Phytopathology* **65**, 45–49 (1975).

Stark, R. W., and F. W. Cobb. Smog injury, root diseases and bark beetle damage in ponderosa pine. *Calif. Agric.* **23**(9), 13–15 (1969).

Tanaka, K., and K. Sugahara. Role of superoxide dismutase in the defense against SO_2 toxicity and induction of superoxide dismutase with SO_2 fumigation. *Res. Rep. Natl. Inst. Environ. Stud. (Japan)* **11**, 155–164 (1980).

Taylor, G. S., and S. Rech. Ozone injury to tobacco in the field influenced by soil treatments with benomyl and carboxin. *Phytopathology* **64**, 814–817 (1974).

Taylor, O. C., and F. M. Eaton. Suppression of plant growth by nitrogen dioxide. *Plant Physiol.* **41**, 132–135 (1966).

Thompson, C. R., E. G. Hensel, G. Kats, and O. C. Taylor. Effects of continuous exposure of navel oranges to NO_2. *Environ. Sci. Technol.* **5**, 1017–1019 (1970).

Thompson, C. R., G. Kats, and J. W. Cameron. Effects of ambient photochemical air pollutants on growth, yield and ear characteristics of two sweet corn hybrids. *J. Environ. Qual.* **5**, 410–412 (1976).

Thorne, L., and G. P. Hanson. Species differences in rates of vegetal ozone absorption. *Environ. Pollut.* **3**, 303–312 (1972).

Ting, I. P., and W. M. Dugger. Factors affecting ozone sensitivity and susceptibility of cotton plants. *J. Air Pollut. Control. Assoc.* **18**, 810–813 (1968).

Tingey, D. T. Ozone-induced alterations in the metabolic pools and enzyme activities of plants. In M. Dugger (ed.) *Air Pollution Effects on Plant Growth.* Chapter 4. American Chemical Society, Washington, D.C., 1974.

Tingey, D. T., and G. E. Taylor, Jr. Variation in plant response to ozone: a conceptual model. In M. H. Unsworth and D. P. Ormrod (eds.) *Effects of Gaseous Air Pollution in Agriculture and Horticulture.* Chapter 6. Butterworth Scientific, London, 1982.

Tingey, D. T., W. W. Heck, and R. A. Reinert. Effect of low concentrations of ozone and SO_2 on foliage, growth and yield of radish. *J. Am. Soc. Hort. Sci.* **96**, 369–371 (1971).

Tingey, D. T., R. C. Fites, and C. Wickliff. Ozone alteration of nitrate reduction in soybean. *Physiol. Plant* **29**, 33–38 (1973).

Tingey, D. T., C. Standley, and R. W. Field. Stress ethylene evolution: a measure of ozone effects on plants. *Atmos. Environ.* **10**, 969–974 (1976).

Treshow, M. The impact of air pollutants on plant populations. *Phytopathology* **58**, 1108–1113 (1968).

Treshow, M. (ed.) *Air Pollution and Plant Life.* Wiley, Chichester, 1984.

Turner, N. C., P. E. Waggoner, and S. Rich. Removal of ozone from the atmosphere by soil and vegetation. *Nature* **250**, 486–489 (1974).

Unsworth, M. H., and D. P. Ormrod. *Effects of Gaseous Air Pollution in Agriculture and Horticulture.* Butterworth Scientific, London, 1982.

Valley, S. L. (ed.) *Handbook of Geophysics and Space Environment.* McGraw-Hill, New York, 1965.

Walmsley, L., M. R. Ashmore, and J. N. B. Bell. Adaptation of radish *Raphanus sativus* L. in response to continuous exposure to ozone. *Environ. Pollut.* (Ser. A) **23**, 165–177 (1980).

Weber, D. E., R. A. Reinert, and K. R. Barker. Ozone and sulfur dioxide effects on reproduction and host-parasite relationships of selected plant–parasitic nematodes. *Phytopathology* **69**, 624–628 (1979).

Wellburn, A. R. Effects of SO_2 and NO_2 on metabolic function. In M. H. Unsworth and D. P. Ormrod (eds.) *Effects of Gaseous Air Pollution in Agriculture and Horticulture.* Chapter 8. Butterworth Scientific, London, 1982.

Wukasch, R. T., and G. Hofstra. Ozone and *Botrytis* interactions in onion-leaf diebacks open top chamber studies. *Phytopathology* **67**, 1080–1084 (1979).

Zahn, R. Untersuchungen über die Bedeutung kontinuierlich und intermittier-ender Schwefeldioxideinwirkung für die Pflanzenreaktion. *Staub* **23**, 343–352 (1963).

Hydrogen Fluoride and Sulfur Dioxide

D. C. McCUNE
Boyce Thompson Institute for Plant Research
Cornell University
Ithaca, New York

1. INTRODUCTION

Before one can generalize the current status and then extrapolate to the future needs of research into the action of air pollutants on vegetation, one ought to have in mind the goals of this research. If the mission of research is the development and application of scientific knowledge to the betterment of environmental quality, there would be three principal goals:

1. The diagnosis and evaluation of pollutant-induced effects that seek to answer the question: To what degree is it likely that exposures to air pollutants are responsible for the effects observed?
2. The prevention of pollutant-induced effects through formulation of a series of air quality criteria, guidelines, and standards that seek to answer the question: For what exposures are the risks of an adverse effect upon the public welfare no greater than a particular value?
3. The prediction of possible or potential pollutant-induced effects, commonly attempted in environmental impact statements, that addresses the question: What are the risks of certain effects, given exposures that may or do occur?

It is obvious that these goals are interrelated and any effective program of research would contribute to the achievement of all three, although to a different extent for each. One of the common characteristics of these goals is that they have a well defined utilitarian value: Research should seek to solve problems of practical significance. Another is that they are phrased in probabilistic terms, which is denoted by the idea of risk or likelihood (it is fair to assume that nothing is certain), and imply the idea of probable cost or harm (to be compared with probable benefit).

Study of the interaction of gaseous pollutants with vegetation has had as its major focus of concern the action of pollutants on plants. But as for the action of plants on pollutants, there would appear to be two kinds, and both may have the same compound effect: a change in the path and form by which the pollutants would otherwise reach the soils and waters of the earth. First, the presence and structure of vegetation can affect the movements of parcels of air that contain the pollutants. Secondly, vegetation can remove pollutants from the air as it passes. The immediate significance of the deflection and interception of pollutants by plants is that care should be taken in the placement of air monitoring equipment and the interpretation of the results. A corollary to this is that models for dispersion should not treat a vegetated surface as a perfect reflector nor as a complete absorber. In view of the effort expended in the development of models for the atmospheric dispersion of pollutants, perhaps a greater burden should be placed on research in the botanical sciences to improve the definition of this area.

There has been much concern with the effects of vegetation on sulfur dioxide (SO_2) and hydrogen fluoride (HF) once they come in contact with the plant, for two reasons. First, there is the concern with sulfur as a nutrient: to what degree it is assimilated, how it is distributed, and what is the fate of sulfate. Secondly, the potential toxicity of fluoride to herbivores has made consideration of its accumulation by plants a matter of practical significance. The more intimate details of the chemical changes undergone by these pollutants in the plant have been more or less relegated to studies on their modes of action.

Perhaps the advent of interest in the problem of acidic precipitation will promote more interest in the action of plants on pollutants and a view of an air mass and plant community as interacting components of a single system. It may also direct more interest towards the problem of the interaction of plant and pollutant at or on the foliar surface. Nevertheless, the major preoccupation of researchers and the remaining portion of this discussion concerns the action of pollutants on vegetation as the former pass through the latter.

A review of the effects of HF or of SO_2 is not needed, as that has been done satisfactorily for the former by the National Academy of Sciences (1971) and Weinstein (1977), and for the latter in criteria issued by the National Research Council of Canada (1977) and the National Academy of Sciences of the United States (1978). A comparison of both pollutants was done by Guderian (1977). However, some review of the results of experimentation on the joint action of these two gases will be made as it illustrates some further aspects of their action. In general, the principal differences between HF and SO_2 or other sets of pollutants lie in their sources and chemistries. One could view ozone (O_3) and HF as representing the ends of a spectrum of phtotoxic gases with SO_2 somewhere in the middle. That is, O_3 is a relatively transient chemical species after absorption by the plant, whereas fluoride is accumulated by the plant as a distinct and potentially toxic chemical species. Sulfur dioxide is like O_3 in its transience but more like NO_2 in its being subject to metabolic transformations (indeed as a nutrient under some conditions) and somewhat like HF in that sulfate (or sulfur) may be accumulated. Unlike photochemical oxidants, the sources of SO_2 and HF in North America are usually readily identifiable points and not areas, and their transformations in the atmosphere are more significant with reference to the depletion than to the formation of the pollutants (except for the problem of SO_2 and acidic precipitation). Thus, if one takes cognizance of the particulars, one could speak analogically about the actions of HF and SO_2 and their interactions in plants and deal with certain persistent problems that are common to all air pollutants.

2. JOINT ACTION OF HF AND SO_2

Published accounts on the joint action of HF with other gaseous air pollutants are relatively scarce. Some qualitative comparisons have been

TABLE 1. A Summary of the Responses of Several Species of Plants to the Joint Action of HF and SO_2

Exposure	Concentration ($\mu g/m^3$) and Duration: HF	Concentration ($\mu g/m^3$) and Duration: SO_2	Species and Cultivar of Plant	Response
1. Consecutive	2.5 or 9.0; 6 or 8 days	5240; 1, 2, 4, or 8 h	*Gossypium hirsuitum*	Foliar lesions: pre-exposure to HF increases susceptibility to SO_2-induced injury, but no injury induced by HF
2. Consecutive	3.1 or 11.3; 17 days	(Same as above)	*Helianthus annuus* "Teddy Bear"	Foliar lesions: chlorosis induced by HF; but pre-exposure decreases susceptibility to SO_2-induced injury
3. Consecutive	5.8 or 19.7; 10 days	(Same as above)	*Helianthus annuus* "Teddy Bear"	Foliar lesions: same result as above
4. Consecutive	3.0 or 3.3; 184 or 209 h	2620; 8 h/day, 1 or 2 days	*Pinus strobus*	Foliar lesions: no evidence for nonadditivity
5. Concurrent	3.9; 24 h/day, 8 days	1050; 8 h/day, 5 days	*Lycopersicon esculentum*	Foliar lesions: induced only by HF; no evidence for nonadditivity
			Gossypium hirsuitum	Foliar lesions: none
			Gladiolus sp.	Foliar lesions: induced by SO_2 or HF; no evidence for nonadditivity
6. Concurrent	10; 1 day	2620; 8 h	*Gladiolus* "White Friendship"	Foliar lesions: induced by either SO_2 or HF alone, but HF prevented SO_2-induced injury
7. Concurrent	10; 4 days	1310; 7 days	*Glycine max* "Amsoy"	Foliar lesions: induced by HF but not SO_2; no evidence for nonadditivity
8. Concurrent	4; 4 h/day, 5 days	262; 4 h/day, 5 days	*Pinus strobus*	Foliar F: no effect of SO_2 on F content of needles
9. Concurrent	10; 4 h/day, 5 days	3200; 4 h/day, 5 days	*Gladiolus* sp.	Foliar F: decreased by SO_2 by 34%
			Lolium perenne	Foliar F: decreased by SO_2 by 29%
			Zea mays	Foliar F: no effect of SO_2

10. Concurrent	10; 4 h/day, 5 days	1300; 4 h/day, 5 days	*Gladiolus* sp. *Lolium perenne* *Zea mays*	Foliar F: decreased by SO_2 by 35% Foliar F: decreased by SO_2 by 28% Foliar F: no effect of SO_2
11. Concurrent	2; 100 h	270; 100 h	*Gladiolus* sp. *Lolium perenne* *Zea mays*	Foliar F: no effect of SO_2 Foliar F: no effect of SO_2 Foliar F: no effect of SO_2
12. Concurrent	4.4; 4 h/day, 1, 2, 3, or 4 days	1050; 4 h/day, 1, 2, 3, or 4 days	*Lolium perenne*	Foliar F: decreased by SO_2 by 13%
13. Concurrent	10; 4 h/day, 1, 2, 3, or 4 days	1050; 4 h/day 1, 2, 3, or 4 days	*Gladiolus* sp. *Lolium perenne* *Zea mays*	Foliar F: decreased by SO_2 by 37% Foliar F: no effect of SO_2 Foliar F: no effect of SO_2
14. Concurrent	2; 4 h/day, 3 weeks	1050; 4 h/day, 3 weeks	*Phaseolus vulgaris*	Foliar F: no effect of SO_2 Elongation: no effect of HF, increased by SO_2 Mass of pods: (tendencies) HF gave decrease, SO_2 gave an increase, and no evidence of nonadditivity
15. Concurrent	0.56; 9 days	217; 9 days	*Zea mays* "Marcross" & "Surecross"	Foliar lesions: greater than additive Foliar F: no effect of SO_2 Growth: no effect on fresh or dry mass
16. Concurrent	0.50; 64 or 71 days	170; 64 or 71 days	*Zea mays* "Marcross"	Foliar F: decreased by SO_2 Growth: no effect on height, fresh or dry mass, number and mass of kernels, number of ears
17. Concurrent	5–6; 24 h/day, 2, 4, or 7 days	105; 3 h/day, 2, 4, or 7 days	*Medicago sativa* "Mesa Sirsa," "Kansa," & 212-H	Foliar F: SO_2 depressed rate in "Mesa Sirsa" Stomatal resistance: increased by SO_2 3× in Mesa Sirsa and 2× in other cultivars Foliar temperature: increased by SO_2 (not known if additive)
18. Concurrent	1.0; 24 h/day, 6 weeks	370; 6 h/day, 6 weeks	*Medicago sativa* "Iroquois" & "Saranac"	Foliar F: decreased 31% by SO_2 Foliar S: decreased 19% by HF Yield: SO_2 less than others, and no SO_2 effect when HF present
19. Concurrent	1.0; 24 h/day, 5 weeks	650; 6 h/day, 5 weeks	*Medicago sativa* "Iroquois" & "Saranac"	Foliar F: decreased 34% by SO_2 Foliar S: decreased 16% by HF Yield: no effect

made on symptomatology, effects on growth and reproduction, and relative susceptibility of forest species in areas subjected to HF, SO_2, or HF and SO_2 (Pollanschütz, 1968; Bohne, 1972; Roques et al., 1980) and on crops and ornamentals (Bohne, 1970; Buckenham et al., 1982). Literature on the experimental investigation of the joint action of HF and other gases in controlled exposures comprises several published accounts (Amundson et al., 1982; Brandt, 1981; Hitchcock et al., 1962; Mandl et al., 1975, 1980; Matsushima and Brewer, 1972; Solberg and Adams, 1956). This is not an extensive catalog of effects upon which to draw if one wishes to determine what characterizes the joint action of HF with other pollutants. Thus, for the purposes of discussion, a summary of unpublished results from several years of experimentation on the joint action of HF with SO_2 at Boyce Thompson Institute is presented in Table 1.

The medley of experimental results presented in the literature and in this table has no readily discernible theme, which is not too extraordinary inasmuch as the experiments comprise many different sets of exposures, many species of plants, and three kinds of measures of response. Moreover, experiments were carried out in the field plots as well as in controlled environment chambers and a variety of experimental designs were used that included not only concurrent but also consecutive exposures to the two gases. The most general and least satisfactory conclusion is that whether an interaction of HF with other pollutants occurs and whether it is less than or greater than additive depends upon the receptor, pollutant, response, and variables of exposure. Nevertheless, some common features of these results allow the construction of more specific conclusions on rather slender experimental and theoretical supports if one starts an analysis of the data with the most readily interpretable response: the accumulation of fluoride (F) or sulfur (S) by the foliage.

2.1 Accumulation of Fluoride

The accumulation of fluoride by the foliage or other tissues of the plant is usually reckoned as the difference in concentration of fluoride between plants that were exposed to atmospheric fluoride and those that were not. Because the concentration is a ratio of mass of fluoride to mass of tissue, one can achieve a different end-point by changes in either variable. For example, a change in concentration will be the consequence of changes in the mass of tissue per unit of sorptive surface, the rate of uptake of atmospheric fluoride, and the rate of loss of sorbed fluoride through elution, weathering, abscission of older leaves containing greater amounts of fluoride, or addition of new leaves containing lesser amounts of fluoride.

The most consistent effect in the joint action of HF and SO_2 was evident on the accumulation of F and S by alfalfa (*Medicago sativa* L.) (Mandl et al., unpublished). In field exposure chambers, the presence of SO_2 significantly decreased the concentration of F in both crops and the presence of HF also decreased the concentration of S. The means by which SO_2 decreased the accumulation of F was partially elucidated by controlled environment exposures to HF alone or with SO_2 present. The diffusive resistance of all cultivars was increased by SO_2, and by a greater amount in Mesa Sirsa than in the two other cultivars. (Foliar temperatures were also increased in the presence of SO_2.) The regression coefficient of foliar F on dose (calculated as $\mu g\ F/m^3\ h$) was less in the presence of SO_2 for Mesa Sirsa but not statistically different for the other cultivars. Thus, on the basis of comparisons across cultivars and treatments, the SO_2-induced decreases in the accumulation of F from HF were associated with SO_2-induced increases in stomatal diffusive resistance. The HF-induced decreases in S accumulation may be caused by a potentiation by HF of the SO_2-induced decrease in diffusive resistance or by HF-induced increases in the translocation or loss of S from the leaves.

Results with three species of monocots—perennial ryegrass (*Lolium perenne* L.), gladiolus (*Gladiolus* sp.), and corn (*Zea mays* L.)—in short-term exposures add to the complexity of the HF–SO_2 interactions. There is a threshold concentration of SO_2 for its effect on the accumulation of F, and this threshold may increase as the concentration of HF increases. With gladiolus, the relative effect of SO_2, when it occurred, was independent of its concentration and independent of foliar F within a certain range of both these variables. The lack of effect of SO_2 in continuous exposures of gladiolus to both gases may be due to a subliminal concentration of SO_2, but because three dark periods intervened and exposures were functionally if not operationally discontinuous, it is possible that sorption of HF during the dark masked an effect of SO_2 in the light. The lack of interactive effects on corn may have resulted because low levels of F (3–20 ppm) in the foliage did not allow detection of moderate relative differences owing to the presence and magnitude of analytical and sampling errors as well as biological variability (Mandl et al., 1975).

The problem is that the less the occurrence of an effect is understood, the more extensive and varied the explanations become for its absence. Thus these or other hypotheses may be advanced to explain the lack of an effect of SO_2 on the accumulation of F by pine (*Pinus strobus* L.) or bean (*Phaseolus vulgaris* L.) foliage. Environmental factors may partially determine the occurrence, kind, and degree of effect. Concurrent exposures of millet (*Setaria italica* L.) to HF and SO_2 resulted in a decreased F accumulation at 75% relative humidity but in an increased F accumulation at 95% relative humidity (Brandt, 1981). An SO_2-induced decrease in F accumulation by leaves of *Citrus* did not become manifest until a certain length of exposure had passed (Matsushima and Brewer, 1972). This may result from the necessity for foliar fluoride levels to reach a certain level for statistically significant differences to be manifest. It may also indicate that the threshold for an SO_2-induced decrease in uptake of HF increases with an increase in the level of fluoride in the foliage. In sweet corn, a sufficiently long duration (≥4 weeks) of exposure may have been necessary for the concurrence of all factors that contributed to an SO_2-induced decrease in F accumulation in the field (Mandl et al., 1980).

2.2 Induction of Foliar Lesions

If the effect of a pollutant were determined solely by its uptake by the leaf, the joint action of HF and SO_2 on the induction of foliar lesions should follow the same pattern as their effects on the accumulation of F or S. Thus, in that domain where SO_2 decreases the accumulation of F, it should also reduce the occurrence of F-induced foliar lesions. In the domain where HF alters stomatal responses to SO_2, it should also alter the occurrence of SO_2-induced lesions. Because in most species of plants HF and SO_2 induce different types of foliar symptoms, it is usually possible to differentiate the action of each pollutant when both are used.

Conceptually, a study of the consecutive action of the two pollutants may be easier to interpret than their concurrent action in the sense of one pollutant altering the susceptibility of the plant to the other pollutant. Prior exposure to HF increased the susceptibility of cotton and decreased the susceptibility of sunflower to SO_2 as judged by three parameters relating dose of SO_2 to response (percentage of leaves with lesions): maximum level of effect; level of effect at lowest dose used; and, dose at which maximum effect was achieved. Some possible explanations for these differences are: the species were different; HF-induced chlorosis was present in sunflower, but no HF-induced symptoms occurred in cotton and this foliar injury may alter the effect of SO_2; and the response is not monotonic with concentration of HF and relatively lower doses of HF (less than those that induce foliar lesions) increased susceptibility to SO_2 whereas relatively higher doses decreased it. Greater insight into modes of joint action would have been provided had the responses been measured with respect to developmental stage of the leaf (instead of being pooled), given the effect of this factor on inherent susceptibility to either pollutant and the real exposure received by each leaf during the course of HF treatment.

Some results tend to favor a mechanism involving a change in susceptibility rather than in uptake as an explanation of the joint action of HF and SO_2. With an increase in the concentration of both HF and SO_2, the response of gladiolus moves from a region in which both pollutants induce foliar lesions without interaction to one where HF suppresses the expression of SO_2-induced lesions. As the latter phenomenon occurs at exposures under which SO_2 reduced the accumulation of F in gladiolus, an effect on uptake may not be a determining factor, unless the presence of fluoride increases the SO_2-induced decrease in stomatal diffusion beyond that which would occur in the absence of HF (Mandl et al., unpublished). Because lesions atypical of HF or SO_2 appeared on the foliage of sweet corn exposed to either pollutant alone or to both in combination, one could not therefore interpret effects on frequency of occurrence, which was much greater than additive, in terms of an effect of HF on SO_2-induced effects or vice versa. But relative to the total number of lesions produced, fewer occurred on the older leaves and more appeared on the younger leaves when HF was present with SO_2. Thus the relative incidence of lesions (with both HF and SO_2 present) increased on leaves more sensitive to the effects of fluoride (Mandl et al., 1975). When experiments were repeated in the field, the same kinds of atypical, bifacial elliptical lesions developed as in previous tests under controlled environmental conditions. But the symptomatology was not so simple because lesions were present not only on all treated plants (although to different degrees) but also on other cultivars of sweet corn in the same area (Mandl et al., 1980). When a response that is sensitive to the joint action of HF and SO_2 can also be induced by an unknown environmental factor, interpretations of the joint action of HF and SO_2 become even more problematic and raise the question whether one mode of joint action of pollutants may be to sensitize the plant to a heretofore ineffective environmental factor.

When field experiments on the joint action of HF and SO_2 were undertaken in open-top chambers, the intrusion of ambient air and photochemical oxidants could occur. The incidence of oxidant-induced injury on sweet corn appeared to be greater in SO_2 and SO_2 plus HF treatments (on 67% and 59% of the plants, respectively) but was unaffected by HF (41% of the plants).

2.3 Effects on Growth and Yield

Effects on growth and yield are the most relevant and practical aspect of the joint action of pollutants, but they are supported by fewer data and are more difficult to interpret than those effects manifested as changes in uptake or foliar lesions, to which they have a theoretical and causal yet uncertain empirical relation.

In *Citrus*, the effects of HF and SO_2 on growth were additive (Matsushima and Brewer, 1972). Effects on biomass of alfalfa were not consistent, but when an effect of SO_2 occurred, the HF plus SO_2 treatment was the same as HF alone or the control. That is, SO_2 decreased yield but not when HF was present, and this effect on yield is concordant with the HF-induced decrease in foliar sulfur from SO_2 (Mandl et al., unpublished). Effects of HF and SO_2 on the growth and yield of corn are difficult to interpret in data available from three experiments under controlled environmental conditions and two field experiments (Mandl et al., 1975, 1980). No effects of the pollutants alone or combined were found in exposures where greater than additive effects on the occurrence of foliar lesions were observed. One can conclude that there was no relation of foliar symptoms to growth, or if there were, it was below the sensitivity of measures of growth to detect it. In a second field experiment, there were no effects on yield but there were on growth: It was increased by HF or SO_2 alone but not by the combination. The interpretation of interactions in 2×2 factorial experiments is difficult in the absence of extrinsic information on dose–response relations. Departures from additivity may not indicate that interactive effects are present but that the relationship of response to exposure is neither linear nor monotonic. Without some estimate of these two characteristics, the interest in and interpretation of nonadditivity as an interaction seems of limited significance.

3. MODES AND MODELS OF INTERACTION

It would appear that the simplest and most general model for the joint action of HF with SO_2 would be a surface, as depicted in Figure 1; the action of one pollutant on the effectiveness of another is shown as having the potential of inducing responses that are greater than additive, additive, or less than additive for those responses to either pollutant by itself. These

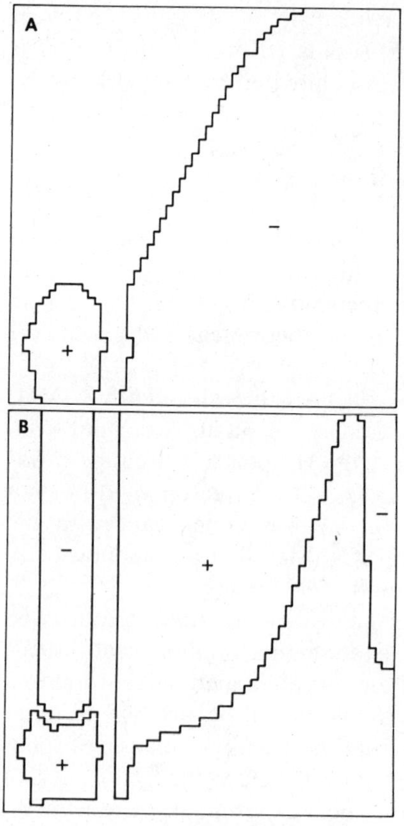

FIGURE 1. The reciprocal effects of HF and SO_2 on plants are shown with reference to (*A*) the effect of exposure to SO_2 (horizontal axis) on the apparent effectiveness of HF (vertical axis); (*B*) the effect of exposure to HF on the apparent effectiveness of SO_2. Regions of exposure in which one pollutant reduces or increases the effect of another are denoted by − or +, respectively.

regions lie in different areas of the surface whose coordinates are an exposure variable for SO_2 (wherein concentration and duration and frequency of exposure enter in an unspecified manner) and an exposure variable for HF (which may also be some function of fluoride accumulated in the tissue). The bases of this model are accumulation of pollutants and induction of foliar lesions wherein the separate effects of HF and SO_2 can be identified. The assumptions in this model are as follows:

1. SO_2 may increase stomatal diffusivity at low exposures and decrease it at greater exposures (Unsworth and Black, 1981).
2. SO_2-induced increases in diffusivity are increased by a low exposure to HF and blocked by higher exposures.
3. SO_2-induced decreases in diffusivity are also blocked by higher exposures to HF but this can be overcome by increasing exposure to SO_2.
4. With sufficiently great exposures to both pollutants, HF may potentiate the SO_2-induced decrease in diffusivity.

The model implicitly assumes that reciprocal effects of the pollutants on susceptibility of the tissues of the plant parallel their effects on accumulation but this assumption is more tenuous than the others. However limited is the experimental evidence for such a model, it does indicate that experimental designs that use very few levels of exposure may give apparently contradictory results concerning the responses of plants to the joint action of HF and another pollutant.

Studies on the modes or mechanisms or actions of air pollutants have fallen on hard times in their perceived importance and hence their likelihood of funding. The reasons for this are understandable and perhaps partially justified as there has been no readily apparent connection of the metabolic or physiologic response of plants with altered agricultural productivity. Nevertheless, studies of mechanisms should be raised from their present position of neglect and resumed in investigations of joint action. Because mechanism or mode implies that which is concerned with a pathway rather than an end point, such studies can aid in the process of extrapolation or interpolation and supplement the studies on productivity. Knowledge of mode of action may more efficiently provide an answer to the question of what would happen with a different exposure, receptor, or environment than would more field experiments, which are long and expensive. The paradigm in this sort of venture has been made in physiologic studies on the reciprocal effects of pollutants on stomatal functions. Through these studies, a means is at hand for estimating the effects of climate and other pollutants and for using a great amount of basic information from physiologic research. Perhaps studies on the metabolic fate of SO_2 will be as successful for this pollutant, but it may be too early to tell.

The occurrence of an interaction (or the interpretation of the results in terms of interactive effects) depends upon what models are assumed for the dose–response relationship and what scale of measurement is used for expression of dose and response. For example: a greater-than-additive response may reduce to a linear additive model if log dose is used; a less-than-additive response may reduce to independent action if response is transformed by a binomial distribution. Probit analysis often assumes both transformations. What are needed are general models that explain and unify apparently disparate results. Such models should have features that are invariant and independent of scale. That is, they should be useful for short- as well as long-term exposures and responses, be applicable to populations of cells or plants, and deal with foliar lesions or morbidity of mature conifers. They should be quantitative and capable of reduction to simpler sets and scales of measurements. A consideration of the joint action of HF and SO_2 and the mode of action of individual pollutants leads to the conclusion that the response of a plant can be viewed as the resultant of two sets of processes: those by which the pollutant accumulates in the receptor and those by which the accumulated pollutant affects the receptor.

Thus, a useful point of attack in the building of such a model would be the statistical and pharmacological structure developed by Hewlett and Plackett (Hewlett and Plackett, 1959, 1964; Plackett and Hewlett, 1967) for quantal response to mixtures of drugs. On the theoretical and phenomenological level, two kinds of classifications of joint action were proposed:

1. In noninteractive joint action, one component (pollutant) does not influence the biological action of the other, but in interactive it does.
2. In dissimilar joint action, the sites of primary action are different, but in similar, they are the same.

Hence, there are four different types of joint action:

	Similar	*Dissimilar*
Noninteractive	Simple similar	Independent
Interactive	Complex similar	Dependent

Mechanistically, the following elements are contained in the model:

1. An external compartment and an internal compartment. Thus for pollutant A, its concentration in the external compartment is denoted by C_A and in the internal compartment by C'_A with $C'_A \leq C_A$.
2. These two compartments are connected by what could be called a gate-and-channel system, which determines the relationship of C'_A to

C_A. The action of this system may involve more than just a restriction of flow between the external and internal compartments. Within this system, a pollutant may be inactivated or changed to a less active form, conducted away from the site of action, stored, or transformed to a more active form.

3. Within the internal compartment is located a receptor–effector system, and if $C'_A > C^*_A$, then a response occurs and C^*_A is termed the action tolerance or threshold.
4. One can consider the same systems as existing for pollutant B, with the variables C_B, C'_B, and C^*_B.
5. The possibilities for interaction therefore lie in the effect of A on the gate-channel or receptor–effector systems for B, and vice versa.

The authors of the model also included certain variations, such as overlapping sites of action and reciprocal side effects. The latter would occur when the primary sites of action are distinct, but one pollutant may have some degree of activity at the site of action of the other.

Such a model seems suitable for use in air pollution because it has general features that can be somewhat invariant with the scale of application:

1. *Response*. Although graded are more common than quantal responses, a graded response can be expressed as a quantal one when some level of response is specified as a definite end-point (critical level) and the response is then expressed as $\geq$ or $<$ that level.
2. *Biological unit responding*. The cell, leaf, plant, or hectare of cropland would appear to be amenable to treatment in this way, as long as the kind of response is appropriate to the receptor.
3. *Functional compartments where C_A and C'_A are to be measured*. There would be at least five possibilities depending upon the responding unit such as the atmosphere in the region of geostrophic wind, the boundary layer of the leaf, the substomatal cavity, the aqueous phase of extracellular space in the mesophyll, and the cellular milieu. There is the possibility of more than one gate-channel system on the macroscale (where there are spatially distinct sources of each pollutant), the microscale (penetration through the plasmalemma), and the leaf where cuticular as well as stomatal absorption is functionally significant.

The proposed model has the advantage of also being stochastic. One formulation for the gate-and-channel system is that $C'_A = \mu_A C_A^N$ where parameters μ_A and N_A vary from individual to individual and have their own distributions over the population. The tolerances of receptors, C^*_A, also have their distribution function and, important to the model, C^*_A and C^*_B

have a joint distribution with significant consequences when their correlations approach −1 or 1.

Liabilities of such a model may be numerous and sufficient to exclude it from future consideration. It may focus attention too much on mode or mechanism of action if such studies tend to center more upon themselves than on extending the extrapolability of the results to other situations. It may be impractical as the data are too few and too poor to judge its applicability to the real world (perhaps more a judgment on the data than the model). There may be complexities in the temporal interactions of pollutants, time-dependent distributions of parameters, and the presence of a stimulation at doses less than those that inhibit. There is also the problem of the form of the pollutant at its primary site of action. With HF, one can assume it is the fluoride ion. But with SO_2, it could be sulfite, sulfate, or a product produced by the oxidation or reduction of the pollutant in the foliar tissues. Finally, no more can be said about the interactive effects of pollutants in combination than is known about the effects of the individual constituents. If the relationship of dose to response is not well described for the individual effects of the constituents (for a range within which both act together), their joint action cannot be interpreted.

Whatever kind of model of joint action is developed, the objective of research should not continue to be to determine in isolated systems and simple designs whether nonadditivity is obtained. As sufficient results accrue, one could define the domain in which nonadditivity does or does not occur (the latter is as important as the former), then develop a quantitative means of expressing joint action that would be suitable for the needs of environmental quality, and ultimately arrive at the means for evaluating the joint action of air pollutants and other abiotic or biotic agents that stress plants or reduce agricultural productivity.

4. PROBLEMS IN EXPERIMENTATION

4.1 Characteristics of Exposures

There are two basic experimental approaches to studies of the action of SO_2 or HF on plants. One is to place plants in an ambient atmosphere near and within the effective vicinity of a source and then monitor exposures as they occur at each site. The other is to add a pollutant, where it does not occur, to the atmosphere of the plant and then regulate the amount and duration of exposure. Variations on these approaches are to enclose the plant in a chamber and exclude the ambient pollutant by different degrees or to add to the ambient pollutant so that exposures greater than the ambient occur.

Whether used to exclude or add a pollutant, enclosure by field exposure chambers can alter the growth and yield of crops. But given the irregular nature and occurrence of these effects, one cannot yet explain or predict them. It is also questionable whether unenclosed plots are normal with respect to what one would obtain in a real field: The limited size of the plots and the presence of nearby structures could alter the microclimate from what might be called normal. Nevertheless, the major question is not so much what is normal but whether enclosure by a chamber exerts an effect or alters the responses of plants to air pollutants. In most experiments, one does not have a direct test for the separate effects of enclosure and its interaction with the pollutant. The use of chamber-free methods, such as the linear gradient or zonal air pollutant systems, eliminates the problems of enclosure but does not resolve the problem of providing experimental plots with realistic exposures.

A well-known and still unmanageable fact is that at any site the concentration of a pollutant will change with time. This time-dependent variation results from fluctuations in source strength, effects of weather on dispersion (which may also include the transformation as well as the transportation of pollutants), micrometeorological conditions, sampling error, and analytical error. The consequence is that one must resort to some kind of statistical description for an exposure. A number have been proposed (the reader is referred to Kornreich, 1974) and there is some divergence as to the forms of the distribution functions, but it appears that with certain averaging (or sampling) periods, empirical data favor a log–normal distribution of concentrations and a Weibull distribution for maxima. One should perhaps have a series of distributions where the statistics and the function itself change with averaging time. When one deals with averaging periods less than 24 h, the empirical distribution often contains occurrences of zeros, especially in the vicinity of a single source. Thus an exposure may be described as a population of episodes—discrete clusters of nonzero events—rather than as collections of means for uniform averaging periods. The best known episode is the daily march of ozone where three or more parameters are needed to describe peak height, duration, and the time-dependent variation of concentration in this episode. If one were to examine a sufficiently large population of episodes, it might turn out that such descriptive statistics are not independently distributed. Indeed, one might even find that these fall in discernible sequences or clusters. The statistics of air quality might then consider not only extreme values of single events (Roberts, 1979) but also the recurrence of certain sequences of events.

There is one great drawback to this approach: It involves sifting a great deal of air monitoring data through a numerical sieve to win a few nuggets of information. Moreover, the air monitoring data itself may not be readily at hand or of sufficient fineness for this purpose. For example, data on HF is based preponderantly on 24- or 12-h averaging periods and the kinds of frequency distributions that can result have a relatively few, high values superimposed on a positively skewed distribution of lower values (McCune et al., 1976). The inherent sensitivity of commonly used monitoring methods, given the concentrations that occur, do not allow sampling periods shorter than two to four hours. Thus with HF, perhaps the most significant advance in description of its occurrence would come from the development of a monitoring system that would be almost instantaneous and as sensitive and precise over the range of phytotoxic concentrations as the present methods for SO_2. Without this development, the characterization and simulation of realistic exposures is restricted. For the joint action of HF with another pollutant, such as SO_2, an exposure to both pollutants should be described with sufficient numbers of parameters in a joint distribution function. On the other hand, photochemical oxidants might be cast in the role of environmental factors owing to their scale of occurrence, patterns of dispersion, and difficulty of control relative to HF, This is probably more a difference in degree than in kind, and with increased specification of conditions in a model there may be no difference.

Even when one can characterize an ambient exposure, it is difficult to interpret its relationship to effects and to transfer these results (exposure–effect relationships) to other sites and different exposures. One could give a rather general qualitative exposure–effect relationship on the principles that the degree and likelihood of an effect are increased:

1. The greater the concentration, duration of exposure, or frequency, because these variables determine the amount of pollutant received.
2. The greater the duration or frequency of exposure, because this increases the chance of exposure during a period of heightened sensitivity.
3. The greater the frequency and recurrence of episodes, because this decreases the countervailing effect of the processes of recovery or regeneration.

A botanically significant expression of exposure might be derived from a weighted sum of the hourly averages where each weight or coefficient for a particular hourly period is a function of the inherent sensitivity of the plant, that varies with ontogenetic and circadian patterns; climatically induced changes in sensitivity; and the effects of previous exposures to the pollutants. The most frequently adopted method is to assign a weight of one to all concentrations with a certain daytime period and a weight of zero to all others. The problem may be that there is no single, botanically significant, quantitative measure of an exposure. With HF, one must also recognize that the effects of repeated exposures are cumulative not only with respect to the effect produced by each but also with respect to accumulation of the pollutant.

It is equally difficult to apply the results of experimental exposures to ambient conditions because in the universe of all possible exposures, only certain sets may be realized in the real world. Unfortunately, these tend not to be those used in experimental exposures. For example, unrealistic concentrations are employed, or realistic but constant concentrations are used continuously for several days. Such exposures are useful in determining the possibility of a certain effect, but they should be succeeded by realistic exposures in determining its likelihood. The supplementation of filtered or unfiltered ambient air with O_3 by a constant amount for a limited period during the daily cycle of photochemical oxidant creates another problem: the regular progression of temporal changes may be altered as the occurrence of peak concentrations is shifted towards earlier hours. This creates an exposure that is not typical of the real world and makes problematic the comparison of air with supplemental O_3 to air without it.

The use of periodic and constant (square wave) exposures, however unreal, may continue for two compelling reasons: ease of generation and control and simplicity of arithmetic expression as dose. On the other hand, the simulation of actual episodes falls into the same problems of interpretation as ambient exposures. In the design of regimes for exposures to pollutants, any experimenter is therefore caught between the Scylla of using an unreal series of exposures which can be interpreted in terms of dose–response or factorial effects, and the Charybdis of simulating realistic regimes, which cannot yet be expressed or interpreted in simple quantitative measures.

4.2 Distributions of Effects

There are often unconscious attitudes that should, from time to time, be brought to mind. One is the tendency to regard a plant or small set of plants as representative of a population. This is most habitual when one deals with crops where a great amount of phenotypic homogeneity is present. Another is to regard an error or residual variance as a bothersome statistic that, were it not so great, would not stand in the way of proving some hypothesis. Consequently, estimates of dispersion tend to be ignored. But they are descriptive statistics with as much potential importance as estimates of location in the evaluation of the effects of air pollutants.

A pollutant-induced change in dispersion can be of interest as indicating an increased homogeneity or heterogeneity of the population. The former would be of significance in natural populations or multiline cropping systems where one is concerned with resistance to biotic or abiotic stress and the consequences of pollutant-induced increase in homogeneity which could be as serious as a shift in median tolerance. On the other hand, it is also possible that a pollutant-induced decrease in the uniformity of a stand or maturity of a crop may prove more commercially significant (in harvest) than an effect on the average yield. It is also important to determine whether some environmental or biological factor affects the variability or distribution of tolerances as well as the median tolerance for a pollutant in a population of plants. There has also been the tendency to neglect statistics, such as correlations or covariances (where more than one characteristic of a plant is measured), of multivariate distributions and only examine each variate independently of the others. In so doing, one loses information on the "shape" of the effect by merely considering its size. Pollutant-induced changes in the associations of traits that influence stress-resistance may prove important, and altered relationships among measures of growth and yield may indicate the ways by which developmental processes are affected.

One of the major goals of experimentation has been to derive dose–response relations, formally and generally expressed as $Y = f(A, B, C) + E$, in which a pollutant-induced response Y is an explicit function of some dose A and the characteristics of the receptor B and environment C, plus a variable component E, which arises from errors in measurement of dose and response and from the variations in receptors and environments that go unobserved. The practical application (e.g., crop-loss assessment) of the dose–response function has centered on its evaluation for certain sets of values of the independent variates, that is, on the conditional expectation of Y given A, B, and C:

$$f(a, b, c) = \mathrm{Ex}[Y \mid A = a, B = b, C = c]$$

However, other practical problems require knowledge of the conditional distributions of responses. For example, the problem of evaluation may be to determine the probability that the effect lies in some range Y_1, Y_2, where Y_1 may be equal to zero, given certain exposures, receptors, and environments:

$$\Pr\{Y_1 \leq Y < Y_2 \mid A = a, B = b, C = c\}$$

Another example would be the relation of one kind of effect to another, such as a reduction in growth and foliar injury, and the problem of estimating the likelihood that a reduction in growth greater than some value occurs given the presence of some amount of foliar injury. In measures for the protection of environmental quality, such as air quality standards, to reduce the risk of undesirable effects to an acceptable level, one needs to know for some upper bound on exposures A_1, the risk of an effect greater than or equal to some level Y_1:

$$\Pr\{Y_1 < Y \mid A_1 > A, B = b, C = c\}$$

Efforts to derive air quality guidelines on the stochastic properties of exposures and effects have been initiated by Schwela and Junker (1978), Shoji and Tsukatani (1973), and Larsen and Heck (1976). In the preparation of environmental impact assessments, one may merely wish to determine the likelihoods of certain sets of outcomes, given certain ranges of the variates. There are also times when the mean or median effect is not the critical one. If the median amount of foliar injury produced is 5%, in what proportion of individuals will it be greater than 10%? These relationships between exposure and effect must consider the distribution of effects as a better basis for extrapolation than a determination of mean response to some exposure. This is particularly important in the problem of thresholds that involves an extrapolation of dose–response at one (lower) end of the curve, because the lowest point on the response curve estimates the threshold of the most susceptible individual in the sample.

5. CONCLUSION

In this discussion of the present status of research on the interaction of SO_2 and HF with plants, the major emphasis has been that the real and the practical should be the determinants of future efforts. The greatest contribution of studies on the effects of HF and SO_2 on yield and quality of crops may come about if greater use is made of models of plant growth and more measurements are made on the individual components of yield and their changes in time. There are also contributions to be made in the study of physiologic responses if undertaken in connection with two problems:

1. What is the nature of the relationship of dose to response, especially in the range of liminal doses?
2. How far apart may two exposures occur to be said to act independently?

In the gathering of new information, future research should be mindful of the need to reevaluate critically the old, not as a series of bad examples but as a source of rather expensively gained information that could still have utility in a new context.

REFERENCES

Amundson, R. G., L. H. Weinstein, P. van Leuken, and L. J. Colavito. Joint action of HF and NO_2 on growth, fluorine accumulation, and leaf resistance in Marcross sweet corn. *Environ. Exp. Bot.* **22**, 49–55 (1982).

Bohne, H. Fluorides and sulfur oxides as causes of plant damage. *Fluoride Quart. Rept.* **3**, 137–142 (1970).

Bohne, H. Klarung eines Rauchschadenfalles bei Keifernbestanden im Ruhrgebiet. *Mitt. Forst. Bundesversuchsanstalt.* **97**, 141–140 (1972).

Brandt, C. J. Wirkungen von Fluorwasserstoff auf *Lolium multiflorum*. *Landesanst. Immissionsschütz Landes Nordrhein-Westfalen Ber.* **14**, 140 pp. (1981).

Buckenham, A. H., M. A. J. Parry, and C. P. Wittingham. Effects of aerial pollutants on the growth and yield of spring barley. *Ann. Appl. Biol.* **100**, 179–187 (1982).

Guderian, R. *Air Pollution: Phytotoxicity of Acidic Gases and Its Significance in Air Pollution Control*. Springer Verlag, New York, 127 pp., 1977.

Hewlett, P. S., and R. L. Plackett. A unified theory for quantal responses to mixtures of drugs: non-interactive action. *Biometrics* **15**, 591–610 (1959).

Hewlett, P. S., and R. L. Plackett. A unified theory for quantal responses to mixtures of drugs: competitive action. *Biometrics* **20**, 566–575 (1964).

Hitchcock, A. E., P. W. Zimmerman, and R. R. Coe. Results of ten year's work (1951–1960) on the effect of fluorides on gladiolus. *Contrib. Boyce Thompson Inst.* **21**, 303–344 (1962).

Kornreich, L. D. (ed.) *Proceedings of the Symposium on Statistical Aspects of Air Quality Data, November, 1972*. EPA-650/4/74-038, U.S. Environmental Protection Agency, Research Triangle Park, North Carolina (1974).

Larsen, R. I., and W. W. Heck. An air quality data analysis system for interrelating effects, standards, and needed source reduction. Part 3. Vegetation injury. *J. Air Pollut. Control Assoc.* **26**, 325–333 (1976).

McCune, D. C., D. C. MacLean, and R. E. Schneider. Experimental approaches to the effects of airborne fluoride on plants. In T. A. Mansfield (ed.) *Effect of Air Pollutants on Plants*. (Soc. Exp. Biol. Seminar Ser. 1), pp. 31–46, Cambridge Univ. Press, 1976.

Mandl, R. H., L. H. Weinstein, and M. Keveny. Effects of hydrogen fluoride and sulphur dioxide alone and in combination on several species of plants. *Environ. Pollut.* **9**, 133–143 (1975).

Mandl, R. H., L. H. Weinstein, M. Dean, and M. Wheeler. The response of sweet corn to HF and SO_2 under field conditions. *Environ. Exp. Bot.* **20**, 359–366 (1980).

Matsushima, J., and R. F. Brewer. Influence of sulfur dioxide and hydrogen fluoride as a mix or reciprocal exposure on citrus growth and development. *J. Air Pollut. Control Assoc.* **22**, 710–713 (1972).

National Academy of Sciences. *Fluorides*. Committee on Biologic Effects of Atmospheric Pollutants. Washington, D.C., 195 pp., 1971.

National Academy of Sciences. *Sulfur Oxides*. Committee on Sulfur Oxides, Board on Toxicology and Environmental Health Hazards. Washington, D.C., 209 pp., 1978.

National Research Council of Canada, NRC Associate Committee on Scientific Criteria for Environmental Quality. *Sulphur and Its Inorganic Derivatives in the Canadian Environment*. Publ. No. NRCC 15015, Ottawa, Canada: 426 pp., 1977.

Plackett, R. L., and P. S. Hewlett. A comparison of two approaches to the construction of models for quantal responses to mixtures of drugs. *Biometrics* **23**, 27–44 (1967).

Pollanschütz, J. Beobachtungen über die Empfindlichkeit verschiedener Baumarten gegenüber Immissionen von SO_2, HF und Magnesitstaub. pp. 371–377. In *Air Pollution, Proc. 1st Europ. Congr., Wageningen*. Centre for Agricultural Publishing and Documentation (Pudoc), Wageningen, 1968.

Roberts, E. M. Review of statistics of extreme values with applications to air quality data. Part II. Applications. *J. Air Pollut. Control Assoc.* **29**, 733–740 (1979).

Roques, A., M. Kerjean, and D. Auclair. Effets de la pollution atmospherique par le fluor et le dioxyde de soufre sur l'appareil reproducteur femelle de *Pinus sylvestris* en foret de Roumare (Seine-Maritime, France). *Environ. Pollut., Ser. A.* **21**, 191–201 (1980).

Schwela, D., and A. Junker. Derivation of air quality standards on the basis of risk considerations. *Wat. Air Soil Pollut.* **10**, 255–268 (1978).

Shoji, H., and T. Tsukatani. Statistical model of air pollutant concentration and its application to air quality standards. *Atmos. Environ.* **7**, 485–501 (1973).

Solberg, R. A., and D. F. Adams. Histological responses of some plant leaves to hydrogen fluoride and sulfur dioxide. *Am. J. Bot.* **43**, 755–760 (1956).

Unsworth, M. H., and V. J. Black. Stomatal responses to pollutants. In P. G. Jarvis and T. A. Mansfield (eds.) *Stomatal Physiology*, pp. 187–263, Cambridge Univ. Press, 1981.

Weinstein, L. H. Fluoride and plant life. *J. Occup. Med.* **19**, 49–78 (1977).

SYNTHESIS

EFFECTS OF GASEOUS AIR POLLUTANTS ON TERRESTRIAL VEGETATION

Walter W. Heck
U.S. Department of Agriculture/
North Carolina State University
Raleigh, North Carolina

S. B. McLaughlin
Oak Ridge National Laboratory
Oak Ridge, Tennessee

INTRODUCTION

Scope of Problem

On a global scale, terrestrial vegetation is exposed intermittently to gaseous air pollutants. Under some conditions the exposure may be continuous, but generally the exposure concentrations are variable.

Similarly, the response of vegetation is variable depending upon factors both external and internal to the plant. Under some conditions, the effects may be beneficial or benign; under others, the effects may be harmful or disastrous. The effects may be acute or chronic, reversible or irreversible, direct (via foliar absorption) or indirect (via deposition on the soil and uptake by roots). The effects may be a direct result of a specific chemical species or of various combinations of chemical species (e.g., $SO_2 + O_3 + NO_2$*). Growth and yield responses can occur without classical injury symptoms and may occur without any appearance of foliar abnormalities. Complex interactions between air pollutants and both biotic and abiotic environmental factors also occur. Specific*

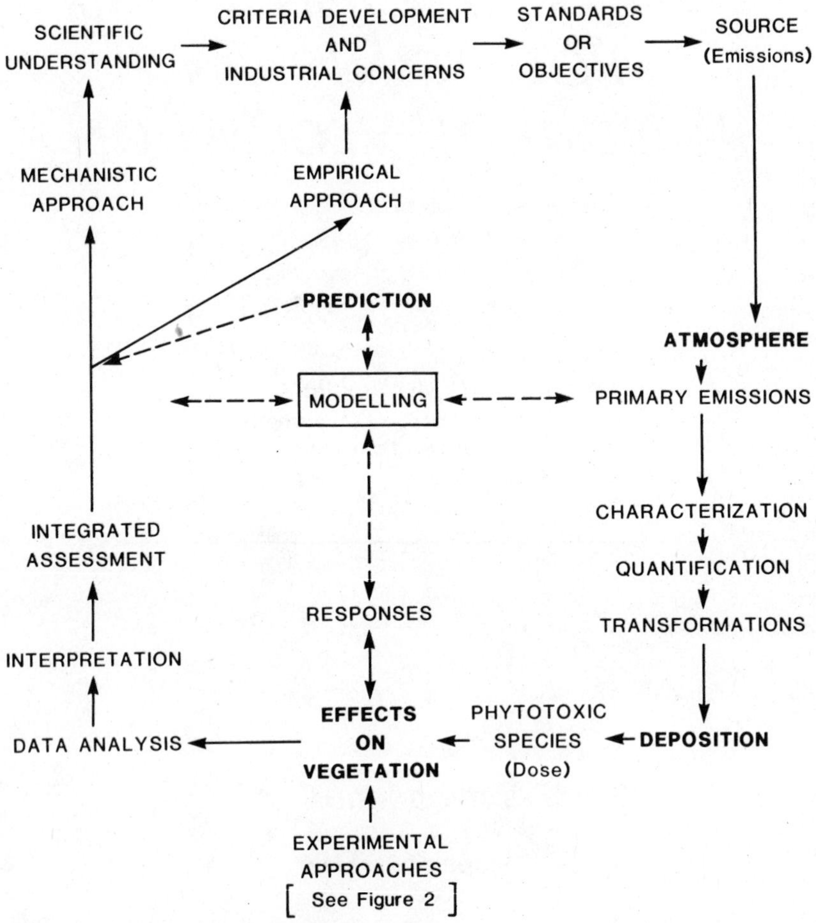

FIGURE 1. Information flow relative to the effects of gaseous pollutants on terrestrial vegetation.

concerns of plant effects researchers include:

1. *The difficulty in translating laboratory data to field situations.*
2. *The difficulty in characterizing the response of plants to pollutant concentrations that simulate ambient conditions.*
3. *The difficulty in expressing a pollutant dose in a form that has consistent biological significance.*
4. *The difficulties of designing experiments that provide the needed understanding of pollutant effects on biological systems which can also be extrapolated to a range of environmental conditions. Parameters of response best suited to meet the objectives of research include, but are not limited to, growth and yield, visible symptoms, reproductive effects, and physiological, biochemical, and cytological changes. Translation of these data into ecological or economic terms is an additional important challenge to research on plant effects. The important interfaces required in translating experimental approaches to ecological or economic assessment are highlighted in Figure 1.*

Section Overview

This section addressed many of the problems mentioned above in the hope that the discussion would help the physical scientists, whose primary involvement is with atmospheric chemistry and meteorology, understand the interrelationships between their research and the effects of air pollutants on the vegetative component of terrestrial ecosystems. This section also highlighted the results and research needs relating to the gaseous air pollutants. The approach to this research is conceptually presented in Figure 2.

In terms of the base of knowledge presently available, the gaseous components of the atmosphere, rather than the aerosols and particulate matter, comprise those chemical species that are most toxic to vegetation, and account for up to 90% of the known adverse effects. Not all of the concerns of plant scientists are reviewed, because many of them are covered in other sections of this book. The following list highlights the major areas of concern.

1. *The basic objective of an effects research program; mandated goals versus scientific goals—an international perspective.*
2. *Dose–response as a primary experimental approach.*
   a. *Purpose of dose–response designs, predictive capability.*
   b. *Effective dose and its relationship to ambient air quality.*
   c. *Characterization of dose to reflect biological response.*
   d. *The occurrence of multiple pollutant interactions.*
   e. *Other phytotoxic chemical species.*

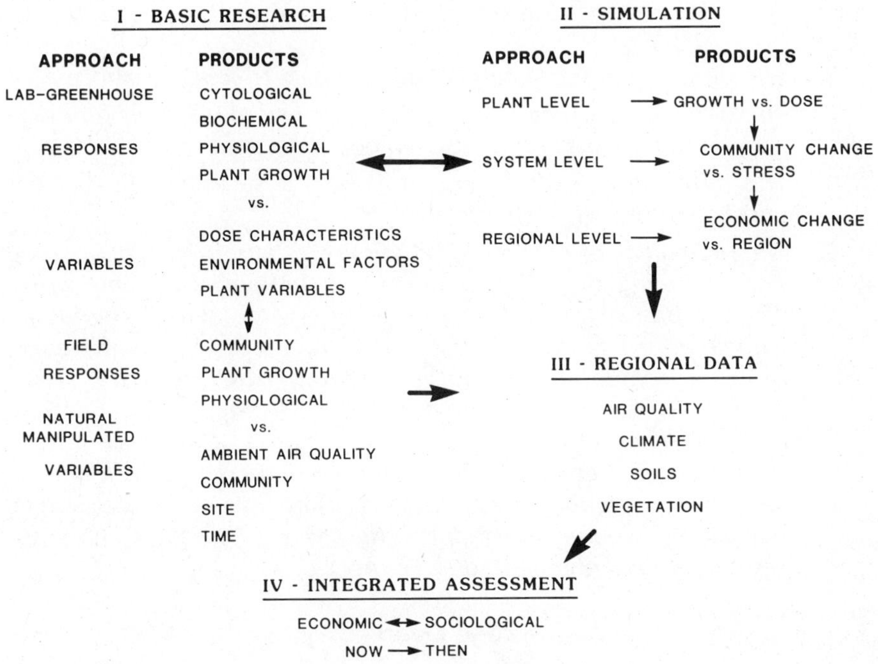

FIGURE 2. Approach and flow of information for research on terrestrial vegetation.

3. *The comparative value of laboratory, greenhouse, and field experiments.*
4. *Additional areas of concern that must be understood for a complete assessment of effects on vegetation.*
   a. *Interaction of pollutants with other pollutants, biotic factors, and abiotic factors of the environment.*
   b. *Effects on growth without visible injury.*
   c. *Biological monitoring—can it be quantified?*

Special areas of interest were highlighted by several scientists and are included in the Special Reports section together with a summary of their discussion. These are separated from the discussion section because their primary purpose was to present new information.

The discussion section is a synthesis of our understanding of the session. Individual contributors are not identified. We hope we have represented the consensus of the conference in this synthesis, but we are responsible for its accuracy and clarity. We have developed the synthesis around the list of highlights (shown above) since it is our opinion that these areas are crucial to our understanding of the effects of gaseous pollutant stress on plant systems.

The review paper by Runeckles highlights areas of research where some information is available. His discussion ranges from pollutant uptake to effects on plant communities. By implication he highlights major areas where additional research and new approaches to research are needed if we are going to develop predictive capabilities. The paper by McCune addresses some specific aspects of needed research. He poses some stimulating questions and ideas, relating to the design and interpretation of studies of pollutant interactions, that should become a part of our thinking process. Neither reviewer has attempted to do the usual literature review or to list research needs. The ideas they have presented, although couched in terms of specific gases (O_3, SO_2, HF), are relevant for any gaseous air pollutant.

It is our conviction that the photochemical oxidants (primarily O_3) comprise the primary air pollution stress within the United States, parts of Canada, parts of Europe, and in some of Africa, Australia, and Asia; this may be true for some areas of all industrialized nations. However, SO_2 has received considerable attention in North America and is currently the pollutant of primary concern to plant scientists in Europe and in many other parts of the world. This fact, and the fact that European and North American scientists generally view the problem in a different context, encouraged us to allow the discussion to focus on SO_2 research. The synthesis, however, is generally applicable to all gaseous pollutants. We have listed a number of the gaseous pollutants that are being studied in several of the countries represented by this conference in Table 1.

SPECIAL REPORTS

1. DETERMINATION OF THE RELEASE OF SULFUR FROM PLANTS USING THE $^{32}S/^{34}S$ RATIO

H. R. Krouse, The University of Calgary, Alberta, Canada

Krouse believes that their $^{32}S/^{34}S$ ratio data prove that significant quantities of sulfur can be emitted from vegetation near sour-gas processing plants. The enrichment of ^{34}S in pine needles (^{32}S was preferentially released) was presumed to be related to isotopic fractionation in the process of release of S compounds by the needles to the atmosphere.

2. EFFECTS OF LOW-LEVEL CONTINUOUS EXPOSURE OF NORWAY SPRUCE TO SO_2

T. Keller, Swiss Forest Research Institute, Zurich, Switzerland

In support of the concept of chronic damage to low SO_2 levels, Keller showed that Norway spruce exposed continuously during bud break and

TABLE 1. Research on the Effects of Gaseous Air Pollutants[a] on Terrestrial Vegetation.[b]

Chemical Species or Complex	Nations Represented						
	United States	Canada	United Kingdom	Switzerland	Germany	Netherlands	Poland
Oxidants (O_3)	a	a	b	c	b	a	d
SO_2	a	a	a	a	a	a	a
NO	d	d	c	d	d	d	d
NO_2	b	c	c	c	c	b	c
Ambient mixtures	a	b	a	b	a	b	a
Controlled mixtures[c]	b	c	b	c	b	a	c
HF	b	b	c	b	c	a	b
CO_2	b	c	d	d	d	d	d
Hydrocarbons (including ethylene)	c	c	d	d	c	b	b
H_2S	c	c	d	d	d	d	d
NH_3	c	d	d	d	d	c	c
Pesticides	c	c	d	d	d	d	d
Cl_2	c	d	d	d	c	c	d
HCl	c	d	d	c	c	d	d
Organic vapors	c	d	d	d	c	d	d

[a]These phytotoxic gases are present principally as a result of industrial activities. They may occur as a direct result of these activities (primary pollutants) or indirectly through atmospheric transformations of primary pollutants (secondary pollutants).

[b]Table shows the relative intensity of research activities within countries: a = high, b = medium, c = low (only occasional work), d = no known ongoing research.

[c]These include various combinations of O_3, SO_2, NO_2, and HF.

needle elongation for 10 weeks to 0.05, 0.1, or 0.2 ppm of SO_2 showed a reduction in growth at all three SO_2 concentrations. Visible injury was not seen at the 0.05 ppm SO_2. Photosynthetic activity was also reduced in spruce exposed for three months (October–December, or January–March) to 0.025, 0.075, or 0.225 ppm of SO_2; the effects in the first exposure persisted through April. Previous exposures also sensitized needles to subsequent frost injury.

3. RELATIONSHIP OF POWER PLANT EMISSIONS TO AIR QUALITY AND THE USE OF THESE DATA IN EXPERIMENTAL DESIGNS FOR EFFECTS RESEARCH

H. Jones, Tennessee Valley Authority, Muscle Shoals, Alabama

A full understanding of power plant emission data is needed so that managed experimental designs can be more realistic. Experimental exposure systems, such as the Zonal Air Pollution System (ZAPS), will provide realistic data for power plant assessment only if frequency distributions of peak concentrations are similar to those occurring around power plants.

DISCUSSION

The discussion which follows is developed around the topic areas highlighted in the session overview. In following this outline we have focused on major areas of discussion voiced by participating scientists in the session. These concepts are not inclusive in terms of research needs, but they do highlight critical needs.

1. OBJECTIVE(S) OF AN EFFECTS PROGRAM

The conference consensus was strongly balanced toward an understanding of effects on vegetation. The participants felt that an understanding of air-pollutant-induced effects is essential for adopting control policies. An assessment of the ultimate response (usually a measure of yield) or other measures of economic or social values in relation to ambient air quality is needed for standard setting. This assessment should include some cost–benefit analysis. This, however, is only a stop-gap approach and is not consistent with the need to predict long-term impacts on agricultural or forest ecosystems.

Most regulatory agencies in the countries represented (with the possible exception of the United Kingdom and Switzerland) do not support research that would lead to a mechanistic understanding of the effects. There are some agencies or groups that give some support to mechanistic studies, but general support or interest in these studies is not high.

2. DOSE–RESPONSE AS A PRIMARY EXPERIMENTAL APPROACH

We have defined dose–response as an experimental design that uses a series of pollutant doses (concentrations and/or time series) and regression analysis.

a. Purpose of Dose–Response Designs, Predictive Capability. *This experimental approach permits the study of plant responses across a range of pollutant concentrations and durations of exposure (doses). Dose–response curves developed from these data are amenable to regression analysis, or the points on the curve can be separated by means comparison tests. The design requires a background or zero concentration for the control and a series of pollutant concentrations (over variable or standard durations of time) that include expected ambient concentrations. Maximum concentrations and/or durations should be such as to ensure at least a 50% response to ascertain a reliable characterization of the response curve. Exceptions to this occur when the concentrations needed are excessive; the determination of excessive concentration is a responsibility of the researcher.*

Under laboratory, greenhouse, and manipulated field designs, the dose concept is practical. In natural systems the dose concept is equally important but is confounded by numerous environmental variations since distance from a source (point or regional) is the only way to develop the dose concept, assuming air monitoring is done across the same distance. The dose–response approach is an essential ingredient of any predictive concept that involves plant effects.

b. Effective Dose and Its Relationship to Ambient Air Quality. *Biologists have noted for years that plants exposed to the same pollutant concentration may respond quite differently, depending on the plant species, growth stage, physiological condition, and a variety of environmental factors such as temperature, light, and relative humidity. Such differences in response are due either to (1) greater uptake of pollutants by plants, or (2) greater sensitivity of plants to a given amount of the pollutant under these conditions. With the development of exposure chamber systems, with which pollutant uptake into plant leaves may be measured, the term effective dose has been introduced (see discussion by Runeckles in this volume). The concept, which is equivalent to flux into the leaf, stimulated considerable controversy stemming mainly from doubts that (1) a unit quantity of pollutant inside the leaf (the context in which the term was introduced) has an effect "equal" in all situations, and (2) biochemical definitions of dose would not be compatible with monitoring data with which the environmental analyst must work. Nevertheless, utilizing the term to relate physiological responses to the quantity of pollutant actually received inside the* *leaf was generally supported. It was supported from the standpoint of mechanistic studies aimed at quantifying cause–effect relationships under experimental conditions and as a means of reducing the variability among data obtained in different laboratories. The term internal dose was suggested as a means of more accurately describing the amount of pollutant entering a leaf. It was also suggested that this term be reported (when possible) in relation to the external dose upon which broader scale analysis must be based.*

c. Characterization of Dose to Reflect Biological Response. *Differences exist between Europe and North America in expected frequency distributions of SO_2 concentrations. The European scientists generally are of the opinion that long-term average concentrations adequately characterize ambient concentrations for the purposes of biological assessment. This reflects their belief that SO_2 constitutes a severe regional problem with relatively little variation over time. Thus, they make extensive use of 24-h continuous fumigations at constant concentrations.*

The North American scientists generally believe that the peak concentrations are most important (with episodic exposures of $\leq$3-h averages) under their meteorological and source pattern conditions. Thus, from point sources there can be at least two orders of magnitude difference between the maximum hourly peak and the annual mean concentration. Even in a multiple-source urban area (e.g., Chicago), the maximum hourly peak may be 10 to 20 times the annual mean. In the single-source case, a given location may be free of measurable SO_2 80% or more of the time. In the regional-source case, a given location may be free of measurable SO_2 from 15 to 30% of the time. The North American scientists are more interested in trying to understand the most biologically significant features of ambient exposures and are using this information to both describe and interpret ambient field exposures and as a tool to design controlled exposures.

There is much interest in North America in understanding frequency distributions of hourly concentrations above 0.25 ppm, in developing peak/mean ratios, and in considering the probable log–normal or other distributions of concentration over time (using hourly or shorter averages). The North Americans still have no consensus as to the best way to express dose so that it adequately measures a biological impact. In general, they are reporting both concentration and duration of exposure (to include the length of individual exposures, the number of exposures per day or week, and the number of weeks over which the controlled exposure was conducted).

There is a need for more comprehensive statistical reduction of air quality monitoring data from various stations for use in assessing field effects and in designing controlled experiments. Some data are tabu-

lated in agency reports and should be made more available. Particular emphasis should go toward tabulation of pollutant mixture data. The theoretical derivation of short-term (30-min averages) concentration frequency distributions and relating these to longer-term averages has been suggested. This higher order of resolution will assist the plant scientist.

No attempt is made to characterize dose for other pollutants, although it is recognized that oxidants generally pose regional problems. Various ways of summarizing dose have been used by various investigators, but no consensus is possible.

North American scientists generally do not expose plants to pollutants during the dark because of the consensus that stomates are generally closed and the plants are generally resistant to dark exposures. Nitrogen dioxide, hydrogen fluoride, and hydrogen chloride may be exceptions to this concept.

d. The Occurrence of Multiple Pollutant Interactions. *There was considerable discussion on the importance of studying multiple pollutants. Ambient air is always composed of a mixture of pollutants, and the importance of this concept is well accepted. Ozone is normally absent from a point-source stack plume but reoccurs (possibly at a peak) once the NO in the plume is gone. There is always* SO_2 *and* NO_2 *in a power plant plume, and this combination can be important. The presence of* CO_2 *in the plume may be protective, but too little is known for much speculation.*

The best way to design multiple pollutant exposures in manipulated experiments is not sufficiently understood. However, multiple pollutant interactions can and should be studied. Scientists responsible for ambient air quality monitoring should develop data sets of pollutant mixtures showing the frequency of occurrence of pollutant concentrations, the relative concentrations, and the average duration of exposure.

e. Other Phytotoxic Chemical Species. *The conferees discussed the possible occurrence of other phytotoxic chemical species. The chemists suggested that the biologists should identify those chemicals that are phytotoxic so the chemist could look for them, but the biologists are not able to do that until they know what chemicals might be present. Even then, screening studies would be required to determine if plants are sensitive to the chemicals and to rank the relative sensitivities. This is an area where the chemists and biologists should work more closely together.*

The importance of the OH radical and possible free radical formation by photolysis of O_3, NO_2, *and* SO_2 *at the leaf surface are of major interest to the plant scientists. However, until procedures are developed to measure and generate these transient chemical species, it will not be possible to study their effects on biological systems.*

3. THE COMPARATIVE VALUE OF LABORATORY, GREENHOUSE, AND FIELD EXPERIMENTS

Basic concerns were raised about the relative importance of results obtained from laboratory, greenhouse, and field experiments and their bearing on an integrated assessment of air pollution effects on vegetation. This includes the comparability of data obtained from the varied experimental and analytical systems employed, the utility of mechanistic studies not accompanied by measurements of changes in plant growth and yield, the applicability of plant growth models to real world situations, and the adequacy of some field systems designed to simulate exposure of plants to SO_2 exposure from coal-fired power plants.

Studies in the laboratory, greenhouse, or field should be viewed as integral parts in understanding the mechanisms of and in measuring and predicting the effects of pollutants on vegetation. Basic biochemical, cytological, and physiological studies increase our understanding of the means by which pollutants influence plant growth and behavior. They also may provide the analytical tools to allow us to develop physiological indicators of stress. Such indicators should be determined when growth effects are concurrently measured in the laboratory. These indicators may then be used in the field to identify and characterize specific pollutant stress episodes that contribute to measured yield effects. This higher resolution analysis in field studies should increase the potential for extrapolating these results to other locations and to other years.

The value of developing physiological models of pollutant effects was challenged on utilitarian grounds. The process is generally viewed by physiologists as a way of integrating information on the multiple environmental and plant factors that influence plant response to pollutant stress and not as an end product in itself. Such models may ultimately enable researchers to extrapolate measured field responses to other locations with different environmental conditions.

The optimum cultural conditions under which plants are generally grown for laboratory or greenhouse exposures are considered by some to enhance their potential sensitivity to air pollutants, thus making quantitative estimates of yield loss under these conditions unreliable. Others point out that such estimates may be low because they exclude interactive effects of environmental stresses normally present in a plant's growing environment. Also, one cannot judge the significance of X% reduction due to a pollutant versus the Y% that may be reduced by all other ambient factors. These arguments highlight the need for combined field and laboratory studies to arrive at reliable quantitative estimates of plant response.

Two concerns pertain to the use of field chambers designed to measure either effects of ambient levels of pollutants or those produced by simulating point source (e.g., power plant) exposure regimes by *controlled release of SO_2. In the former situation, particular care should be taken to ensure no loss of the pollutant in the often long sampling lines; in the latter, the frequency distribution of short-term peak concentrations should resemble those of typical emitters, or the results may not accurately reflect yield responses to be expected around these sources.*

Evidence has been provided, in both Europe and the United States, for the evolution of resistance of SO_2 by herbaceous species. The possibility that this potentially may lessen the ultimate effects of pollutant stress on plant populations, though real, has not been broadly tested and is generally regarded by biologists as an undesirable alternative to pollution control. Evolution of resistance to anthropogenic stress may result in the alteration of competitive relationships between plant species and alter resistance to other natural stresses. Known alterations in community structures around uncontrolled point sources and in some industrialized areas of the United States argue that adaptation is not rapid enough to be effective in these situations.

4. ADDITIONAL AREAS OF CONCERN

a. Interaction of Pollutants. *Vegetation is sensitive to many stresses. Thus the effect of any given pollutant is affected by other pollutants, biological factors, and various abiotic environmental factors. Many of these have been well studied under laboratory conditions and have permitted us to acquire a much broader understanding of the influence of many environmental factors.*

b. Effects on Growth Without Visible Injury. *The concept of reduced growth of vegetation at low concentrations of air pollutants, in the absence of visible injury, has been defended by European scientists based on published data from England, France, Germany, Japan, and Switzerland. Thresholds for SO_2 growth reductions in England were 0.05 to 0.10 ppm of continuous exposure. Sensitivity of plants to acute injury was not well correlated with yield reductions from chronic exposures. Physiological responses were noted below 0.05 ppm.*

c. Biological Monitoring—Can It Be Quantified?. *According to certain scientists, biological monitoring has intrinsic scientific value, while others feel it should be studied to determine if it could form a practical basis for assessment of economic loss. Questions remain as to whether this type of correlation could be found.*

PART SIX

TERRESTRIAL VEGETATION–AIR POLLUTANT INTERACTIONS: NONGASEOUS AIR POLLUTANTS

The effects of fine and coarse particulates on vegetation are poorly understood compared with those of gaseous air pollutants. Little is known about the relationship between the atmospheric concentrations of particulates and the damage they cause to plants. There are major uncertainties about the chemistry of particulates deposited upon plants and about the interactive effects of mixtures of gaseous and particulate pollutants.

It is important to recognize that a pollutant may be deposited by both dry and wet deposition processes, the two combining to produce at the leaf surface a total burden of pollutant whose impact will be determined by its chemical and physical nature. This is particularly important in respect to acid rain, which, in reality, is the sum total of acid deposition of sulfur, nitrogen, and other ions.

Further attention must be given to the development of laboratory techniques which can be extrapolated into the field with a high degree of accuracy, so that we can predict and assess impacts upon terrestrial ecosystems with confidence. These impacts will involve both direct effects upon vegetation and indirect effects upon plant growth, and the like, mediated by changes in the physics, chemistry, and microbiology of soils. Similarly, we need to achieve a better understanding of the nature of runoff and drainage waters in polluted ecosystems and its effects upon the quality and biota of aquatic environments.

These and other topics, including research needs, are addressed in this section.

Terrestrial Ecosystems: Particulate Deposition

R. GUDERIAN
University of Essen
Essen, West Germany

1. INTRODUCTION

The injury of plants by gaseous atmospheric pollutants is far greater than injury by the nongaseous pollutants. This general statement is based on the fact that gaseous atmospheric pollutants have a widespread distribution and high phytotoxicity when considering the net economic and ecological effects on vegetation. Notwithstanding, nongaseous emission components have recently received increased attention both in air quality research as well as in air pollution control practices, principally for the following reasons:

1. Particulate atmospheric pollutants are more widely distributed in the upper atmosphere today.
2. The number of toxic components in particulates has increased, especially those associated with the increasing emissions of heavy metals.

It seems appropriate to analyze and evaluate the present status of this aspect of air quality research and environmental protection and then develop suggestions for further research needs. This goal assumes that, apart from the definition of nongaseous atmospheric pollution, a classification and evaluation of the resultant effects is required.

2. CLASSIFICATION OF THE EFFECTS

Particulate atmospheric pollutants consist of solid particles, of any shape, structure, or composition which are dispersed in the atmosphere. These particles are deposited either faster or slower due to the gravitational field of the earth. Particulates with diameters between 0.01 and 50 μm are characterized as follows:

Coarse particles	$>10\ \mu$m	rapid deposition
Medium particles	0.5–10 μm	slow deposition
Fine particles	$<0.5\ \mu$m	deposition velocity approaching 0

The particle size differences are of great importance in determining the behavior of the particular matter during transport in the atmosphere and their rate of deposition.

It is not possible to separate totally the effects of particulate matter and aerosols, because they are both present in the atmosphere. There are, however, solid or liquid particles which, due to their small mass, are not subjected to gravity and thus behave like a gas. These aerosols, with sizes between 0.01 and 1.0 μm, have definite surfaces onto which gaseous substances can be absorbed, and from which they can be given off to the atmosphere. Particles of this size range should be able to pass through stomata. Godzik and Sassen (1978) have, however, through examination by scanning electron microscopy, excluded the stomata as an entry point for particulates in a number of conifers.

2.1 General Indirect Effects

The residence time of particulate matter in the atmosphere is a function of particle size. The particulate components, together with the gaseous pollutants and aerosols, can form a haze cloud. This type of pollution can decrease net global incoming radiation through reflection and absorption of solar energy. As an example, a decrease in incoming radiation of up to 20% was caused in the Ruhr region (Faerber et al., 1959) and in Hamburg (Gräefe, 1963; Gräefe and Schlunk, 1965). As a possible consequence of the decrease in radiation, we should consider the decrease in photosynthetic performance.

Another mediating effect resulting from the problem of particulate matter is the fact that particles can act as condensation nuclei (Kreutz and Walter, 1956) which favor fog formation. For example, as a result of increased mist formation there were ice formations on spruce trees on the ridges of the Erz Mountains. These were directly related to the intensity of the air pollution and caused the collapse of the entire stand (Pelz, 1960).

The last illustration of the mediating effects of particulates to be considered is the deposition of particulate matter on coldframes and greenhouses. Fortmann (1963) showed that, as a result of reduced light intensity due to particulate deposition, the growth period of plants was lengthened and the crop quality and yield were reduced. The requirements for frequent cleaning of the dusty glass, as well as increased corrosion of the metal parts, gave rise to additional costs.

2.2 Direct Effects on Above-Ground Plant Parts

To consider the direct effects of particulates on above-ground plant parts, it is appropriate to divide the particles into inert and toxic particulate matter. Inert particulate matter has mainly a physical effect on plants, while the toxic particles have both chemical and physiological effects.

2.2.1 Inert Particulate Matter

Deposition of particulate matter on above-ground plant organs can give rise to a reduction of the radiation received, an increase in temperature especially in the leaves, and a blockage of the stomata. These factors can lead to a modification in the dry matter production of the plants.

Eller (1974) as well as Eller and Brunner (1975) have shown in their research that road dust deposition influenced the diffuse spectral reflectivity, transmissibility, and absorbtivity of leaves. As a result, such modified spectral characteristics of leaves could adversely affect the vitality of the plants. Measurements of leaves of *Hedera helix* L. (Eller and Willi, 1977) with road dust accumulations showed a 3- to 4-fold increase in absorption between 750 and 1,350 nm wavelengths when compared with dust-free leaves. The amount of absorption in the visible range, from 400 to 750 nm, was only insignificantly influenced. The increase in the amount of absorption in the infrared region by the upper leaf surface could be accounted for by equal decreases in the amount of reflection and transmission. Calculation of the amount of energy absorbed from the incoming global radiation gave an increase of about 30% for a dirty leaf in the 400–1,350-nm wavelength range when compared with a clean leaf. If only the infrared region of the spectrum was considered, that is, from 750 to 1,350 nm wavelength, an increased value of about 137% for the dirty leaf was found.

The increased amount of energy from the absorption of global radiation can lead to significant increases in leaf temperature. Leaves of *Rhodendron catawbiense* that were coated with road dust showed a temperature increase of up to 4°C (Eller, 1977), while leaves of *Populus nigra* covered with foundry dust had a temperature increase of 8.7°C (Steinhüebel and Halas, 1967). Decreased photosynthetic ability and increased respiration rates could result from temperature differentials as small as 2.5°C between polluted and nonpolluted leaves (Steinhüebel and Halas, 1967). With increased outside temperature and increased temperature differentials in the leaf, the dry matter production of the plants decreased. In the leaves of deciduous plants, dust deposition caused higher temperature increases than in evergreens (Steinhüebel, 1966). The generally thicker leaves of the evergreen plants evidently facilitate more efficient heat dissipation.

Dust layers on the leaves can exert different influences on the leaves as a function of light intensity. As was found by Steinhüebel (1963) both inert solid particles and the particulate matter emitted by a foundry had an influence on starch production. Under diffuse light conditions in the morning and evening or with cloudiness, the starch production was reduced, while during intense solar radiation a shade-loving plant exhibited increased photosynthetic performance.

With deposition of road dust, according to findings by Eller and Willi (1977), the photosynthetically active radiation, between 400 and 750 nm, was not affected. These findings indicate that less emphasis should be placed on the effect of light conditions.

2.2.2 Particulate Matter with Toxic Effects

The particulate atmospheric pollutants represent a mixture of physically and chemically heterogeneous substances. Apart from the phytotoxicity of the particles, it is particle solubility that is the controlling factor for direct foliar injury of plant tissue. The factors which determine the injurious effects of a given particulate load are the particle size, particle size distribution, particle shape, particle surface characteristics, the time-dependent deposition rate, the type of plant species, the species specific resistance, the developmental stage of the leaves, and the morphology of the leaves. Suitable examples will be used to explain the type of effects toxic particles have. With these examples as a basis, selected methods can be understood and evaluated.

Particulate Matter from Cement Factories. An analysis of the literature concerning the effects of cement dust on plants, until recently, has presented two distinctly separate ways of thinking (Darley, 1966; Lermann and Darley, 1975). Most authors of the "old world" (Sorauer, 1898; Haselhoff, 1908; Ewert, 1917, 1926; until Scheffer et al., 1961) approached the question of the effect of cement dust from the origin of the dust. Inadequate methods were used to apply known quantities of particulates per unit surface area to experimental plants in the field or in greenhouses. No attention was paid to the meteorological conditions or their effects on the particulates during transmission, nor was attention given to the microspheres at the plant surface. As a result of these factors, no direct damaging effects were determined, even by applications of unrealistically high dust quantities.

This approach was opposed by a group of researchers (Steffeck, 1896; Peirce, 1910; Parish, 1910; Anderson, 1914) who considered the question of the effects of cement dust from the viewpoint of the plants which are the recipients of the dust. This was the standard approach of Czaja (1960, 1961, 1962a, b), and today it is the accepted way of thinking. (Darley deserves the honor of having made these data available to a wide sphere in North America.)

The chemical composition of toxic particulates is the primary factor determining their effects. With the dust from cement factories we are primarily concerned with calcium particulates, such as CaO and $CaCO_3$, as well as with the water soluble cement. If an abundance of moisture is available at the surface of the plant organ, such as standing water on the surface, very durable crusts are formed. These grey-white crust coverings were observed by Parish (1910) and Peirce (1910) and can be one- or two-layered (Czaja, 1960, 1961, 1962a, b), depending on the quantity of deposited dust.

A strong glossy crystalline layer of about 30 μm is adhered adjacent to the leaf surface and appears rough on the upper surface. On the reflective and shiny undersurface in direct contact with the cuticle of the leaves, microscopic details of the cell surfaces can be distinguished. With the combination of atmospheric humidity and the water originating from transpiration, sufficient moisture is available to allow for the hydration of the four standard components in the deposited cement: tricalcium silicate, dicalcium silicate, tricalcium aluminate, and tetracalcium aluminum ferrite. The outer layer of the cement crust consists of cement particles hardened together with one another, but leaves their individuality still abundantly clear. The moisture which is available here from the surrounding atmosphere is frequently not sufficient for the complete hydration process.

This type of crust injures plants in a physical manner. These crusts affect the leaf light regime and the leaf temperatures as well as the gas exchange and transpiration rates of the plants. The chemical reactions that take place during the setting of the deposited cement particles should especially be considered. The liberation of calcium hydroxides creates an oversaturated solution, with pH values of up to 12, which can penetrate into the epidermis and mesophyll. This penetration can result in hydrolysis of the lipids, coagulation of the protein compounds, as well as plasmolysis of the leaf tissue, resulting in a reduction of growth and quality of the plants (Czaja, 1960, 1962a, b). According to Holobrady and Toth (1968) this type of effect can also be attained without cement encrustment of the leaves.

In a manner similar to gaseous pollutants, particulate emissions can also have an effect on the structure and function of plant communities. Research by Brandt and Rhoades (1972) has shown that *Quercus coccinea*, *Quercus velutina*, or *Tilia americana* can disappear as the important species in an Oak–Chestnut association and *Liriodendron tulipifera*, *Acer saccherum*, and possibly *Quercus muehlenbergii* can become the dominant species. With magnesium emissions this leads to a reduction in the number of species in the phytocoenoses and a qualitative change in the association structure (Kaleta, 1972). Apart from the very high load, which totally disrupts plant communities, there is an underlying parallel degeneration which is a spontaneous or human-supported process in which both the original or adaptive resistant members of the existing community and new arrivals form a secondary succession (Guderian and Kueppers, 1980). As further possible effects of lime- and cement-containing emissions we can name: disturbance of pollination, especially in fruit trees, in which the dust emissions soak up the stigmatic fluid; reaction changes in the stigmatic secretion; and impaired pollen germination and pollen growth. The resistance to phytopathogenic fungi also appeared to be increased with lime- and cement-containing atmospheric pollutants (Schönbeck, 1960; Czaja, 1961).

Heavy Metal-Containing Particulates. The effect of heavy metals on plants is especially dependent on the type of plant resistance and the environmentally controlled resistance, as well as the phytotoxicity of the particular compound, the amount of the compound taken up, and the site of uptake. A number of aspects pertaining to the latter factors will be briefly discussed.

To determine the concentrations of heavy metal pollution in plant tissue we must have a knowledge of the normal concentration range of heavy metals in plant tissue from unpolluted areas. The concentration of elements in plant tissue is highly variable and depends on the origin of the plant material, the type of experiment, and the analytical methods used. The increasing anthropogenic dispersal of heavy metals into ecosystems makes it more and more difficult to determine the normal concentrations.

Because of the difficulty of ascertaining quantitative relationships between the concentration of heavy metal-containing particulates in the atmosphere and their effects on plants, the following question remains very important: How far can the increase in concentration of harmful components within the plants, through uptake from the polluted atmosphere, be attributed to the degree in which the plants are endangered? (Compare this also with the explanation in Section 4.) One should proceed cautiously for the following reason: It is important to determine whether the substance is actually taken up by the leaf or whether it is an active substance adhering to the leaf surface. This distinction has already been brought to our attention by Wittwer and Teubner (1959). With the aid of special washing techniques, most of the particulates attached to the surface can be removed (Krause, 1974).

The amount of injury to plant tissue from incorporated heavy metals is a function of the site of uptake. Heavy metals that penetrate the aboveground plant parts are frequently less damaging than heavy metals that are taken up by the roots from the soil. According to nutrient solution and sand culture experiments, concentrations of as low as 500 ppm zinc in the plant dry weight can cause damage (Hunter and Vergnano, 1953; Carroll and Loneragan, 1969). In contrast, the very high concentrations of up to a few thousand ppm zinc in the plant dry weight, taken up through the leaves, still do not cause plant damage (Buchauer, 1973; Little, 1973; Krause, 1974; Guderian et al., 1977). The same is true for cadmium. According to the research of Krause (1974), several hundred ppm cadmium in the dry weight through application to the leaves did not cause any visible symptoms, while much less than 5 ppm cadmium in the dry weight taken up through the roots gave rise to plant damage (Allaway, 1968; Turner, 1973). The lack of external damage symptoms makes it difficult to recognize human and animal toxicologically critical concentrations in food crops and animal fodder plants (refer to Section 2.4).

With the introduction of chemical analysis for the detection of these dangerous chemical levels in a plant, it must be remembered that the increase in pollutant concentration in the plant varies greatly with the plant species. Anatomical and morphological differences of the leaves are the main reasons for these differences. The very low heavy metal concentrations found in some experiments with kohlrabi plants (*Brassica oleracea* var. *gongylodes*) (Guderian et al., 1977) is certainly the result of its thicker cuticle, through which metal ion penetration was significantly retarded. Dust particles are also removed more easily by rain and other environmental mechanisms from leaves with thick cuticles (Rentschler, 1973, 1977; Evans et al., 1977). The up to 20-times-higher accumulation of heavy metals in radishes, tomatoes, and broad beans when compared with kohlrabi can be traced back to other morphological characteristics of the leaves. The pubescence of the leaves not only favors the capture of the dust particulates, but also favors ion uptake. This could be attributed to the basal cells of the trichomes having an increased ectoderm density.

At given deposition rates the amount of pollutant uptake is mainly dependent on the solubility of the respective pollutant components on the surfaces of the plant parts. Readily soluble compounds, such as sulfates, pose an increased hazard. Most atmospheric pollutants, however, are compounds that are difficult to dissolve in water. To provide practical emission protection, it is important to know if and in what quantity these pollutants are taken up by plants. According to dusting experiments it is possible that both pure heavy metals (Krause, 1974) and their oxides (Guderian et al., 1977) can end up within the plants. In bushbean leaves this uptake was always about 10% of the dusted metallic zinc or cadmium applied (Krause, 1974). With experiments in the vicinity of heavy-metal emitting industries, it has been determined (Schönbeck, 1974) that both metallic zinc and metallic cadmium can be taken up directly through the leaf surface. The solubility of the dust layer was raised, both through the carboxylic acid secreted by the leaves and through the water film on the leaf surfaces in which CO_2 had been dissolved (Krause, 1974). Sulfur dioxide concentrations of 0.2 mg/m^3 in air decreased the heavy metal accumulation in bean plants dusted with zinc and cadmium. However, it raised the heavy metal specific damage to the leaves and stems. The cadmium and zinc absorbed by the above-ground plant parts is translocated acropetally as well as basipetally. This aspect must also be considered when we try to evaluate the contamination in food crops and animal fodder.

The transmission electron micrographs that have been examined so far have shown that the chloroplasts are the most sensitive of all the cell organelles to heavy metal-containing emissions. After treatment with Fe_2O_3 about one third of the chloroplasts exhibited beaklike processes (so-called protuberances), the tonoplast showed numerous cracks, and there were numerous plastids and mitochondria in the vacuole. The action of PbO_2 led to a noticeable shrinking of the tonoplasts. The combined action of Fe_2O_3, CuO, and PbO_2 can result in the membranes of the thylakoid system moving further apart, thus reducing starch formation. The increase in number and size of the plastoglobuli is an early indication of aging of the leaves.

Gas-exchange measurements showed that when plants were dusted they had a distinct lowering of the light compensation point. When heavy metals were applied, after dilution with quartz dust, the resulting dust layer caused a 15% decrease in light available to the leaf. The plants consequently adapted to the decreased light intensity; they had already shown a positive CO_2 balance under reduced light intensities. In contrast, millet plants fumigated with SO_2 and HF showed a distinct increase of the light compensation point (Brandt, 1980).

2.3 Indirect Effects via the Soil

Indirect effects on plants after the soil has been contaminated by atmospheric pollutants have until recently been viewed as a problem mainly confined to the immediate vicinity of the emission source. For this reason, greater emphasis should be placed on control of heavy metals and their emissions. The concentrations of heavy metals in the soil, however, decrease very rapidly with an increase in distance from the emission source, both with a point source such as metal production or metal processing facilities (Ratsch, 1974; MAGS, 1977) and with strip emitters such as highways (Kloke, 1974). When considering the effects above the ground, attention has until now mainly been given to the effective area of the source.

New evidence, however, shows from the emission measurements as well as from the chemical analysis of plant and soil samples, that there is also more than local stress in the area of air pollution control. Krämer (1976) found contributions of heavy metals above the levels that are considered tolerable, while carrying out research on soils of pastures and crop lands in a 200-km^2 area of the western Ruhr region. These data are shown in Table 1. The soil samples from pastures, which were taken at a soil depth of 5 cm, showed distinctly higher concentrations than those from the cropland. The latter samples were taken at a depth of 16–32 cm, depending on the thickness of the cultivation horizon. A dilution effect is obviously brought about by cultivation in the croplands.

Chemical analyses of aerosol samples (Buck and Ixfeld, 1976) as well as analysis of plant samples (Scholl and Schönbeck, 1975; Prinz and Scholl, 1975) showed that heavy metals are widespread as atmospheric pollutants. Huttunen (1979) even considers heavy metal particulate atmospheric pollutants in boreal forest ecosystems as one of the stress factors.

The heavy metal concentration in the soil should receive special attention because it leads to different types of effects when it is above specific

TABLE 1. Heavy Metal Concentrations in Pastures and Croplands of the Ruhr Area

Element	Tolerable Amount (ppm in the soil as dry weight)	Percentage Contribution in Values Above the Tolerable Amount	
		Pasture ($n = 199$)	Croplands ($n = 99$)
Lead	100	62	13
Zinc	300	56	8
Cadmium	5	22	14
Copper	100	17	2

From Krämer (1976).

limits. Microflora and microfauna can be affected, with the result that the raw humus addition is not broken down and the mineralization is disturbed (Folkeson, 1976; ECE, 1976; Greszta et al., 1979). The concentration of plant-available nutrients and their uptake can also be reduced. Ultimately the growth and quality of food crops and animal fodder can be reduced through heavy metal contamination and resulting necrotic appearance.

Analyses of soil samples from different horizons have shown that concentrations of lead and zinc, for example, are limited in the upper soil layer to a few centimeters in pastures or meadows, and to the plowing depth in croplands (Schönbeck, 1974; Krämer, 1976). In an area surrounding a zinc smelter it has occasionally been seen that individual turnip plants with their roots in the deeper layers of the soil not enriched with heavy metals apparently grew normally. However, the majority of plants had their roots restricted to the cultivated zone. These plants were severely damaged or even dead. Significant leaching of heavy metals to deeper soil horizons apparently does not take place. After certain concentrations of heavy metals are reached in the soil, these concentrations can remain the same for tens or even hundreds of years after the emission source has been removed. Consequently, it is not sufficient in a time of severe increases in heavy metals emissions to simply determine the tolerable concentration of heavy metals in the soil when the goal is to guarantee normal plant growth and to avoid contamination in food crops and animal fodder. If we want to protect the soil and its various functions, we shall have to set limits on the yearly input of heavy metals, so that the concentration considered to be tolerable will not be reached over a predetermined time period.

2.4 Indirect Effects via the Nutrient Chain

Particulate atmospheric pollutants are of increasing importance when considering practical emission control because they can have an indirect effect via the nutrient chain. Loading of forage and foodstuffs with heavy metals can pose a danger to domestic animals and create a human health hazard.

The inert dusts are of lesser importance. They can cause qualitative decreases in such items as leafy vegetables, fruits, and cereals. Decreases in cattle production are also possible by ingestion of contaminated forage. If the current valid guidelines for "no-risk dust deposition" in the Federal Republic of Germany, which are set at 35 mg/m^2 day, are followed, we would not be confronted with significant damaging concentrations.

An increasing number of cases are coming to the surface, however, concerning the endangerment of humans and animals by exposure to compounds with potentially toxic effects, especially heavy metals such as lead, zinc, or cadmium (Ratsch, 1974; Vetter, 1974; MAGS, 1975, 1977; Umweltbundesamt, 1976, 1977). A spectacular case of plant injury as well as danger to humans and animals via the nutrient chain, by thallium-contaminating emissions from cement factories in Nordrhein–Westfalen, was disclosed in the spring of 1979 (Prinz et al., 1979).

Human health is endangered when high concentrations of heavy metals are consumed in foodstuffs for extended periods. Therefore, attention must be paid not only to single foodstuffs with elevated heavy metal concentrations, but also to the total heavy metal uptake from all foodstuffs over an extended period of time. This requires both analysis of food preferences and quantities consumed, and long-term continued supervision of the ingredients of the different foodstuffs, such as all animal and plant products.

The products of plant origin are of special interest. As an example, according to the research of Vetter (1974), in the region of a lead smelter at Nordenham, about 92% of the lead taken up by the population originated from plant foodstuffs. When choosing plant samples to be studied it is important to note that, with a given load of heavy metals, the concentration of heavy metals in the plant sample varies greatly not only with the plant species and the plant part but also with the manner in which the plant product is prepared before human consumption. Surface adhering pollutants, for example, are almost totally removed with peeling, while the percentage removed by washing depends on the nature of the surface. By knowing the heavy metal concentration in the food samples chosen, as well as the quantity of foodstuffs consumed, one can calculate the total uptake of heavy metals. A comparison with the recommended objectives set by WHO (1972) gives information about the future dangers to man.

Farm animals, house pets, and wild animals also have the potential of being exposed to ever-increasing quantities of heavy metals with increasing industrial activity. The primary danger here is the increased uptake of the heavy metals with foodstuffs. A direct effect via inhaled air is of secondary importance.

The increased exposure of humans and animals to toxic components requires systematic control, especially in the form of chemical analysis of plants. These chemical analyses are important because the above-ground plant parts show lower phytotoxic effects when the heavy metals are taken up from the air, than when they are taken up from the soil by the roots (compare with this statements in the section on Heavy Metal-Containing Particulates and in Section 4). Thus, plants have largely lost their usefulness as biological indicators of dangerous contaminants for humans and animals. The pollutant content determined through chemical analysis of plants in food crops and animal fodder provides a foundation for setting allowable limits for cultivation practices in the vicinity of emission sources of particulates with potentially toxic effects (MAGS, 1975, 1977).

3. EFFECTS OF POLLUTANT COMBINATIONS

Air pollution today can best be described as the simultaneous occurrence of many gas and particulate components. The important question which arises then, is what effect does this complex mixture of pollutants have on plants? Are dose–effect relationships of particular importance when considering air pollution control (Persson, 1971)?

While numerous experiments have been conducted on combined effects of gaseous pollutants (compare to this the review of Reinert et al., 1975), we have only recently started to be concerned with the effects of gaseous and particulate components in combination (Krause, 1974; Guderian et al., 1977; Krause and Kaiser, 1977). I shall present a brief review of the results of my own experiments in this area to explain the problems that arise and to illustrate approaches for their possible solution.

3.1 Research Methodology

From the results of air quality measurements (Buck and Ixfeld, 1976) and the determination of plant effects (Scholl and Schönbeck, 1975; Prinz and and Scholl, 1975) it has been shown that, in the airsheds of the Rhein and the Ruhr, sulfur dioxide, hydrogen fluoride, and heavy metal-containing particulates occur together over a widespread area. The pollutant components, including fumigation concentrations as well as the amount of dust applied in the experiments, were derived from these data and are shown in Table 2.

The selected pollutants were applied both as single components and in all possible combinations. The heavy metals were applied as oxides, since it is known that they occur in this form and are widely distributed in the Ruhr area. The concentrations of gaseous pollutants and the quantity of heavy-metal particulates applied to the plants represent concentrations that are

TABLE 2. Present-Day Background Concentration and Deposition Data for Selected Atmospheric Contaminants in the Rhein and Ruhr Regions[a]

Component	Concentration ($\mu g/m^3$ air)	Deposition (pure metal) (mg/m^2 week)
Sulfur dioxide (SO_2)	110–170	
Hydrogen fluoride (HF)	0.75–1.90	
Cadmium oxide (CdO)		±1
Cupric oxide (CuO)		±10
Lead oxide (PbO_2)		±100
Ferric oxide (Fe_2O_3)		±300

[a] Research was based on these values.

known to occur in heavily loaded locations in the Ruhr. The fumigation with sulfur dioxide and hydrogen fluoride was carried out continuously, while the dusting with heavy metals was only done once a week.

A variety of agricultural and garden species were used as research plants. The height of the plants as well as the morphology of their leaves were important considerations in the choice of the experimental plant species.

The research concentrated on experiments in small greenhouses. I refer to my earlier work (Guderian, 1977) with regard to the methods employed in the dosage and measurement of SO_2 and HF. The reproducible applications of heavy metals followed the methods and equipment developed by Krause (1974) and Krause and Kaiser (1977).

A 3.1-mm^3 dispensing chamber was filled with the selected heavy metal dust, using a storage cone which was heated and equipped with a stirring rod. An electronically controlled system was responsible for transferring the dispensing chamber contents, which were under a pressure of 1.5 atm, to the exposure chamber containing the experimental plants. An air nozzle, which was located opposite the heavy metal input port, provided an airstream which, in connection with a deflection plate, dispersed the dust–air mixture into a horizontal plane. To obtain a uniform coating of heavy metal dust over the entire surface area of the experimental plants, the plant material was rotated through the dust–air mixture.

The amount deposited on the slant surface was determined by simultaneously exposing microscope slides with the plant material in the fumigation/dusting chamber. With the aid of a mist generator in the chamber, a thin water film formed on the above-ground parts. This improved the adhesion of the heavy metal dust and also dissolved the heavy metal dust so it could penetrate into the leaf.

Apart from the greenhouse experiments, a number of growth chamber experiments were conducted to determine the effects of the pollutants on the gas-exchange capability of the plants. A portable gas-exchange system was used to measure the CO_2 uptake of the plants treated in the dusting chamber and those growing in the open fields (Guderian and Thiel, 1973; Guderian, 1977).

The portable gas-exchange system consisted of four 30-L plexiglass cylinders, which could be housed in a growth chamber of average size. With the assistance of an infrared gas analyzer and the use of a time clock and solenoid valves, the CO_2 gas exchange of the plants in the individual plexiglass cylinders could be monitored continuously. Apart from the CO_2 gas exchange, growth rate and external appearance were used as criteria for the evaluation of pollutant effects. The determination of pollutant content by chemical analysis of individual plant organs not only gave information about the accumulated quantity and percentage of the applied pollutant in the plant organs, but also allowed an evaluation of the dangers to humans and animals via the nutrient chain. The transmission electron microscopic examination of structural changes in the cell organelles (Masuch et al.,

1973) as well as the gas-exchange measurements, helped in understanding the overall effect, which was expressed as a change in growth production.

3.2 Reactions to Gas and Dust Components

The reaction of different agricultural and garden plant species to gaseous and particulate atmospheric pollutants will be illustrated using three series of experiments.

The data obtained from experiments conducted in 1975 are shown in Table 3 (Guderian et al., 1977). All six plant species were treated with SO_2 and with CdO, CuO, and PbO_2 as single components as well as all of their possible combinations during a period of 19–75 days.

Figure 1 presents the average yield of the research plants as a result of exposure to SO_2, CdO, and PbO_2 as single components and in combination. The yield for the various treatments are shown as a percentage of the untreated control. A comparison of the group average values, $\bar{X}_1$ to $\bar{X}_4$, shows a decrease in yield with an increase in the number of components. While treatment with one component had an average yield reduction of 17% relative to the control plants, the decrease in yield production rose to 22% with two components, 26% with three components, and 30% with four components. All four group averages are significantly reduced when compared with the controls. Yield reduction occurred without the appearance of visible damage.

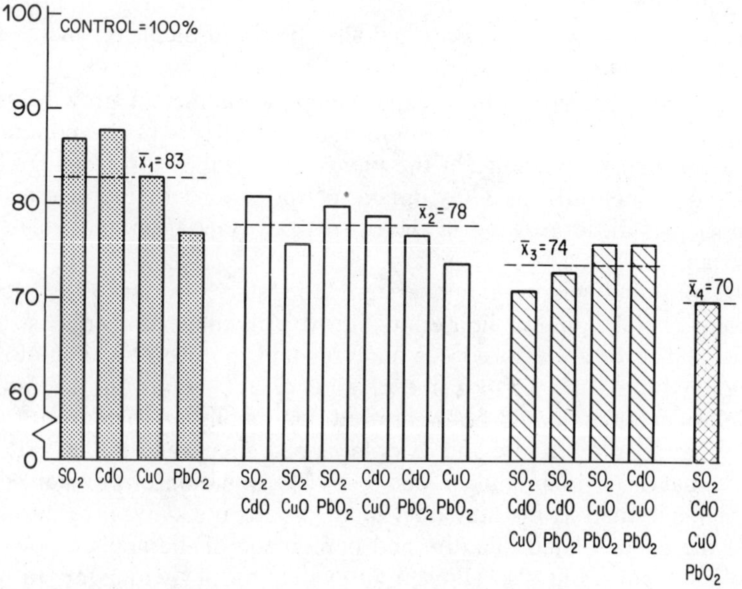

FIGURE 1. Average yield of various plant species after exposure to SO_2, CdO, CuO, and PbO_2 alone and in combination.

TABLE 3. Background Data for the Completion of the Experiments

	Development Stage		Exposure		Sulfur Dioxide			Dusting		
Plant Species	Experiment Start	Experiment End	Dates	No. of Days	$[c]$ (mg/m^3 air)	$[c \cdot t]$[a] ($mg\ h/m^3$)	Number	Total Cd	Amount Cu	(mg/m^2) Pb
Cress (*Lepidium sativum*), Hilds big-leaved	Small-leaves	Growth height 15 cm	6/27–7/16	19	0.13	46.4	3	4.4	29.8	405
White clover (*Trifolium perenne*) NFG Giant	After first cutting	Start of flowering	6/27–7/25	28	0.13	75.7	4	6.1	41.4	543
Kohlrabi (*Brassica oleracea* var. *gongylodes*), Regglis white	Primary-leaf stage	Market ripe	6/27–7/30	33	0.13	89.8	5	7.7	53.0	681
Tomatoes (*Lycopersicon esculentum*), Rentita GS	2-leaf stage	Start of fruit ripening	7/2–9/15	75	0.12	178.5	11	11.5	98.7	1211
Radish (*Raphanus sativus* var. *radicula*) Saxa race Primetta	2-leaf stage	Market ripe	8/27–9/13	17	0.11	40.2	3	1.8	17.0	213
Broad bean (*Vicia faba minor*) Ackerperle	2-leaf stage	Green feed ripe	9/12–1/27	45	0.11	97.0	6	7.4	67.5	798

[a]In the calculation of the product $c\,.\,t$ the interruption of the continuous SO_2 exposure by heavy-metal dusting has been taken into account.

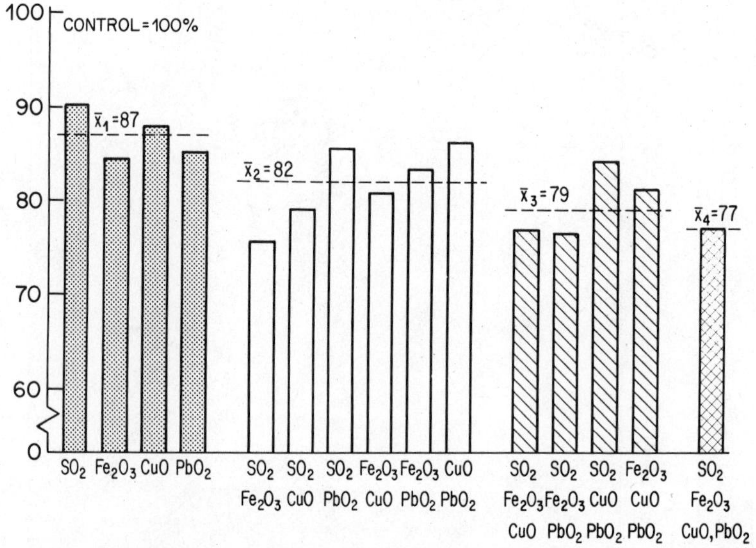

FIGURE 2. Average yield of various plant species after exposure to SO_2, Fe_2O_3, CuO, and PbO_2 alone and in combination.

Figure 2 illustrates a further series of experiments using red clover (*Trifolium pratense*), lettuce (*Lactuca sativa*), radish (*Raphanus sativus* var. *radicula*), and spinach (*Spinacia oleracea*) in which iron oxide was applied instead of cadmium oxide. Here the same tendency manifested itself as was shown in the first experiments illustrated in Figure 1: an increased combined effect with an increase in the number of single components. As was shown previously, the combined effect of the pollutants was less than the sum of the individual effects.

In the next experiment with seven plant species, spinach (*Spinacia oleracea*), red clover (*Trifolium pratense*), tomato (*Lycopersicon esculentum*), radish (*Raphanus sativus* var. *radicula*), lupine (*Lupinus luteus*), lettuce (*Lactuca sativa*), and a violet (*Viola tricolor*)), the plants were exposed to two gases, SO_2 and HF, and two heavy metals, CdO and PbO_2. These data are shown in Figure 3. Once again there is a similarity with the previous results. The experiments illustrated in Figures 1–3 were conducted between 1975 and 1977. The variations in climatic conditions during the three experiments have apparently had no overriding effects on plant response.

It must be mentioned in this connection that with long duration exposure to SO_2, even at concentrations close to the air quality standards at that time in the Federal Republic of Germany (0.14 mg SO_2/m^3) and with an HF concentration well below the current long-term value of 2 $\mu g/m^3$ (TAL,

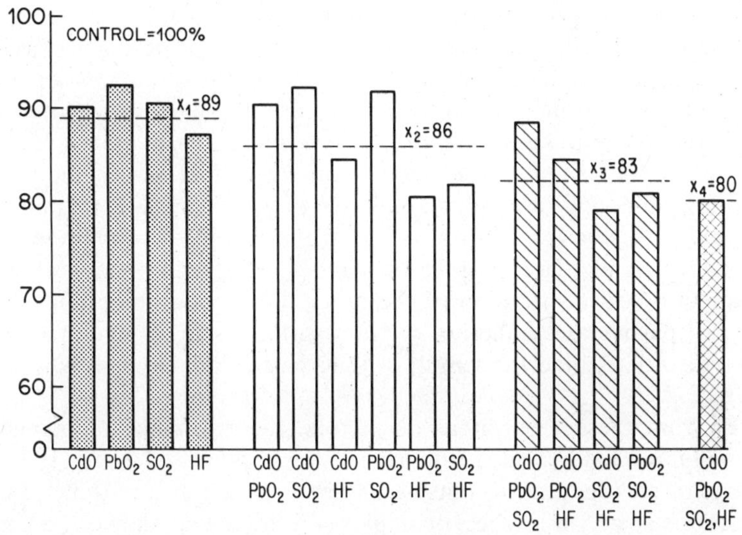

FIGURE 3. Average yield of various plant species after exposure to CdO, PbO_2, SO_2, and HF alone and in combination.

1974), significant growth reduction still appeared. This is further evidence to confirm that, even at the current allowable emission values, there is still a significant risk of damage to the vegetation.

Refer to the section on Heavy Metal-Containing Particulates for a discussion of a number of aspects concerned with the uptake of heavy metals by above-ground plant parts and for their methods of action.

4. DOSE–EFFECT RELATIONSHIPS

The quantitative relationships between the pollutant concentration in the air and its effects on the object that needs protection must be known for preventive air quality measures to be effective. Because they have been studied extensively and for many years, established air quality criteria are available for important gaseous components; similar values for particulate pollution, however, are lacking. We shall now briefly examine the difficulties associated with dose–effect relationships.

The morphology and chemical composition of a particulate atmospheric pollutant are of primary importance for the determination of damaging effects. Particle size, particle size distribution, shape, and surface characteristics are important factors in determining the adhesion of a particulate to the above-ground plant parts, while the solubility of a particulate is mainly dependent on its chemical composition. In general it can be said that the action of a specific element increases with decreasing particle size and increasing solubility. The first requirement for the determination of the dose–effect relationship is, therefore, a careful characterization of the particles according to structure, size, and solubility. Particulate solubility is not a fixed value but is dependent on such factors as the specific meteorological conditions and the presence of other atmospheric pollutants. Thus Kabata-Pendias (1979) has reported the increased mobility of heavy metals and their increased uptake by leaves during acid precipitation. The type, amount, and distribution of the precipitation determine not only the quantity of particulates adhering to plant surfaces with a given particulate load but also determine the length of time particulates remain adhering to the surface.

Apart from these environmental factors, the resistance of plants to a given pollutant load is also important. According to Levitt (1972) the resistance of an organism to outside influences, such as atmospheric pollutants, consists of two stress complexes: stress avoidance and stress tolerance.

The term stress avoidance refers to those mechanisms which reduce the uptake or the degree of toxicity of the pollutant that has been taken up, while stress tolerance refers to those mechanisms that lead to a cellular resistance against atmospheric pollutants. The topic of stress avoidance has already been touched upon briefly. It was indicated, for example, that the morphology of the plant organ was very important in determining the pollutant uptake from the air. When considering genetically controlled plasmatic resistance, we must refer especially to our knowledge about the effects of heavy metals via the soil (Ernst, 1966). The great susceptibility of the pinto bean (*Phaseolus vulgaris*) to lead, for example, is dependent on the sensitivity of the nodulating bacteria to the lead (Garber, 1970). As the experiments of Bradshaw (1952) have indicated with different populations of *Agrostis tenuis*, there exist even within one plant species distinct lead-resistant strains. For the development of air quality criteria it is important to know to what extent agreement exists in resistance relationships of plants to increasing pollutant loads between the soils via the roots and the air via the above-ground plant organs.

Special consideration must be given to the criteria used for determining the effects of particulates on plants during the development of air quality criteria for particulate atmospheric pollutants. Two types of questions, to a large extent, determine the choice of effects criteria: What site in the ecosystem will be subject to pollutant effects, and what functions of the objects to be protected are endangered?

The increase of pollutants in the soil caused by industrial emissions must be quantified in order to determine the magnitude of the indirect effects attributable to the industrial sources. Possible toxic effects on the plants which result in changes in growth and yield or influence nutrient availability or nutrient uptake are also questions that need to be considered, as are the effects on microflora and microfauna. Reference has already been made in Section 2.3 to the necessity of preserving the full production capacity of the soil through limiting the yearly input of accumulating substances.

When considering the direct effects of airborne pollutants, we are concerned with the effects on growth and reproduction as well as with the quality of the plant product. For plants utilized as food crops or animal fodder, the increase in toxic components is frequently the determining criterion, while with ornamental plants, leaf vegetables, or fruits it is usually the outward appearance. In the case of forest trees, the determination of wood production is often not enough. The end use of the plant, therefore, determines the method of analysis used for characterizing the pollutant effects.

Under actual field conditions there are numerous difficulties in separating the proportion of direct or indirect effects from the total effect. With the assistance of the modified capture-plant method of Sorauer (1911) and the test-plant method of Schönbeck (1963), it is possible to make delineations between direct and indirect effects. Using this approach, comparable experiments with contaminated and uncontaminated soils were conducted within and outside an area influenced by industrial emissions.

The characterization and quantification of dose–effect relationships is further complicated by the fact that particulate atmospheric pollutants occur both as deposited particulates and as aerosols. Through the measurement of pollutant deposition on the one hand and pollutant concentration on the other, the total pollutant load can be determined, and the effects can be characterized in an adequate manner. Finally, it is important to characterize the particulates that adhere to the plant organs (Vaughn, 1976).

The chemical and physical characteristics of specific particulate atmospheric pollutants as well as their mode of action must naturally be considered when choosing the approach for determining their effects. Depending on the research questions asked, experiments should be conducted under natural conditions in the region where the emission source is located, under as close to natural conditions as is possible in various exposure systems, or under closely controlled and reproducible field-simulated conditions in growth chambers. In many cases, especially in the determination of the dose–response relationships, it will be necessary to conduct combinations of experiments to satisfy the analytical constraints as well as the practical requirements of air quality control (Guderian and Kueppers, 1979). The transferability of data obtained with the dusting installation in the laboratory to field conditions is much more difficult than for gaseous components. Field research, therefore, requires special emphasis.

5. CONCLUSION

Particulate atmospheric pollutants are so widely distributed that it is essential that increased attention be paid to them when determining the effects of air pollution on terrestrial ecosystems. I shall now refer to some very important and pressing questions that need to be answered.

Particulate atmospheric pollutants represent an extraordinary collection of physically and chemically heterogenous substances. Their effects on plants range from a purely physical effect, such as large quantities of inert dust, to the very toxic low concentrations of heavy metal-containing compounds. Because it is impossible to examine such a large number of particulate substances at the same time, we should compile a list of substances to be tested in their order of priority. As criteria for this list of priorities, we should use pollutant toxicity, the pollutant's site of action in an ecosystem, and the pollutant's spatial extent and intensity.

According to the air quality principle, it is valid to determine the dose–effect relationship for setting air quality standards. Because of the chemical–physical characteristics and effects of particulate atmospheric pollutants on biological systems, we must pay special attention when selecting our research methods and objectives.

Atmospheric pollutant loads have to be quantified by their deposition rate or their concentration measurements. The choice of the organism to be studied must take into consideration the possible effects not only on the producers, the consumers, and the decomposers, but also on the soil. The selection of the criteria for the assessment of air pollution effects is dependent on the kind of effects. The plants must be chosen for their economic or ecological functions as well as for their genetic and environmentally controlled resistance to the pollutants.

The development of such dose–response relationships includes the elucidation of the mode of entry into the organism and of the mode of action in an organism. To protect the productivity of the soil, the annual input of accumulating substances, such as heavy metals, must be limited to such an extent that, over a prescribed time span, a tolerable concentration is not exceeded.

The increased danger to humans and animals via the nutrient chain requires not only an analysis of their eating habits, but also a systematic and long-term monitoring of the pollutants contained in various foodstuffs, especially plant products. Very important in this regard is the development of reproducible methods for the evenly distributed application of particulates, down to doses of less than 1 mg/m^2. The previously cited characteristics of particulates and the pronounced dependence of pollutant uptake on meteorological conditions makes it difficult to determine the validity of results obtained in laboratory experiments. Investigations under natural conditions, therefore, require special emphasis. By choosing a specific method or a combination of specific methods it can be guaranteed that results will be obtained which will satisfy the requirements of cause–effect analysis as well as the goal of environmental protection.

ACKNOWLEDGMENT

Appreciation is expressed to E. M. van Zinderen Bakker and A. H. Legge for translating this manuscript from the original German.

REFERENCES

Allaway, W. H. Agronomic controls over the environmental cycling of trace elements. *Adv. Agron.* **20**, 235–274 (1968).

Anderson, P. J. The effect of dust from cement mills on the setting of fruits. *Plant World* **17**, 57–68 (1914).

Bradshaw, A. D. Populations of *Agrostis tenuis* resistant to lead and zinc poisoning. *Nature* **169**, 1098 (1952).

Brandt, C. J. Untersuchungen über Wirkungen von Fluorwasserstoff auf *Lolium multiflorum* und andere Nutzpflanzen. Dissertation, Bonn, 1980.

Brandt, C. J., and R. W. Rhoades. Effects of limestone dust accumulation on the composition of a forest community. *Environ. Poll.* **3**, 317–325 (1972).

Buchauer, M. J. Contamination of soil and vegetation near a zinc smelter by zinc, cadmium, copper and lead. *Environ. Sci. & Technol.* **7**, 131–135 (1973).

Buck, M., and H. Ixfeld. Immissionsüberwachung im Lande Nordrhein–Westfalen. *Schriftenr. der Landesanstalt für Immissions- und Bodennutzungsschultz des Landes NW* **38**, 43–110 (1976).

Carroll, D. M., and J. F. Loneragan. Response of plant species to concentrations of zinc in solution. II. Rates of zinc absorption and their relation to growth. *Aust. J. Agric. Res.* **20**, 457–463 (1969).

Czaja, A. T. Die Wirkung von verstäubtem Kalk und Zement auf Pflanzen. *Qual. Plant. et Mat. Veg.* **7**, 184–212 (1960).

Czaja, A. T. Zementstaubwirkungen auf Pflanzen: Die Entstehung der Zementkrusten. *Qual. Plant. et Mat. Veg.* **8**, 201–238 (1961).

Czaja, A. T. Über das Problem der Zementstaubwirkungen auf Pflanzen. *Staub* **22**, 228–232 (1962a).

Czaja, A. T. Die Beeinflussung der Pflanzen durch Luftverunreinigungen besonders durch Kalk- und Zementstaub. In *Fortschritte der biologischen Aerosol-Forschung in den Jahren 1957–1961*, pp. 88–98, Schattauer-Verl., Stuttgart, 1962b.

Darley, E. F. Studies on the effect of cement-kiln dust on vegetation. *J. Air Poll. Contr. Ass.* **16**, 145–150 (1966).

Economic Commission for Europe (ECE). The leaded–unleaded petrol dilemma. Preliminary note prepared by the Secretariat. ENV/R. 49, 20. October 1976.

Eller, B. M. Strassenstaub heizt Pflanzen auf. *Umschau* **74**, 283–284 (1974).

Eller, B. M. Beeinflussung der Energiebilanz von Blätten durch Strassenstaub. *Angew. Botanik* **51**, 9–15 (1977).

Eller, B. M., and U. Brunner. Der Einfluss von Strassenstaub auf die Strahlungsabsorption durch Blätter. *Arch. Met. Geoph. Biokl. Ser. B* **23**, 137–146 (1975).

Eller, B. M., and P. Willi. Globalstrahlungsabsorption von *Hedera helix* L. unter Strassenstaubimmissionen. *Gartenbauwissenschaft* **42**, 49–53 (1977).

Ernst, W. Ökologisch-soziologische Untersuchungen an Schwermetall-Pflanzengesellschaften Südfrankreichs und des östlichen Harzvorlands. *Flora* (*Jena*) **B156**, 301–318 (1966).

Evans, L. S., N. F. Gmur, and F. DaCosta. Leaf surface and histological perturbations of leaves of *Phaseolus vulgaris* and *Helianthus annuus* after exposure to simulated acid rain. *Am. J. Bot.* **64**, 903–913 (1977).

Ewert, R. Untersuchungen über den Einfluss des Zementstaubes auf den Pflanzenwuchs. *Zement* **6**, 299–300, 307–309 (1917).

Ewert, R. Der Einfluss des Zementstaubes auf die Vegetation. *Zement* **15**, 39–42, 61–64, 83–85, 103–106, 128–130, 148–150, 168–170, 203–206 (1926).

Faerber, K. P., A. Hoffman, and G. Schmitz. Untersuchungen zum Nachweis schädigender Einflüsse von Luftverunreinigungen auf die Gesundheit des Menschen an grösseren Bevölkerungsgruppen. *Öffentl. Ges. Dienst* **20**, 493–511 (1959).

Folkeson, L. B. Särskilt avgasbly i den terrestra miljön. *SNV PM* **794**, 1–103 (1976).

Fortmann, H. Beitrag zur Frage der Staubsedimentation und ihrer Wirkungen im Gartenbau. Raucheinwirkungen im Gartenbau. *Forschung und Beratung, Reihe C, H. 5*, 69–77 (1963).

Garber, K. Erganzende Untersuchungen über die Auswirkungen der Luftverunreinigung auf die Vegetation in den Industriegebieten der Hansestadt Hamburg, *Jahresbuch f. Angewandte Botanik* **87/88**, 207–235 (1970).

Garber, K. Schwermetalle als Luftverunreinigung—Blei-, Zink-, Cadmium-Beeinflussung der Vegetation. *Staub-Reinh. d. Luft* **34**, 1 (1974).

Godzik, S., and M. M. A. Sassen. A scanning electron microscope examination of *Aesculus hippocastanum* L. leaves from control and air-polluted areas. *Environ. Pollut.* **17**, 13–18 (1978).

Gräefe, K. Globalstrahlungsunterschiede als lufthygienisches Kriterium für Grossstadt und Umgebung. *Städtehygiene* **14**(4), 68–74 (1963).

Gräefe, K., and C. Schlunk. Globalstrahlenmessung als Beitrag zu lufthygienischen Problemen. *Gesundheits-Ing.* **86**, 54–60 (1965).

Greszta, J., S. Braniewski, K. Marczynska-Galkowska, and A. Nosek. The effect of dusts emitted by non-ferrous metal smelters on the soil, soil microflora and selected tree species. *Ekol. po.* **27**, 397–426 (1979).

Guderian, R. *Air Pollution. Phytotoxicity of Acidic Gases and its Significance in Air Pollution Control* (Ecological Studies 22), Springer-Verlag, Berlin, Heidelberg, New York, 1977.

Guderian, R., and K. Kueppers. Problems in determining dose–response relationships as a basis for ambient pollutant standards. *Symposium on the Effects of Air-borne Pollution on Vegetation. August 20–24, Warsaw* (1979).

Guderian, R., and K. Kueppers. Responses of plant communities to air pollution. *Intern. Symp. on Effects of Air Pollutants on Mediterranean and Temperate Forest Ecosystems. June 22–27, Riverside, California* (1980). General Technical Report PSW-43, Pacific Southwest Forest and Range Experiment Station, Berkeley, Cal., 187–199 (1980).

Guderian, R., and K. Thiel. Versuchsanlage zur Ermittlung immissionsbedingter Kombinationswirkungen an Pflanzen. *Schriftenr. der Landesanstalt für Immissions- und Bodennutzungsschutz des Landes NW* **29**, 61–64 (1973).

Guderian, R., G. H. M. Krause, and H. Kaiser. Untersuchungen zur Kombinationswirkung von Schwefeldioxid und schwermetallhaltigen Staüben auf Pflanzen. *Schriftenr. der Landesanstalt für Immissionsschutz des Landes NW* **40**, 23–30 (1977).

Haselhoff, E. Versuche über die Wirkung von Flugstaub auf Gras. Landw. *Versuchsstationen* **69**, 477–482, Berlin (1908).

Holobrady, K., and J. Toth. Beitrag zu den Einwirkungen von festen aus Zementfabriken emittierten Parkikeln auf die Vegetation. Die VI. *Internationale Arbeitstagung forstlicher Rauschschadensachverständiger 9–14 Sept. in Kattowice* (1968).

Hunter, J. G., and O. Vergnano. Trace-element toxicities in oat plants. *Ann. Appl. Biol.* **40**, 761–777 (1953).

Huttunen, S. The integrative effects of air-borne pollutants on the performance of boreal forest ecosystems. *Symposium on the Effects of Air-borne Pollution on Vegetation. August 20–24, Warsaw* (1979).

Kabata-Pendias, A. Effects of inorganic air pollutants on the chemical balance of agricultural ecosystems. *Symposium on the Effects of Air-borne Pollution on Vegetation. August 20–24, Warsaw* (1979).

Kaleta, M. Die Wirkung von Magnesit-Immissionen auf die Änderung von Pflanzengemeinschaften. *Mitteil. d. Forstl. Bund. Vers. anst., Wien* **97**, 569–584 (1972).

Kloke, A. Blei–Zink–Cadmium. Anreicherung in Böden und Pflanzen. *Staub-Reinh. d. Luft* **34**, 18–21 (1974).

Kozel, J. Die Verstaubeng der Spaltöffnungen der landwirtschaftlichen Früchte und der Waldgehölze in Industriegebieten. *Verdecke prace Vyzkumneho ustavu melioraci ve Zbraslavi*, 211–226 (1971).

Krämer, F. Erste Untersuchungen zur Erstellung eines Bodenbelastungskatasters (Pb, Zn, Cd, Cu) im Raume Duisberg-Dinslaken. *Schriftenr. der Landesanstalt für Immissions- und Bodennutzungsschutz der Landes NW* **39**, 45–48 (1976).

Krause, G. H. M. Zur Aufnahme von Zink und Cadmium durch oberirdische Pflanzenorgane. Dissertation, Rheinische Friedrich-Wilhelms-Universität, Bonn (1974).

Krause, G. H. M., and H. Kaiser. Plant response to heavy metals and sulfur dioxide. *Environ. Pollut.* **12**, 63–71 (1977).

Kreutz, W., and W. Walter. Der Wind als Träger von Zementstaub und dessen Ablagerung auf Boden und Pflanzen. *Gartenbauwissenschaften* **21**(15), 151–164 (1956).

Lermann, L., and E. F. Darley. Particulates. In *Responses of Plants to Air Pollution*: Physiological Ecology, pp. 141–156. Academic, New York, London, 1975.

Levitt, J. *Responses of Plants to Environmental Stresses*. Academic, New York, London, 1972.

Little, P. A. Study of heavy metal contamination of leaf surfaces. *Environ. Pollut.* **5**, 159–172 (1973).

MAGS. Umweltprobleme durch Schwermetalle im Raum Stollberg. Minister für Arbeit, Gesundheit und Soziales des Landes Nordrhein–Westfalen, Dusseldorf (1975).

MAGS. Umweltbelastung im Raum Datteln. Minister für Arbeit, Gesundheit und Soziales des Landes Nordrhein–Westfalen, Dusseldorf (1977).

Masuch, G., R. Guderian, and H. Weinert. Wirkung von Chlorwasserstoff auf die Ultrastruktur der Chloroplasten von *Spinacea oleracea* L. *Proc. 3rd International Clear Air Congress*, pp. A160–A163. VDI-Verlag, Dusseldorf, 1973.

Parish, S. B. The effect of cement dust on citrus trees. *Plant World* **13**, 288–291 (1910).

Peirce, G. J. An effect of cement dust on orange trees. *Plant World* **12**, 283–288 (1910).

Pelz, E. Die phytotoxische Wirkung von Staubemissionen und Massnahmen zu ihrer Verhutung. *Technik* **15**, 549–555 (1960).

Persson, G. A. Massnahmen zur Reinhaltung der Luft. Staub Reinh. *Luft* **31**, 283–284 (1971).

Prinz, B., and G. Scholl. Erhebungen über die Aufnahme und Wirkung gas- und partikelförmiger Luftverunreinigungen im Rahmen eines Wirkungskatasters. *Schriftenr. der Landesanstalt für Immissions- und Bodennutzungsschutz des Landes NW* **36**, 62–86 (1975).

Prinz, B., G. H. M. Krause, and H. Stratmann. Thalliumschaden in der Umgebung der

Dyckerhoff Zementwerke AG in Lengerich, Westfalen. Staub-Reinh. *Luft* **39**, 457–462 (1979).

Ratsch, H. C. Heavy-metal accumulation in soil and vegetation from smelter emissions. EPA, Roap/Task 21BCJ-01, Oregon (1974).

Reinert, R. A., A. S. Heagle, and W. W. Heck. Plant responses to pollutant combinations. In *Responses of Plants to Air Pollution. Physiological Ecology*, Academic, New York (1975).

Rentschler, I. Die Bedeutung der Wachstruktur auf den Blättern für die Empfindlichkeit der Pflanzen gegenüber Luftverunreinigungen. *Proc. 3rd Internat. Clean Air Congr. Düsseldorf 1973*, pp. A139–142, VDI-Verlag, Düsseldorf, 1973.

Rentschler, I. The suitability of plants as indicators of air pollution. *Proc. 4th Internat. Clean Air Congress, Tokyo*, pp. 99–102 (1977).

Scheffer, F., E. Premeck, and W. Werner. Untersuchungen über den Einfluss von Zementofen-Flugstaub auf Boden und Pflanze. *Staub* **21**, 251–254 (1961).

Schönbeck, H. Beobachtungen zur Frage des Einflusses von industriellen Immissionen auf die Krankheitsbereitschaft der Pflanze. *Berichte d. Landesanstalt für Bodennutzungsschutz des Landes NW*, 89–98 (1960).

Schönbeck, H. Beispiel für Untersuchungen von Raucheinwirkungen im Freiland durch Anwendung einer Modifikation des Fangflanzenverfahrens. *Forschung und Beratung, Reihe C*, 47–56 (1963).

Schönbeck, H. Nachweis schwermetallhaltiger Immissionen durch ausgewählte pflanzliche Indikatoren. *VDI-Berichte* **203**, 75–85 (1974).

Scholl, G., and H. Schönbeck. Erhebungen über Immissionsraten und Wirkungen von Luftverunreinigungen im Rahmen eines Wirkungskatasters. *Schriftenr. der Landesanstalt für Immissions- und Bodennutzungsschutz des Landes NW* **33**, 73–80 (1975).

Sorauer P. Jahresber des Sonderausschusses für Pflanzenschutz 1897. *Arb. Deutsche Landw. Gesellsch.* **29**, 79–81 (1898).

Sorauer, P. Die makroskopische Analyse rauchgeschädigter Pflanzen. Sammlung von Abhanlungen über Abgase und Rauchschaden. H 7 (1911).

Steffeck, H. Der Schutz gegen Fluorschädigungen durch gewerbliche Einwirkungen. *Deutsche Landw. Gesellsch.* **30**, 27–35 (1896).

Steinhüebel, G. Zmeny v škrobových rezervách listov cezminy po umelom znecisteni pevnym popraskom (Veränderungen in den Stärkereserven der Blätter der gemeinen Stechpalme nach einer künstlichen Verunreinigung durch Staub). *Biologia* **18**, 23–33 (1963).

Steinhüebel, G. Ucinok prachovej vrstvy na prehrienvanie listovej cepele pripriamej insolach (Wirkung einer Staubschicht auf die Überwärmung der Blattspreite bei direkter Insolation). *Biologia*, **21**(4) 277–294, (1966).

Steinhüebel, G. Immergrüne Laubgehölze in verunreinigter Atmosphäre. In *Einführung in die ökologische Sempervirenz*. Vydavatelstvo Slovenskej akademic vied, Praha, 1967.

Steinhüebel, G., and L. Halas. Poruchy v tvorbe sušiny pri zvyšenych teplotach vyvolanych v listoch drevin prasnon imision (Störungen der Trockensubstanzbildung bei durch Staubimmissionen hervorgerufenen höheren Temperaturen in den Blättern von Gehölzen) *Lesnicky Casopis* **13**(40), 365–382 (1967).

Turner, M. A. Effect of cadmium treatment on cadmium and zinc uptake by selected vegetable species. *J. Environ. Quality* **2**, 118–119 (1973).

Umweltbundesamt. *Luftqualitätskriterien für Blei*. Umweltbundesamt, Berlin, 1976.

Umweltbundesamt. *Luftqualitätskriterien für Cadmium*. Umweltbundesamt, Berlin, 1977.

Vaughn, B. E. Suspended particle interactions and uptake in terrestrial plants in atmosphere–surface exchange of particulate and gaseous pollutants, 1974. ERDA Symposium Series 38, CONF-740921, 228–243 (1976).

Vetter, H. Belastungen und Schaden durch Schwermetalle in der Nahe einer Blei- und Zinkhutte in Niedersachsen. *Staub-Reinh. d. Luft* **34**, 10–11 (1974).

World Health Organization (WHO). Evaluation of certain food additives and the contaminants mercury, lead and cadmium. Sixteenth Report, WHO, Technical Report Series 505, Geneva (1972).

Wittwer, S. H., and F. G. Teubner. Foliar absorption of mineral nutrients. *Ann. Rev. Plant Physiol.* **10**, 13–32 (1959).

Terrestrial Ecosystems: Wet Deposition

DAVID S. SHRINER
Environmental Sciences Division
Oak Ridge National Laboratory
Oak Ridge, Tennessee

1. INTRODUCTION

One of the most rapidly growing areas of research into the effects of air pollutants on terrestrial ecosystems during the past decade has been wet deposition of pollutants and the subsequent effects of those precipitation-deposited pollutants on vegetation and soils. The objectives of this paper are to review the state of our collective knowledge on the effects of wet deposition of pollutants, to identify where research to date has left gaps, and to present our views on the future research needs in this area. After a decade of rather intensive research into the effects of wet deposition of pollutants on vegetation, a number of conclusions can be drawn about the way in which wet deposition affects vegetation.

Rain, fog, dew, and other forms of wet deposition that are not contaminated with pollutants play an important role for vegetation and soils (beyond providing water) as sources of nutrient inputs and as mechanisms of removal from vegetation of mineral nutrients, amino acids, carbohydrates, and growth regulators by leaching (Tukey, 1975). Monitoring has made us aware of the role precipitation plays in scavenging pollutants from the atmosphere and of changes in the character of wet deposition over the last 30 years. Considerable speculation and attention have been centered on the effects that increases in ionic strength of sulfate, nitrate, and hydrogen ions scavenged by rain might have on receptor organisms. The effects of this form of wet deposition, popularly referred to as acidic rain, are interesting because, in contrast to most of the gaseous pollutants which we are called upon to investigate, there are no documented cases of direct, visible injury to vegetation in the field situation as a result of so-called acidic rains [with the exception of a single instance of volcanic origin (Kratky et al., 1974)]. The knowledge we have accumulated is the result of a wide variety of approaches involving, in most instances, the use of some form of rain simulation.

This review does not present an account of all of the individual research efforts which comprise our corporate knowledge; instead, it concentrates on those common conclusions we may draw as well as on those inconsistencies which have appeared thus far. On the basis of those commonalities and inconsistencies, perhaps a clearer picture of research needs will emerge.

2. WHAT DO WE KNOW ABOUT TERRESTRIAL VEGETATION–WET DEPOSITION INTERACTIONS?

Wet deposition is believed to affect a wide range of plant species. These effects can be categorized broadly as either direct, resulting in some directly measurable effect on the target plant itself, or indirect, resulting in alteration of the manner in which the target plant interacts with some other factor of its environment.

2.1 Direct Effects on Vegetation

The most commonly described process of pollutant scavenging by rain is that which results in the acidification of rainfall and other forms of wet deposition. The chemistry leading to the creation of so-called acidic rain, as well as its occurrence and distribution, has been reviewed frequently in recent years, and is not further belabored in this paper. Instead, for the purpose of our discussion, let us accept as given the fact that, due to the removal of sulfur and nitrogen compounds from the atmosphere primarily by precipitation, rainfall deposits from 10 to 100 times the quantities of hydrogen ions expected under unpolluted circumstances and proportionally higher than expected concentrations of sulfates, nitrates, and insoluble particles.

These increased inputs of H^+, SO_4^{2-}, and NO_3^- have both positive and negative effects on soil–plant systems. Concentrations of hydrogen ion, equivalent to that measured in more acidic rain events ($\leq$pH 3), are observed experimentally to cause tissue injury in the form of necrotic lesions to a wide variety of plant species under greenhouse and laboratory conditions. The threshold for the occurrence of this visible injury is reported as being between pH 3.0 and 3.6. The various types of direct effects which have been reported are shown in Table 1. Such results must be interpreted with caution, however, because the growth and morphology of leaves in a greenhouse are often atypical of field conditions.

The most common form of direct injury, small necrotic lesions, appears to be the result of the collection and retention of water on plant surfaces and the subsequent evaporation of these water droplets. Once a lesion occurs, collection of water is further enhanced by the depression formed by the lesion. Other reported effects include the alteration of surface features (such as epicuticular waxes) and altered physiological processes (notably, carbon fixation and allocation).

Let us consider the direct effects of wet deposition in terms of the following questions:

1. What components of wet deposition are responsible for any effects observed to date?
2. What do we know about dose-response?
3. What characteristics of a plant impart sensitivity to wet deposition of acidic substances?
4. What characteristics of a precipitation event dominate the type and degree of effect involved?
5. To what extent are site characteristics involved in plant sensitivity?

TABLE 1. Types of Direct, Visible Injury Attributed to Acidic Wet Deposition

Injury Type	Species	pH Range	Reference	Remarks
Pitting, curl shortening, death	Yellow birch	2.3–4.7	Wood and Bormann (1974)	
1-mm necrotic lesions, premature abscission	Kidney bean, soybean, loblolly pine, E. white pine, willow oak	3.2	Shriner et al. (1974)	
Cuticular erosion	Willow oak	3.2	Shriner (1978a) Lang et al. (1978)	
Chlorosis	Sunflower, bean	2.3–5.7	Evans et al. (1977)	
(a) Small, shallow circular depressions: slight chlorosis	Sunflower, bean	2.7	Evans et al. (1977)	More frequent near veins. (a)–(d) represent sequential stages of lesion development, through time, up to 72 h (one 6-min rain event daily for 3 days)
(b) Larger lesions, chlorosis always present, palisade collapse	Sunflower, bean	2.7	Evans et al. (1977)	
(c) 1-mm necrotic lesions, general distortion	Sunflower bean	2.7	Evans et al. (1977)	
(d) 2-mm bifacial necrosis due to coalescence of smaller lesions, total tissue collapse	Sunflower, bean	2.7	Evans et al. (1977)	

Wrinkled leaves, excessive adventitious budding, premature abscission	Bean	1.5–3.0	Ferenbaugh (1976)	
Incipient bronzed spot	Bean	2.0–3.0	Hindawi et al. (1980)	After first few hours
Bifacial necrotic pitting	Bean	2.0–3.0	Hindawi et al. (1980)	After 24 h (reported pooling of drops = more injury)
Necrotic lesions, premature abscission	E. white pine, scotch pine, spinach, sunflower, bean	2.6–3.4	Jacobson and van Leuken (1977)	Injury associated with droplet location within 24–48 h
Marginal and tip necrosis	Bean, poplar, soybean, ash birch, corn, wheat	Submicron H_2SO_4 aerosol	Lang et al. (1978)	
Galls, hypertrophy, hyperplasia	Hybrid poplar	2.7–3.4	Evans et al. (1978)	
Dead leaf cells	Soybean	3.1	Irving (1979)	
Necrotic lesions	Citrus	0.5–2.0	Heagle et al. (1978)	

2.1.1 What Components of Wet Deposition Are Responsible for the Effects Observed to Date?

The major components of wet deposition which are of greatest concern in terms of vegetation effects are dissolved gases (e.g., SO_2), sulfate, nitrate, and the hydrogen ion concentrations associated with them. Experiments have shown that hydrogen ion concentration may have negative effects on vegetation in terms of growth, yield, and quality. No evidence at the present time suggests any positive or beneficial effects of hydrogen ion inputs *per se*. Recent research has identified the extremely important role of sulfur and nitrogen inputs in precipitation to the ultimate response of vegetation to acid deposition. Jacobson et al., (1980) investigated the impact of simulated acidic rain on the growth of lettuce at acidities of pH 5.7 and 3.2. At pH 3.2, solutions were compared with $NO_3:SO_4$ mass ratios of 20:1, 2:1, and 1:7.5. For those growth parameters that showed a significant treatment difference (root dry weight, apical leaf dry weight), the high-nitrate pH 3.2 treatments were not different from the pH 5.7 controls, but the results were significantly less than those from the low-N, high-S treatment. These observations suggest that perhaps sulfur was a limiting factor in the nutrition of these plants, resulting in a sulfur response that overwhelmed the hydrogen ion effect. Irving (1979) reported that when field-grown soybeans were exposed to simulated acid rains, an apparent positive growth response was obtained from the sulfur and nitrogen inputs in simulated acid rain. Wood and Bormann (1974) and Abrahamsen and Dollard (1979) also presented data suggesting positive growth responses to nitrogen and sulfur in simulated rain treatments with forest tree species. Turner and Lambert (1980) presented evidence of positive growth response in Monterey pine to sulfur deposition in ambient precipitation in Australia.

It is clear from the evidence available that wet deposition of pollutants may result in a range of direct effects on vegetation, depending on (1) precedent and antecedent conditions (e.g., soil nutrient status, plant nutrient requirements, plant sensitivity, and growth stage) and (2) the total loading or deposition of critical ions (H^+, NO_3^-, SO_4^{2-}).

2.1.2 What Do We Know About Dose–Response?

Several investigators have reported results of dose–response experiments with a variety of plant species. It is appropriate also to discuss the comparability of results including those factors that influence the extent to which results in the literature can be compared.

Thresholds for direct, visible injury to foliage in greenhouse studies with acid rains typically lie above pH 3.1 (Jacobson et al., 1980; Shriner et al., 1974). However, when the same investigator worked with the same species and the same treatment solutions under both greenhouse and field situations, there were significant shifts in species sensitivity. Shriner (1978a), for example, observed visible injury and significant growth reduction at pH 3.2

in greenhouse studies; however, similar doses resulted in no visible injury and no significant growth impact in field experiments. Irving (1979), working with field-grown soybeans exposed to rainfall simulant of pH 3.1, also observed no apparent visible injury, although significantly higher numbers of dead leaf cells did occur. A. S. Heagle (personal communication), also working with soybeans in field experiments, found no visible injury from simulated rain applications as low as pH 2.8.

Data from Linskens (1952), suggesting that cuticle development under light, temperature, and humidity conditions in greenhouses is markedly different from that under similar conditions in the field, may provide an explanation for the difference in sensitivity between greenhouse and field-grown plants with respect to acid rain. Experiments in our own laboratory, where cuticle thickness was evaluated by means of transmission electron microscopy of transverse sections of the cuticle, suggest that the cuticle plus epicuticular wax layer on field-grown loblolly pine (*Pinus taeda*) needles may be as much as 75 times thicker than that on needles of the same age, grown under greenhouse conditions. The surface characteristics of leaves have critical bearing not only on the effects of wet deposition but also on the way in which wet deposition interacts with dry particles and gases deposited on leaf surfaces.

It is clear from most of the investigations conducted thus far that leaf morphology and leaf surface characteristics play an important role in understanding why lesions occur where they do on a leaf surface (Evans et al., 1977). Evans et al. found that initial injury to beans and sunflowers occurs near trichomes and stomata after exposure to simulated acidic rain. They hypothesized that cuticle thickness near these structures may be a factor in lesion development. Rentschler (1973) discussed the importance of the superficial wax layer on leaves in relation to plant sensitivity to air pollutants. The presence of the wax greatly reduces the wettability of leaves and, therefore, the extent to which water is retained on the leaf surfaces. Rentschler also mentions that in certain species (e.g., lupin, clover) "the neighborhood of the stomate is free of wax," supporting the hypothesis of Evans et al., (1977) that components of the cuticle may develop differentially near trichomes and guard cells.

Martin and Juniper (1970) and Purnell and Preece (1971) described physical changes in the structural appearance of epicuticular waxes as a direct result of weathering agents (rain, wind, dust, abrasion with foreign objects). Shriner (1974) observed that, in addition to obvious weathering of these superficial waxes by the impaction of simulated rain drops, there appeared to be differential rates of weathering as a function of the treatment when treatments of differing acidity were used (pH 5.6 versus 3.2). Shriner was unable to postulate a chemical mechanism for the differences observed on *Quercus prinus* and *Liriodendron tulipifera*. However, Hoffman et al. (1980) proposed a mechanism by which precipitation acidity can act as an additional, chemical factor in the weathering of epicuticular waxes.

TABLE 2. Thresholds for Visible Injury and Growth Effects Associated With Experimental Studies of Wet Deposition of Acidic Substances

Effect	Species	Threshold[a]	Reference	Remarks
Foliar lesions, decrease in growth	Yellow birch	pH 3.1	Wood and Bormann (1974)	Greenhouse
Foliar aberrations, decrease in growth	Bean	pH 2.5	Ferenbaugh (1976)	Greenhouse
Foliar lesions	Bean, sunflower	pH 3.1	Evans et al. (1977)	Greenhouse
Foliar lesions	Bean	pH 3.2	Shriner (1978a)	Greenhouse
Foliar lesions	Hybrid poplar	pH 3.4	Evans et al. (1978)	Greenhouse
Foliar lesions	Sunflower	pH 3.4	Jacobson and van Leuken (1977)	Greenhouse
Reduced growth	Bean	pH 4.0	Longnecker and Shriner (unpublished)	Greenhouse
Foliar symptoms, no reduced growth	Soybean	pH 3.0	Jacobson (1980b)	Greenhouse
Increased growth, increased/decreased nutrient content	Lettuce	pH 3.0, 3.2	Jacobson (1980b)	Greenhouse (varied with SO_4^{2-} and NO_3^-)

Reduced growth	Pinto bean	pH 3.1	Jacobson (1980b)	Greenhouse
Reduced yield	Pinto bean	pH 2.7		
Reduced growth	Soybean	pH 3.1		
Reduced yield	Soybean	pH 2.5		
Increased yield	Soybean	pH 3.1		
Foliar symptoms	Tomato	pH 3.0	Jacobson (1980b)	Greenhouse
Reduced growth	Tomato	pH 3.0		
Reduced yield	Tomato	pH 3.0		
Reduced quality	Tomato	pH 3.0		
No foliar symptoms, or effects on growth	Soybean	pH 3.1	Irving (1979)	Field
No foliar symptoms, but	Soybean	pH 2.8	Jacobson (1980b)	Field, low ozone
(a) decreased growth, yield	Soybean	pH 2.8		Field, high ozone
(b) increased yield	Soybean	pH 2.8		Field, low ozone
No effect on growth, yield	Tomato	pH 3.0	Jacobson (1980a)	Field
Reduced quality	Tomato	pH 3.0		Field

After Jacobson (1980a,b).

[a]Highest pH to elicit a negative response, or lowest pH to elicit a positive response.

They point out that the wax composition, as polymeric structures of condensed long-chain hydroxy carboxylic acids, may result in an imperfect wax matrix in which the uncondensed sites containing hydroxy functional groups are more readily weathered. Strong acid inputs to such a system would oxidize and release a wide range of carbon chain acids from the basic waxy matrix, conceivably yielding the type of change in weathering rate observed by Shriner. More recently, Shriner and Cowling (1980) discussed the potential implications of such weathering in terms of the plant's ability to respond to a variety of external stresses.

Wettability appears to be an important factor in the response of plants to acid deposition. This can be seen in the work of Evans et al. (1977), Jacobson and van Leuken (1977), and Shriner (1978a), who variously report a threshold of between pH 3.1 and 3.5 for development of foliar lesions on beans. The cultivars of *Phaseolus vulgaris* L. used in the above studies are all relatively nonwaxy and therefore fairly easily wettable. By comparison, in studies with the very waxy leaves of citrus, Heagle et al. (1978) reported a threshold for visible symptoms to be near pH 2.0. Waxy leaves apparently minimize the contact time for the acid solutions, thus accounting for the greater than 400-fold increase in H^+ ion concentration required to induce visible injury. Table 2 summarizes the thresholds, including range, species sensitivity, concentration, and time, for visible injury associated with experimental studies of wet deposition of acidic substances.

Physiological responses of vegetation to wet deposition have received limited attention, although Tukey and Morgan (1963) and Tukey (1970, 1975, 1980) reviewed extensively the role of precipitation in the loss, by leaching, of nutrients from above-ground plant parts. Sheridan and Rosenstreter (1973) reported marked reduction of photosynthesis in a moss exposed to increasing H^+ ion concentrations. Sheridan and Rosenstreter (1973), Ferenbaugh (1976), and Hindawi et al. (1980) reported reduced chlorophyll content as a result of tissue exposure to acid solutions. In the case of Ferenbaugh (1976), the significant reductions in chlorophyll content did not occur at pH 2.0, and chlorophyll content at pH 3.0 may have been slightly increased. Irving (1979), working with field-grown soybeans, also reported higher chlorophyll content of leaves exposed to simulated rain at pH 3.1. Hindawi et al. (1980), on the other hand, reported an 8% reduction in chlorophyll content at pH 3.0, decreasing to 15% at pH 2.5 and 31% at pH 2.0. They also observed no changes in the ratio of chlorophyll *a* to chlorophyll *b* as a result of the treatments.

Ferenbaugh (1976) determined photosynthesis and respiration rates for bean plants exposed to acid treatments, and found not only a slight, significant increase in the rate of respiration but also a large (nearly four-fold) increase in photosynthetic rate (as determined by oxygen evolution) at pH 2.0. Ferenbaugh concluded that because biomass accumulation and sugar and starch concentrations of the treated plants were significantly reduced, photophosphorylation was somehow being uncoupled by the acid treatments.

Irving (1979) reported increased photosynthetic rates in soybeans, and attributed them to increased nutrition by the sulfur and nitrogen components of the acid simulant apparently overcoming any negative effect of the pH 3.1 treatment in the field.

Jacobson et al. (1980) reported a shift in partitioning of photosynthate from vegetative to reproductive organs as a result of acid rain treatments of pH 2.8 and 3.4. In other experiments designed to look at the effects of acid treatments on photosynthate allocation, Leopold (unpublished) found that when $^{14}CO_2$-tagged bush beans were stressed for 21 days prior to the tag and for 4 days after the tag with rain at pH 3.2, 4.0, and 5.6, a significant shift in carbon allocation occurred, with the tagged carbon remaining in the foliar portions of the plants at the expense of allocation to the roots. Earlier work by Shriner and Johnston (1981) on the effects of simulated acid rain on nodulation of legumes also suggested that pH 3.2 simulated rain caused a shift in photosynthate allocation, resulting in reduced energy available for root nodule formation.

To summarize our knowledge of dose–response:

1. Visible injury thresholds for acid precipitation lie between pH 2.0 and 3.6, depending on species, and may vary from pH 3.0 to 3.6 within the same species (e.g., *Phaseolus vulgaris*).
2. Species-to-species variability in sensitivity may be as great as two orders of magnitude in hydrogen ion concentration.
3. The threshold for growth effects in the absence of visible injury is reported to lie between pH 3.5 and 4.0, but sulfur and nitrogen inputs in the precipitation may result in a net positive growth impact, depending on soil nutrient status, buffer capacity, other growth conditions, and plant nutrient requirements.
4. Total dose of hydrogen ions appears to be most clearly related to visible injury.

2.1.3 What Characteristics of a Plant Impart Sensitivity?

As mentioned in the discussion of dose–response relationships, certain characteristics of the plant are critical to its sensitivity to injury from wet deposition of acidic substances. One of the most significant of these characteristics is wettability. The less wettable the leaf surface, the less contact time a water droplet has with the leaf surface and the lower the resulting dose. Also, on less wettable leaves, dry deposits of particles can be dislodged more readily by precipitation.

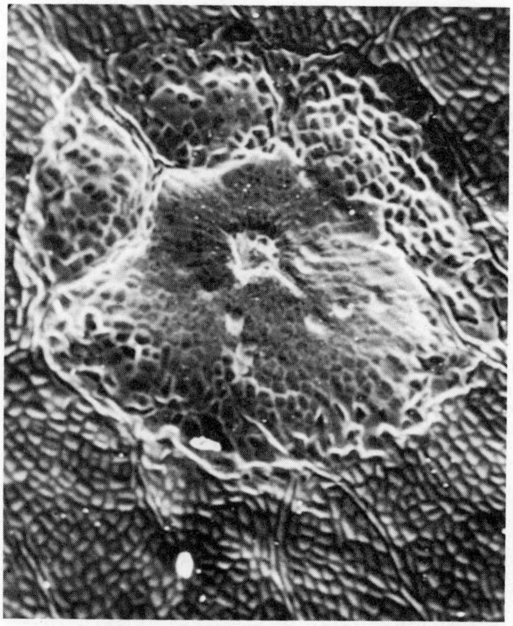

FIGURE 1. Scanning electron micrograph of red kidney bean leaf surface, illustrating a typical necrotic lesion induced by acidic precipitation of pH 3.2 or lower. Note trichome at center of lesion. Actual diameter of lesion, approximately 1 mm.

Another characteristic of leaf surfaces influencing sensitivity to visible injury is the presence or absence of sensitive morphological features such as trichomes (Fig. 1) and/or stomatal subsidiary cells (Evans et al., 1977). General leaf morphology is also important. Patterns of venation, leaf shape, and so forth which favor the pooling of water along veins or at the margins of leaves will tend to favor formation of local lesions at those sites (Evans et al., 1977).

Related to the inherent sensitivity of a plant species due to morphological advantage or disadvantage is the role that tissue age plays in plant sensitivity. Several studies (Evans et al., 1977; Evans and Bozzone, 1977; Shriner, 1978a) showed that sensitivity varies with life stage which plays an important role in the actual threshold of effects observed in higher plants and ferns, and in host–parasite relationships. For instance, Evans et al. (1977) found that the relationship between the degree of injury and plant age was inverse in *Helianthus annuus*. In *Phaseolus vulgaris*, however, young leaves before full expansion were not as sensitive to acidic rain as older leaves.

2.1.4 What Characteristics of the Event Control the Effect?

From the work of Jacobson and van Leuken (1977), it appears that the following characteristics of wet deposition events control the ultimate plant response:

1. *pH*. Foliar injury increases with decrease in pH, with a threshold for visible injury between pH 3.0 and 3.6 for most species.
2. *Duration*. Total dose of hydrogen ion is critical. Threshold for injury may range from 1 min at pH 2.6 to 9 h of exposure at pH 3.4.
3. *Frequency*. Repetition of treatments increases the frequency of foliar lesions. One 3-h treatment resulted in 25% leaf injury; two and three 3-h treatments resulted in 48 and 74% injury, respectively.

Jacobson and van Leuken (1977) and Shriner (1978b) suggested that additional characteristics of the rain event (intensity, droplet size, and changes in acidity within the rainfall event) also have significant influences on plant response, as do those factors affecting drying time (temperature, humidity, airflow, and sunlight).

2.1.5 Are Site Characteristics Involved in Plant Response?

A number of site characteristics are important factors in the long-term impacts of wet deposition of pollutants in soil–plant systems. Attention is drawn especially to the influence of soil properties such as cation exchange capacity, base saturation, soil depth, and sulfate adsorption (Reuss, 1978; Norton, 1980; McFee, 1980). Reuss (1978) concluded that soils well supplied with bases are most susceptible to base loss, and as base saturation and lime potential fall to low levels, acid precipitation may cause increased leaching of H^+ and Al^{3+} ions rather than bases. This effect was reported by Cronan (1980) and by European workers. There is little information on the role of soil fertility levels on plant response to acidic deposition, but it would seem reasonable at this point to consider carefully those ecosystems where cation nutrients are in short supply, and where both soil and foliar leaching of those ions are accelerated by acid deposition (Reuss, 1978). Generally, those sites most likely to experience effects of acid precipitation on vegetation will be those of marginal productivity in terms of sulfur and/or nitrogen nutrition, cation nutrients, and base exchange capacity. In the absence of fertilization and/or liming, the primary productivity of marginal sites might be reduced over a time scale of several decades or longer (Reuss, 1978). Should soil acidification occur at specific sites, then a wide range of micronutrient deficiency–toxicity problems could potentially arise.

2.2 Indirect Effects on Vegetation

2.2.1 What Types of Indirect Effects Does Wet Deposition of Pollutants Have on Vegetation?

Wet deposition of pollutants may result in a range of indirect effects on plants and may play a role in the individual plant's relative fitness or ability to survive or otherwise compete at the community or ecosystem level. These indirect effects fall into two distinct categories:

1. Biotic—alteration of the plant's physiology to such a degree as to result in detectable shifts in physiological process rates or biotic interactions.
2. Abiotic—function of the plant as a reaction surface for interactions between wet and dry deposits of particles and gases.

Both these areas are deserving of critical consideration because they have received somewhat limited attention. They are, for the most part, processes or interactions which are extremely difficult to characterize.

Biotic Effects. Acid deposition effects on physiological processes were discussed earlier with regard to direct effects. An additional effect, not previously discussed, is the uptake and loss of nutrients by foliar surfaces of plants. This discussion is included under indirect effects because leaching is largely regarded as a passive process. Tukey (1970, 1975, 1980) and Tukey and Morgan (1963) extensively reviewed the leaching of substances from plants as well as the foliar uptake or absorption of nutrients from water films on plant surfaces. Leaching involves exchange reactions on the leaf surface in which cations at exchange sites of the cuticle are exchanged for hydrogen from leaching solutions (Tukey, 1970). Because leaching is a sufficiently effective mechanism for removing nutrients from foliage, nutrient deficiencies can occur (Jacobson et al., 1980) in the absence of efficient cycling (subsequent reabsorption by adjacent foliage or uptake from soil pools) mechanisms. It is important to note that:

1. The rate of cation leaching is influenced strongly by the chemical composition of the plant, the composition of the leaching solution, and especially the concentration of hydrogen ions in the solution.
2. Regardless of the cause of injury to foliage (such as the necrotic spotting discussed under direct effects), injured foliage is more susceptible to leaching than is healthy foliage, and leaching may accentuate the harmful effects of injury to plants (Tukey and Morgan, 1963).

Saunders (1971), Purnell and Preece (1971), and Shriner (1978a) observed the effects of wet deposition alone, or its effects in the presence of gaseous and particulate pollutants, on some aspect of the growth of leaf-surface saprophytes and pathogenic microorganisms. In certain instances, these effects appeared to be directly on the microorganisms, resulting in an indirect benefit or injury to the host plant. In other instances, effects on the host plant, in the form of necrotic lesions acting as infection courts for pathogenic microorganisms (Shriner, 1978a) or altered host physiology breaking down host resistance to the pathogen (McLaughlin and Shriner, 1980), were identified.

There are a number of other aspects of leaf-surface microfloral populations which deserve mention with regard to effects of wet deposition. The chemical environment of the leaf surface has long been recognized as a critical factor in the development of many plant diseases (Blakeman, 1971). Exudates from within the leaf may serve as sources of nutrients for germinating fungal spores on leaf surfaces, or, in contrast, may be fungistatic or fungitoxic. Purnell and Preece (1971) and Blakeman and Atkinson (1976) presented evidence for water-soluble spore germination inhibitors occurring in leaf-surface waxes. At the present time, one can only speculate that the wet deposition of acidic substances, or the mobilization of dry-deposited trace metals by wet-deposited acidic substances, might result in significant changes in leaf-surface host–parasite interactions. By the same token, the deposition of such pollutants on leaf surfaces may influence the population dynamics of leaf-surface microfloral populations, thus affecting the balance between saprophytes and pathogens competing for substrate (Fokkema, 1976; Hudson, 1971).

In some forest ecosystems, nitrogen fixation by phylloplane-inhabiting nitrogen-fixing bacteria may represent a significant source of input of combined nitrogen to the forest (Jones, 1976; Dennison, 1973). In such instances, the potential of acid deposition to alter rates of N_2 fixation by these microorganisms could represent an important consequence of the pollutant deposition. Closely related research with nitrogen-fixing lichens of a Douglas fir forest canopy exposed to acid solutions (Dennison et al., 1976) showed a reduction in N_2 fixation by the lichens below a threshold between pH 4 and pH 2. Considering the importance of biological N_2 fixation to the nitrogen economy of many natural ecosystems, this area deserves additional research in connection with our evaluation of long-term impacts of acid deposition at the ecosystem level.

Because of the widespread occurrence of wet deposition of acidic substances in the eastern United States and in northern Europe, vegetation is commonly exposed not only to gaseous phytotoxicants common to urban (and even some rural) areas but also to so-called acidic rain. Little information is available which will permit a conclusive evaluation of the potential for interaction effects of wet- and dry-deposited pollutants as far as plant response is concerned. However, preliminary work by Shriner (1978b), Irving (1979), and Jacobson et al. (1980) suggests that interactions may occur. In controlled-exposure greenhouse experiments, Shriner

(1978b) reported no significant interaction between multiple rain exposure (pH 4.0) and four SO_2 exposures [3-ppm peak, 1 h; see method described by McLaughlin et al. (1979)] as far as growth of bush beans was concerned. Irving (1979), working with field-grown soybeans, found that simulated acid precipitation (pH 3.1) apparently diminished a photosynthetic decrease that resulted from 17 exposures to 0.19 ppm of SO_2 during the growing season. Shriner (1978b) also exposed plants to 0.15 ppm ozone (four 3-h exposures) in between four weekly exposures to rainfall of pH 4.0, and observed a significant growth reduction at the time of harvest. Jacobson et al. (1980), using open-top field chamber-grown soybeans, compared growth and yield in exposure treatments of three pH levels of simulated rain (pH 2.8, 3.4, and 4.0) and two levels of ozone (<0.03 and ≤ 0.12 ppm). Results demonstrated that ozone depressed both growth and yield of soybeans with all three rain treatments, but that the depression was greatest with the most acidic rain. The aforementioned investigations clearly point to the need for research in this area, because ozone levels equal to or greater than those used in the study are relatively common in most of the areas of the eastern Unites States where acid deposition is also a problem (Jacobson et al., 1980). Mechanisms of interaction are currently unknown and should be investigated.

Abiotic Effects. The particular mechanisms involved in what I have termed *indirect abiotic effects* are discussed from the standpoint of the difficulty of measurement by Lindberg and McLaughlin (this volume). However, this is a very important subject which deserves additional consideration here. The two major abiotic phenomena may be classed as processes which take place at the leaf surface, and which use the leaf surface as a reaction surface. The leaf surface itself may be either an active or a passive participant in these processes.

2.2.2 What Are the Wet Deposition–Gaseous Pollutant Interactions?

During significant portions of the growing season, the foliage of vegetation will be wetted by rain, fog, or dew in midlatitude temperate climates. During these periods one would expect leaf surfaces to become more efficient sinks for the dry deposition of gases and perhaps become more efficient at particle retention due to surface-tension effects. Unsworth and Fowler (1976), in describing SO_2 fluxes to a wheat canopy, defined surface resistance of the crop canopy. By their definition, if that crop surface resistance term should be zero, then the crop is a perfect sink for SO_2. During the course of their experimentation, they found that when the leaves of the crop were wet with dew, the surface resistance was very small (near or equal to zero). As SO_2 dissolved in the film of moisture on leaf surfaces, they found that the pH of the dew decreased, suggesting the potential for direct acid-type effects on wet surfaces of vegetation, until the pH-dependent SO_2 solubility limits the process.

Probably a more important role of wet foliage is the potential for increased rate of absorption of ambient SO_2 in the surface films, foliar uptake of that sulfur, and its subsequent availability to the metabolic pools within the plant. There is also a potential loss of sulfur from labile pools by leaching. Wet surface reactions may also be effective detoxification mechanisms, permitting the neutralization of H_2SO_4 by alkaline particulate matter on the leaf surface. Under these conditions, foliar uptake of sulfur may be as sulfur rather than though the direct introduction of SO_2 into the substomal cavity, as observed under dry conditions.

2.2.3 What Are the Wet Deposition–Particle Interactions?

Vaughn (1976) lists a number of extrinsic factors operating at the leaf surface which can significantly modify the subsequent sorption and translocation processes associated with particle deposition. Those factors include humidity, precipitation, light cycle, mass loading of the leaf, and absolute size of the aerosol particle. The leaf surface is involved in dynamic interaction with the atmosphere. Vaughn (1976) stated that mass loading in excess of ~1 $\mu g/cm^2$ of leaf surface area significantly inhibits translocation from the leaf to the root by impairing gas exchange. Data of Lindberg et al. (1979) suggest that if the 1-μg figure is accurate, quite large areas of deciduous forest canopies may be operating under conditions of impaired gas exchange during at least portions of the growing season, due to particle mass loading.

Total mass loading of the leaf surface is influenced by many factors. Research (Lindberg et al., 1979) indicates that the pattern of residual particles on leaf surfaces may be strongly influenced by the drainage patterns of wet deposition passing from the leaves as throughfall, with collection occurring along major veins and leaf margins (Fig. 2). An additional significant factor in particle retention during short periods in the growing season is the presence of excretions of biological origin, such as aphid honeydew, which results in the leaves becoming sticky and efficient collection surfaces.

Few data are available from which to draw conclusions regarding the total role of interactions between wet- and dry-deposited pollutants on the leaf surface. However, Hosker and Lindberg (1979) reviewed a number of considerations which must be mentioned in the context of the potential fate and effects of these pollutants in soil/plant systems. The interaction of particles with vegetation is similar in some respects to that with gases. Exceptions arise due to the following processes: (1) aerodynamic effects, (2) diffusion of small particles, (3) gravitational sedimentation of large particles (>1 μm), and (4) particle bounce leading to resuspension from the receptor surface. Physical entrapment by structural features of the leaf surface (leaf hairs, irregularities associated with vein patterns, hyphae of leaf surface microflora, etc.) may also be an important factor, but it is virtually uncharacterized.

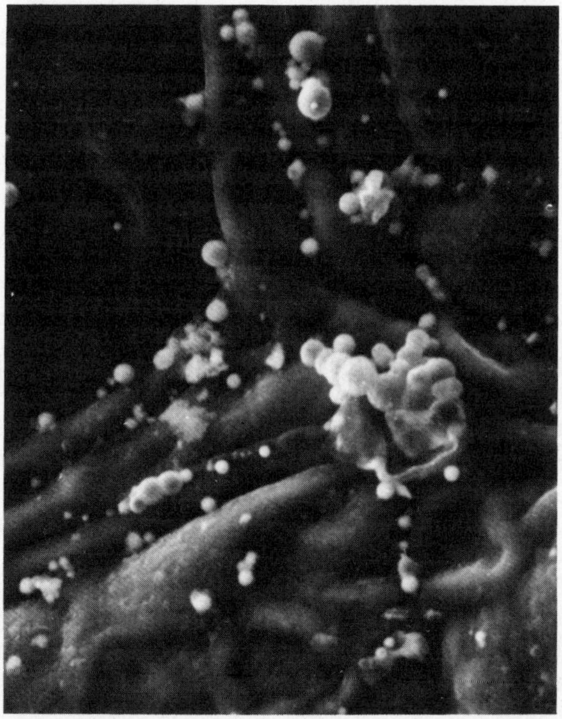

FIGURE 2. Scanning electron micrograph of chestnut oak leaf surface, illustrating the collection of fly-ash particles in an area of surface roughness at the juncture of two veins.

Chamberlain (1975) discussed the role that wet leaf surfaces play in dampening the particle bounce mentioned above, suggesting increased particle retention by dry deposition to previously wetted surfaces. However, impaction of water droplets during wet deposition may effectively dislodge and resuspend dry-deposited particles, resulting in secondary movement of the particles through the canopy and ultimately to the soil surface either in throughfall or stemflow.

Hosker and Lindberg (1979) also listed (1) particle size, density, and shape, (2) chemical composition, especially solubility in water or degree of hydration, and (3) electrical charge as properties likely to affect the interaction of particles with the leaf surface. Of these, the first two would appear to be of special significance in the potential interaction of liquids and particles on the leaf surface. Carlson et al. (1976), looking at wind reentrainment versus rainwash of lead aerosol particles deposited on plant leaves, found that wind reentrainment was a minor factor in comparison to rainwash. Simulated rainfall removed up to 95% of the topically applied lead (Pb). Of the Pb removed in rainwash, dissolution or suspension and subsequent runoff appeared to predominate over splashoff from leaf surfaces.

An additional factor influencing particle retention on leaf surfaces is cuticle weathering. As discussed earlier, rainwash of leaf surfaces results in weathering of epicuticular waxes which, in turn, causes increased wettability of the leaf surface. This increased wettability will result in increased retention time by the leaf of both moisture films and particles. Hosker and Lindberg (1979) suggested that this retention time is probably "the most critical parameter" in terms of effects. This may result in the potential for particles to interact with wet-deposited pollutants or dry-deposited gases.

Particles in residence on the leaf surface include soluble compounds which, when mobilized, may penetrate via natural flaws in the cuticle, as well as across cell membranes of trichomes and other specialized cells devoid of epicuticular waxes. Direct cuticular penetration is possible, with rates dependent on compound concentration, cuticle thickness, and the affinity of cuticular constituents for the solutes involved (Hosker and Lindberg, 1979). It is interesting to speculate that the cuticle could represent a significant sink for nonpolar compounds deposited on the leaf surface.

Particles on a leaf surface, in essence, increase the reactive area of that leaf surface, potentially increasing the sorptive capacity of the surface for gases, as well as serving as a source of soluble constituents to be leached by dew, fog, and rain. The roles that strong and weak acids in precipitation or organic acids derived from the vegetation play in the rates of solubilization of nutrients and/or toxic trace elements from dry-deposited particles on leaf surfaces are largely unknown. Lindberg (personal communication) has evidence for increased deposition of sulfate by such a mechanism. He found increased deposition of sulfate to be apparent through absorption of SO_2 by surface moisture followed by oxidation (possibly catalyzed by Mn) to SO_4^{2-}.

3. SUMMARY AND CONCLUSIONS

The major components of research on the effects of wet deposition of pollutants on vegetation have been discussed. The following sections will review what we do not know about wet deposition and its interactions with dry deposition in relation to effects on vegetation.

3.1 How Are Effects of Wet Deposition Caused?

Negative growth impacts and direct injury associated with wet deposition of pollutants appeared to be linked primarily with the associated hydrogen ion concentrations. However, the mechanisms of action involved in shifts in photosynthate allocation, for example, are not yet understood. It is quite clear that beneficial effects may accrue to plants as a result of wet deposition of nutrient elements and especially of nitrogen and sulfur. One of the most critical parameters we must identify, however, is the threshold point for a tradeoff between the negative effects of hydrogen ions and the positive effects of nitrogen and sulfur in wet deposition. Not only will the threshold need to be identified but, in addition, the effects of (1) species sensitivity, (2) N:S ratios in rain, soils, and tissue, (3) environmental variables (light, temperature, humidity, soil parameters), and (4) presence of other pollutants at the location of the threshold must be understood before predictive estimates can be made from some generalized model of plant response. Given that thresholds for direct injury of cultivars within the same species can vary by more than an order of magnitude, the task is substantial indeed.

We must also determine the magnitude of effects at the individual organism level which are necessary to produce effects also at the population, community, and ecosystem levels. Some evidence suggests that thresholds for visible injury may have little or no meaning in terms of significant effects at higher levels of system organization.

3.2 Is There Some Common Indicator of Plant Sensitivity to Wet Deposition?

Research has been singularly unsuccessful in the search for common denominators amongst responses, denominators which might be used to screen species for characteristics of sensitivity to wet deposition of pollutants. Leaf wettability holds promise because it can be measured easily as a function of the contact angle of water droplets with the leaf surface. It is admittedly an indirect measurement of sensitivity, but it does influence greatly two important variables, (1) contact or residence time and (2) total volume or mass of pollutant in contact with the leaf surface.

3.3 What Are the Effects of Within-Event Variations in the Composition of Wet Deposition?

Many workers suggest that substantial biological significance may be associated with short-term fluctuations in the composition of wet deposition. Sequential sampling within rain events has confirmed that short-term variations do occur within a storm, but it is difficult to establish regular patterns in such events. There is a need to evaluate the biological significance of within-event fluctuations of wet deposition. Can plants respond to short-term fluctuations, or do they effectively integrate the dose over the course of the event?

3.4 Can "Sensitive" Sites Be Identified Relative to Vegetation and Soil Microbial Effects of Wet Deposition?

Certain soil parameters (e.g., base exchange capacity, cation exchange capacity, pH, fertility levels, sesquioxide content) permit the identification of broad soil classifications which are potentially sensitive identifiers in terms of the effects of plants (Norton, 1980; McFee, 1980).

Research on the effects of wet deposition and wet–dry deposition interactions is perhaps at the same stage of development as that of gaseous pollutant effects 15–20 years ago. Recognition of the important role that wet deposition plays in pollutant transport, fate, and effects has only come within the last decade. As each of the areas is studied in greater detail, working hypotheses will give way to clearer understanding and, hopefully, to the predictive capability necessary to evaluate the long-range agricultural and environmental consequences of an energy-constrained, industrial society. Wet deposition effects research is at a point where demand for precise quantification through crash programs could lead to oversimplified analysis. Such an occurrence would be indeed unfortunate under the current adversary climate of environmental regulation. The problems dealt with and the products of research may have social implications so profound that only a scientific case of unprecedented credibility and completeness will be able to balance the complex costs of regulatory compliance with an acceptable level of environmental quality.

ACKNOWLEDGMENT

Research sponsored in part by Interagency Agreement DOE No. 40-740-78, EPA No. 79-D-X0533, and in part by U.S. Department of Energy Office of Health and Environmental Research under contract W-7405-eng-26 with the Union Carbide Corporation; Publication No. 1526, Environmental Sciences Division, Oak Ridge National Laboratory.

REFERENCES

Abrahamsen, G., and G. J. Dollard. Effects of acidic precipitation on forest vegetation and soil. In G. Howells (ed.) *Ecological Effects of Acid Precipitation* EPRI SOA77-403. Electric Power Research Institute, La Jolla, California, 1979.

Blakeman, J. P. The chemical environment of the leaf surface in relation to growth of pathogenic fungi. In T. F. Preece and C. H. Dickenson (eds.) *Ecology of Leaf Surface Microorganisms*, pp. 255–268, Academic, New York, 1971.

Blakeman, J. P., and P. Atkinson. Evidence for a spore germination inhibitor co-extracted with wax from leaves. In C. H. Dickenson and T. F. Preece (eds.) *Microbiology of Aerial Plant Surfaces*, pp. 441–449, Academic, New York, 1976.

Carlson, R. W., F. A. Bassaz, J. J. Stukel, and J. B. Wedding. Physiological effects, wind reentrainment, and rainwash of Pb aerosol particulates deposited on plant leaves. *Environ. Sci. and Technol.* **10**, 1139–1142 (1976).

Chamberlain, A. C. The movement of particles in plant communities. In J. L. Monteith (ed.) *Vegetation and the Atmosphere, Vol. I, Principles*, pp. 155–203, Academic, London, 1975.

Cronan, C. S. Consequences of sulfuric acid inputs to a forest soil. In D. S. Shriner, C. R. Richmond, and S. E. Lindberg (eds.) *Atmospheric Sulfur Deposition: Environmental Impact and Health Effects*, pp. 335–344, Ann Arbor Science Publishers, Ann Arbor, Michigan, 1980.

Dennison, R., B. Caldwell, B. Bormann, L. Elred, C. Swanberg, and S. Anderson. The effects of acid rain on nitrogen fixation in Washington coniferous forests. In Proc., First Int. Symp. on Acid Precipitation and the Forest Ecosystem. USDA Forest Service General Technical Report NE-23 (1976), pp. 933–949.

Dennison, W. C. Life in tall trees. *Sci. Am.* **228**(6), 74–80 (1973).

Evans, L. S., and D. M. Bozzone. Effect of buffered solutions and sulfate on vegetative and sexual development in gametophytes of *Pteridium aquilinum*. *Am. J. Bot.* **64**, 897–902 (1977).

Evans, L. S., N. F. Gmur, and F. DaCosta. Leaf surface and histological perturbations of leaves of *Phaseolus vulgaris* and *Helianthus annuus* after exposure to simulated acid rain. *Am. J. Bot.* **64**, 903–913 (1977).

Evans, L. S., N. F. Gmur, and F. DaCosta. Foliar responses of six clones of hybrid poplar to simulated acid rain. *Phytopathology* **68**, 847–856 (1978).

Ferenbaugh, R. W. Effects of simulated acid rain on *Phaseolus vulgaris* L. (Fabaceae). *Am. J. Bot.* **63**, 283–288 (1976).

Fokkema, N. J. Antagonism between fungal saprophytes and pathogens on aerial plant surfaces. In C. H. Dickenson and T. F. Preece (eds.) *Microbiology of Aerial Plant Surfaces*, pp. 487–506, Academic, New York, 1976.

Heagle, A. S., W. W. Heck, W. M. Knott, J. W. Johnston, E. P. Stahel, and E. B. Cowling. Responses of citrus to acidic rain from simulated SRM fuel exhaust mixtures and exhaust components, North Carolina State Univ., Department of Botany, Internal Report, 11 pp. Raleigh, North Carolina (1978).

Hindawi, I. J., J. A. Rea, and W. L. Griffis. Response of bush bean exposed to acid mist. *J. Bot.* **67**, 168–172 (1980).

Hoffman, W. A., S. E. Lindberg, and R. R. Turner. Precipitation acidity: The role of the forest canopy in acid exchange. *J. Environ. Qual.* **9**, 95–100 (1980).

Hosker, R. P., and S. E. Lindberg (eds.) Atmospheric Transport, Deposition, and Plant Assimilation of Airborne Gases and Particles, Proc., Second Ecology-Meteorology Workshop, Pellston, Michigan, Aug. 16–20, 1976. NOAA/ATDL, Oak Ridge, Tennessee, 27 pp. (1979).

Hudson, H. J. The development of the saprophytic fungal flora as leaves senesce and fall. In T. F. Preece and C. H. Dickenson (eds.) *Ecology of Leaf Surface Microorganisms*, pp. 447–455, Academic, New York, 1971.

Irving, P. M. Response of field-grown soybeans to acid precipitation alone and in combination with sulfur dioxide. Ph.D. Dissertation, Univ. of Wisconsin, Milwaukee (1979).

Jacobson, J. S. Experimental studies on the phytotoxicity of acidic precipitation: The United States experience. In T. C. Hutchinson and M. Havas (eds.) *Effects of Acid Precipitation on Terrestrial Ecosystems*, pp. 151–160, Plenum, New York, 1980a.

Jacobson, J. S. The influence of rainfall composition on the yield and quality of agricultural crops. In Proc. International Conference on the Ecological Impact of Acid Precipitation, Sandefjord, Norway, March 1980, pp. 11–14, (1980b).

Jacobson, J. S., and P. van Leuken. Effects of acid precipitation on vegetation. In Proc., Fourth Int. Clean Air Congress, Tokyo (1977), pp. 124–127.

Jacobson, J. S., J. Troiano, L. J. Colavito, L. I. Heller, and D. C. McCune. Polluted rain and plant growth. In T. Y. Toribara, M. W. Miller and P. E. Morrow (eds.) *Polluted Rain, 12th Rochester International Conference on Environmental Toxicity*, pp. 291–306, Plenum, New York, 1980.

Jones, K. Nitrogen fixing bacteria in the canopy of conifers in a temperate forest. In C. H. Dickenson and T. F. Preece (eds.) *Microbiology of Aerial Plant Surfaces*, pp. 451–463, Academic, New York, 1976.

Kratky, B. A., E. T. Kukunage, J. W. Hyline, and T. T. Kakano. Volcanic air pollution: Deleterious effects on tomatoes. *J. Environ. Qual.* **3**, 138–140 (1974).

Lang, D. S., S. V. Krupa, and D. S. Shriner. Injury to vegetation incited by sulfuric acid aerosols and acidic rain. In *Proc. 71st Ann. Mtg. of the Air Pollut. Control Assoc.*, Houston, Texas, 78-7.3, 1978, pp. 1–15.

Lindberg, S. E., R. C. Harriss, R. R. Turner, D. S. Shriner, and D. D. Huff. Mechanisms and rates of atmospheric deposition of selected trace elements and sulfate to a deciduous forest watershed, ORNL/TM-6674. Oak Ridge National Laboratory, Oak Ridge, Tennessee (1979), 519 pp.

Linskens, H. F. Uber die Anderung der Benetzbarkeit von Blattoberflachen und deren Ursache. *Planta* **41**, 40–51 (1952).

McFee, W. W. Sensitivity of soils to acidification by acid precipitation. In D. S. Shriner, C. R. Richmond, and S. E. Lindberg (eds.) *Atmospheric Sulfur Deposition: Environmental Impact and Health Effects*, pp. 495–506, Ann Arbor Science Publishers, Ann Arbor, Michigan, 1980.

McLaughlin, S. B., and D. S. Shriner. Allocation of resources to defense and repair. In J. G. Horsefall and E. B. Cowling (eds.) *Plant Disease: An Advanced Treatise, Volume V*, pp. 407–432, Academic, New York, 1980.

McLaughlin, S. B., D. S. Shriner, R. K. McConathy, and L. K. Mann. The effects of SO_2 dosage kinetics and exposure frequency on photosynthesis and transpiration of kidney beans (*Phaseolus vulgaris* L.). *Environ. Exp. Bot.* **19**, 179–191 (1979).

Martin, J. T., and B. E. Juniper. *The Cuticles of Plants*, 247 pp., St. Martins, New York, 1970.

Norton, S. A. Sensitivity of aquatic systems to atmospheric deposition. In D. S. Shriner, C. R. Richmond, and S. E. Lindberg (eds.) *Atmospheric Sulfur Deposition: Environmental Impact and Health Effects*, pp. 521–532, Ann Arbor Science Publishers, Ann Arbor, Michigan, 1980.

Purnell, T. J., and T. F. Preece. Effects of foliar washing on subsequent infection of leaves of swede (*Brassica napus*) by *Erysiphe cruciferarum*. *Physiol. Plant Pathol.* **1**, 123–132 (1971).

Rentschler, I. Significance of the wax structure in leaves for the sensitivity of plants to air pollutants. In *Proc. Third Int. Clean Air Congress, Dusseldorf*, ORNL-TR-4120 (translation). Oak Ridge National Laboratory, Oak Ridge, Tennessee (1973).

Reuss, J. O. Simulation of nutrient loss from soils due to rainfall acidity, EPA-600/3-78-053, 21.5. Environmental Protection Agency, Washington, D.C., (1978) 44 pp.

Saunders, P. J. W. Modification of the leaf surface and its environment by pollution. In T. F. Preece, and C. H. Dickenson (eds.) *Ecology of Leaf Surface Microorganisms*, pp. 81–90, Academic, New York, 1971.

Sheridan, R. P., and R. Rosenstreter. The effect of hydrogen ion concentrations in simulated rain on the moss *Tortula ruralis* (Hedw.) Sm. *Bryologist* **76**, 168–173 (1973).

Shriner, D. S. Effects of simulated rain acidified with sulfuric acid on host–parasite interactions. Ph.D. Thesis, North Carolina State Univ., Raleigh, North Carolina (1974), 79 pp.

Shriner, D. S. Effects of simulated acidic rain on host–parasite interactions in plant diseases. *Phytopathology* **68**, 213–218 (1978a).

Shriner, D. S. Interactions between acidic precipitation and SO_2 or O_3: Effects on plant response. *Phytopathol. News* **12**, 153 (1978b).

Shriner, D. S., and E. B. Cowling. Effects of rainfall acidification on plant pathogens. In T. C. Hutchinson and M. Havas (eds.) *Effects of Acid Precipitation on Terrestrial Ecosystems*, pp. 435–442, Plenum, New York, 1980.

Shriner, D. S., and J. W. Johnston. Effects of simulated, acidified rain on nodulation of leguminous plants by *Rhizobium* spp. *Environ. Exp. Bot.* **21**, 199–200 (1981).

Shriner, D. S., M. Decot, and E. B. Cowling. Simulated acid rain caused direct injury to vegetation. *Proc. Am. Phytopath. Soc.* **1**, 112 (1974).

Tukey, H. B., Jr. The leaching of substances from plants. *Ann. Rev. Plant Physiol.* **21**, 305–324 (1970).

Tukey, H. B., Jr. Regulation of plant growth by rain and mist. *Proc. Int. Plant Prop. Soc.* **25**, 403–406 (1975).

Tukey, H. B., Jr. Some effects of rain and mist on plants, with implications for acid precipitation. In T. C. Hutchinson and M. Havas (eds.) *Effects of Acid Precipitation on Terrestrial Ecosystems*, pp. 141–150, Plenum, New York, 1980.

Tukey, H. B., Jr., and J. V. Morgan. Injury to foliage and its effect upon the leaching of nutrients from above-ground plant parts. *Physiol. Plant.* **16**, 557–563 (1963).

Turner, J., and M. J. Lambert. Effects of atmospheric sulfur deposition on forest growth. In D. S. Shriner, C. R. Richmond and S. E. Lindberg (eds.) *Atmospheric Sulfur Deposition: Environmental Impact and Health Effects*, pp. 321–334, Ann Arbor Science Publishers, Ann Arbor, Michigan, 1980.

Unsworth, M. H., and D. Fowler. Field measurements of sulfur dioxide fluxes to wheat. In *Atmosphere—Surface Exchange of Particulate and Gaseous Pollutants, 1974*. ERDA Symposium Series 38, CONF-740921, National Technical Information Services, Springfield, Virginia (1976), pp. 342–353.

Vaughn, B. E. Suspended particle interactions and uptake in terrestrial plants. In *Atmosphere–Surface Exchange of Particulate and Gaseous Pollutants*, 1974. ERDA Symposium Series 38, CONF-740921, National Technical Information Services, Springfield, Virginia (1976), pp. 228–243.

Wood, T., and F. H. Bormann. The effects of an artificial acid mist upon the growth of *Betula alleghaniensis* Britt. *Environ. Pollut.* **7**, 259–268 (1974).

SYNTHESIS

TERRESTRIAL VEGETATION–AIR POLLUTANT INTERACTIONS: NONGASEOUS AIR POLLUTANTS

P. J. W. Saunders
Natural Environment Research Council
Swindon, Wilts, England

S. Godzik
Polish Academy of Sciences
Institute of Environmental Engineering
Zabrze, Poland

BACKGROUND

In this section existing knowledge is reviewed and priorities for research on the effects of wet and dry deposition of nongaseous pollutants upon terrestrial vegetation are identified. The topic covers coarse and fine particulate matter. Consideration is given to both direct and indirect (via soil) effects on vegetation and terrestrial ecosystems.

Aerosol pollution in industrial areas alters the spectrum of radiation and reduces the net total photon flux to plants by up to 5%. Such changes have been shown experimentally to have adverse effects upon plant growth, but no such effects have been detected in the field.

There is a substantial body of knowledge on the effects of gaseous pollutants upon vegetation. Nongaseous pollutants are infinitely more diverse in physical and chemical form and, although perhaps less significant in terms of direct injury, they may have many more subtle and long-term indirect effects through their influences upon soil processes.

Impacts

Two categories of impact have been identified:

1. *Regular and accidental emissions of particulates from single industrial sources. The local pathways and effects of such emissions, including those of coarse particles and those including toxic metals and fluoride, have been studied in some detail. There remains a need, however, to study the processes of uptake and toxicity of individual pollutants and mixtures in plants and soils.*
2. *Fine particulate matter is dispersed over a wide area resulting in elevated ambient levels of pollution. This category includes pollutants removed from the atmosphere by wet deposition in rain, dew, and so forth. Our perception of the hazards associated with this category of pollutants varies from region to region. Thus, concern is focused on acidic rain and ozone in North America, on acidic rain, ozone, and sulfur dioxide in western Europe, and on chemically inert particulates and sulfur dioxide in central Europe.*

Generally, somewhat different emphasis is placed upon single- and multiple-source pollution and on the problems of incident and ambient pollution in the regions of the world. This leads naturally to some differences in both approaches to and priorities for research.

With the exception of emissions in category 1 (above), there is little evidence of visible injury and loss of yield in agricultural vegetation exposed to ambient levels of fine particles and wet-deposited pollutants in the field. Nevertheless, there can be some changes in the chemical composition and physiological function of plants in such areas.

Long-term exposure to low levels of nongaseous pollutants is probably of greatest significance in forests and other natural and seminatural systems of marginal productivity. The accumulation of toxic pollutants (e.g., metals) and changes in soils (e.g., acidification due to acidic deposition) in such areas may not become apparent for 20 or more years. It is important, therefore, to take a long-term view of research and to concentrate on improving our understanding of the processes of deposition, uptake, and effects of particulate pollutants on plants in the context of plant communities and ecosystems. The results of such work will improve our ability to predict the effects of long-term exposures to low levels of pollution and of incidental exposures to industrial emissions, and will provide a rational basis for sound management of the environment.

Pollutants and Vegetation

Vegetation is an effective scavenger of gaseous and particulate pollutants and a sink for all forms of atmospheric pollution deposited at ground level. The size and shape of a particle influences its rate of deposition and may also influence its impaction, adherence to, and uptake by the plant. Indeed, there is some evidence to imply that aerosols of less than 1 μm in diameter may enter foliage directly via stomata.

The ultimate fate of a nongaseous pollutant and its impact upon vegetation depend upon

1. *the physical and chemical properties of the pollutant,*
2. *the structure and properties of the receptor, especially its surface,*
3. *the presence of other pollutants, biogenic materials, and so forth, and*
4. *the various influences of the environment upon deposition, interactions with other pollutants, and receptor condition.*

In most cases, no useful purpose is served by distinguishing between dry and wet deposition once the pollutant reaches the surface of the receptor. Wet deposition may, however, enhance certain processes in the absence of surface moisture. The principal receptors of nongaseous pollutants in terrestrial ecosystems are leaves and the soil. Attention is focused here mainly on the leaf, although reference is made to soils.

The Leaf Surface

Physical and chemical studies of pollution often regard the leaf as an inert surface on which deposition occurs at random and without interference by the plant. However, the surface of the leaf is structured and may bear trichomes or hairs and be covered with a waxy layer, depending on the species of the plant. It is also a habitat for microorganisms, including plant pathogens. Gas and moisture are exchanged with the atmosphere, mainly through the stomata, and sugars, organic acids, and other materials may exude or seep onto the surface.

A wettable, hairy leaf is especially efficient as a scavenger of dry particulate matter in the atmosphere and at retaining dry and wet deposits. There is evidence of nonrandom distribution of particles around hairs and other surface structures. The wax is eroded naturally by weathering processes, which are accelerated in the presence of gaseous and particulate pollutants, thus reducing the leaf's natural protection against injury from pollution, pests, and so forth. Experiments demonstrate the leaching of sugars and other natural compounds, as well as pollutants, from the leaf surface by precipitation. Concentrations of sulfur, nitrogen, and other elements in precipitation may be increased by a factor of two or even three after passing through plant canopies. This process may be accelerated by acidic precipitation.

The presence of films of moisture on the surface of the leaf enhances the absorption of gaseous pollutants and the retention and dissolution of soluble particulate matter. The chemical composition of such films is complex, but the ionic strength of the solution is probably low and the pH regime unstable. There is thus great potential for a wide variety of chemical interactions on the leaf surface as a prelude to uptake. Little is known about these interactions, which may have great influence upon the relationship between the effective dose of a nongaseous pollutant and its ultimate effect upon the plant.

There is a need for research on

1. *interactions between various pollutants,*
2. *interactions between pollutants and the leaf surface, including any resulting injury to the leaf surface,*
3. *effects of pollutants on the leaf surface microflora and fauna, especially pests and pathogens of plants,*
4. *the mobility of soluble pollutants in relation to their uptake through the cuticle and stomata, and*
5. *the substantiation of reports of relatively minor emissions of pollutants and their derivatives in gaseous and particulate form (e.g., H_2S, SO_2, HF, Pb) from the leaf surface.*

These studies require the development of sophisticated equipment and techniques for microphysical and microchemical investigations of the leaf surface.

Most evidence of visible injury to plants by acidic rain and particulate matter is obtained by experiments in the laboratory using simulated deposits, often under highly artificial conditions. There is no supporting evidence of such effects from observations in the field, although changes in chemical composition are reported. It is possible that the discrepancies between laboratory and field studies are the result of poor simulation of deposition processes. More attention should be paid to this question, including the accurate reproduction of rain events and soil conditions.

The Soil

Soil is also a receptor of dry and wet deposits of nongaseous pollutants. Several studies in central and western Europe and in North America have revealed the role of vegetation as a scavenger and integrator of atmospheric pollutants. Their derivatives are washed from plant canopies by precipitation, thus augmenting direct inputs. The spatial distribution of such leachates beneath a forest canopy is especially complex.

Forest soils are naturally acidic and it is doubtful that inputs from the atmosphere will affect most soils and, hence, tree growth to a marked degree in the short term. The longer-term hazards to base-poor soils, however, should not be overlooked. Despite claims to the contrary, surprisingly little is known about the processes involved in the transfer of acidic precipitation through the soil into ground and surface waters. These processes ought to be studied, if only to quantify the net contribution of acidic rain to the acidification of lakes and rivers in North America and northern Europe.

Metals and fluorides are also deposited in soils. Most are locked into the soil by chemical processes and are not available to plants. Metals tend to accumulate in the litter and upper horizons where, at high concentrations, they may inhibit decomposition. Fluoride is also not taken up by plants, except to a limited extent in soils of low calcium content. There is a need to determine effective doses for individual metals in soils around specific industrial sites.

The Ecosystem

Few studies have successfully identified significant effects of nongaseous pollutants upon populations, communities, and ecosystems. Most of these studies have focused on industrialized areas which are polluted heavily by deposits of chemically inert as well as active dusts. Reductions in growth, yield, flowering, fruit production, and so forth have been recorded, as have differences in the sensitivities of individual species. The latter result in changes in the composition and structure of natural ecosystems. Some plants may become predisposed to attacks by pests and pathogens.

Similar changes have not been observed in communities and ecosystems exposed to typical, ambient conditions of fine particulate pollution and of acidic rain. Field studies have also been unsuccessful, perhaps because of inadequate design for taking into account natural variability and the long time needed to observe any effects. Consequently, the Scandinavian approach of long-term field monitoring of wet and dry deposition and of biological change at selected sites, is correct.

The results of short-term laboratory and greenhouse tests on in-

dividual species exposed to acidic rain should not be extrapolated to communities and ecosystems unless the experimental and field conditions are directly comparable. This does not mean that short-term work on processes should be discontinued; the results give a valuable insight into the mechanisms of uptake and toxicity.

Evidence from studies with gaseous and metal pollutants indicates that some plants exhibit a wide range of sensitivity within a population, and that this property allows the natural selection of tolerant individuals. It follows that natural selection may occur and probably has occurred in many ecosystems exposed to relatively low levels of pollution. This process deserves further research to enhance our understanding of the capacity of terrestrial systems to withstand stress and to identify prospects for the breeding of tolerant strains of plants for use in agriculture and so forth.

CONCLUSION

It is impossible to define clearly what constitutes an effective dose for any nongaseous pollutant–plant interaction. The leaf surface is clearly an important factor and deserves intensive investigation, if only to understand the processes involved in the uptake of pollutants. Generally, it appears that more emphasis should be given to longer term research on processes which will enhance our predictive capabilities, and less effort should be devoted to ill-advised, short-term studies of visible injury and loss of yield, notably in agriculture and forestry. Field studies should focus on the longer-term implications of pollution, especially for forestry, grasslands, and so forth, on soils of marginal productivity. Such areas have less immediate economic value but are of immense value in terms of natural resources for conservation, future exploitation, and recreation.

PART SEVEN

SOIL–AIR POLLUTANT INTERACTIONS

Soil acts as a source and a sink for air pollutants. Acidification of soils due to pollutant impaction has become a serious concern. Several studies have shown alterations in nutrient cycling in areas impacted by pollutant load. This in turn has important implications in long-term ecosystem function and character. In describing soil–air pollutant interactions, one needs to consider various mechanisms involved in their deposition.

The section considers the interactions from the points of view of both physical and biological sciences. Additionally, evaluation of current studies on point-source emission impacts on soils are discussed, and the future needs of such studies are examined.

Soil–Air Pollutant Interactions

T. CRAIG WEIDENSAUL AND JAMES R. McCLENAHEN
Laboratory for Environmental Studies
Ohio State University
Ohio Agricultural Research and Development Center
Wooster, Ohio

1. INTRODUCTION

There are several literature reviews on the effects of individual air pollutants on vegetation, but few on the subject of air pollutant–soil interactions. This review should not be considered exhaustive, as it was our intention only to review pertinent publications that, it is hoped, represent the philosophy as well as the state of knowledge of pollutant–soil interactions. There are not many accounts of actual pollutant effects on soils and *vice-versa*, as most report the influences of various soil parameters on pollutant sorption (absorption and/or adsorption). Several papers allude to the capabilities of various soils to reduce atmospheric levels of various pollutants. The concept of soil as an effective pollutant sink is no doubt most appropriately viewed on a regional or on a global basis (Inman and Ingersoll, 1971).

2. GENERAL BACKGROUND

Virtually any element or compound, if in an undesirable abundance, can be classified as an air pollutant, but those considered important include SO_2, HS, H_2SO_4, mercaptans, dimethyl sulfide, dimethyl disulfide, CO, N_2O, NO, NO_2, hydrocarbons such as C_2H_4, Cl_2, O_3, NH_3, peroxyacetyl nitrate (PAN), peroxybenzyl nitrate (PBN), peroxypropionyl nitrate (PPN), HCl, HF, and transition metals such as copper, chromium, zinc, lead, nickel, and cadmium. Of these materials, a relatively small number have been investigated with respect to soils. These include nitrous and nitric oxides, nitrogen dioxide, hydrogen sulfide, carbon disulfide, sulfur dioxide, ethylene, and carbon monoxide. Very little information is available regarding ozone. Studies cited in this paper involve these pollutants and their relationships to the chemistry and microbiology of soils. There is a great lack of information regarding soil interactions with the oxidants chlorine, ozone, PPN, PBN, PAN, and phosphorus-containing gases. The heavy-metal group will not be considered in this discussion since it has been covered by Elseewi, Page, and Straughan (this volume).

The general consensus of most investigators is that there are apparently no drastic or potentially catastrophic short-term pollutant effects on soils, but there is a lack of information on the effects of long-term pollutant exposures on temporary and permanent changes in soils. It is not known which or what proportions of the atmospheric gases react with soil components or what are the rates of reaction if they occur. Another void in our knowledge is the effects of air pollutants on soil chemistry and microbial activity and whether or not microbes affect pollutant detoxification. Similarly, there is very little information regarding effects on nitrogen fixation.

Much has been published on the subject of precipitation acidity effects on vegetation and on cultivated agricultural soils. Much has been written on theoretical effects of acid deposition on soils and ecosystems, but conclusive hypothesis testing concerning naturally occurring precipitation poses difficulties that have yet to be satisfactorily resolved. Few accounts involve sound, complete data; many are somewhat contrived or unrealistic; and the conclusions drawn are frequently tenuous or hypothetical.

Studies regarding soil–pollutant interactions must consider many biotic, physical, and chemical factors that influence this relationship. These are air moisture, soil temperature, soil type as related to its buffering and exchange capacities, microorganism influences, soil organic matter, oxygen and other gases in the atmosphere and soils, transition metal ions, duration of exposure of soils to pollutants at various concentrations, micrometeorological conditions, and the acid-titratable basicity of soils.

The bulk of the literature is occupied by soil uptake of air pollutants. In a review of this subject, Bohn (1972) reported that soils generally sorb organic gases faster and in greater amounts as their molecular weights increase. He also noted that absorption rates increase as other functional groups, such as nitrogen, phosphorus, oxygen and sulfur are attached to organic gases. The lower-molecular-weight and less-substituted organic gases depend on the buildup of microbial populations for sorption. It was concluded that organic gases generally involve microbial influences for uptake by soils and that the absorption of inorganic pollutants by soils is primarily attributed to chemical and/or physical parameters.

The above generalizations are substantiated by the work of Abeles et al. (1971), in which these investigators examined the nature of sorption of ethylene, nitrogen dioxide, and sulfur dioxide by soils. It was pointed out that there are only two known mechanisms for ethylene degradation in the atmosphere. These are ethylene oxidation by ozone and a photochemical reaction with nitrogen oxides. Ethylene sorption did not occur in the absence of oxygen, and C_2H_4 uptake was inhibited in soils sterilized by ethylene oxide or autoclaving. On the other hand, the initial SO_2 level of 100 ppm (266,484 $\mu g/m^3$) was reduced to 8 ppm (21,319 $\mu g/m^3$) in 15 min in the presence of autoclaved soil. Abeles et al. (1971) also observed a similar phenomenon with NO_2, although the NO_2 sorption rate was slower than that of SO_2. The initial NO_2 level of 100 ppm (191,271 $\mu g/m^3$) dropped to 3 ppm (5,741 $\mu g/m^3$) after 24 h in unsterile soil and to 13 ppm (24,878 $\mu g/m^3$) in autoclaved soil.

These results suggest that most SO_2 and NO_2 is removed by soil by some chemical reaction since sorption rates for each were quite similar in sterile and unsterile soils. Other researchers found that NO_2 uptake by soils is actually retarded by the nitrates already in soil (Blackmer and Bremner, 1976). One might assume that the absorption of NO_x by soils is proportional to the acid-titratable basicity of soils. However, Eberhardt (1972) has shown that the NO_x sorption capacity of soils greatly exceeds their acid-titratable basicity because of additional NO_x retention by chemisorption and physisorption. He observed no differences between NO and NO_2 reactions. The end product of the sorption reaction was NO_3–N in both cases. He further observed that additional nitrate increases in soil due to NO_x sorption were correlated with decreases in pH, and moisture tended to increase the NO_x sorption capacity. As Bohn (1972) suggested, organic matter seems to play an important role in the sorption of nitrogen oxides. This was confirmed by Eberhardt (1972) who determined that nitrates formed in soil remain as nitrates in the absence of organic matter. When 8% barley straw was added under aerobic conditions, a 60% gaseous loss of nitrates occurred. Additions of $(NH_4)_2SO_4$ or KNO_3 fertilizers with the straw increased the total nitrogen loss from nitrates to 90%. He further reported that NO_x had no influence on ammonia immobilization. These results indicate that bare soils with a minimum of organic matter are probably the most efficient in sorbing and retaining nitrogen oxide gases, and the net result for soil as an atmospheric NO_x sink may be somewhat limited.

Apparently, only a small amount of SO_2 is sorbed by acid soils, but more basic soils sorb SO_2 fairly rapidly and convert it to H_2SO_4 at moderate soil moisture contents (Bohn, 1972). It has been postulated by Miyamoto et al. (1974) that calcareous soils could be useful for removing SO_2 from waste gases, particularly when SO_2 concentrations are low. In agricultural soils, acidity is easily controlled by liming, and therefore SO_2 sorption involves conversion to sulfates, which contribute to good plant nutrition. Conversely, hydrogen sulfide sorption is slower in soils with a high moisture content. This may be due to much of the soil pore space being filled with water.

Although the forms of the absorption curves for NO_2 and SO_2 are similar, soil apparently removes a greater percentage of SO_2 from the air than it does NO_2 (Ferenbaugh et al., 1979). They introduced SO_2 and NO_2 simultaneously into an air stream to which soils were exposed. The sorption characteristics for each were unaffected by the other's presence. This apparent lack of competition for absorption sites suggests that the uptake mechanisms for the two gases are different or that the sorption capacity exceeded the input. On a loam soil at pH 7.2, Labeda and Alexander (1978) observed that continuous exposure to 5 ppm (9,569 $\mu g/m^3$) NO_2 inhibited ammonium disappearance, caused greater rates of nitrate formation, and resulted in nitrate accumulation. The same soil was not affected by continuous exposure to 0.5 ppm (1,332 $\mu g/m^3$) or higher levels of SO_2 for brief periods.

Although Eberhardt (1972) indicated that organic matter influenced nitrogen losses from soil, Nelson and Bremner (1970) determined that other binding sites, such as minerals and metallic cations, do not promote decomposition of nitrates in soils. Nor do they play a significant role in processes leading to gaseous loss of nitrogen from soils due to chemodenitrification.

Just as soil moisture is important in gaseous sorption, water is also important in the air itself. Soil sorption of NO_2 from dry air reached a maximum within 2.5 min in air-dry soils (Prather et al., 1973a), and the sorption capacity increased as the NO_2 concentration increased. However, the sorption capacity was essentially unaffected by the NO_2 concentration in moist air. It was found that under dry conditions NO_2 absorption is slow and is related to soil surface area but not to acid-titratable basicity. The same relationships found for NO_2 sorption by soils (Prather et al., 1973b) hold true for nitric oxide, but the NO_2 sorption capacity is approximately twice that of NO. The sorption of NO by calcareous soils reached a maximum within 15 min, regardless of soil type, aggregate size, or NO concentration (Prather et al., 1973b). These investigators assumed that, due to the short time involved and large amounts sorbed, the NO sorption process is physiochemical. It was concluded by Prather et al. (1973a) that calcareous soils could be used as NO_x sorbents for waste gas treatments, but the sorption capacity is not renewable.

Although certain environmental factors influence the uptake of various gases by soils, Smith and Mayfield (1978) concluded that because 95% of NO_2 administered to soils was sorbed after a 15-min exposure period, environmental factors, such as temperature, oxygen, and soil moisture will probably exert little influence on its long-term sorption even though decreases in soil moisture result in an increase in absorbance.

Soil differs in its capacity to sorb different gases. For example, Smith et al. (1973) observed that soils have a capacity to sorb substantial amounts of SO_2, H_2S, and methyl mercaptan (CH_3SH), but smaller amounts of CO, acetylene (C_2H_2), and ethylene (C_2H_4).

Soil type influences the magnitude of SO_2 sorption. Terraglio and Manganelli (1966) observed that in calcareous soils SO_2 sorption increased markedly in the presence of moisture in either the airstream or soil. However, the sorption rate was less in neutral and acidic soils under the same moisture regimes. Yee et al. (1975) feel that, although such increased sorption rates may be due to the moderate water solubility of SO_2, the effect of water was probably to hasten SO_2 oxidation in the reaction of its oxidation process with soil bases.

Just as there are variations in gaseous uptake rates in soils, they differ in their susceptibility to changes induced by acidic precipitation. Malmer (1976) suggests that soils most affected would be acid soils with hydrated aluminum compounds present because of their low buffering capacity. It was postulated by Mayer and Ulrich (1977) that more than half of the nitrogen and sulfur acids deposited by precipitation are neutralized by basic pollutants (e.g., CaO). Similarly, sulfur and nitrogen filtered from the atmosphere by trees are subsequently washed out by rain and can contribute to acidity but are mostly neutralized by basic air contaminants or plant metabolites. In discussing the effects of acidic precipitation on soil pH and base saturation of exchange sites, McFee et al. (1976) pointed out that, due to the resistance of most soil systems to pH changes, there is a small likelihood that rapid soil degradation due to acidic precipitation will occur.

One of the more interesting soil–pollutant interactions involves CO uptake. Carbon monoxide is apparently removed from the air at a linear rate by dry soil, but in moist soil the rate is exponential (Heichel, 1973). The optimum soil moisture of forest soils for CO sorption was found to be near 55% (w/w). Carbon monoxide uptake by field soil increased up to a soil moisture content of approximately 5%, but did not increase further as moisture was increased to 15%. These data indicate that CO uptake by forest soils is greater than in field soils. Stirring or tilling soil increased the CO absorption. However, slowly mixing air over forest soil decreased CO sorption by as much as 40–50%. Other investigators (Inman and Ingersoll, 1971) indicated that sterile soil does not favor CO removal. Nonsterile soil exposed to 120 ppm (139,820 $\mu g/m^3$) CO in air reduced the CO level to almost zero (in chamber air) within 3 h. The activity of a variety of California soils was related to high organic matter and low pH, and many varied in their capacity to remove CO. Inman and Ingersoll (1971) observed that CO uptake by soils could be inhibited by treating soils with 10% sodium chloride. Soil incubation under anaerobic conditions for 5 days prior to testing also inhibited uptake. This strongly suggests that microorganisms are responsible for CO absorption. It was concluded that most soils tested had great capacities to remove CO from the atmosphere (Inman et al., 1971). The rate of entry of CO into soil is opposed by a given diffusive resistance of soil that varies with soil type, water content, air turbulence, and, of course, the CO level (Heichel, 1973). Heichel also noted a CO removal from soil which increased in rate between CO concentrations of 2 (2,330 $\mu g/m^3$) and 40 ppm (44,606 $\mu g/m^3$). The maximum rate of sorption was 0.20 mg CO/dm^2 h.

Temperature, apparently, is also critical for CO sorption. Uptake rates fall off sharply on either side of a peak at 30°C (Inman and Ingersoll, 1971). These same investigators noted that when sterile soil was amended with nonsterile soil (2.8 : 1), the uptake activity for CO was the same as for naturally inoculated soil after 36 days. Inman et al. (1971) observed the average sorption rate of soils tested to be 8.44 mg of CO/h m^2, equivalent to approximately 191.1 metric tons/yr m^2.

Not only do different variables in air and soil influence the rate of uptake of various gases, but different gases can be sorbed at different rates under the same conditions. For example, Abeles et al. (1971) observed that soil removes SO_2 and NO_2 faster than it does ethylene. Most SO_2 is removed within 15 min in the presence of soil, and NO_2 requires approximately 24 h for a similar amount. Alway et al. (1937) had data that suggested an approximate rate of sorption by bare soil in kilograms of sulfur per hectare. The method of calculating this rate is $0.55 \times SO_2$ concentration in micrograms per cubic meter. This estimate is similar to the sorption capacity of leaves for SO_2.

While studying flux measurements of atmospheric ozone over land and water, Aldax (1969) concluded that soils are effective reductants of O_3. Macdowall (1974) predicted that diffusivity within the plant canopy allows the strong sorbent capacity of soil to shift the O_3 concentration gradient downward and thus play a protective role. Turner et al. (1973) observed that autoclaving soil decreased its resistance to ozone removal, and therefore assumed that microbes were not involved. However, they did not rule out the possibility that microorganisms may provide O_3 reduction sites. Turner et al. concluded that bare soil can be an important sink for atmospheric O_3; that dry, cultivated soil removes more O_3 than wet, compacted soil; that the resistance to O_3 uptake by a dry, cultivated field soil seems to be due largely to the resistance of the boundary layer of still air near its surface; and that the decrease in atmospheric O_3 after sunset can be attributed in part to O_3 removal by soils.

In addition to the previously listed factors influencing direct soil uptake of various pollutants, vegetation can transfer large quantities of sorbed airborne pollutants to the soil via throughfall and stem flow and in litter. Wainwright (1978) studied the distribution of sulfur oxidation products in soils and observed that the acidity of polluted soils was greater beneath the canopy of *Acer pseudoplatanus* than 2 m away; at an unpolluted site, there was no difference.

3. FATE OF POLLUTANTS IN SOIL

Fluorides, which are among the most abundant compounds in the crust of the earth, accumulate in soils. Rao and Pal (1978) observed that an increase in fluoride concentration in litter and soil was paralleled by an accumulation of organic matter in the surface horizons. They suggested that increased fluoride levels in soil reduced microorganism activity, and hence the rate of organic matter decomposition decreased. Their reported increase in soil organic matter near the fluoride source could also be interpreted as being due to more rapid decomposition and incorporation of surface litter; however, they provided no information on surface organic matter per se or on the type of vegetative cover, soils, and so forth.

Both gaseous and particulate fluorides accumulate in soils. Thompson et al. (1979) observed that damage to vegetation and fluoride levels in soil humus were inversely related to the distances from a phosphorus plant. Available fluorides ranged between 3.8 and 58 ppm (w/w), and total fluorides ranged between 16 and 908 ppm in soils near the phosphorus plant. Few plants accumulate significant amounts of fluorides via root absorption.

In a survey by McClenahen and Weidensaul (1977), maximum soil fluoride levels also occurred nearest a large aluminum smelter and in a downwind direction. Where there was low atmospheric fluoride exposure, fluorides increased with soil depth, but the highest levels were in the surface (0–2 in.) layer closer to the alumina reduction facility. They concluded that the fluoride source affected the distribution of soil fluorides by (1) increasing the general fluoride content of nearby soils, and (2) causing highest fluoride levels near the surface. This latter finding confirms the general contention that soils tend to quickly immobilize fluorides, although immobilization is generally enhanced by high pH and increasing calcium, phosphorus, and clay content (Bohn, 1972; Jacobson et al., 1966).

Bohn (1972) reported in his literature review that as the oxidation state of nitrogen oxides decreases, these compounds are progressively less reactive chemically, unless water soluble. Nitric oxide sorption is enhanced by transition metal ions in soil. The rate of nitrous oxide (N_2O) sorption is slower than that for the more highly oxidized states, NO and NO_2. Nitrous oxide uptake by soils was found to be promoted by anaerobic conditions and organic substances that promote growth of soil microorganisms. Bohn also mentions that soil can rapidly inactivate fluorides, the degree of inactivation increasing with increasing soil alkalinity.

Not only do soils sorb NO_x gases, but due to chemical reaction in the soil, some nitrogenous gases are released as well. Soils released N_2, N_2O, O, NO, and some NO_2, but compared with N_2, the amount of oxides released is small and mostly N_2O (Bohn, 1972).

Sulfur dioxide sorbed by nonsterile soil was found to be rapidly oxidized to SO_4^{2-}–S. Similar results in gamma-irradiated soil were observed, but oxidation of SO_2 to SO_4^{2-} was fairly slow in autoclaved soil.

Ghiorse and Alexander (1976) noted that more nitrogen was ultimately recovered as NO_3^-–N in nonsterile soils than had been introduced originally as NO_2. The excess nitrogen, according to the investigators, probably resulted from the stimulation of soil nitrifiers when exposed to NO_2. Their data suggest that nitrifying microbes are responsible for nitrite oxidation, and that it was plausible that microorganisms may passively sorb SO_2 by providing organic matter that reacts with SO_2. They postulated that ignited soil sorbed more SO_2 (subsequently oxidized to SO_4^{2-}) than air-dried soil because there were more sorption sites. When NO_2 was added to soil at the rate of 51 μg N/g soil, the nitrite form was oxidized logarithmically, but the final microbe count was too low to account for the amount of nitrite metabolized. Discrepancies between the predicted bacterial numbers and nitrifying activity were also noted in soils amended with nitrite. Ghiorse and Alexander (1977) concluded that soils may contain large numbers of unrecognized or poorly understood nitrite-oxidizing populations. They based this partly on the fact that when 106 μg N (as NO_2)/g soil was administered, the nitrite form was oxidized, but the number of autotrophic nitrite oxidizers did not increase.

When a neutral loam soil was fumigated continuously with 5 ppm (9,569 $\mu g/m^3$) NO_2, ammonium disappearance was inhibited and nitrites accumulated.

Oxidation of elemental sulfur in soils was investigated by Nor and Tabatabai (1977), who observed that after an incubation period of 70 days at 30°C, elemental sulfur oxidation to sulfate in some Iowa soils ranged from 39 to 75 mg S/g soil. Oxidation rates were more rapid in alkaline than in acid soils and lower in air-dried than in moist field soils.

Studies by Nyborg et al. (1977) showed that the sulfur content of snow is so low in Alberta that the snowpack has a deposition of less than 1 kg S/ha, even downwind from large SO_2 sources. Rain is seldom abnormally acidic, but contributes about four times as much sulfur as snow near SO_2 sources. The Canadian soils sorb a considerable amount of sulfur from these emissions, but much is in nonsulfate form. Acidification of the Canadian soils is thought to occur very slowly as a result of SO_2 deposition, and the rate is estimated at approximately one pH unit decrease in 10–20 yr. At this rate of acidification, even though the authors mention a slow change in pH, such soils will be unable to support the present vegetation within 60 yrs.

In studying the deposition rate of NO, Prather et al. (1973b) noted that sorbed NO oxidized almost completely to NO_3^- and caused some reduction of organic nitrogen. The percentage change in organic nitrogen was high only in low-capacity soils, however. Other investigators (Smith and Mayfield, 1978) observed relatively high concentrations of soil nitrite nitrogen after exposure to NO_2. This suggests that NO_2 reacts with cations and organic matter and existed in interparticle spaces in gaseous form, or dissolved in water to form HNO_2 because high levels of NO_3^-–N were not immediately detected.

Microorganisms have been associated primarily with soil uptake of organic-type pollutants rather than with materials such as SO_x and NO_x. Abeles et al. (1971) suggested that microorganisms are the major sink for C_2H_4 and other hydrocarbons. Soil (or its components) is very effective in reducing atmospheric levels of C_2H_4.

Whereas SO_2 and NO_2 uptake have been considered in terms of physical or chemical absorption, N_2O uptake by soils is thought to be a microbial process involving the reduction of N_2O to N (Blackmer and Bremner, 1976). Nitrous oxide uptake is promoted by anaerobic conditions and by organic substances that promote the growth of soil microorganisms. Soil uptake of N_2O, C_2H_4, and the mercaptans is thought to be dependent upon microbial populations (Bohn, 1972). Bohn also reports that soil microbes are the component which sorbs unburned hydrocarbons, automobile exhaust, and C_2H_4 from air. Carbon monoxide removal by soil, although relatively slow, is extremely important because its transformation to CO_2 by soil microorganisms is the only known mechanism by which this pollutant is removed from the earth's atmosphere.

Inman and Ingersoll (1971) observed in their studies that all the CO in the air (120 ppm or 139,820 $\mu g/m^3$) could be removed by nonsterile soil in 3 h. Sterile soil removed less CO. Their tests with cycloheximide, high-temperature, anaerobic conditions, antibiotics, and high salinity all decreased CO uptake and assumed that CO sorption was due to biological mechanisms. These same investigators found 16 fungi capable of sorbing CO from the atmosphere. They included *Penicillium digitatum*, *Penicillium restrictum*, four species of *Aspergillus*, *Mucor hiemalis*, *Haplosporangium parvum*, and *Mortierella vesiculata*.

Compared with CO and C_2H_4, SO_2 was rapidly absorbed from air by both sterile and nonsterile soils. The investigators (Ghiorse and Alexander, 1977) assumed that microorganisms were not directly involved in chemical transformations after sorption. Sulfates formed from SO_2 in both sterile and nonsterile soils. It is felt that the role of soil microorganisms in the fate of NO_2 is in the conversion of nitrite to nitrate.

Although microorganisms are not thought to be instrumental in soil sorption of SO_2, Wainwright's results indicate that microbial sulfur oxidation is actively occurring in soils exposed to heavy atmospheric pollution. Soils exposed to SO_2 pollution in excess of 0.05 ppm (125 $\mu g/m^3$) contained greater numbers of thiobacilli and sulfur-oxidizing fungi than did similar, but unpolluted soils. Polluted soils contained more fungi but fewer heterotrophic bacteria (Wainwright, 1979).

Turner et al. (1973) determined that microbial involvement is not required for the sorption of ozone by soils, but they have not ruled out the possibility that microorganisms may provide active reduction sites. Similarly, Smith and Mayfield (1978) reported that microorganisms are not involved in the oxidation of NO_2^-–N to NO_3^-–N. In fact, NO_2 at a concentration of 100 $\mu g/mL$ is bacteriocidal for both aerobic and anaerobic bacteria. Nitrogen dioxide severely inhibited the nitrogenase enzyme system at concentrations that did not inhibit the growth of nitrogen-fixing microorganisms. They, like others, found that microorganisms did not affect the rate of NO_2 sorption by soils.

Substantiating the work of others, Smith et al. (1973) demonstrated in studies with steam-sterilized soils that microorganisms are responsible for sorption of CO, C_2H_2, and C_2H_4 in moist soils, but play no significant part in the sorption of SO_2, H_2S, or CH_3SH.

The question of gas influence on extractable cations has been raised. Smith and Mayfield (1978) stated that it is unlikely that atmospheric levels of NO_2 commonly encountered will significantly reduce the fertility of soil in exposed areas. Supporting this contention, but for SO_2, Labeda and Alexander (1978) note that soluble levels of K, Mg, Ca, Mn, Fe, and Al were not increased as a result of soil treatment with SO_2. Continuous fumigations of acid soil (pH 5.0) with 10 ppm SO_2 (26,648 $\mu g/m^3$) decreased nitrification, but continuous fumigation with 1 ppm (2,665 $\mu g/m^3$) increased the quantity of soluble Mn and Fe.

The highest concentration of sulfate and tetrathionate were found in polluted soils while more thiosulfate was found in unpolluted soils (Wainwright, 1979). Johnson and Henderson (1979) reported that highly weathered surface soils in a mixed deciduous forest had no detectable sorbed sulfate, but were a substantial reserve of water-soluble SO_4^{2-}–S. Surface soils apparently were very capable of sorbing more SO_4^{2-}–S into forms not extractable with water or phosphate solution. The high proportion of water-soluble SO_4^{2-} appears to play an imporant role in determining SO_4^{2-} concentrations in headwater streams, as during peak storm flow periods, SO_4^{2-} concentrations increased in headwater streams. The investigators hypothesized that this, although based on circumstantial evidence, is due to the leaching of SO_4^{2-}–S from the surface horizons.

It has been postulated (Labeda and Alexander, 1978) that nitrification in certain soils could be inhibited in areas acutely polluted with SO_2 and NO_2. Other studies in Iowa (Blackmer and Bremner, 1976) reported that surface soils demonstrated a greater capacity for N_2O uptake under conditions favorable to denitrification of nitrates than their capacity to release NO_2.

Concern has arisen recently in the United States and Canada regarding the acidification of atmospheric deposition and the related potential effects on terrestrial and aquatic ecosystems. Long-term monitoring of precipitation quality has been conducted in the Scandinavian countries. Similar monitoring programs emphasizing deposition quality have been initiated in Canada and the United States, in which several factors other than acidity are considered. Much of the emphasis placed on effects of atmospheric deposition on various ecosystems, particularly terrestrial ecosystems, is based on potentially harmful effects. Little attention has been given to the beneficial effects of precipitation components. For example, there are many elements and ions present in precipitation that are essential for growth and development of normal healthy green plants. It should be pointed out that in terms of nitrate and ammonia nitrogen contributions, more than 10% of the nitrogen required for corn production is contributed as soluble N via precipitation. Similarly, about 20% of the available nitrogen required by the most demanding coniferous trees is also in precipitation. Most people recognize that normal liming of agricultural soils easily offsets any acidity contributed by precipitation. Frink and Voigt (1976) stated that the acidity produced by biological cycling of nitrogen and sulfur compounds in forest stands exceeds that found in rain, and, unless a substantial increase in the acidity of precipitation occurs, it should pose no serious threat to forest soils of the northeastern United States. Similarly, Johnson and Cole (1976) studied sulfate mobility in soils in western Washington and observed that in a second-growth Douglas fir ecosystem there was the ability to buffer and moderate cation losses due to heavy loads of sulfuric acid over short periods. These researchers suggested that sulfate interacts strongly with soils that have appreciable sesquioxide contents. Characteristics of this interaction must be understood and employed in evaluating and modeling cation losses due to strong acid deposition.

The fate of hydrogen and sulfur ions associated with precipitation have been studied (Mayer and Ulrich, 1977). Soil acidification by precipitation was greatest within 1 or 2 cm of the surface. Therefore, a disturbance of rooting near the surface might be expected. However, these investigators found that in calcareous soils the total amount of hydrogen ions is neutralized, and in acid soils aluminum forms buffer systems for the soil solution, and still, about 80% of the hydrogen ions are neutralized. They also reported that the hydrogen ion concentration and sulfur fluxes in a spruce forest were greater than in a beech forest. Therefore, they postulated that the effect of acidic precipitation would be more accentuated in a spruce forest ecosystem.

Sulfur dioxide in precipitation increases the extractability of hydrogen and aluminum and decreases the exchangeable bases, particularly calcium and magnesium (Baker et al., 1976). In related work, Norton (1976) reported that impoverishment of soils by acid rain is unlikely for certain rock types initially high in certain nutrients, but that many rock types are not so invulnerable. Norton further indicates that absolute nutrient losses may be negligible over the short term, but that increases in the amount of dissolved elements may have a significant impact in aquatic systems, particularly with respect to the mobility of Fe, Mn, and an obvious increase in the mobility of Al. The cation exchange capacity may also be altered over a long period of time. Norton points out some possible effects of increased precipitation acidity on soils. These include (1) increased mobility of most elements, primarily the monovalent, divalent and trivalent cations; (2) increased loss of existing clay minerals; (3) changes in the cation exchange capacity (increases or decreases may result); (4) proportionate increases in the rate of removal of all soil cations; and (5) an increased flux of nutrients through ecosystems contiguous to soil and through aquatic ecosystems below the soil zone. Tamm (1976) has speculated that soil contamination could result in decreases in pH and lead to effects on carbon and nitrogen mineralization as well as influences on cation availabilities to forest trees. The most serious adverse effects of precipitation acidity would be expected in soil types transitional between brown earths and podsols (Wiklander, 1973–1974).

It has been postulated that poor growth of trees such as Scots pine and Norway spruce can be attributed to acidification by atmospheric pollution in susceptible soils (Jonsson, 1977). Others (Wood and Bormann, 1976) noticed that tree seedling productivity increased significantly with simulated rain (distilled H_2O acidified with sulfuric, nitric, and hydrochloric acids). Using pH values between 2.3 and 5.6, they noticed that the least growth occurred at a pH of 5.6 and the most at pH 2.3. They state that nitrate fertilization probably caused this trend and that the accelerated growth at pH 2.3 may be a short-term phenomenon. They observed significant leaching of Mg and Ca, which increased steadily with increased rain acidity through the pH range 5.6–2.3. Potassium losses did not increase until the pH was lowered to 2.3–3.0. Declines in exchangeable K, Mg, and Ca were measured at pH's of 3.0 and less. Greatly depleted levels of exchangeable cations occurred at a pH of 2.3.

McClenahen (1979) has conducted element cycling studies in two mixed mesophytic forest stands on similar sites in the Ohio River Valley. One site was subjected to a high level of industrial emissions from chemical industries, aluminum reduction, and coal-fired electric generating plants. The second stand was rather remote from industries or cities and was designated as a low-exposure site. Deposition, throughfall, and leachate movement out of the top 15 cm of soil were measured for Cl^-, F^-, NO_3^-–N, SO_4^{2-}–S, Mg^{2+}, Ca^{2+}, and K^+. These studies indicate that deposition rates of plant nutrients and potentially phytotoxic elements on forest soil can be increased through industrial air emissions, and that the rate of movement of at least several nutrient elements through the surface of the soil can likewise be greater in forests exposed to air pollutants. Such losses appear related to direct enhancement of wet and dry deposition and perhaps pollutant-induced acceleration of litter decomposition. Evidence from laboratory experiments shows that exposure of soil to SO_2 could also enhance nutrient mobilization. The net impact of these and possibly other pollutant-related effects on the nutrient budget of forest ecosystems is not clear. It does appear that the pollutant regime is enriching the soil nutrient pool. However, potentially toxic elements may also be accelerating the element cycling/loss rates.

Rates of breakdown of yellow buckeye and sugar maple leaf litter and element mineralization were also studied in the two stands. Weight loss of maple leaves was greater after 2 months in the high-exposure stand (McClenahen, 1979). Further, rates of element loss from litter tended to be greater in the high than in the low-exposure stand, particularly for Ca. Notably, the amount of fluorides in litter exposed in the high-exposure stand increased to over eight times that in the low-exposure stand. More mesic conditions near the soil surface of the high-exposure stand seem to have been fostered by dense herbage and may at least account for the faster litter breakdown and element mineralization. A dense herb layer appears to be the result of pollutant-induced stand structural changes (McClenahen, 1978). Thus, air pollutants may be both a direct and an indirect causal factor in accelerated litter decomposition and element flux.

In soil fumigation studies involving forest soil microcosms, at lower SO_2 concentrations (less than 0.25 ppm), cation losses via soil leachate tended to remain similar to those of the unfumigated control (McClenahen, 1979). Increasing SO_2 concentrations above 0.25 ppm (666 $\mu g/m^3$) resulted in increasingly greater cation losses, with a maximum occurring at 1.0 ppm SO_2 (2,665 $\mu g/m^3$). Nitrate losses were generally reduced below those of controls at the intermediate SO_2 exposures, but reached or exceeded those of the control at 1.0 ppm SO_2. Losses of SO_4^{2-}–S increased linearly with increased SO_2 concentrations.

4. RESEARCH CONSIDERATIONS FOR THE FUTURE

One of the major shortcomings in research planning regarding the effects of gaseous air pollutants and atmospheric deposition has been the predominant approach which has assumed that most or all effects are adverse. There are many beneficial effects associated with compounds and materials identified as air pollutants, and they should certainly not be disregarded. It is important that we weigh environmental concerns and insults against economic and social costs in light of tradeoffs and future land use patterns when planning research. We are now at a stage in our research maturation process at which we must not be purely academic in our outlook. We must consider cost–benefit comparisons of harmful and beneficial effects from gaseous pollutants and components of precipitation in terms of their stability in soils and their effects on productivity. We must consider the influence of past, present, and future agricultural and forest management techniques on soil nutrient depletion and rejuvenation. Research is definitely needed in the area of the fate of gases and dry deposition components.

We would be wise to consider the likelihood that there will be great changes in the types of fuels and rates of consumption for both mobile and stationary sources. Pollutants resulting from combustion of materials such as hydrogen and alcohol for mobile transportation facilities will be different from those we experience from the combustion of fuel oil and gasoline. Sulfur pollutants will probably be abundant for a long time and evaluations of their stability and mass transport over land capable of sorbing them should be investigated. The reactivity of sulfur oxides with wastes from hydrogen, alcohol, and methane combustion should be considered. In addition to effects of nonradioactive gases, nuclear waste effects on the productivity of agricultural and forest soils should be probed. We feel that a good beginning has been made toward understanding many of the basic factors regarding the interaction of soils and air pollutants. The general approach until now has been one involving mostly piecemeal research of specific pollutants and soil responses, often under highly controlled laboratory conditions. Although the soil system is an entity that deserves continued investigation as such, the ultimate concern is for the impact of ambient air pollutants on the larger ecosystem—be it agronomic or forest. Future research should therefore seek to incorporate soil studies in an ecosystem context, one that will require the integration of both field and laboratory studies.

We offer the following list of subjects for consideration of effects research relative to soils and gaseous pollutants in addition to atmospheric deposition. They are not arranged in any particular order of priority, but represent areas where there are deficiencies in our present knowledge.

1. Long-term studies on the effects of precipitation components are appropriate relative to nutrient cycling and productivity of agricultural and forest soils.
2. Research is needed on the adsorptive capacity of surface soil horizons, in terms of sulfate and nitrate adsorption and the ac-

companying percolation rates of sulfates to the subsoil when the adsorptive sites in upper soil horizons become saturated.

3. In-depth studies of nitrite oxidation by microorganisms are appropriate since NO_2 sorption by soils, although considered a physical or chemical phenomenon, relates directly to the oxidation of nitrogenous compounds by microorganisms.
4. The influence of soil microorganisms on the fate of pollutants such as CO, SO_2, CH_2H_4, and NO_2 in soil should be investigated, as well as microbes as reduction sites for photochemical oxidants. Such studies should address oxidation–reduction reactions as a result of microorganism activity, especially where pollutant sorption is thought to be chemical or physical in nature.
5. Soil resistance factors involved in pollutant sorption should be studied. Factors such as microenvironmental influences related to laminar flow and residence time of ozone near suitable receptor soils should be investigated.
6. Some important, and certainly more insidious, effects of pollutants on soils will involve long-term exposures of soils to low atmospheric concentrations of gases and precipitation components. Soil changes or responses such as leaching and nutrient cycling should be studied.
7. A general area of research that should be addressed before any substantial understanding of pollutant–soil interactions occurs is that of modeling various pollutant pathways in surface and subsurface soil horizons.

In summary, it is our opinion that more research planning should be based on future fuels, emission source control technology, future land uses, and changes in life styles that will induce changes in the production of goods and services. We advocate long-range studies that are essential to the understanding of real-world effects of atmospheric contaminants on soils and terrestrial ecosystems in general. Some blue-sky research is certainly in order; however, in the past we have concentrated on pollutants that have caused obvious symptoms on vegetation, presently exist in abundance, and cause adverse effects to human and other animal health. Foresight and planning are critical to the understanding of long-term effects of pollutants that are not obvious today.

Let us not busy ourselves studying soil chemical reactions that have long been known and understood, but not necessarily published in the context of environmental protection or degradation. A large part of the understanding of chemical reactions in soils as they pertain to sulfurous and nitrogenous compounds, especially, is available in agronomic, forestry, and horticultural literature. In many cases we are simply dealing with a different source of materials input. This is particularly important as we begin a major effort in effects research involving atmospheric deposition. Taking advantage of knowledge already gained is expeditious in terms of both time and available funds.

REFERENCES

Abeles, F. B., L. E. Craker, L. E. Forrence, and G. R. Leather. Fate of air pollutants: removal of ethylene, sulfur dioxide, and nitrogen dioxide by soil. *Science* **173**, 914 (1971).

Aldax, L. Flux measurements of atmospheric ozone over land and water. *J. Geophys. Res.* **74**, 6943 (1969).

Alway, F. J., A. W. Marsh, and W. J. Methley. Sufficiency of atmospheric sulfur for maximum crop yields. *Soil Sci. Soc. Am. Proc.* **2**, 229 (1937).

Baker, J., D. Hocking, and M. Nyborg. Acidity of open and intercepted precipitation in forests and effects on forest soils in Alberta, Canada. In L. S. Dochinger and T. A. Seliga (eds.) *Proceedings of the First International Symposium on Acid Precipitation and the Forest Ecosystem. Ohio State University, May 12–15, 1975*, pp. 779–790, USDA, Forest Service General Technical Report Ne-23, Upper Darby, Pennsylvania, 1976.

Blackmer, A. M., and J. M. Bremner. Potential of soil as a sink for atmospheric nitrous oxide. *Geophys. Res. Letters* **3**, 739 (1976).

Bohn, H. L. Soil absorption of air pollutants. *J. Environ. Qual.* **1**, 372 (1972).

Eberhardt, P. J. J. Absorption of nitrogen oxides and influence on nitrogen transformations in soil. Ph.D. Dissertation, Univ. of Arizona, 1972, 55 pp.

Ferenbaugh, R. W., W. S. Gaud, and J. S. States. Pollutant sorption by desert soils. *Bull. Environ. Contam. Toxicol.* **22**, 681 (1979).

Frink, C. R., and G. K. Voigt. Potential effects of acid precipitation on soils in the humid temperate zone. In L. S. Dochinger and T. A. Seliga (eds.) *Proceedings of the First International Symposium on Acid Precipitation and the Forest Ecosystem. Ohio State University, May 12–15, 1975*, pp. 685–709, USDA, Forest Service General Technical Report Ne-23, Upper Darby, Pennsylvania, 1976.

Ghiorse, W. C., and M. Alexander. Effect of microorganisms on the sorption and fate of sulfur dioxide and nitrogen dioxide in soil. *J. Environ. Qual.* **5**, 227 (1976).

Ghiorse, W. C., and M. Alexander. Effect of nitrogen dioxide on nitrite oxidation and nitrite-oxidizing populations in soil. *Soil Biol. Biochem.* **9**, 353 (1977).

Heichel, G. H. Removal of carbon monoxide by field and forest soils. *J. Environ. Qual.* **2**, 419 (1973).

Inman, R. E., and R. B. Ingersoll. Uptake of carbon monoxide by soil fungi. *J. Air Poll. Cont. Assoc.* **21**, 646 (1971).

Inman, R. E., R. B. Ingersoll, and E. A. Levy. Soil: a natural sink for carbon monoxide. *Science* **172**, 1229 (1971).

Jacobson, J., L. H. Weinstein, D. C. McCune, and A. Hitchcock. The accumulation of fluorine by plants. *J. Air Poll. Contr. Assoc.* **16**, 412 (1966).

Johnson, D. W., and D. W. Cole. Sulfate mobility in an outwash soil in western Washington. In L. S. Dochinger and T. A. Seliga (eds.) *Proceedings of the First International Symposium on Acid Precipitation and the Forest Ecosystem. Ohio State University, May 12–15, 1975*, pp. 827–835, USDA, Forest Service General Technical Report Ne-23, Upper Darby, Pennsylvania, 1976.

Johnson, D. W., and G. S. Henderson. Sulfate adsorption and sulfur fractions in a highly weathered soil under a mixed deciduous forest. *Soil Sci.* **128**, 34 (1979).

Jonsson, B. Soil acidification by atmospheric pollution and forest growth. *Water Air Soil Pollut.* **7**, 497–510 (1977).

Labeda, D. P., and M. Alexander. Effects of SO_2 and NO_2 on nitrification in soil. *J. Environ. Qual.* **7**, 523 (1978).

McClenahen, J. R. Unpublished Data, Ohio Agricultural Research and Development Center, Wooster (1979).

McClenahen, J. R. Community changes in a deciduous forest exposed to air pollution. *Can. J. For. Res.* **8**, 432 (1978).

McClenahen, J. R., and T. C. Weidensaul. Geographic distribution of airborne fluorides near a point source in southeast Ohio. Research Bulletin No. 1093, Ohio Agricultural Research and Development Center, Wooster, (1977), 29 pp.

Macdowall, F. D. H. Importance of soil in the absorption of ozone by a crop. *Can. J. Soil Sci.* **54**, 239 (1974).

McFee, W. W., J. M. Kelly, and R. H. Beck. Acid precipitation effects on soil pH and base saturation of exchange sites. In L. S. Dochinger and T. A. Seliga (eds.) *Proceedings of the First International Symposium on Acid Precipitation and the Forest Ecosystem. Ohio State University, May 12–15, 1975*, pp. 725–735, USDA, Forest Service General Technical Report Ne-23, Upper Darby, Pennsylvania, 1976.

Malmer, N. Acid precipitation: chemical changes in the soil. *Ambio* **5**, 231 (1976).

Mayer, R., and B. Ulrich. Acidity of precipitation as influenced by the filtering of atmospheric sulfur and nitrogen compounds—its role in the element balance and effect on soil. *Water Air Soil Pollut.* **7**, 409 (1977).

Miyamoto, S. A., A. W. Warrick, and H. L. Bohn. Land disposal of waste gases: I. Flow analysis of gas injection systems. *J. Environ. Qual.* **3**, 49 (1974).

Nelson, D. W., and J. M. Bremner. Gaseous products of nitrate decomposition in soil. *Soil Biol. Biochem.* **2**, 203 (1970).

Nor, Y. M., and M. A. Tabatabai. Oxidation of elemental sulfur in soils. *Soil Sci. Soc. Am. J.* **41**, 736 (1977).

Norton, S. A. Changes in chemical processes in soils caused by acid precipitation. In L. S. Dochinger and T. A. Seliga (eds.) *Proceedings of the First International Symposium on Acid Precipitation and the Forest Ecosystem. Ohio State University, May 12–15, 1975*, pp. 711–724, USDA, Forest Service General Technical Report Ne-23, Upper Darby, Pennsylvania, 1976.

Nyborg, M. J., J. Crepin, D. Hocking, and J. Baker. Effect of sulfur dioxide on precipitation and on the sulfur content and acidity of soils in Alberta, Canada. *Water Air Soil Pollut.* **7**, 439 (1977).

Prather, R. J., S. Miyamoto, and H. L. Bohn. Sorption of nitrogen dioxide by calcareous soils. *Soil Sci. Soc. Am. Proc.* **37**, 860 (1973a).

Prather, R. J., S. Miyamoto, and H. L. Bohn. Nitric oxide sorption by calcareous soils. I. Capacity, rate, and sorption products in air dry soils. *Soil Sci. Soc. Am. Proc.* **37**, 877 (1973b).

Rao, D. N., and D. Pal. Effect of fluoride pollution on the organic matter content of soil. Plant and Soil **49**, 653 (1978).

Smith, E. A., and C. I. Mayfield. Effects of nitrogen dioxide on selected soil processes. *Water Air Soil Pollut.* **9**, 33 (1978).

Smith, K. A., J. M. Bremner, and M. A. Tabatabai. Sorption of gaseous atmospheric pollutants by soils. *Soil Sci.* **116**, 313 (1973).

Tamm, C. O. Acid precipitation: Biological effects in soil and forest vegetation. *Ambio* **5**, 5 (1976).

Terraglio, F. P., and R. M. Manganelli. The influence of moisture on the adsorption of atmospheric sulfur dioxide by soil. *Air Water Pollut. Inst. J.* **10**, 783 (1966).

Thompson, L. K., S. S. Sidhu, and B. A. Roberts. Fluoride accumulations in soil and vegetation in the vicinity of a phosphorus plant. *Environ. Pollut.* **18**, 221 (1979).

Turner, N. C., S. Rich, and P. E. Waggoner. Removal of ozone by soil. *J. Environ. Qual.* **2**, 259 (1973).

Wainwright, M. Distribution of sulfur oxidation products in soils and on *Acer pseudoplatanus* L. growing close to sources of atmospheric pollution. *Environ. Pollut.* **17**, 153 (1978).

Wainwright, M. Microbial S-oxidation in soils exposed to heavy atmospheric pollution. *Soil Biol. Biochem.* **11**, 95 (1979).

Wiklander, L. The acidification of soil by acid precipitation. *Grundforbattring* **26**, 155 (1973–1974).

Wood, T., and F. H. Bormann. Short-term effects of a simulated acid rain upon the growth and nutrient relations of *Pinus strobus* L. In L. S. Dochinger and T. A. Seliga (eds.) *Proceedings of the First International Symposium on Acid Precipitation and the Forest Ecosystem. Ohio State University, May 12–15, 1975*, pp. 815–825, USDA, Forest Service General Technical Report Ne-23, Upper Darby, Pennsylvania, 1976.

Yee, M. S., H. L. Bohn, and S. Miyamoto. Sorption of sulfur dioxide by calcareous soils. *Soil Sci. Soc. Am. Proc.* **39**, 268 (1975).

Soil–Coal Fired Power Plant Effluent Interactions: An Evaluation

A. A. ELSEEWI AND A. L. PAGE
Department of Soil and Environmental Sciences
University of California
Riverside, California

I. R. STRAUGHAN
Research and Development
Southern California Edison Co.
Rosemead, California

1. INTRODUCTION

Coal is an abundant energy resource capable of expanding the world's energy supply. The total world coal resources are estimated at 11,500 billion tons, which in terms of energy are equivalent to about 50,000 billion barrels of oil or six times the estimated world reserve of oil (Griffith and Clarke, 1979). In 1974, estimates of coal resources in the United States which were mineable by current technology and economics were approximately 434 billion tons (Averitt, 1975).

The majority of the coal mined is used for electric power generation. For example, out of about 596 million tons of coal produced in the United States in 1975, nearly 369 million tons, or more than 61%, were used by the electric power utilities (U.S. Bureau of Mines, 1977). In 1976, coal accounted for almost 43% of the total electric power produced in the United States, while by the year 2000 it is estimated that more than 51% of the total U.S. electric power production will be derived from coal, requiring an additional 515 new coal-fired power plants (Berman, 1979). Griffith and Clarke (1979) approximate an annual world demand for coal in the year 2000 of at least 1.6 billion tons.

There are several major issues associated with coal resource development which have to be addressed to ensure that the resource is utilized in an environmentally safe manner. Important among these issues are the global effects of increased atmospheric CO_2 on climate, effects of acid deposition on aquatic and terrestrial flora and fauna, release of potentially hazardous trace elements in vapor and condensed phases, release of naturally occurring radionuclide and polynuclear aromatic hydrocarbons, and the reclamation (or lack of) of regions where surface mining is practiced. The present report concerns itself mainly with the mobilization and the fate of trace elements released during the coal combustion processes and examines their subsequent interaction with terrestrial ecosystems.

2. EMISSIONS AT THE COAL-FIRED POWER PLANT

2.1 Coal Composition

A comprehensive chemical analysis of U.S. coals has been completed by the U.S. Geological Survey (Swanson et al., 1976). Several hundred samples representing some 150 major producing mines in the U.S. were analyzed for 36 major, minor, and trace elements, in addition to a number of other proximate and ultimate analyses. Occurrence and distribution of 60 elements in coals from the Illinois Basin, mainly, and from other parts of the United States, have also been reported (Gluskoter et al., 1977).

Analyses by the U.S. Geological Survey indicated averages of 10% moisture, 11% ash, 2% sulfur, and 11,180 Btu/lb for all coal in the United

TABLE 1. Concentrations of Trace Elements in Coal, Rocks, and Soil

	Coal[a]			Rocks[a]		Soil[a]	
Element	USA	Australia	Europe	Igneous	Sedimentary	Typical	Range
As	15	3	6.3	2	1–3	6	0.1–40
B	50	60	—	10	20–100	10	2–100
Be	2	1.5	—	3	1–3	6	0.1–40
Cd	1.3	0.2	3.0	0.2	0.03–3	0.06	0.01–7
Cr	15	6	26.4	100	10–100	100	5–3000
Cu	19	15	41	55	5–45	20	2–100
Hg	0.18	0.10	0.5	0.08	0.03–14	0.03	0.01–3
Mn	100	150	80	1000	50–1100	850	100–4000
Mo	3	1.5	1.7	1.5	0.2–3	2	2–5
Ni	15	15	47	75	2–70	40	10–1000
Pb	16	10	—	12.5	5–20	10	2–200
Sb	1.1	≤20	1.2	0.2	0.05–1.5	6	0.6–10
Se	4.1	0.79	1.3	0.05	0.1–1	0.2	0.01–2
Ti	800	900	600	—	—	5000	—
V	20	20	43	135	20–150	100	20–500
Zn	39	<100	70	70	10–100	50	10–300

From Straughan et al. (1978).

[a]Concentrations are in micrograms per gram.

States. They also showed that more than 60% of S is in a sulfide form (pyritic sulfur), and that silica is the most abundant element in coal ash, followed by aluminum and iron. Calcium and magnesium oxides were found to be 2–3 times higher in lignite and subbituminous coals than in bituminous and anthracite coals, and cadmium, lead, and zinc showed highest average concentrations in bituminous coals.

On a whole-coal-weight basis, Gluskoter et al. (1977) observed elemental concentrations of western U.S. coals to be lower than their eastern counterparts; coals from the Illinois Basin were intermediate. Elements that showed a wide range of concentration were associated with the mineral matter of coal, while those with a narrow range occurred mostly in the coal organic fraction. Compared with the earth's crust, Gluskoter et al. (1977) suggested that only boron, chlorine, selenium, and arsenic are enriched in coal. They also indicated that boron was only concentrated in the Illinois Basin coals, presumably as a result of a greater marine influence.

Straughan et al. (1978) summarized concentrations of 16 trace elements in various coals, rocks, and soils (Table 1). They indicated that, on the average, Se, As, B, Cd, Hg, and Mo are enriched in coal relative to soil.

The above-mentioned studies, and others, clearly demonstrate that coal contains every naturally occurring element, most of them occurring at concentrations comparable to other geologic materials, such as rocks, soil, and shale. Sulfur, together with a number of trace elements, may on the average be higher in coal. The studies also demonstrate significant variation in coal composition depending upon rank, location, bed, and so forth. For example, Swanson et al. (1976) cautioned against over generalization of coal composition since, according to their study, every sample analyzed represented one billion tons of coal in the United States, which could hardly be a statistically valid basis for evaluating the quality of all coal in the United States.

2.2 Coal Combustion

In boilers of modern coal-fired power plants, temperatures average around 1500°C. Various combustion residues, some in the vapor phase and others in the form of solid particles, are produced. Elements and their compounds which are easily volatilized at the prevailing temperature of combustion remain in the vapor phase (unless otherwise captured by additional emission control devices) and are discharged to the atmosphere. Matrix elements such as silicon, aluminum, calcium, and iron form a melt which ultimately solidifies in a spherical form as a result of continuous gas bubbling. These spheres vary in size, with diameters ranging from <1 to >100 μm and are carried along in the flue gas. Most of the larger size particles, together with unburned and partially burned coal particles, are collected at the furnace bottom, forming what is known as boiler slag or bottom ash. Upon leaving the combustion zone in the flue gas, elements are exposed to lower temperatures and tend to condense somewhat uniformly on the surface of particles. Since smaller size particles possess larger surface area per unit weight than larger size particles, greater quantities of the element in question will condense at the surface of the smaller size particles.

The inverse relationship between element concentration and particle size of coal combustion residues has been examined by a number of investigators (Davison et al., 1974; Linton et al., 1977; Coles et al., 1979; Ondov et al., 1979; Campbell et al., 1978; Smith et al., 1979; Gladney et al., 1976; Kaakinen et al., 1975; Lee et al., 1975). In these studies, attempts have been made to group the elements into three major categories with respect to their enrichment (or lack of) in small-size particles:

1. Group I, elements generally showing no enrichment in small-size particles.
2. Group II, elements exhibiting marked enrichment with decreasing particle size.
3. Group III, elements which show an intermediate behavior with respect to particle size.

There seems a general agreement between various researchers that Al, Ba, Ca, Ce, Co, Eu, Hf, Fe, La, Mg, Mn, K, Rb, Sm, Sc, Sr, and Ti belong to Group I; Sb, As, Cd, Cu, Ga, Pb, Mo, Se, and Zn fall under Group II; and Be, Cr, Ni, Na, U, and V belong to Group III.

The flue gas stream with its suspended particulates (fly ash) passes to emission control equipment such as electrostatic precipitators (or bag filters), where particles with diameters generally greater than 5 μm are removed (Paulson and Ramsden, 1970). The residue so collected is referred to as fly ash but should be more correctly called precipitator or filter ash. In situations where high-sulfur coals are burned, additional scrubbing devices may be installed to remove sulfur gases from the flue gas. Various calcium and basic salts are employed in such scrubbers and the resultant discarded residue is referred to as flue gas desulfurization sludge (FGD sludge). In addition to removing the majority of sulfur in the flue gas, fine ash particulates not removed upstream are also trapped by this technique. The remaining ash and uncaptured gases are discharged via tall stacks to the surrounding environment.

The National Ash Association (1975) estimated that in 1974 the U.S. electric utilities had burned approximately 363 million tons of coal, which resulted in the collection of about 54 million tons of ash. The bulk of this ash is collected as precipitator/filter ash (fly ash). The remainder, bottom ash and associated residues, accounts for 10–15% of the total ash. Estimates of the amount of FGD sludge collected are variable. Bern (1976) projected that when approximately two-thirds of all generating capacity have FGD systems in operation, about 37 million tons of FGD sludge will be produced annually.

Schwitzgebel et al. (1975) calculated the amount of fly ash released through stacks of three coal-fired power plants without scrubbers and found it to range from 1800 to 44,000 tons/yr/1000 MW. Reports by the U.S. Environmental Protection Agency (1978) indicate that particulate emission from electric generation accounts for 63.9% of total particulate emission from stationary combustion systems in the United States. This amounts to about 4.1 million tons annually. According to their estimates, the amount of oxides of sulfur emitted by the electric utilities was about 14.5 million tons/yr. If we assume 4% of ash from coal combustion used in power generation is emitted as suggested by Gladney et al. (1976), during 1975, 1.6 million tons of particulates were emitted from this source. Campbell et al. (1978) pointed out that particulate emissions from coal combustion in the United States will reach approximately 5 million tons/yr by the year 2000.

Emissions of mercury are unique because the majority of the amount of this element entering with coal is released as vapor to the atmosphere. Billings and Matson (1972) and Billings et al. (1973) estimated that about 90% (by weight) of the mercury released from a 755-MW furnace appears to be in the vapor phase, and 10% remains with the furnace residual ash. Based upon their estimates, the amount of Hg emitted annually to the atmosphere equals 1.2 tons/1000 MW. Anderson and Smith (1977) estimated the atmospheric discharge of Hg from a 1200-MW facility at 97% in the form of vapor. Klein et al. (1975) concluded that most Hg in coal is discharged at the rate of 0.1 g/min or 1.8 tons Hg/1000 MW/yr from a 290-MW unit. This corresponds reasonably closely to the estimates derived from Billings et al. (1973). Schwitzgebel et al. (1975) estimated the fraction of total Hg entering with coal and discharged into the atmosphere at three midwestern power plants at 86.8, 97.9, and 96.1%.

Amounts of mercury released by coal combustion on a worldwide basis vary widely. The variation is due to the difference in the mean concentration of mercury in the total coal combusted. Joensuu (1971) estimated a mean concentration of 1 ppm Hg in coal; and, using this figure and assuming a total quantity combusted of 3 billion tons, annual release amounted to 3000 tons of mercury. Bertine and Goldberg (1971), on the other hand, estimated the mean concentration of Hg in coal as 0.012 ppm. The estimated world value for Hg released from coal combustion, according to these two sources, therefore varies from 3.6 to 3000 tons annually.

Estimates of the rate of atmospheric discharge per ton of coal combusted, for 23 elements, summarized from literature data, are shown in Table 2. These estimates were based on various mass balance studies and, for the most part, measurement of flue gas element concentration and flow rate were taken in the stack or somewhere near the end of the residue stream. Table 2 shows that emissions vary greatly between elements with regard to the amount released and the fraction emitted relative to input from coal. Sulfur shows the greatest amount emitted and cadmium the least.

TABLE 2. Estimates of Atmospheric Emission of Elements at the Coal-Fired Power Plant[a]

	Range of Atmospheric Emission	
Element	(g/ton coal)	(% of element in coal emitted)
Sulfur	4100–12,270	58–96
Calcium	6–2481	0.2–19
Iron	20–1300	0.4–18
Aluminum	18–1000	0.2–14
Magnesium	17–600	1–17
Fluoride	3–52	2–97
Boron	2–43	5–64
Titanium	3–27	0.3–27
Chromium	0.2–14	1–47
Barium	0.2–10	0.1–2.4
Zinc	0.2–5	2–66
Manganese	0.1–10	0.4–13
Nickel	0.05–4.5	0.4–82
Vanadium	0.3–4	1–28
Copper	0.04–3	0.6–29
Arsenic	0.002–1.6	0.1–21
Selenium	0.06–0.6	3–36
Lead	0.1–0.6	4–67
Antimony	0.01–0.27	1–75
Cobalt	0.01–0.4	0.5–56
Uranium	0.02–0.25	1–17
Beryllium	0.01–0.05	3–8
Cadmium	0.003–0.09	3–48

[a]These estimates were calculated from data given by Schwitzgebel et al. (1975) for three power plants; Lyon (1977) for one plant sampled twice in two consecutive years; and Andren et al. (1974) for three plants combined.

The large range of elements emitted is mainly due to differences in emission control devices in operation at any one plant. For example, data included in the calculations shown in Table 2 involved a North Dakota power plant which was equipped only with a mechanical cyclone particulate collector (Schwitzgebel et al., 1975). The fraction of total ash exiting this station with the flue gas was 12.9%, compared with 0.3% and 0.7% calculated for the other two power plants reported in that study.

Thus various amounts of each element contained in coal are released to the atmosphere. The amount released, however, depends upon the characteristics of the element and its compounds, coal composition, power plant configuration, and the emission control devices available.

2.3 Deposition on Land

Particulate and gaseous products emitted from stacks of coal-fired power plants eventually are deposited on land and on fresh or marine waters. Two methods have been employed to evaluate the extent to which the deposition of materials exiting the stacks may enrich the surrounding environment. These include those which model the process involved and case studies of actual measurements.

2.4 Modeling

Various attempts have been made to model ground-level, surface-air concentrations of particulates emitted from coal-fired power plants (Vaughan et al., 1975; Lyon, 1977; Jurinak et al., 1977). Models used in these studies can generally be classified into two major categories. Those which account for a variable source strength (emission rate) due to continuous wet and dry deposition are called source-depletion models. Models which account for depletion only at the earth's surface due to the action of soil and vegetation are called surface-depletion models.

No attempts will be made to explain the mathematical derivation and the various details contained in each model. Instead, we shall present the major assumptions and findings of three modeling studies for coal-fired power plants in the St. Louis, Missouri (Vaughan et al., 1975) and Memphis, Tennessee (Lyon, 1977) areas, and a proposed power plant located in southern Utah (Jurinak et al., 1977).

The model used by Vaughan et al. (1975) is a source-depletion model and distinguishes clearly between wet and dry deposition. The model was applied to a 1400-MW power plant in the St. Louis, Missouri area, assuming an emission rate of 1 g/s and a deposition velocity of 1 cm/s. Surface air concentrations for summer (July, August, September) and winter (January, February, March) months were predicted by the model at various distances from the source (S) for various release heights. Mass deposition rates computed from air concentration are presented in Figure 1, which shows that deposition decreases as a function of distance from the source. Because of predominantly southerly winds in the summer, maximum deposition rates extend to the north (Figure 1a). In Figure 1b, the effect of west–northwest wind is reflected on the pattern where it is skewed to the east–southeast. The increase in the release height (Figure 1c) resulted in much less deposition near the source. The authors acknowledged the fact that concentrations shown are very low because of their assumption of only 1 g/s for the emission rate. Assuming a 40-yr exposure time, the authors calculated

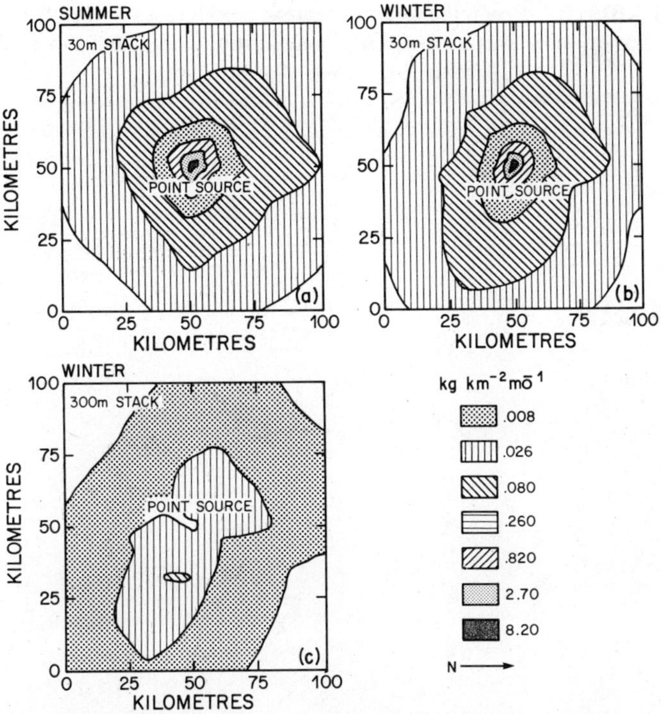

FIGURE 1. Modeled mass deposition of fly ash around a 1400-MW coal-fired power plant in the St. Louis, Missouri area (adapted from Vaughan et al., 1975). Emission rate = 1 g/s; particle size = <15 μm; deposition velocity = 1 cm/s.

the increase in total and available trace element content of soil arising from the deposition of coal-fired power plant effluent (Table 3). These calculations were based upon 30-cm soil depth and a soil bulk density of 1.5 g/cm^3. The data indicate quite small changes in total concentrations of elements. Because of natural variability, in all probability, in practice it would not be possible to measure the changes predicted by the model. However, if all deposited elements were considered available, a number of elements show marked increases in concentration. Notably among these elements are Cu, Hg, and Mo. Although the authors suggested that toxicity problems to plants and animals could arise from increased concentrations of Cu and Ni, in our opinion the probability is extremely small.

Lyon (1977) developed a model for particulate emission and deposition originating from the Allen steam plant in Memphis, Tennessee. The plant capacity is 870 MW and emission was set at 3.2×10^3 g/s, which corresponds to 0% emission efficiency. All particles were assumed to have a diameter of 2 μm, and dry deposition velocity was set at 1 cm/s. Although the model also takes into account both wet and dry deposition, a fixed amount of wet deposition was assumed. The rate of deposition derived from

TABLE 3. Predicted Trace Element Enrichment in Soil from the Deposition of Particulates Originating from Coal-Fired Power Plant in the St. Louis, Missouri Area[a]

Element	Increase over Total Soil Concentration (%)	Increase over Available Soil Concentration[b] (%)
Ag	0.03	0.3
As	0.02	0.3
Cd	0.2	0.51
Co	0.002	0.4
Cr	0.0002	0.8
Cu	0.06	63.0
Hg	2.0	68.0
Mn	0.001	0.4
Mo	0.1	247.0
Ni	0.01	4
Pb	0.02	0.3
Sb	0.01	0.4
Se	0.1	0.3
Sn	0.04	10.0
Tl	0.03	0.08
W	0.3	0.3
Zn	0.03	0.4

From Vaughan et al. (1975).

[a] Calculations were based upon 20-cm soil depth and a soil bulk density of 1.5 g/cm^3.

[b] Assumes all deposited material is in a plant-available form.

their data, presented in Figure 2, indicates little change in deposition rate with compass direction, although maximum deposition at a given distance from the source would be to the west. Soil enrichment was calculated for the 595 kg/km^2 month area, assuming complete mixing of particulates with the upper 1, 3, and 10 cm of soil. The results are shown in Table 4.

Thus, if an enrichment of about 50% were detected in soil for an element, it would mean that concentration of this element in the deposited particulates is $\geq$100 times its normal concentration in soil. Very few elements in the fly ash produced at the power plant had concentrations greater than 10 times that found in soil. Also, since emission controls are usually better than 90%, and 0% was assumed for the model, Lyon (1977) indicated that the probability of observing an enrichment would be even more remote.

A more comprehensive model was developed by Jurinak et al. (1977) for

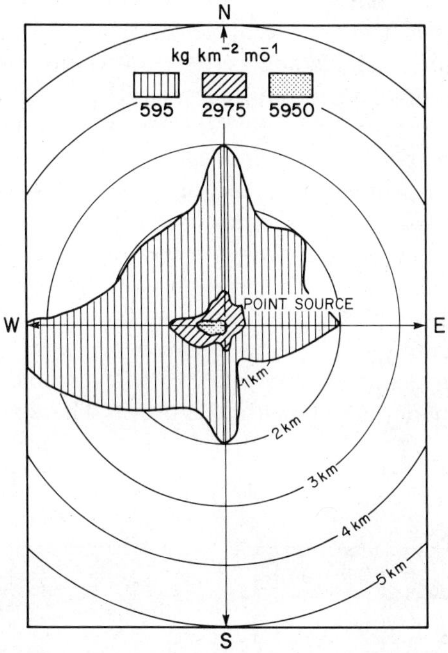

FIGURE 2. Modeled mass deposition of fly ash around an 870-MW coal-fired power plant in the Memphis, Tennessee area (adapted from Lyon, 1977). Emission rate = 3.2×10^3 g/s (0% emission control); particulate size = 2 μm; deposition velocity = 1 cm/s. Mass deposition given by the original authors was estimated for 14 yr of exposure and has been normalized to kg/km^2 month.

TABLE 4. Predicted Soil Enrichment[a] of Trace Elements Discharged from a Coal-Fired Power Plant in Memphis, Tennessee Area

Ratio of Particulate Concentration to Soil Concentration	Depth of Soil-Particulate Layer (cm)		
	1	3	10
1	1.00	1.00	1.00
5	1.02	1.01	1.00
10	1.04	1.01	1.00
100	1.39	1.13	1.04
1000	4.98	2.33	1.40

From Lyon et al. (1977).

[a]Values are ratio of enriched to normal soil concentration.

a 3000-MW coal-fired power plant that was proposed for southern Utah. The model belongs to the surface-depletion group and takes into account variable stack (plume) heights and vertical diffusivity profiles according to atmospheric conditions and terrain features. The model thus is said to be specific diffusivity-terrain corrected. In this model, deposition velocities varied according to particle size and distance from the source. For small particles (≈5 μm in diameter), deposition velocities were estimated in the range of 1 to 10 cm/s. Larger size particles were treated as gravitationally deposited, and a mean particle diameter was assumed. The model assumed a total particulate emission of 1.05 tons/h (292 g/s), assuming 99.1% precipitator plus scrubber particulate removal efficiency. Furthermore, of the total particulates emitted, 3% were assumed to have a mean diameter of 50 μm, 5% 25 μm diameter, and 10% 15 μm diameter. Deposition rates predicted by the model are shown in Figure 3. The changes in deposition

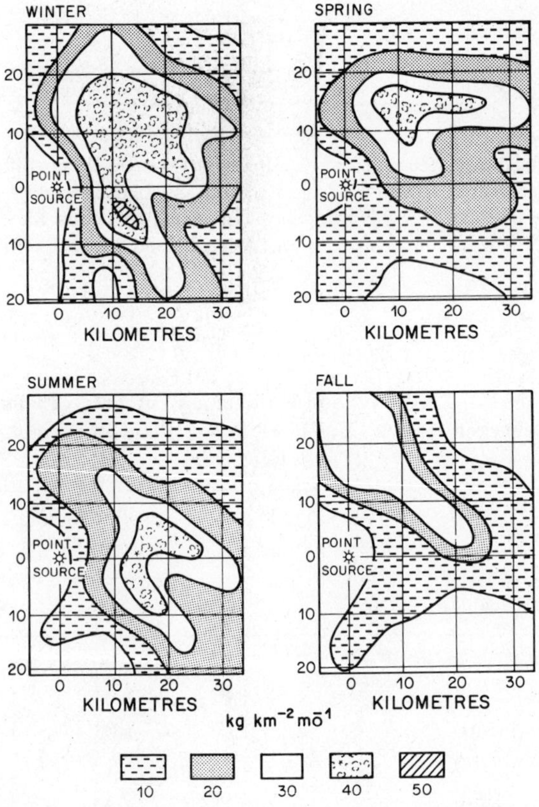

FIGURE 3. Modeled mass deposition of fly ash around a 3000-MW coal-fired power plant in southern Utah (adapted from Jurinak et al., 1977). Emission rate = 292 g/s (99.1% emission control); particulate size = <5–50 μm; deposition velocity = 1–10 cm/s.

pattern with season reflect corresponding changes in the wind speed and direction, atmospheric stabilities, and depths of mixed layer. The greatest deposition occurred on a plateau in all seasons with highest deposition in summer and lowest in fall. Based upon maximum deposition rates (50 kg/km^2 mo), Page et al. (1979) estimated soil enrichment assuming a particle size of <3 μm with composition similar to that given by Ondov et al. (1976) and a 35-yr exposure. Calculation of percent enrichment of soil is shown in Table 5.

Based upon data shown in Table 5, soil enrichment, except for Se, appears negligible over the 35-yr exposure period. Thus, trace element enrichment in soil from particulate emission of coal combustion predicted by models developed for diverse conditions seems quite low and, in many cases, not measurable by current technology.

TABLE 5. Estimated Maximum Deposition of Trace Elements onto Soil Adjacent to a Coal-Fired Power Plant

	Amount Deposited (mg/kg)		Common Concentrations (mg/kg)		Soil Enrichment
Element	Annual[a]	Lifetime[b]	Typical	Range	(%)
As	0.00039	0.014	6	0.1–40	0.2
Cd	0.000014	0.0005	0.06	0.01–7	0.8
Pb	0.00082	0.029	15	2–200	0.2
Mo	0.00015	0.0052	2	0.2–5	0.3
Se	0.00059	0.021	0.2	0.01–2	10.5
U	0.000087	0.003	1	0.1–10	0.3
Zn	0.00163	0.057	50	10–300	0.1
Sb	0.000061	0.0021	6	2–10	0.04
Be	0.00003	0.0011	6	0.1–40	0.02
Cr	0.00019	0.0066	100	5–3000	0.007
Co	0.000062	0.0022	8	1–40	0.03
Cu	0.00041	0.0144	20	2–100	0.07
Ga	0.00053	0.0182	30	0.4–300	0.06
Ni	0.00012	0.0042	40	10–1000	0.01
Th	0.0009	0.0032	5	0.1–12	0.06
V	0.00097	0.034	100	20–500	0.03

[a]Derived from deposition data of Jurinak et al. (1977) and concentration data of Ondov et al. (1976): 3000-MW power plant, western U.S. coal, precipitator removal efficiency 99%.

[b]Assuming 35-yr lifetime.

2.5 Case Studies

Field studies to monitor power plant emissions utilize a general concept in which evidence of contamination is suggested where concentrations of a particular element in soil or vegetation in regions surrounding the emission source are inversely related to distance from the source. The major problem with case studies, however, is natural variation and lack of reliable background levels. Bradford et al. (1978) measured 22 elements in soil and vegetation samples taken on a grid pattern at 6.4-km intervals up to 19 km in eight compass directions from a 1500-MW power plant in Nevada. Progressive decreases in the concentration of Ca, Mg, B, Ba, and Sr in saturation extracts of surface soils were observed as the distance from the source was increased in one compass direction. For the same compass direction, only Ca, B, and Sr in vegetation samples showed the inverse relationship with distance from the source. They concluded, however, that the distance–concentration relationship was due to natural changes in soil properties and that four years of the power plant operation resulted in no measurable contamination of either soil or vegetation in the region studied.

Gough and Erdman (1976) noticed an increase in Se, F, and Sr in lichen (*Parmelia chlorochroa*) as the distance from a western 750-MW coal-fired power plant, which had been in operation for about 16 yr before sampling, was decreased. They emphasized the potential use of this species as a monitor of power plant element emission, particularly in the Northern Great Plains region. Conner et al. (1976a) used sagebrush (*Artmesia tridentata*) to monitor element enrichment from emissions from the same power plant. Their data suggested low-level contamination of Se, Sr, U, and V and possible contamination of Co, Li, Ti, and Zn originating from the power plant.

Conner et al. (1976b) also sampled soil (surface and subsurface) and vegetation (Indian rice grass, *Oryzopsis hymenoides*) in the downwind direction from a major coal-fired power plant in the southwestern United States. Distance-related geochemical variability indicated that out of 38 elements analyzed in plants, nearly half exhibited statistically significant variations associated with distance, with essentially all trends decreasing away from the power plant. Trends in soils data were fewer and showed opposite directions. The authors concluded that Se, Fe, and Sr contamination originating from the coal-fired power plant was most likely because all exhibited significant distance-related decreases in vegetation near the operating power plants. They further concluded:

> *A large number of elements have been implicated as suspect pollutants. Most soil elements so implicated appear to be better interpreted as reflecting natural geochemical changes in the soil substrate (i.e., geologic control), but this explanation seems less satisfactory for the interpretation of distance-related plant chemistry.*

Thus, Conner et al. felt that vegetation shows greater promise than soil for monitoring contamination from coal combustion.

Klein and Russell (1973) noticed that soils around a 650-MW power plant on the eastern shore of Lake Michigan, which has been in operation for about 10 yr and has burned 10^7 tons of Ohio coal, were enriched in Ag, Cd, Co, Cr, Cu, Fe, Hg, Ni, Ti, and Zn compared with background soils. Vegetation (native grass, maple leaves, and pine needles) were enriched in Cd, Fe, Ni, and Zn. Mercury in soil, however, was only slightly enriched and could not be correlated with the Hg content of coal. The authors suggested the reason mercury was only slightly enriched is because mercury discharged to the atmosphere by coal combustion is much more widely dispersed than other metals similarly discharged.

Anderson and Smith (1977) studied dynamics of mercury around a 1200-MW coal-fired power plant which had been in operation for 6 yr prior to sampling of soils, lake sediments, fish, macrophytes, and ducks. Their estimates indicated that from 26 to 70% of the Hg emitted was incorporated into soil within a 19.3-km radius of the plant. They could not, however, account for the remaining Hg but indicated that it might have dispersed over longer distances, have revolatilized by some unknown mechanism from precipitated particulate matter, or may have found its way into the surrounding lake and other aquatic environments. Lake sediments only accounted for 1% of Hg emitted, and fishes contained unusually low concentrations of Hg. They concluded that their findings failed to implicate mercury as a serious pollutant in the lake and the surrounding agricultural land.

Crockett and Kinnison (1979) also investigated the fate of mercury around a 2100-MW power plant in the southwestern United States. They indicated that, despite the fact that this power plant was emitting 1–2% of all the Hg released by U.S. coal-fired power plants, the soil Hg levels within a 30-km radius around the plant were low and were not significantly different from background levels. Near the fly ash ponds, however, Hg concentrations in soil were slightly elevated but not unusual for soil in general. They concluded that the fate of mercury emitted by the plant is not yet known.

Owing to repeated observations on Sr enrichment of soils and vegetation in regions adjacent to coal-fired power plants (Bradford et al., 1978; Gough and Erdman, 1976; Conner et al., 1976b), we examined the isotopic ratio $^{87}Sr/^{86}Sr$ in soils and plants treated with coal ash (Straughan et al., 1981). The objectives were to develop an index by which atmospheric deposition of residues from coal-fired power plants could be monitored. Significantly different Sr isotopic ratios were found between fly ash and soil, but plants treated with fly ash exhibited ratios similar to that of fly ash. Further studies, however, are needed to examine these ratios under field conditions.

3. SOIL–COAL ASH INTERACTIONS

Addition of coal combustion residues such as fly ash to soil has been shown to exert major changes in the chemical properties of the recipient soils, thereby affecting their agronomic and engineering characteristics. Furthermore, such changes may also reflect upon the quality of surface and groundwaters in contact with the affected soils. Changes in soil pH, salinity, and trace element content are discussed in the following sections.

3.1 Effect on Soil pH

Coal combustion products in contact with water produce solutions which are generally alkaline. A few ashes, however, may generate acidic solutions, and these are primarily derived from eastern and southeastern U.S. coals (Furr et al., 1977). Although this is not known with certainty, the acidic and basic behavior of coal combustion residues appears to be related at least in part to the relative proportions of Ca and Fe oxides contained in the residue (Theis and Wirth, 1977).

When mixed with soils, alkaline fly ash induces an elevation in the soil pH which is dependent upon the buffering capacity of the recipient soil, the amount of fly ash added, and the time of contact. Calcareous soils, which are dominant in the arid and semiarid southwestern United States, are highly buffered compared with acid soils. They show an initial increase in their pH upon coal ash addition, which diminishes with time to levels comparable to background (Page et al., 1979; Elseewi et al., 1978a). Fly ash amended soils, however, tend to maintain pH levels well above background levels for extended periods of time.

Effects of small atmospherically dispersed solid particles from coal combustion sources on soil pH have not been investigated. Greenhouse experiments conducted by the authors with various size fractions of fly ash derived from western U.S. coal sources added to a slightly acid soil and planted with tall fescue (*Festuca elatior*) and alfalfa (*Medicago sativa*) have clearly demonstrated that all size fractions added produce an increase in soil pH (Figure 4). In fact, the experiment showed that the smaller the particle size, the larger the increase in soil pH.

Thus, in situations where high-Ca coals are burned, one might expect that the deposition of fine particulates emitted from Ca-rich coal would exert a pH effect on soil similar to that depicted in Figure 4. In fact, Prince and Ross (1972) noted that rain in East Germany downwind from power plants burning lignite was alkaline, presumably from washout of basic particulates. Primary oxides of S and N which have been implicated in acid rain formation induce an effect on soil pH opposite to that observed for small solid particulates.

Acidification of soils from the latter source is undoubtedly dependent upon the buffering capacity of soils, which in turn is determined by various

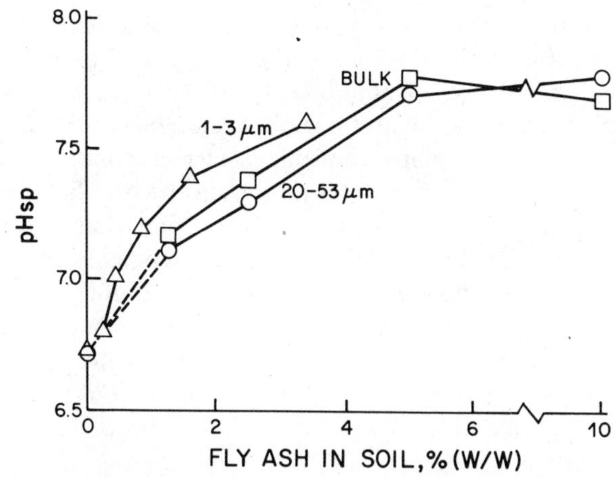

FIGURE 4. Effect of fly ash particle size on pH of Josephine loam soil. Soil was cropped to tall fescue under greenhouse conditions for one year.

mineralogical, physical, chemical, and biological properties of the soil system. For example, calcareous and clayey soils are highly buffered compared with the poorly buffered sandy soils. Thus, despite the numerous reports in the literature showing increased acidity of rain correlated with increasing acidity of surface waters and deleterious effects on higher and lower plants as well as on various forms of aquatic life, no such correlations have been established with respect to soil acidity. Malmer (1976) concluded that failure to observe such changes in soil could probably be related to the fact that only a short amount of time has elapsed since the problem was first recognized. In the United States, data on pH of precipitation for the entire United States were collected only between 1964 and 1966 (Likens, 1976). The problem is further complicated by the fact that causes of acid precipitation seem to be global in nature. More than 77% of the amount of anthropogenic sulfur over Sweden is thought to have originated from distant and densely populated industrialized areas in England and the Ruhr Valley (Likens, 1976). Therefore, although the potential for soil acidification from anthropogenic sources exists, it is rather difficult to assess at this time.

3.2 Effect on Soil Salinity and Trace Element Soluble and Exchange Ions

Residues from coal combustion when dissolved in water produce solutions which contain appreciable amounts of soluble salts. Salinity of such solutions is usually expressed in terms of electrical conductivity units. Fresh British coal ashes were reported to have conductivities as high as 13 mmho/cm (Townsend and Hodgson, 1973). Similarly, water saturation extracts of fly ash derived from western U.S. coal sources were found to exhibit electrical conductivity values as high as 12 mmho/cm (Elseewi et al., 1978a; Page et al., 1979). Unpublished data by the authors show that saturation extracts from fly ash and bottom ash collected simultaneously at a western coal-fired power plant showed electrical conductivity values of 11.1 mmho/cm for the fly ash and 1.1 mmho/cm for the bottom ash. Lower salinity of the bottom ash is attributed to the fact that it is made up of coarse-textured ash and coal particles that are virtually chemically inert. The high salinity of the fly ash is due mainly to solubility of Ca and OH ions. This $Ca(OH)_2$ system is characteristically unstable, as insoluble $CaCO_3$ is being formed upon contact with CO_2. Salinity of fresh fly ash is thus greater than that of weathered or aged fly ash, as has been noticed by a number of investigators (Hodgson and Holliday, 1966; Page et al., 1979; Adriano et al., 1980).

Fly ash amended soils are consequently enriched in their soluble salt content (Mulford and Martens, 1971; Elseewi et al., 1978a; Page et al., 1979). Such enrichment may interfere with the normal growth of a number of agronomic crop species which are sensitive to salts. As was the case with the influence of particle size on soil pH, soil salinity is also increased with decreasing particle size of the added ash (Figure 5). At equal rates of addition, the 1–3-μm-sized fraction produced higher residual salinity in soil than the 20–53-μm-sized fraction. The unclassified fly ash (bulk) showed an integrated effect which is intermediate between the largest and the smallest size fraction used (Figure 5).

The increase in solution Ca in soils amended with fly ash brings about a new equilibrium between soluble and exchangeable cations in the soil system, with a resultant increase in base saturation. Phung et al. (1978)

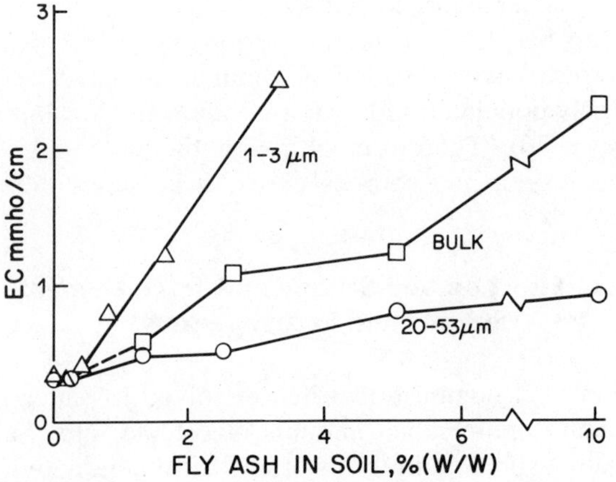

FIGURE 5. Effect of fly ash particle size on salinity (EC) of Josephine loam soil. Soil was cropped to tall fescue under greenhouse conditions for one year.

demonstrated a dramatic increase in concentrations of exchangeable Ca in an acid soil amended with variable amounts of fly ash. Such an increase was associated with moderate increases in exchangeable Mg and drastic reductions in exchangeable Al and Mn. From the standpoint of fly ash use as a soil amendment, their data demonstrated the potential of fly ash as a liming material for acid soil, capable of increasing the pH and exchangeable Ca and reducing the incidence of Al and Mn toxicities to plants. Furthermore, one of the most common effects of acid rain is the mobilization of increased quantities of available Al and Mn in soils. This effect is thus counteracted in situations where alkaline solid particulates from coal combustion are added to soil.

Fly ash treated soils are also significantly enriched in sulfates in proportion to the amount of ash added to soil. Although the majority of S input from coal is discharged in the flue gas and would ultimately be atmospherically emitted unless captured by additional scrubbing devices, substantial quantities remain with the residual ash. The content of total S in fly ash ranges from <0.1 to $>1.0\%$ (Natusch et al., 1975; Bickelhaupt, 1974; Block and Dams, 1975), and is mostly in a biologically available form (Elseewi et al., 1978b). The concentration is also inversely related to particle size. Studies conducted by the authors showed that concentrations

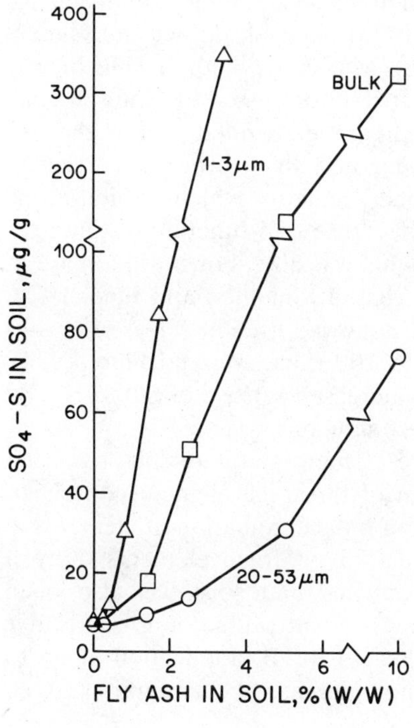

FIGURE 6. Effect of fly ash particle size on the SO_4–S content of Josephine loam soil. Soil was cropped to tall fescue under greenhouse conditions for one year.

of SO_4–S increased from 0.18% in coarse fly ash (10–53 μm) to 1.1% in fine fly ash (1–3 μm). The amount of S extracted from soil using NH_4OAc was dramatically increased in proportion to the amount of ash added to soil and as the particle size decreased (Figure 6). Consequently, plants grown on soils amended with ash are greatly enriched in S and, as a result, they often show positive growth response to ash additions to soil (Elseewi et al., 1978b; Page et al., 1979).

The increased availability of a major plant nutrient such as S should not be overlooked as a positive credit to coal-fired power plant effluents. Increased emission controls which regulate the amount of S released to the atmosphere has, among other results, created a situation of near sulfur starvation in plants (Coleman, 1966). Further discussion of this observation is found in a later section of this report.

3.3 Trace Element Enrichment of Ash-Amended Soils

Solubility studies with ash systems indicate that the amount of trace elements extracted is dependent upon pH, surface predominance of the element in question, amount of ash in solution, and time of contact between the ash and the extracting solution. Decreasing the pH of the extracting media has been shown to result in increased release of As, B, Pb, Cu, Cr, Ni, Mn, Zn, and Cd (Phung et al., 1979; Theis and Wirth, 1977). Dreesen et al. (1977) emphasized the importance of surface predominance of Mo, F, Se, B, As, and Cd on the ash particles, which makes them more available for mobilization. Similarly, Natusch et al. (1975) showed surface enrichment and extractability in water for Cd, Co, Li, Mn, P, Tl, and Zn. Distinction between surface and matrix elements was further discussed by Natusch et al. (1975). Matrix elements and those which are embedded within the ash matrix exhibit low initial solubility in water and show progressive release with time in contrast to surface enriched elements which exhibit rapid maximum solubility followed by gradual decrease as a function of time.

Extractability of a number of ash elements was also shown to increase as the amount of fly ash in solution was decreased (Shannon and Fine, 1974; Elseewi et al., 1980). Calcium, Mg, and Fe showed greater release in water as the water–ash ratio increased from 5 to 100 (Shannon and Fine, 1974), while Elseewi et al. (1980) demonstrated increased release of SO_4, Ca, and B with increasing dilution of various fly ash aqueous systems.

The effect of ash residues on the trace element status in soils has been investigated either by direct measurement of the trace element content of the ash-amended soil or via measurement of bioaccumulation of elements in plants grown on treated soils. Soil attenuation of trace elements derived from interactions between ash pond leachates and soils has also been evaluated to predict the impact of ponding of coal residue on groundwater quality. Phung et al. (1979) found that, except for B, application of up to 1% fly ash to three different soils did not increase the amount of trace elements extracted in water or in DTPA extracts. In fact, extraction of Fe, Mn, Ni, Co, and Pb in DTPA was reduced upon addition of the fly ash to soils. They indicated that the alkalinity of the ash plays a significant role in regulating the availability of trace elements in the treated soils. Studies by the authors showed that soils treated with up to 10% western fly ash are enriched in available B and Mo (Page et al., 1979; Elseewi et al., 1978a; Jones and Straughan, 1978).

Plants grown on soils treated with fly ash from various sources have been found to accumulate a number of trace elements, including As, B, Mo, Se, Zn, Cu, Fe, Hg, I, Ni, and Sb (Page et al., 1979; Elseewi et al., 1978a; Furr et al., 1976, 1978; Plank and Martens, 1974; Schnappinger et al., 1975; Romney et al., 1977; Gutenmann and Lisk, 1979). In most cases, the accumulation is proportionate to the amount of fly ash added to soil. In situations where highly alkaline fly ash is added to soil, the availability of Zn and Mn to plants may be curtailed (Page et al., 1979). Bioaccumulation of trace elements in aquatic organisms grown in drainage basins of settling ash ponds were also found to concentrate a large number of trace elements in concentrations greater than those of the water effluent (Guthrie and Cherry, 1979). These authors indicated that, although accumulation by aquatic plants offers an efficient mechanism for trace element removal, a potential hazard exists for higher organisms feeding on these plants.

Trace element enrichment of subsurface water seeping from ash-settling ponds situated near power plants has also been demonstrated (Theis and Marley, 1979). Theis et al. (1978) indicated that the enrichment is sensitive to ash loading rates and procedures. They also indicated that soil attenuation of the trace elements occurred for most metals near the pond itself. Theis and Richter (1979) detailed the attenuation by showing that the most important factors influencing attenuation were pH, iron and manganese oxides, and concentrations of sulfates. They showed that adsorption on the oxides was the major solubility control for Cd, Ni, and Zn, while precipitation was the mechanism responsible for the removal of Cr, Cu, and Pb from solution.

4. GENERAL DISCUSSION AND CONCLUSIONS

Effluents from coal combustion are grouped, for convenience, into two categories, namely, captured and uncaptured residues. The fate of the uncaptured residues or aerosols is largely unknown. In particular, knowledge concerning interactions between these atmospherically discharged particles and soils is very limited, and in most situations completely absent. We therefore resort to theoretical modeling to predict such interactions. Models previously discussed were developed for diverse environmental conditions and, interestingly, they show that soil enrichment of trace elements originating at the coal-fired power plant is rather negligible.

Information obtained by such studies should thus be fully utilized to avoid unneccesary duplication.

The same, however, does not apply to sulfur and other gaseous emissions whose magnitude is tremendously greater than that of solid particulates. Nevertheless, the sulfur problem may be viewed as two-fold with respect to trace element mobilization. On the one hand, restricting the amount of sulfur emission will undoubtedly reduce the potential for acid rain formation and subsequent soil acidification. On the other hand, acid rains have the ability to mobilize trace elements in soil, probably in much the same way as residues from sulfur trapping will enrich the trace element content in soil when disposed into the environment. This tradeoff brings about a question of prime importance, that is; the biogeochemical cycle of elements. Basic understanding of such a cycle is crucial for the successful control of any potentially hazardous impacts from coal resource development.

The research outlined in the previous sections indicates that little has been done in the area of residue–*soil* interactions. These residues vary greatly in composition (Furr et al., 1977; Adriano et al., 1980), and their disposal occurs under various environmental conditions with respect to soil, climate, vegetation, water resources, and so forth. Research emphasis thus far has been given to the residues themselves and little, if any, research has been done with regard to their interactions with the various environments in which they are discharged. For example, despite the long history of coal resource development in the eastern and midwestern United States, virtually no studies have been conducted to evaluate interactions between these residues and soils of these regions. Atmospheric conditions in these regions are vastly different from those of the western United States, and may result in the production of atmospheric residues different from those developed under the western arid climatic conditions. In the western United States, where the shift to coal as an energy source is at its inception, soil–residue interaction studies are under way; few were cited in this report. They show that with the exception of boron, molybdenum, selenium, and salinity buildup in soils, which can be controlled successfully, rather positive impacts from the residues are expected. For example, improved crop yields have been demonstrated when moderate amounts of these residues are applied to agricultural lands. The use of these residues as soil amendments may thus offer a potential for recycling them beneficially.

A concept of integrated approach should thus be adopted to approach the consequences of any resource development. Adopting such a concept would lead to avoiding errors which have been frequently committed in the past. A case in point is the unilateral introduction of solid particulate control technology, which resulted in reduced emissions of particulates that might have otherwise counteracted the acidic effects of S and N oxides. The latter pollutants were later controlled by the introduction of scrubbers. It would have been far more realistic if a holistic approach had been adopted initially to tackle the two problems simultaneously instead of following the "pollutant of the month" syndrome. Our research guidelines should never be driven by such an approach.

In conclusion, the entire biosphere reserve area of the total environment has to be considered for the assessment of the potential impact of effluents from coal combustion.

REFERENCES

Adriano, D. C., A. L. Page, A. A. Elseewi, A. C. Chang, and I. R. Straughan. Utilization and disposal of fly ash and other coal residues in terrestrial ecosystems: A review. *J. Environ. Qual.* **9**, 333–344 (1980).

Anderson, W. L., and K. E. Smith. Dynamics of mercury at coal-fired power plant and adjacent cooling lake. *Environ. Sci. Technol.* **11**(1), 75–80 (1977).

Andren, A. W., B. G. Blaylock, E. A. Bondietti, C. W. Francis, S. G. Hildebrand, J. W. Huckabee, D. R. Jackson, S. E. Lindberg, F. H. Sweeton, R. I. Van Hook, and A. P. Watson. Ecology Research. In *Ecology and Analysis of Trace Contaminants*. pp. 61–104. National Science Foundation, ORNL-NSF-EATC-11, 1974.

Averitt, P. Coal resources of the United States. January 1, 1974. *U.S. Geological Survey Bull.* **1412**, 119 (1975).

Berman, I. M. New generating capacity: When, where, and by whom. *Power Engineering*, **83**(4), 54–64 (1979).

Bern, J. Residues from power generation: Processing, recycling, and disposal. In *Land Application of Waste Materials*. pp. 226–248. Soil Conservation Society of America, Ankeny, Iowa, 1976.

Bertine, K. K., and E. D. Goldberg. Fossil fuel combustion and the major sedimentary cycle. *Science* **173**, 233–235 (1971).

Bickelhaupt, R. E. Electrical volume conduction in fly ash. *J. Air Poll. Control Assoc.* **24**(3), 251–254 (1974).

Billings, C. E., and W. R. Matson. Mercury emissions from coal combustion. *Science* **176**, 1232–1233 (1972).

Billings, C. E., A. M. Sacco, W. R. Matson, R. M. Griffin, W. R. Coniglio, and R. A. Harley. Mercury balance on a large pulverized coal-fired furnace. *J. Air Poll. Control Assoc.* **23**(9), 773–777 (1973).

Block, C., and R. Dams. Inorganic composition of Belgian coals and coal ashes. *Environ. Sci. Technol.* **9**, 146–150 (1975).

Bradford, G. R., A. L. Page, I. R. Straughan, and H. T. Phung. A study of the deposition of fly ash on desert soils and vegetation adjacent to a coal-fired generating station. In D. C. Adriano and I. L. Brisbin, Jr. (eds.) *Environmental Chemistry and Cycling Processes*. pp. 383–393. U.S. Dept. of Energy Symposium Series 45, CONF-760429 (1978).

Campbell, J. A., J. C. Laul, K. K. Nielson, and R. D. Smith. Separation and chemical characterization of finely-sized fly-ash particles. *Anal. Chem.* **50**(8), 1032–1040 (1978).

Coleman, R. The importance of sulfur as a plant nutrient in world crop production. *Soil Sci.* **101**, 230–239 (1966).

Coles, D. G., R. C. Ragaini, J. M. Ondov, G. L. Fisher, D. Silberman, and B. A. Prentice. Chemical studies of stack fly ash from a coal-fired power plant. *Environ. Sci. Technol.* **13**(4), 455–459 (1979).

Conner, J. J., J. R. Keith, and B. M. Anderson. Trace-metal variation in soils and sagebrush in the Powder River Basin, Wyoming and Montana. *J. Research U.S. Geol. Survey* **4**(1), 49–59 (1976a).

Conner, J. J., B. M. Anderson, J. R. Keith, and J. G. Boerngen. Soil and grass chemistry near the Four Corners power plant. In *Geochemical Survey of the Western Energy Regions*, pp. 112–120. U.S. Geol. Survey, Open File Report 76-729, 1976b.

Crockett, A. B., and R. R. Kinnison. Mercury residues in soil around a large coal-fired power plant. *Environ. Sci. Technol.* **13**(6), 712–715 (1979).

Davison, R. L., D. F. S. Natusch, J. R. Wallace, and C. A. Evans, Jr. Trace elements in fly ash, dependence of concentration on particle size. *Environ. Sci. Technol.* **8**(13), 1107–1113 (1974).

Dreesen, D. R., E. S. Gladney, J. W. Owens, B. L. Perkins, C. L. Wienke, and L. E. Wangen. Comparison of levels of trace elements extracted from fly ash and levels found in effluent waters from a coal-fired power plant. *Environ. Sci. Technol.* **11**(10), 1017–1019 (1977).

Eaton, J. S., G. E. Likens, and F. H. Bormann. The input of gaseous and particulate sulfur to a forest ecosystem. *Tellus.* **30**(6), 546–551 (1978).

Elseewi, A. A., F. T. Bingham, and A. L. Page. Growth and mineral composition of lettuce and Swiss chard grown on fly ash-amended soils. In D. C. Adriano and I. L. Brisbin (eds.) *Environmental Chemistry and Cycling Processes*. pp. 568–581. U.S. Dept. of Energy Symposium Series 45, CONF-760429 (1978a).

Elseewi, A. A., A. L. Page, and F. T. Bingham. Availability of sulfur in fly ash to plants. *J. Environ. Qual.* **7**, 69–73 (1978b).

Elseewi, A. A., A. L. Page, and S. R. Grimm. Chemical characterization of fly ash aqueous systems. *J. Environ. Qual.* **9**(3), 424–428 (1980).

Furr, A. K., W. C. Kelly, C. A. Batche, W. H. Gutenmann, and D. J. Lisk. Multielement uptake by vegetables and millet grown in pots on fly ash amended soils. *J. Agric. Food Chem.* **24**(4), 885–888 (1976).

Furr, A. K., T. F. Parkinson, R. A. Hinrichs, D. R. Van Campen, C. A. Bache, W. H. Gutenmann, L. E. St. John, I. S. Pakkala, and D. J. Lisk. National survey of elements and radio-activity in fly ashes. Absorption of elements by cabbage grown in fly ash-soil mixtures. *Environ. Sci. Technol.* **11**(13), 1104–1112 (1977).

Furr, A. K., T. F. Parkinson, W. H. Gutenmann, I. S. Pakkala, and D. J. Lisk. Elemental content of vegetables, grains and forages field-grown on fly-ash amended soils. *J. Agric. Food Chem.* **25**(2), 257–359 (1978).

Gladney, E. S., J. A. Small, G. E. Gordon, and W. H. Zoller. Composition and size distribution of in-stack particulate material at a coal-fired power plant. *Atmospheric Environ.* **10**, 1071–1077 (1976).

Gluskoter, H. J., R. R. Ruch, W. G. Miller, R. A. Cahill, G. B. Dreher, and J. K. Kuhn. Trace elements in coal; occurrence and distribution. *Ill. State Geological Survey Circular* **499**, 154 (1977).

Gough, L. P., and J. A. Erdman. Elements in lichen near the Dave Johnston power plant. In *Geochemical Survey of the Western Energy Regions*, pp. 22–29. U.S. Geological Survey Open File Report 76-729, 1976.

Griffith, E. D., and A. W. Clarke. World coal production. *Sci. Am.* **240**(1), 38–47 (1979).

Gutenmann, W. H., and D. J. Lisk. Absorption of selenium from coal fly ash-amended soil by *Astragalus racemosus*. *Bull. Environ. Contam. Toxicol.* **23**, 104–106 (1979).

Guthrie, R. K., and D. S. Cherry. The uptake of chemical elements from coal ash and settling basin effluent by primary producers. I. Relative concentrations in predominant plants. *Science Total Environ.* **12**, 217–222 (1979).

Hodgson, D. R., and R. Holliday. The agronomic properties of pulverized fuel ash. *Chem. and Ind.*, 785–790, May 14 (1966).

Joensuu, O. I. Fossil fuels as a source of mercury pollution. *Science* **172**, 1027–1028 (1971).

Jones, D. G., and I. R. Straughan. Impact of solid discharges from coal usage in the southwest. *Environ. Health Perspective* **27**, 275–281 (1978).

Jurinak, J. J., W. J. Grenney, G. L. Woldridge, J. P. Riley, and R. J. Wagenet. A model of environmental transport of heavy metals originating from stack derived particulate emission in semi-arid regions. Research and Development 77-RD-27, Southern California Edison Co., Rosemead, California (1977).

Kaakinen, J. W., R. M. Jorden, M. H. Lawasani, and R. E. West. Trace element behavior in coal-fired power plant. *Environ. Sci. Technol.* **9**(9), 862–869 (1975).

Klein, D. H., and P. Russell. Heavy metals: Fallout around a power plant. *Environ. Sci. Technol.* **7**(4), 357–358 (1973).

Klein, D. H., A. W. Andren, J. A. Carter, J. F. Emergy, C. Feldman, W. Fulkerson, W. S. Lyon, J. C. Ogle, Y. Talmi, R. I. Van Hook, and N. Bolton. Pathways of thirty-seven trace elements through coal-fired power plants. *Environ. Sci. Technol.* **9**, 973–979 (1975).

Lee, R. E., Jr., H. L. Crist, A. E. Riley, and K. E. MacLeod. Concentration and size of trace metal emissions from a power plant, a steel plant, and a cotton gin. *Environ. Sci. Technol.* **9**(7), 643–647 (1975).

Likens, G. E. Acid precipitation. *Chem. Engineer. News.* **54**(48), 29–44 (1976).

Linton, R. W., P. Williams, C. A. Evans, Jr., and D. F. S. Natusch. Determination of the surface predominance of toxic elements in airborne particles by ion micro-probe mass spectrometry and auger electron spectrometry. *Anal. Chem.* **49**(11), 1514–1521 (1977).

Lyon, W. S. *Trace element measurements at the coal-fired steam plant*. CRC Press, Cleveland, Ohio, 1977.

Malmer, N. Acid precipitation: chemical changes in the soil. *Ambio.* **5**, 231–234 (1976).

Mulford, F. R., and D. C. Martens. Response of alfalfa to boron in fly ash. *Soil Sci. Soc. Am. Proc.* **35**, 296–300 (1971).

National Ash Association. Ash collections soar to 59.5 million tons in 1974. *Ash at Work* **7**(3), 1–4 (1975).

Natusch, D. F. S., C. F. Bauer, H. Matusiewicz, C. A. Evans, J. Baker, A. Loh, R. W. Linton, and P. K. Hopke. Characterization of trace elements in fly ash. *International Conference on Heavy Metals in the Environment. Toronto, Ontario, Canada*, Vol. II, part 2,(1975) pp. 553–575.

Ondov, J. M., R. C. Regaini, R. E. Heft, G. L. Fisher, D. Silberman, and B. A. Prentice. Interlaboratory comparison of neutron activation and atomic absorption analyses of size-classified stack fly ash. *8th Materials Research Symposium Methods and Standards for Environmental Measurement. Gaithersberg, Maryland, Sept. 20–24* (1976).

Ondov, J. M., R. C. Ragaini, and A. H. Bierman. Emissions and particle-size distributions of minor and trace elements at two western coal-fired power plants equipped with cold-side electrostatic precipitators. *Environ. Sci. Technol.* **13**(18), 947–953 (1979).

Page, A. L., A. A. Elseewi, and I. R. Straughan. Physical and chemical properties of fly ash from coal-fired power plants with reference to environmental impacts. *Residue Rev.* **71**(2), 83–120 (1979).

Paulson, C. A. J., and A. R. Ramsden. Some microscopic features of fly-ash particles and their significance in relation to electrostatic precipitation. *Atmos. Environ.* **4**, 175–185 (1970).

Phung, H. T., L. J. Lund, and A. L. Page. Potential use of fly ash as a liming material. In D. C. Adriano and I. L. Brisbin, Jr. (eds.) *Environmental Chemistry and Cycling Processes*, pp. 504–515. DOE Symposium Series 45, CONF-760429, 1978.

Phung, H. T., L. J. Lund, A. L. Page, and G. R. Bradford. Trace elements in fly ash and their release in water and treated soils. *J. Environ. Qual.* **8**(2), 171–175 (1979).

Plank, C. O., and D. C. Martens. Boron availability as influenced by application of fly ash to soil. *Soil Sci. Soc. Am. Proc.* **36**(6), 974–977 (1974).

Prince, R. and F. F. Ross. Sulfur in air and soil. *Water, Air and Soil Poll.* **1**(3), 286–302 (1972).

Romney, E. M., A. Wallace, and G. V. Alexander. Boron in vegetation in relation to a coal-burning power plant. *Commun. in Soil Sci. and Plant Anal.* **8**(9), 803–807 (1977).

Schnappinger, M. G., Jr., D. C. Martens, and C. O. Plant. Zinc availability as influenced by application of fly ash to soil. *Environ. Sci. Technol.* **9**(3), 258–261 (1975).

Schwitzgebel, K., F. B. Meserole, R. G. Oldham, R. A. Magee, F. G. Mesich, and T. L. Thoem. Trace element discharge from coal-fired power plants. *International Conference on Heavy Metals in the Environment*, Vol. II, part 2, Toronto, Ontario, Canada (1975) pp. 533–550.

Shannon, D. G., and L. O. Fine. Cation solubilities of lignite fly ashes. *Environ. Sci. Technol.* **8**(12), 1025–1028 (1974).

Smith, R. D., J. A. Campbell, and K. K. Nielson. Concentration dependence upon particle size of volatilized elements in fly ash. *Environ. Sci. Technol.* **13**(5), 553–558 (1979).

Straughan, I., A. A. Elseewi, and A. L. Page. Mobilization of selected trace elements in residues from coal combustion with special reference to fly ash. In D. D. Hemphill (ed.) *Trace Substances in Environmental Health XII*. pp. 389–402. Univ. of Missouri Press, Columbia, Missouri, 1978.

Straughan, I. R., A. A. Elseewi, A. L. Page, I. R. Kaplan, R. W. Hurst, and T. E. Davis. Fly ash-derived strontium as an index to monitor deposition from coal-fired power plants. *Science* **212**, 1267–1269 (1981).

Swanson, V. E., J. M. Medlin, J. R. Hatch, S. L. Coleman, G. H. Wood, Jr., S. D. Woodruff, and R. T. Hildebrand. Collection, chemical analysis and evaluation of coal samples in 1975. U.S. Dept. Int., Geological Survey, Open File Report 76-468, (1976) 503 pp.

Theis, T. L., and J. J. Marley. Environmental consideration for fly ash disposal. *J. Energy Div. ASCE* **104**, 47–61 (1979).

Theis, T. L., and R. O. Richter. Chemical speciation of heavy metals in power plant ash pond leachate. *Environ. Sci. Technol.* **13**(2), 219–224 (1979).

Theis, T. L., and J. L. Wirth. Sorptive behavior of trace metals on fly ash in aqueous systems. *Environ. Sci. Technol.* **11**(12), 1097–1100 (1977).

Theis, T. L., J. D. Westrick, C. L. Hsu, and J. J. Marley. Field investigation of trace metals in groundwater from fly ash disposal. *J. Water Poll. Cont. Fed.* **50**(11), 2457–2469 (1978).

Townsend, W. N., and D. R. Hodgson. Edaphological problems associated with deposits of pulverized fuel ash. In R. J. Hutnik and G. Davis (eds.) *Ecology and Reclamation of Devastated Land, Vol. I*. pp. 45–56. Gordon and Breach, New York, 1973.

U.S. Bureau of Mines. *Minerals Yearbook, Vol. 1, Metals, Minerals and Fuels*, 1550 pp. U.S. Gov. Printing Office, Washington, DC (1977).

U.S. Environmental Protection Agency. *Energy/Environment Fact Book*. 76 pp. EPA-600/9-77-041, 1978.

Vaughan, B. E., K. H. Abel, D. A. Cataldo, J. M. Hales, C. E. Hane, L. A. Rancitelli, R. C. Rutson, R. E. Wildung, and E. G. Wolf. Review of potential impact on health and environmental quality from metals entering the environment as a result of coal utilization. Pacific Northwest Laboratories, Battelle Memorial Institute, Richland, Washington (1975).

SYNTHESIS

SOIL–AIR POLLUTANT INTERACTIONS

S. N. Linzon
Ontario Ministry of the Environment
Toronto, Ontario, Canada

M. A. Tabatabai
Iowa State University
Ames, Iowa

BACKGROUND

Despite the sulfate that is deposited by air pollution and ocean aerosols, sulfur deficiencies have occurred in inland aerobic soils in certain parts of the world. Although the information available indicates that the sorption of some pollutants (e.g., CO, C_2H_4, and mercaptans) by soils involves microbial processes, all the experiments reported involve comparison of rates of sorption of the pollutants by autoclaved and nonautoclaved soils. The possible effect of autoclaving soils, thereby changing the nature of the soil organic matter on sorption of pollutants by soils, requires further investigation.

The information obtained from laboratory studies indicates that soil is an important sink as well as a source (under certain conditions) for a number of air pollutants. However, it is important that such information should not be extrapolated to field conditions, mainly because most soils are covered with forest or agricultural crops that act as receptors for the pollutants, and the soil–air pollutant interaction is minimized. Therefore, generally plants are better indicators of the presence of air pollutants than are soils.

Addition of heavy metals to soils near smelters would lead to inhibition of enzyme-catalyzed reactions in soils. Such inhibitions have been reported for phosphatases in soils contaminated with copper and vanadium. No such inhibitions are expected, however, from trace elements, including heavy metals, added to soils by dry and wet deposition under natural conditions.

Although acidification of soils by acidic precipitation has been considered, it is doubtful if the effect of hydrogen ion concentration in precipitation on soil acidification could be elucidated. Numerous chemical and biochemical reactions in soils produce, in one year, hydrogen ions in concentrations markedly exceeding those added by precipitation in many years. Therefore alterations in nutrient cycling in areas impacted by acidic precipitation are difficult to evaluate.

DISCUSSION

1. Interactions of Gaseous Pollutants with Soils

Most of the information available in the literature concerning gaseous pollutant–soil interactions involves soil as a sink, with emphasis being placed on such pollutants as sulfur dioxide, oxides of nitrogen, carbon monoxide, ethylene, and fluorides. Very little information is available on the interactions of oxidants such as peroxyacyl compounds and ozone with soils. Generally, sorption of inorganic gaseous air pollutants by soils increases as the molecular weights decrease, but organic gases are absorbed at faster rates as their molecular weights increase, and other functional groups such as nitrogen, phosphorus, oxygen, and sulfur are included. Sorption by soils of low molecular weight, less-substituted organic gases is generally thought to depend on the soil microbial population. Sorption of most inorganic gases by soils is attributed primarily to chemical and/or physical parameters of soils.

In many cases, especially under controlled conditions with finite amounts of soil, the initial sorption of gases is very rapid, reducing fairly substantial concentrations to very low values in a matter of minutes to hours. For example, 1 g soil exposed to 100 mL of air containing 100 ppm of SO_2 can result in the SO_2 reduction of air concentration to near zero in *about 30 s to 1 min. Relative to sulfur gases in the atmosphere, it is safe to assume that generally the gases are sorbed by soils in rates related to the gas molecular weight: the lower the molecular weight, the faster sorption by soils. One exception is the inert SF_6, which is not sorbed by soils. With the nitrogen oxides, however, it is generally accepted that the more highly oxidized the nitrogen gas, the more rapidly it is sorbed by soils.*

Under normal air quality, sorption of gaseous pollutants by soils would not affect the chemical and biochemical reactions in soil systems, and current knowledge indicates that gaseous air pollutant interactions with soils have not generally resulted in detrimental effects. Dry and wet deposition of low levels of sulfur compounds on soils seem to indicate beneficial effects to crop production in many areas around the world. The reported beneficial effects of atmospheric sulfur on crop production in the north-central region of the United States constitute one example.

2. Acidic Precipitation

2.1. DIRECT EFFECTS ON SOIL

Reported effects of acidic precipitation on vegetation are based mainly on greenhouse or field experiments using simulated acidic rain. In discussing the direct effects of acidic precipitation on soil, it should be noted that soil is naturally acidic in many parts of the world, especially in forest ecosystems. Rainfall over hundreds and even thousands of years prior to man-made industrial inputs had naturally acidified soil by long-term leaching of neutralizing cations. The decomposition of forest litter and the natural biological processes taking place in soil, such as N mineralization and nitrification, also contributed to soil acidity. In recent times soils have been acidified by agricultural practices in the use of fertilizers such as anhydrous ammonia and other ammonium-bearing compounds.

In acidified soils which occur in forested and agricultural areas and near point sources of industrial emissions, a number of changes can occur. Some of these changes, which also have been induced in soils in experiments using simulated acidic rain, include the leaching of basic cations such as magnesium and calcium, the mobilization of soil-bound metals such as aluminum and iron, and changes in biological activities (e.g., nitrification and nitrogen fixation). However, the contribution of current industrial activities to acidic precipitation and its role in enhancing chemical and biochemical reactions in soil requires further study.

It is generally considered that soils low in cation exchange capacity and base saturation are most sensitive to acidification. However, this concept is challenged by a number of soil scientists. In Sweden, based on laboratory experiments and field studies, it has been suggested that acidified soils with a $pH < 5$ would be only slightly affected by acidic *precipitation, since most of the cation exchange sites are already occupied by H^+ ions. Calcareous soils with a very high buffering capacity would be most resistant, whereas noncalcareous sandy soils with a pH > 6 would be most sensitive.*

According to some, the acidity of lakes in Norway is due not to acidic precipitation but to changes in land use. The change from agriculture to forestry has produced increasing organic matter in soil, and the humic acid leached from humus by rainfall has resulted in the acidity of runoff 2 times greater than the precipitation itself. However, others believe that both wet and dry deposition have resulted in sulfate and H^+ accumulations on vegetation and, with heavy rains, the washing off of these pollutants produces a runoff with an acidity up to 5 times greater than the precipitation itself.

It would appear that both groups of scientists are correct, since both precipitation and vegetation can contribute to lake acidity. However, it has been observed in other parts of the world—for example, in Ontario, Canada—that lakes have been gradually becoming more acidic in forested locations where no change in land use has occurred. The geological conditions, the naturally developed acid soil, the current forested vegetation, and precipitation all have contributed to the acidification of the lakes.

2.2. INDIRECT EFFECTS ON VEGETATION

It has been suggested that certain forests in Sweden had suffered an annual growth loss of 0.3%, attributed to soil impoverishment caused by acidic precipitation. However, these observations have not been confirmed in studies conducted in Norway and Sweden, and no direct proof has been offered to show that the acidity of rain commonly occurring in southern Scandinavia has adversely affected tree growth, since there is a great individual variation in tree growth due to genotype, age, competition from neighboring trees, and site conditions. There are also a number of compensatory factors such as the stimulatory effect of the nitrogenous component of acidic precipitation and nutrient cycling in soil caused by leaf fall and root weathering of bedrock.

There is some controversy as to the need for control of industrial emissions of sulfur dioxide and nitrogen oxides, since it does not appear to some that the acidity of precipitation, soil, or lakes had really increased in the last two decades. Experiments conducted by scientists using artificially acidified solutions have shown that adverse effects can be induced in soil and vegetation, but the acidities of the solutions used were far greater than those occurring in rainfall. There was a general agreement that much more research and field monitoring was needed.

3. *Biogenic Emission of Gaseous Pollutants*

A variety of biological processes are involved in emissions of a number of gaseous pollutants to the atmosphere. Some of the nonindustrial sources of atmospheric sulfur gases are sewage sludge, animal manure, and hot springs. Studies at the University of Calgary showed that SO_2 in the air in the vicinity of hot springs is isotopically similar to the H_2S evolved from the hot springs, indicating that the biogenic H_2S is being very rapidly converted to SO_2. Other studies in France showed that tomato plants, soon after exposure to $^{35}SO_2$ had 98% of the ^{35}S in the form of sulfate, and some of this sulfur was subsequently volatilized as $H_2^{35}S$, especially if the plants were illuminated.

Similarly, production of methyl mercaptan has been reported to occur in river water, rice paddies, soils, spoiled foods, and animal manure. Also, emission of this gas occurs as the product of the decomposition of cruciferous plants and other organic material. Similarly, dimethyl sulfide can be found in a variety of atmospheric environments, with sources being ocean water, paper mill effluents, soils, river water, lake water, and spoiled foods.

Carbon disulfide has been found to be a major product in certain soils; it is also emitted from manure and in very small amounts from seawater. Carbon disulfide is extremely effective at inhibiting nitrification in soils. Carbonyl sulfide has also been detected from manure and soils.

Emissions of N_2O, NO, NO_2, and NH_3 from soils can be significant under certain conditions. The relative contribution of the biogenic sources to atmospheric contents of these gases deserves investigation.

4. *Trace Element Enrichment of Soil*

The combustion of coal results in the release of many trace elements. The concentrations of several elements found in soil are similar to those found in coal—for example, arsenic, cadmium, lead, mercury, chromium, copper, nickel, and zinc. However, coal is relatively more enriched in boron, molybdenum, and selenium. This is also generally true of fly ash.

A number of case studies have attempted to document the contamination of soil surrounding power plants but with little success, because of the high background concentrations of most trace elements in the soil. Models can be used to predict deposition of particles from point sources. By knowing the concentrations of the various trace elements in the particulate matter as they exit the stack, one can compute the extent to which soils may become enriched. In most cases, the amount deposited has been computed to be very small compared with the amount already present in soils.

With respect to the growth of plants on soils amended with fly ash, reductions in yield occur at around 2% addition of fly ash. The addition of fly ash to soil increases pH, since it is highly alkaline, and also increases salinity. Fly ash amendments can result in toxic levels of trace elements such as boron, which can injure vegetation, and molybdenum and selenium, which can make forage unsafe for animals. Conversely, sulfur in fly ash has been found to be beneficial to plants grown in sulfur-deficient soils. Also, fly ash contains about 30% calcium and can be used as a supplemental source of agricultural limestone. At the present time, except for experiments and in isolated instances, fly ash is not being used as a soil amendment, because of the potential toxic effects.

Methods to ameliorate soil contamination around various sources are known but may not be practical. Heavy-metal contamination of soil will persist for long periods of time. A number of organic compounds are complexers of metals and can remove metals from soils in solution, but the drainage may create a problem elsewhere. In many instances contaminated soil is removed to a landfill and is replaced with fresh uncontaminated soil; however, this is very expensive. Research is needed to find inexpensive, practical soil-amelioration methods.

5. Research Needs

A number of questions regarding soil–air pollutant interactions require elucidation. Some of these questions which could be resolved by further research are:

1. *What are the fate and persistence of various pollutants in soil?*
2. *What are the agricultural and forest effects from pollutants in sewage sludge?*
3. *What are the effects of pollutants on microorganisms and on biological and biochemical processes in soil?*
4. *What are the indirect effects of soil pollution on germination and growth of plants?*
5. *What are the permissible levels of pollutants in fruits and vegetables?*
6. *Could the use of stable sulfur and other elemental isotopes help identify sources of pollutants, chemical and biochemical transformations of pollutants in soils, and uptake by plants?*
7. *What is the contribution of biogenic sources to atmospheric pollution?*
8. *How can reentrainment and resuspension of soil pollutants to the atmosphere be assessed?*
9. *What are the sensitivities of various soil types to acidic precipitation?*
10. *What are practical, cost effective measures to ameliorate soil that has been contaminated by pollutants?*

PART EIGHT

DATA ACQUISITION, INTERPRETATION, MODELING, AND APPLICATION IN TERRESTRIAL VEGETATION–AIR POLLUTANT INTERACTION STUDIES

Proper and scientifically sound experimental design and data acquisition are preludes to all subsequent analysis and application of such data. For example, with ozone, at low wind velocities (below 5 mph) a forest canopy is subjected to a higher ozone concentration over time in comparison to a crop canopy. Therefore, appropriate data acquisition is imperative to effects analysis. Similarly, data acquisition relative to pollutant concentrations over time, above and below the canopy, should be addressed with regard to frequency and type of data acquisition based upon the type of effects criteria in question.

A great deal of effort is needed to integrate data acquisition and interpretation in relation to terrestrial ecosystems research. This is particularly true when one relates other disciplines of science such as statistics and reproducibility.

The aforementioned considerations are critical to modeling impacts. A model may be unsatisfactory if the proper types of parameters measured and their frequency are inadequate. Models can be successful if they are verified with actual measurements and derivations of such data.

The verification of mesoscale and local air pollution stress ecosystem-response models is vital to understanding both the productivity and maintenance of agronomic and natural ecosystems. Physical, chemical, and biological processes have both short- and long-term global consequences. The state of the art of application of current models requires refinement.

The objective of this section is to address these and other related issues.

Air Pollutant Interactions with Vegetation: Research Needs in Data Acquisition and Interpretation*

STEVEN E. LINDBERG AND S. B. McLAUGHLIN
Environmental Sciences Division
Oak Ridge National Laboratory
Oak Ridge, Tennessee

*While this paper was originally written in 1980, meaning that at the time of publication the reference list is no longer current, we feel that most of the research needs are still very relevant. Since 1980, the community has begun to address several of the gaps in our knowledge described here, and significant advances are anticipated in this field over the next 10 years.

1. INTRODUCTION

Collection of the data necessary to address the issue of air pollutant–vegetation interactions adequately is not a straightforward task. A formidable challenge is acquisition of data unbiased by the experimental design, uncomplicated by the wide degree of variability associated with natural systems, and uncontaminated by the sampling apparatus. Those who advocate strictly controlled experiments in which variables are manipulated are faced with noncomparability with the real world. On the other hand, those who advocate field studies of pollutant–ecosystem interactions are often accused of being unable to interpret their data clearly in light of the known variability of several uncontrolled, interrelated parameters. The ultimate solution will likely involve an integration of these two approaches.

Another difficult problem is one of scale. The most apparent effects of air pollutants on terrestrial biota, including man, occur on a localized scale in urban and industrial areas where pollutant concentrations are high due to local activity. However, ecosystem-level effects result from pollutant exposures integrated over both time and space. Such effects are often due to chronic exposure to larger portions of the atmosphere where pollutant concentrations are much smaller because of dilution and removal mechanisms. The situation is further complicated by effects related to exposure to pollutant mixtures and to transient pulses of high pollutant concentrations. The distinction between local and regional pollutant concentrations is critical in the design of sampling programs. Because of the considerable decrease in atmospheric concentrations of gases and particles from local to regional and global scales, the necessary sample collection and analytical methodologies are quite different in many cases. Methods useful in monitoring local pollution levels may be inappropriate for monitoring regional background levels. The problem is further compounded by the fact that most constituents of anthropogenic emissions also occur naturally in the atmosphere. This suggests the need for increased sensitivity, precision, accuracy, and time resolution in future monitoring techniques.

The objective of this discussion is to consider these and other problems involved in the acquisition, interpretation, and application of data collected in studies of air pollutant interactions with the terrestrial environment. Emphasis will be placed on a critical evaluation of current deficiencies and future research needs by addressing the following questions:

1. Which pollutants are sufficiently toxic, pervasive, or persistent to warrant the expense of monitoring and effects research?
2. What are the interactions of multiple pollutants during deposition, and how do these influence toxicity?
3. How do we collect, report, and interpret deposition and air quality data to ensure its maximum utility in assessment of potential regional environmental effects?

4. What processes do we study, and how are they measured to describe most efficiently the relationship between air quality dose and ultimate impacts on terrestrial ecosystems?
5. How do we integrate site-specific studies into regional estimates of present and potential environmental degradation (or benefit)?

2. CHARACTERIZING AND MONITORING ATMOSPHERIC CONSTITUENTS

2.1 Which Pollutants Are of Interest?

Monitoring implies field measurements which can sense changes in environmental quality so that corrective measures may be taken before serious deterioration results. Physical measurements of certain atmospheric parameters (temperature, wind speed and direction) can be made with speed and simplicity, so that interpretations of changes in physical parameters can be made nearly simultaneously with the measurements. Some chemical measurements, such as pH and trace gas concentrations (SO_2, NO_x, O_3, CO), can also be monitored continuously using reliable field instrumentation. However, for other critical chemical parameters (toxic metals and organic compounds in either gaseous form, particle-associated, or dissolved in meteoric waters), substantial time delays between sampling in the field and measurement in the laboratory are required.

The decision of which constituents to monitor on local and regional scales is not as straightforward as it may seem. There is, of course, the general physical categorization of gas phase versus particle-associated, and beyond that, organic versus inorganic. Since in many cases scientists have specialized in either particle or gaseous pollutant interactions, this seems a logical division.

2.1.1 Particles

The sources of many of the long-lived particles in the atmosphere are various combustion processes, both natural and associated with human activities. It follows that those elements which exhibit, or whose compounds exhibit, high volatilities might be expected to occur in relatively high atmospheric concentrations. In addition, volatile elements generally are discharged from combustion sources as either gases or associated with submicron particles, both phases being relatively poorly contained by conventional emission control technology and exhibiting relatively long atmospheric residence times (Van Hook and Shultz, 1977; Keyser et al., 1978). Gas-phase reactions with particles result in preferential concentration of volatile elements on particle surfaces (Natusch et al., 1974; Natusch and Wallace, 1974; Keyser et al., 1978), which may increase their effective bioavailability through solubilization reactions (Eisenbud and Kneip, 1975; Lindberg et al., 1979a).

A list of the elements having these characteristics and suspected of or known to cause adverse environmental effects indicates that there is some justification for the monitoring of approximately 25% of the elements in the periodic table. Although it is difficult to develop rigid priorities from such a cursory examination, it is helpful to select a more tractable list of priority constituents of atmospheric particulate matter. We have selected those elements which have been repeatedly identified elsewhere in lists of toxic elements and compounds, elements readily mobilized by human activities, and elements highly enriched in both urban and regional aerosols (Schroeder, 1971; Duce et al., 1975; Zoller et al., 1974; Andren and Lindberg, 1977; Heindryckx, 1976; Lawson and Winchester, 1978; Gordon, 1975; Van Hook and Shultz, 1977; Morgan, 1975; Vaughan et al., 1975; Lim, 1979). This list includes Hg, Cd, Pb, Sb, Se, S (as acid sulfates), and C (as hydrocarbons). Because of the existence of Hg primarily as a vapor in combustion and other industrial processes and in the ambient air (Lindberg and Turner, 1977; Lindberg, 1980), it will be considered in the discussion of gaseous pollutants.

Carbon is included in this list to represent the general class of airborne microcontaminants which are becoming increasingly important because of their ubiquity and toxicity (including carcinogenic and mutagenic activity) (Chrisp et al., 1977; Fisher et al., 1979). Of these compounds, simple, polyaromatic, substituted, and halogenated hydrocarbons may be of most concern (Weisburger, 1979; Bjorseth et al., 1979; Greist and Guerrin, 1979). Because of the limited knowledge concerning their behavior, considerable research is required in the areas of sampling and analytical methodology, atmospheric transport and transformation, and the effects of organic micropollutants on the terrestrial environment.

2.1.2 Gases

The gaseous pollutants which currently occur at phytotoxic levels over broad areas of industrialized countries are SO_2, O_3, and NO_x. The U.S. Environmental Protection Agency (EPA) annually publishes a National Air Quality and Emissions Trend Report (most recently for 1984) to report temporal and spatial variability in levels of these pollutants. Concern over their occurrence and distribution is based primarily on their potential interactions (see Reinert et al., 1975) at the relatively low levels ($\leq$0.10 ppm) at which they occur on a regional basis. Problems of a local scale may also occur around industrial point sources of HF, SiF_4, Cl_2, HCl, NH_4, and H_2S; however, none of these pollutants currently warrants the expense of a national monitoring network.

Future trends in industrial development, transportation patterns, and fossil fuel combustion may alter our air quality monitoring needs. Secondary pollutants such as peroxyacetyl nitrates (PAN and other homologs) may become more significant. Complex organic and sulfur-containing compounds may become important as effluents from the developing coal conversion industry (Gehrs et al., 1980). Their significance will depend heavily on technological choices (sulfur recovery and off-gas flaring) currently being considered.

Increased reliance on coal may make monitoring of volatile trace inorganics an important consideration in evaluating potential long-term impacts on terrestrial ecosystems. Mercury is a potentially phytotoxic pollutant which is emitted primarily in gaseous form during the combustion of coal, remains as a vapor in ambient air, and can be directly absorbed by some crop plants (Lindberg et al., 1979b; Lindberg, 1980). The potential for interactions between subacute levels of mercury and other regional-scale gaseous pollutants is not known.

2.2 What Levels of Concentration Are of Interest at the Source and at the Receptor?

The variability of trace element concentrations in ambient air and in precipitation collected above and below the forest canopy has been reported to be over several orders of magnitude (McMullen and Faoro, 1977; Lindberg et al., 1979a). It is apparent from these reports that some knowledge of expected concentration ranges is necessary to ensure accurate selection of both sampling and analytical techniques prior to the initiation of monitoring. Only then can one be assured of optimum precision and accuracy in reported data without complicating future data analyses with the reporting of "less than" or "below detection" values or data resulting from detector overload or nonlinear response problems.

Although considerable data exist on time-averaged concentrations in various media, there is a need for determination of short-term concentrations at the exposure microsite itself. In the context of this symposium this includes the concentration in the air at the leaf boundary layer, the concentration of dry deposited material on the foliar surface, and the concentration of material in solution on the foliar surface as a result of wet deposition and subsequent interaction with previously dry-deposited material. These data are strongly related to vegetation effects and should prove essential for predictive purposes. An additional need is for analysis of fine-scale temporal variations in concentration both at the source and at the receptor. Studies of temporal variability have proved very useful in the identification of pollutant sources and mechanisms of surface transfer and in the prediction of vegetation effects (Raynor, 1976; Lawson and Winchester, 1978; McLaughlin et al., 1979).

2.3 Sampling and Analytical Methodology

It is widely recognized by those familiar with highly sensitive, multielement analytical techniques employed in air pollution studies that problems associated with sample collection are often the limiting step in achieving the levels of precision and accuracy necessary for unbiased data analysis. The general problem of representative sample collection (as influenced by contamination from collection surfaces or reagents, irreversible loss of some components of the pollutant to the sample container, and postdepositional reactions between several pollutants) often controls detection limits, rather than sensitivity of the analytical instrumentation (Altshuller, 1972; Lindberg et al., 1977).

It is not the intent of this review to describe in detail the wide range of sampling and analytical techniques used in this field (see the discussions by Newman and Gordon in this volume). Rather, we shall review the problems associated with the collection, analysis, and interpretation of particulate pollutants in the atmosphere. The problems associated with sampling and analysis of gases are primarily those of achieving microscale response times of the associated analytical instrumentation and unequivocal separation of gas and particulate species of the same element (e.g., S, N). The former problem is largely limited to the field of micrometeorology where subsecond response times are often necessary for eddy correlation calculations of pollutant fluxes (Hicks and Wesely, 1980). In general, the available technologies afford sufficient response times (seconds to minutes) for the needs of the plant physiologist interested in major gaseous pollutant effects (O_3, SO_2, NO_x). Work is needed, however, on development of subminute response samplers and detectors for trace-level gases (e.g., CS_2, Hg^0). The problem of separation of gas and particulate forms applies equally to particle characterization and will be discussed below.

In our opinion the problems associated with particle collection are more complex and the solutions more urgently needed. While aerosol concentration monitoring is accepted as a necessity in human health effects studies, its relationship to terrestrial effects studies may not be as apparent (National Research Council, 1979). At an EPA-sponsored workshop on dry deposition methodologies (Hicks et al., 1980), several experts in the field advocated the use of monitoring data on suspended-particle concentrations to calculate dry-deposition fluxes. This suggestion reflected the recognition of the inadequacies of dry deposition collectors (discussed further in Section 4.1.2).

Particle sampling procedures are subject to errors related to gravitational, inertial, diffusional, thermal, and electrical forces and to chemical interactions. In addition, the problem of spatial heterogeneity of the dispersed aerosol may, in itself, overwhelm the others. While these problems directly confront the researcher actively involved in sample collection, we are all subject to the problem of improper interpretation and application of data which are collected without careful attention to potential sampling and analytical errors. This discussion is aimed primarily toward those individuals who may not be experts in the field of particle physics, but who need to interpret existing aerosol data in the context of air pollutant deposition and effects studies or to acquire their own data while constrained with limited resources. A more detailed review of aerosol measurement is presented in Lundgren et al. (1979).

Suspended particles are collected primarily by three techniques: filtration, impaction, and impingement, or some combination of these, with filtration and impaction used in more than 75% of the applications. With some exceptions, filtration yields only the total aerosol, while impaction also provides the particle size distribution. In their simplest form aerosol filters consist of an inert filter holder through which air is drawn at flow rates ranging from approximately 1 L/min to approximately 1800 L/min. The most often used impactor, the cascade impactor, consists of a series of orifices of decreasing diameter behind which are mounted the impaction surfaces intended for chemical analysis, and onto which particles in a selected size range are deposited as air is drawn through the device. Impactors also often utilize filters as impaction surfaces or in a backup mode to collect the finest particles not impacted on the upper stages.

Care must be taken in the use of these methods and subsequent analysis and interpretation of the data, particularly when applied in the context of the papers in this volume. Table 1 lists several specific problems inherent in these methods. In general, the specific problems all relate to the collection of a sample representative of material suspended in the atmosphere or the performance of an analysis truly representative of the collected material.

In the case of filtration methods, the analytical problems are often related to the filter itself or to postdepositional changes in particle chemistry following collection. Of the many filters now available, the most often used are glass fiber and cellulose in high volume (~1800 L/min) samplers (Hi-vols), and various membrane filters (cellulose acetate, polycarbonate, Teflon) in addition to glass fiber and cellulose in low volume (~1–30 L/min) and impactor samplers. While glass fiber filters have been the standard in Hi-vol sampling and are known to exhibit high efficiencies for retention of submicron aerosols (Lockhart et al., 1963), they are often unacceptable for trace-level inorganic analysis because of their high and variable trace element content (Neustadter et al., 1975; Sievering et al., 1978). On the other hand, because of their inert nature with respect to organic solvents and their relatively low ability to adsorb many hydrocarbons, particularly polyaromatic hydrocarbons (Brown et al., 1983), glass fiber filters may be the only choice for trace organic analysis of aerosols (Pierce and Katz, 1975). While cellulose filters are consistently lower in trace-element content than are glass fiber filters, they may exhibit lower particle collection efficiencies (Lindekin et al., 1963; Sievering et al., 1978), although there is some question as to the statistical significance of these differences (Neustadter et al., 1975).

The question of filtration efficiency also applies to low-flow samplers employing membrane filters. Large-particle collection efficiencies (>3 μm) decrease under conditions of high filter loading, while efficiencies for

TABLE 1. Typical Problems Encountered in the Characterization of Airborne Particles by Filtration and Impaction Methods

Filters	Impactors
1. High blanks (contamination) (+)[a]	1. High blanks (+)
2. Nonisokinetic sampling (−)[b]	2. Nonisokinetic sampling (−)
3. Small-particle loss (−)	3. Particle bounce (ΔS)[c]
4. Pressure-drop, decreasing flow (−)	4. Wall loss (ΔS, −)
5. Clogging (−)	5. Small-particle loss (−, ΔS)
6. Gas–particle reaction on filter surface (±)	6. Pressure-drop, variable flow rate (ΔS)
7. Incompatibility of filter surface (±)	7. Clogging of backup filter (−, ΔS)
	8. Gas–particle reactions on impaction surfaces (±, ΔS)
	9. Incompatibility of sticky-coated surfaces with extraction or analytical methods (±)
	10. Detailed aerosol behavior obscured by long sampling times (ΔS)
	11. Uncertainty in interpretation of particle sizes on backup filters ("less-than" values)

[a](+), generally results in overestimate of concentration.
[b](−), results in underestimate of concentration.
[c](ΔS), results in a shift in the apparent particle size distribution.

smaller particles (<0.5 μm) may decrease under conditions of very low filter loading (Fan and Gentry, 1978; Skogerboe et al., 1977). Laboratory studies of membrane filter efficiencies have indicated that particle collection efficiency decreases rapidly, from approximately 90 to 20%, as particle size decreases below 1 μm for Nuclepore filters of pore size ranging from 0 to 8 μm and is dependent on pressure drop (Liu and Lee, 1976; Spurny et al., 1969, 1974). For 0.6-μm Nuclepore and 5-μm Teflon membranes, efficiencies ranged between 80 and 100% for particles 0.03–1.0 μm (Liu and Lee, 1976). Further research is needed to develop theory applicable under the conditions of high pressure-drop and compressible gas flow occurring in routine use. Particle retention is also strongly influenced by the design of the sampler intake, its orientation with wind direction, and the use of external rain shields (Pattenden and Wiffen, 1977). A typical filter sampler collects large particles (>10 μm) with an efficiency of <15%. Since particles in this size range are readily removed from the atmosphere by deposition processes, the underestimate of this fraction has important implications to pollutant effects studies.

The choice and application of filter and sampler types in the field must be based on considerations of anticipated loading, particle size distribution, and analytical methodology. Such restrictions may preclude the development of a universally applicable total aerosol sampler. Until this becomes available it is our responsibility to present and interpret particle concentration data with consideration of the limitations of the techniques and to apply these data in effects studies with considerable care.

The quality of air concentration data may be limited by the accuracy of the measurement of total air flow through the sampler. An uncertainty in air flow rates of approximately 25%, for example, results in an equal uncertainty in element concentrations in the atmosphere. Continuous flow monitoring is not always feasible; yet flow rates may vary substantially during the course of a single collection period. The researcher is faced with the task of choosing a sampling interval sufficiently long to be compatible with his analytical sensitivity and yet short enough to prevent significant variations in flow rate.

Postdepositional chemical changes in the aerosol collection surface may result in losses or gains of some constituents during sampling. Berg (1976) reported that sampler design played a significant role in the conversion of particulate chlorine to gaseous form. Such reactions lead to an underestimate of particle concentrations and are enhanced by high filter loading and flow rates and may occur for other halides, some compounds of S and C, and Hg. In addition, in ambient air sampling for particulate sulfates, overestimates may result from absorption of SO_2 and subsequent oxidation to sulfate on the collection surface (Coutant, 1977; Cheney and Homolya, 1979). The effects of gas–particle interactions at sample collection surfaces must be considered in more detail (and for other elements) with the intent of developing techniques which minimize this problem.

Pollutant concentration data reported on a particle size basis may be more useful in effects research because of the influence of particle size on wet and dry deposition processes. In addition to the problems common to filters and impactors, Table 1 presents additional problems related to sampler design and aerosol chemistry which influence impactor operation. Nonisokinetic sampling, which results from sampling without compensation for wind speed and direction effects, leads to low collection efficiencies for the large-particle fraction (>5 μm). Wind tunnel tests of the widely used 8-stage Andersen impactor (28.3 L/min flow rate) indicated collection efficiencies for 5-μm particles of ~70% and for 15-μm particles of 2% (Wedding et al., 1977a). Subsequent modifications to this impactor (now available from the manufacturer) have increased collection efficiencies to near 100% for 5-μm and 50% for 15-μm particles (Wedding et al., 1977b; McFarland et al., 1977). In comparing the size distributions of Pb, Cd, and Zn in aerosols collected in California, Davidson (1977) found a significantly larger mass of these elements in particles >10 μm during isokinetic experiments compared with nonisokinetic experiments performed earlier in the same area. Because of high deposition velocities associated with large particles (Sehmel and Hodgson, 1976), every effort must be made to sample this fraction efficiently.

Wall losses and particle bounce also involve primarily the large-particle fraction. Particle rebound and impaction on surfaces other than those intended for analysis generally result in an underestimate of the large-particle concentration (Cheng and Yeh, 1979). An equally serious consequence is distortion of the entire particle-size spectrum, with large particles being inadvertently deposited on stages designed to collect smaller particles (Dzubay et al., 1976). This phenomenon is relatively simple to detect by inspection of impaction stages by scanning electron microscopy and has been reported for some standard impactors (Dzubay et al., 1976; Lindberg et al., 1979a). These problems can be minimized by isokinetic sampling (Wedding et al., 1977b), by selecting samplers designed to be less subject to particle bounce (Gordon et al., 1973), or by coating the impaction surfaces with a sticky substance such as grease (Dzubay et al., 1976). However, the use of sticky substrates may not be amenable to many extraction and analytical techniques and is inefficient under high temperature conditions (Cheng and Yeh, 1979; Casey, 1975). We have studied the water solubility of size-fractionated aerosols for several trace elements and have found grease-coated impaction surfaces to be incompatible with our extraction methods (Lindberg and Harriss, 1980). In addition, organic substrates are likely to cause serious matrix problems for the extraction and analysis of organic micropollutant aerosols.

Problems related to aerosol characteristics involve gas–particle interactions and the effect of relative humidity on aerosol impaction efficiency. Dingle and Joshi (1974) observed ~25-μm crystals of $(NH_4)_2SO_4$ on the stage of an Andersen impactor designed to collect 0.5-μm particles. Crystal formation, enhanced under high relative humidities, could result in gross overestimates of fine-particle atmospheric sulfate concentrations. Relative humidity has also been found to influence particle adhesion to impactor surfaces, such that at humidities >75% no sticky coating on the impaction surfaces was needed to reduce particle bounce (Winkler, 1974). With decreasing relative humidity the relative composition of the sampled aerosol may change due to preferential deposition of water-soluble material, leading to an erroneous size distribution. Aerosol solubility is extremely important in vegetation effects (see discussion in Section 5.2.1); therefore, it is critical to sample this material accurately.

The collection, interpretation, and application of particulate, and to some extent gaseous, air pollutant concentration data are obviously complex tasks. The preceding review was intended to raise several questions in the mind of the air pollution effects researcher; we would like to raise a few more of our own. The importance of a constant and accurately known flow rate during filter sampling is even more critical in impactor sampling because of the additional dependence of particle-size separation efficiency on flow rate. When no backup filter is used, the designed constant flow rate for any given impactor can be readily maintained. However, in this mode, considerable fine-particle material is not collected. When a backup filter is used, clogging and pressure drop can result in a variable and decreasing flow rate such that the specified particle retention characteristics of the impactor are no longer directly applicable. How is one to interpret the resulting size spectrum?

Another source of ambiguity in impactor data arises because a significant fraction (>50%) of the total air concentration of many important major (S) and minor (Pb) pollutant elements are captured on the backup filter. The aerodynamic cutoff diameter for the backup filter in the standard 8-stage Andersen impactor, for example, is expressed as a "less-than" value (<0.5 μm). In the application of deposition models to these data, should this particle size be interpreted as equal to 0.5 μm, equal to 0.25 μm (midpoint between 0.5 and 0 μm), or in some other manner? This is critical in the estimation of dry deposition fluxes because of the strong dependence of deposition velocities on particle size for particles <0.5 μm (Sehmel and Hodgson, 1976).

Commonly used samplers require collection periods of approximately 12 h to several days. Winchester and co-workers have studied temporal variations in total particulate air concentrations of several elements using a streaker sampler of their own design and proton-induced X-ray emission analysis, yielding a 2-h resolution time (see Lawson and Winchester, 1978). If we hope to understand the mechanism of deposition and its effect on vegetation, we need to develop instrumentation with the ability to provide data on short-term variations in pollutant concentrations and particle-size distributions.

Although considerable research is still needed to address these issues, there have been important developments, some of which are now in general use. The dichotomous sampler has been used in several monitoring networks (EPA, Multi-State Atmospheric Power Production Pollution Study (MAP3S), National Atmospheric Deposition Program (NADP-selected sites) and is a virtual impactor designed to eliminate several typical impactor problems. Because this sampler was originally intended for use in inhalation studies to separate particles into respirable and nonrespirable sizes (Conner, 1966; Dzubay and Stevens, 1975), it is of limited value in studies of deposition to environmental surfaces because of its limited particle-size fractionation (stage 1, ≤~2 μm; stage 2, >~2 μm). These same arguments apply to two-stage respirable aerosol samplers which use Nuclepore filters in series (Parker et al., 1977), although there are other multiple-filter series samplers which have been used under controlled conditions to fractionate aerosols between 0.1 and 1 μm (Melo and Phillips, 1975).

A promising development is a more recently designed low-pressure impactor compatible with ambient air sampling and available analytical techniques (Hering et al., 1978, 1979). The sampler provides highly accurate and efficient separation of aerosols into eight size classes ranging from 0.05 to 4.0 μm and minimizes size distribution distortion caused by particle bounce, reentrainment wall loss, and evaporation. This device, combined with a sampler providing efficient large particle (>4 μm) size classification, would provide the necessary data for application of theory and laboratory studies to the calculation of dry deposition fluxes.

In general, sampling problems outweigh those of analytical sensitivity, precision, and accuracy. For detailed discussions of analytical problems the reader is referred to other papers on the analysis of pollutants in air (Altshuller, 1972), in precipitation (Turner and Lindberg, 1976; Lindberg et al., 1977; Peden et al., 1979), and in environmental samples in general (Tolg, 1975; Morgan, 1975). The application of X-ray and activation analysis to studies in the atmospheric environment was reviewed by Winchester and Desaedleer (1975), while Harris (1976) has reviewed chemical and physical characteristics of atmospheric particles, describing suitable analytical techniques for most of the elements in the periodic table.

A crucial research need with respect to evaluating the exposure of terrestrial biota to air pollutants is chemical and physical speciation analysis. It may no longer be sufficient to know only the concentration of an element in suspended or deposited particles. One might also need to know:

1. The chemical form; for example, trivalent arsenic is considerably more toxic than the pentavalent form in solution (U.S. Atomic Energy Commission, 1973; National Academy of Sciences, 1977).
2. The physical distribution within the particle; for example, trace elements in the atmosphere do not generally occur as discrete particles of Se, but as heterogeneous mixtures of compounds distributed on the surface and/or throughout the interior of particles; (Keyser et al., 1978).
3. The solubility of the various forms of the element associated with the particle (known to influence bioavailability; Lindberg and Harriss, 1980).

3. PREDICTING MESOSCALE TRANSPORT

Transport of a pollutant from source to receptor is controlled by mesoscale meteorological properties. Wind and temperature fields, precipitation patterns, terrain, vegetation cover, and turbulence encountered during transport influence the quantity of a pollutant ultimately delivered to the vicinity of the receptor. Several papers and reviews (Gifford, 1975, 1976; Hanna, 1976; Hosker, 1974; Nappo, 1976) and one workshop (Haugen, 1975) have considered some of the common techniques and inherent problems in modeling the effluent delivery process. The consensus of this symposium is that there are severe limitations in the present modeling capability for regional transport, dispersion, and deposition of atmospheric pollutants. A detailed consideration of data requirements, interpretation, modeling and application to estimates of dispersion, and removal in the context of this symposium is presented in the companion paper (Hosker, this volume).

4. MEASURING MASS TRANSFER TO THE CANOPY

Material is removed from the atmosphere by two general processes: dry deposition (convective diffusion, impaction and turbulent inertial deposition, and sedimentation) and wet deposition (precipitation scavenging occurring both within and below the cloud layer). The importance of dry deposition to the terrestrial environment is now well recognized; dry deposition of "pollutant materials to a forest from ground-based sources appears to be comparable to wet deposition" (Shinn, 1977). Wet deposition is of particular importance, because of both its episodic nature and the fact that pollutants are partly (or wholly) in solution, thereby enhancing the possibility of absorption by vegetation surfaces and generally increasing the mobility of the element in the landscape. The consensus at two workshops on atmospheric deposition was that the wet segment of input is much more addressable with current methodologies than the dry component (Hicks et al., 1980; Shriner et al., 1980); consequently, we shall emphasize the acquisition, interpretation, and application of dry deposition data.

4.1 Dry Deposition

As the crucial link between transport of atmospheric pollutants and their ultimate uptake and effects on vegetation, the processes influencing dry deposition have been examined at levels of resolution ranging from cellular to regional in scale. Deposition rates of gaseous pollutants are controlled by a variety of physical, chemical, and biological factors (Bennett and Hill, 1975; Heck and Brandt, 1977). These include micrometeorological conditions above and within the plant canopy, leaf surface characteristics (shape, degree of pubescence, and surface chemistry), and the chemical and biological properties of individual pollutants (diffusivity, water solubility, and biological reactivity within cell systems). Evaluating the rates and direction of transport of pollutants in terrestrial ecosystems obviously requires generalization of the complex biological and environmental variables inherent in the structural complexity of these systems. Determining the level of generalization acceptable to address relevant questions in the framework of this complexity is a continuing challenge.

Transport of pollutants to the leaf surfaces within a plant canopy is controlled by the rate of air movement and its turbulent properties. With a simple, somewhat homogeneous, crop canopy, three distinct air layers can be distinguished (Bennett and Hill, 1975): (1) a logarithmic wind profile layer (boundary) above the canopy surface, (2) an exponential canopy-eddy layer, and (3) the lowest plant–air layer where direct physical influence of the vegetation and ground surface influence the wind profile. Ultimate movement of pollutants into plant leaves occurs as a result of interactions between micrometeorological and plant physiological processes. The pathway of entry includes the leaf boundary layer, the stomata, and the aqueous layer on substomatal mesophyll cell walls (Bennett and Hill, 1973).

4.1.1 Gases

Studies to determine rates of pollutant flux into plant canopies have been conducted both in the field using principles of momentum, mass, and heat exchange (Thom, 1975) and under controlled laboratory conditions where environmental variation is minimized (Bennett and Hill, 1973). Both approaches have provided valuable insights. In the field, generalizations are required with respect to the geometry of the plant canopy, the turbulence it induces, and the average physiological status of foliage which occurs in multiple layers. Wesely and Hicks (1977) and Hicks and Wesely (1980) have reviewed the processes influencing turbulent transfer and deposition of gases to vegetation. They note the importance of both diurnal and spatial variations on the series of within-canopy resistances determining deposition velocity. A major limitation of most approaches is the assumption of a uniform deposition velocity of 1 cm/s. Eddy flux measurements over a loblolly pine canopy (Hicks and Wesely, 1980) indicate that resistances as high as 10 s/cm for O_3 and 15 s/cm for SO_2 can occur at night as stomatal closure occurs. Compounding this temporal variability for a monospecific forest canopy is the spatial heterogeneity of a typical mixed deciduous forest. Here, quantification of deposition rates to understory shrubs and tree seedlings represents an important consideration if we are to relate pollutant dose (typically measured as an above-canopy value) to uptake and effects of pollutants on the total forest system.

Chamber studies offer intriguing possibilities for examining not only deposition rates but also the associated effects on plant physiological properties. Major contributions in this area have been made by Hill (1971) and Bennett et al. (1973) (see also Bennett and Hill, 1975). Using a standardized plant canopy, Hill (1971) was able to compare uptake rates of HF, SO_2, Cl_2, NO_2, O_3, PAN, NO, and CO. A direct correlation between uptake rate and solubility of these pollutants in water (see Table 2) was shown. This relationship emphasizes the importance of ultimate pollutant uptake by moistened substomatal cell walls. Uptake rate was also inversely related to molecular weight, a factor directly influencing molecular diffusivity. These data suggest a potentially useful approach to screening

TABLE 2. Uptake Rate of Pollutants and Solubility in Water

Pollutant	Uptake Rate (Alfalfa at 1 pphm[a]) (L/min m^2)	Equivalent Deposition Velocity (cm/s m)	Solubility at 20°C (cm^3 gas/cm^3 H_2O)
CO	0.0	0.00	0.02
NO	0.6	0.10	0.05
CO_2	2.0	0.33	0.88
PAN	3.8	0.63	—
O_3	10.0	1.67	0.26
NO_2	11.4	1.90	Disproportionate
Cl_2	12.4	2.07	2.30
SO_2	17.0	2.83	39.40
HF	22.6	3.77	446

After Hill and Chamberlain (1976).

[a]pphm = parts per hundred million.

phytotoxicity of the variety of complex organic gases (i.e., phenols) which may be released in coal conversion processes and about which very little is known.

A number of potentially critical questions remain, which appear particularly well suited to controlled chamber studies. These include the influence of dosage kinetics, frequency, and distribution on rate of uptake; the influence of leaf surface chemistry on absorption and transformation (and possibly detoxification, i.e., $O_3 \rightarrow O_2$ or $SO_2 \rightarrow SO_4^{2-}$) of pollutants; and the influence of pollutant mixtures on uptake of component species. Absorption of a pollutant such as SO_2 into a surface moisture layer may be very rapid but may permit subsequent volatilization and internal absorption as the moisture layer evaporates.

The integration of chamber-derived data into mathematical models of pollutant uptake (Bennett et al., 1973; Steinhardt et al., 1977) offers promise of a tool for calculating uptake rates over larger temporal and spatial scales. Steinhardt et al. (1977) used a single-leaf model to calculate SO_2 uptake on a statewide basis for the Northern Great Plains. Estimates based on projected increases on a more simplistic basis may also be made over larger regions based on assumed uniform cover of a single test crop. Hill (1971), for instance, used uptake data for alfalfa to estimate that total canopy removal of SO_2 would be 562 tons/day in a 3520-km^2 area near Sudbury, Ontario. Potential deposition of NO_2 in the Los Angeles basin was similarly calculated to be on the order of 100 kg/km^2 day. Such calculations may prove valuable for other regions in the future, as both beneficial and detrimental effects of S and N amendments are considered.

The principal utility of measurements of dry deposition rates of gases lies in their potential for relating effects on the physiology, growth, and productivity of the exposed vegetation to the actual dose received (Winner and Mooney, 1980a, b). Unfortunately, very little concurrent data exist on the relationship between external dose, internal dose, and plant effects. The development of advanced-design exposure and uptake chambers by Rogers et al. (1977) should make these data more easily available in the future. While these data may be obtained now in the controlled laboratory situation (see Section 5.1), documentation of responses under field conditions must await development of suitable fast-response instrumentation. In this regard, eddy correlation techniques utilizing concurrent measurements of CO_2, H_2O, SO_2, and O_3 may provide a powerful tool for assessing impacts on primary production processes of natural systems.

4.1.2 Particles

Problems encountered in measuring particle deposition to vegetation include aerodynamic effects, small-particle diffusion, large-particle sedimentation, and particle bounce from the receptor surface. Physical entrapment by structural components of the surface is also a factor. Particularly important aspects requiring attention are the behavior of particles during impaction on foliage and the efficiency of retention. Particle deflection, bounce, rebound, and resuspension from surfaces all contribute to the local air concentration at the same time that deposition is occurring. Thus, particle fluxes may actually occur against a measured gradient. Because of these uncertainties, measured and calculated fluxes are often in poor agreement.

Several experts at a U.S. EPA-sponsored workshop to examine critically the field and laboratory methods used to determine the dry deposition of pollutants to environmental surfaces compiled a list of methods currently or previously in use for estimating dry deposition and discussed in detail their applicability to monitoring and vegetation effects studies, emphasizing their shortcomings, advantages, and the developmental needs (Hicks et al., 1980).

The methods fell into three basic groups: (I) physical measurements performed in the field, (II) field and wind tunnel meteorological techniques, and (III) indirect methods of calculation. These are summarized in Table 3 with an indication of their applicability to gaseous or particulate pollutants, suitability for use in routine monitoring, and utility in recent studies (if available). This discussion will be concerned primarily with methods of groups I and III, since these are generally more applicable to routine and wide-scale use and include methods with which the authors are experienced. The techniques in group II are more applicable in a research mode and are briefly considered in the following paper (Hosker, this volume).

Group III methods are dependent on accurate airborne concentrations

TABLE 3. Methods Used to Estimate Dry Deposition[a]

Method	Pollutant Measured	Suitability	Reference to Recent Application
	I. Physical Measurement in the Field		
1. Surface snow contamination	Particles (and gases?)	Monitoring/research	Barrie and Walmsley, 1978
2. Background atmospheric radioactivity	Particles	Monitoring/research	Hodge et al., 1978; Graustein and Armstrong, 1978
3. Applied tracers	Particles or gases	Research	Garland, 1976
4. Open pots or buckets	Particles (and gases?)	Monitoring	Gibson and Baker, 1979
5. Discs, plates, dishes	Particles (and gases?)	Monitoring/research	Lindberg et al., 1979a; Davidson, 1977
6. Sticky surfaces	Particles	Monitoring/research	
7. Filter paper	Particles (and gases?)	Monitoring	Cawse, 1974
8. Foliage washing	Particles (and gases?)	Monitoring/research	Lindberg et al., 1979a; Parker et al., 1980
	II. Wind Tunnel and Field Meteorological Techniques		
9. Box budget	Particles or gases	Research	See discussions in Kellogg et al., 1972; Whelpdale, 1978

10. Gradient	Particles or gases	Research/monitoring?	Droppo, 1980
11. Eddy correlation	Particles or gases	Research	Hicks and Wesely, 1980
12. Eddy accumulation	Particles or gases	Research	(See discussion in Hicks and Wesely, 1978a)
13. Variance method	Particles or gases	Research	(See discussion in Hicks and Wesely, 1978a)
14. Bowen ratio	Particles or gases	Research	(See discussion in Hicks and Wesely, 1978a)
15. Wind tunnel studies	Particles	Research	Wedding et al., 1975; Klepper and Craig, 1975
16. Chamber mass balance studies	Gases	Research	Hill, 1971
	III. *Indirect Methods of Calculation*		
17. Air concentration	Particles or gases	Monitoring/research	Andren and Lindberg, 1977
18. Calculations based on physical fundamentals	Particles	Research, monitoring?	Davidson, 1977

[a] Adapted from Hicks et al., 1980.

and, additionally, on accurate particle-size determinations in some cases. The method of concentration sampling involves use of an assumed relationship between atmospheric concentration and deposition rate. The ratio between the vertical flux of an element (F) and its mean concentration in the air (c) at some reference height is termed the deposition velocity: $V_d = F/c$ (Chamberlain, 1955). Deposition flux can thus be estimated, given the air concentration and published values of V_d. An advantage of this method is the existence of several networks which currently monitor air particulate concentrations. As we continue to improve our understanding of the relationship between deposition flux and air concentration, we can apply this information to existing data bases. The disadvantages of this technique include those associated with accurate determination of air concentrations and size distributions (Section 2.3). The calculated deposition flux is also limited by the accuracy of the deposition velocity used. Considerable research is needed in this entire field.

We need a more thorough understanding of the influence of particle size on V_d because many published values are given for specific elements without regard to size distribution (Cawse, 1974); we need to define a standard reference height above the deposition surface for all field and laboratory determinations of V_d; and we need to define the relationships between field and laboratory determinations of deposition velocities. Sehmel (1980), reviewing deposition velocities for particulate and gaseous pollutants, commented that even for specific elements, literature values of V_d ranged over several orders of magnitude. This has hindered the generalization of these types of data. Because of the wide-ranging values and lack of control of several variables under field conditions, predicted deposition velocity models based on field data are scarce (Sehmel, 1980). Another knowledge gap is in the reliable application of deposition velocity data in complex terrain situations. While this problem has been considered on a regional basis for SO_2 (Murphy, 1976), it has not been approached for particulate pollutants.

The calculation method based on physical fundamentals also relies on accurate measurements of atmospheric concentrations, but combines them with existing laboratory data on particle fluxes to well-defined, geometrically simple surfaces. Because of the complexity of natural surfaces, several simplifying assumptions are necessary in this application (Davidson, 1977). Considerable research is needed on deposition to complex-geometry surfaces before this method can be applied to general vegetated canopies.

Several group I methods are particularly appealing for use in monitoring networks or in field effects studies. These methods are readily adaptable to most environments, are relatively inexpensive, can be deployed by personnel with a minimum of training, are adaptable to extraction and analytical techniques, are applicable to uniform use in monitoring networks, and yet can be used in a research mode (for purposes of calibration with vegetation, for example). The remaining methods are somewhat more involved technically or analytically (background atmospheric radioactivity and applied tracer methods) or are applicable only under certain conditions (surface snow contamination), making them useful for monitoring in only a limited number of situations, although they are very useful in a research mode.

Consider first this latter subset of methods. Recent analytical and sampling advances have made it possible to quantify accurately trace contaminants in snow. In practice, (1) surface snow is sampled immediately following a snowfall and again several days later, the net increase in chemical content being ascribed to dry deposition, or (2) snow cores are analyzed in vertical sections to yield a measure of temporal variations in bulk deposition over time scales of years (Boutron and Lorius, 1978; Boutron, 1979; Dovland and Eliassen, 1976; Barrie and Walmsley, 1978; Jernelov and Wallin, 1973; Applequist et al., 1978). While this method is relatively straightforward and can provide historical deposition information, the results may be confounded by spatial inhomogeneity in snow chemistry, ease of sample contamination, element migration in snow layers, sublimation of surface snow, and wind redistribution of surface snow layers. Further research is needed to quantify the influence of these complicating factors.

The use of background atmospheric radioactivity to estimate deposition may be applied to closed systems, such as lakes or instrumented watersheds, as well as in smaller-scale studies of particle flux, taking advantage of radioactive fallout as a tracer (e.g., see the symposia proceedings edited by White and Dunaway, 1977; Klement, 1965; Fowler, 1965; Shinn, 1977; Hodge et al., 1978; Graustein and Armstrong, 1978). The technique may require the use of sophisticated counting equipment, but it does result in accurate estimates of radioactive small-particle fluxes. Details of deposition mechanisms or short-term deposition rates are not generally possible, however, because of the often necessarily long averaging times. In investigations of radionuclide accumulation in water bodies, plants, or sediments, it is difficult to differentiate between wet and dry deposition. Research is needed to define more thoroughly the atmospheric particle-size distribution of background radioactivity to enhance our understanding of the responsible deposition mechanisms. This point also applies to the use of tracers in laboratory and field studies. While tracers are most often used as gases such as $^{35}SO_2$ (Noggle, 1980; Garland, 1976; Owens and Powell, 1974), particulate tracers (^{131}I, ^{198}Au, for example) have been used in wind tunnels to measure deposition processes and rates to foliar or inert surfaces (Clough, 1973; Klepper and Craig, 1975; Carlson et al., 1976) and in field studies of deposition and particle retention by vegetation (see review by Chamberlain, 1970). In many applications, the labeled particles are monodisperse or, if polydisperse, are not representative of the size distribution encountered in the atmosphere. Further research is needed using labeled aerosols with realistic size distributions applied on larger spatial scales to facilitate interpretation of data for extrapolation to the field. This method may be very useful in the calibration of artificial surfaces with vegetation, a point considered in the following discussion. Research on this question must be encouraged.

The methods in Table 3 which are most readily adaptable to regional monitoring networks involve surrogate surfaces. Although this was the consensus of opinion at the EPA workshop, it was the only statement regarding inert surfaces which had such support; no other topic generated as much controversy. From a practical standpoint, it must be accepted that surface methods or air concentration measurements (discussed above) hold the only hope, at present, for the monitoring of pollutant dry deposition (if indeed it *can* be monitored). This sentiment was also expressed at the Expert Meeting on Dry Deposition in Sweden (Brosset et al., 1977).

The problems of field measurement of dry deposition by inert surfaces have been considered in detail (Galloway and Parker, 1980; Lindberg et al., 1979a). Since measurement of dry deposition on natural surfaces is particularly difficult, it is not straightforward to relate deposition on an artificial surface to that on living vegetation. With few exceptions, most large collection devices for deposited particles currently in use detect primarily the fall of relatively large, sedimenting particles. Efficient collection of small particles and gases transferred via turbulence and deposited by impaction processes likely will require modified samplers and techniques. Conventional methods suffer not only from artificiality of the surfaces but also from their common use at ground level. For evaluation of the overall mass transfer to the forest canopy, there seems little alternative to the task of direct in-canopy sampling.

The various surrogate surfaces in use include the so-called HASL (Health and Safety Laboratory) bucket, which had been adopted by the NADP network as a standard collection device, perhaps as much for convenience as any other reason because wetfall is collected by the same device (Volchok et al., 1974; Gibson and Baker, 1979; Dana, 1979); discs and cylinders coated with a sticky substance and noncoated discs (White and Turner, 1970); flat Teflon plates or discs (Elias et al., 1976; Davidson, 1977); polycarbonate petri dishes (Lindberg et al., 1979a); filter paper (Cawse, 1974); artificial foliage (Schlesinger et al., 1974); stacks of plastic sheeting (Nihlgard, 1970); and polyethylene screen (Hart and Parent, 1974). These devices have been placed at ground level in the open, 1 m above bare ground, and beneath, within, or above forest canopies. Several methods devised to deal with the complications of unwanted deposition (i.e., bird droppings, rain) include permanent covers, wetfall-activated covers, manual collection prior to rainfall, and anti-bird spikes. It should be understood here that we are considering the use of these devices to measure dry deposition only and not bulk deposition (wet plus dry), which is discussed in Section 4.2.

Since several mechanisms comprise the dry deposition process, no one surface may suffice to collect a representative sample of total dry deposition, each process requiring a different technique with its own inherent problems. To measure the flux of particulate pollutants to the terrestrial system accurately, it is necessary to essentially duplicate the vegetation surface (Galloway and Parker, 1980). Because of its biological and chemical reactivity, its complex microstructure, and its diverse variety of occurrence, the foliar surface is currently nonduplicable. Thus, measurement of dry deposition by any surrogate surface yields only an estimate of deposition flux, obviously limiting our understanding of the deposition process and its effects. Research in the fields of plant surface morphology and pollutant–foliage interactions should be encouraged to develop an inert, reproducible surface more representative of vegetation than those currently in use.

With these qualifying statements firmly in mind, it is useful to consider the advantages, disadvantages, and further needs associated with the surface collection methods currently available. The open plastic bucket, although perhaps the most widely used surface, was also the most assailed at the EPA workshop. Its major advantage is widespread use under standard conditions (e.g., in an upright position, 1 m above bare ground, as in the HASL wet–dry sampler), thus facilitating comparative studies. Its ability to collect relatively large quantities of material facilitates analysis for both chemical speciation and concentration. The major disadvantages lie in the geometry of the bucket with its large side-to-bottom surface area ratio. Material is deposited to all surfaces, both inside and outside the vertical walls, such that the amount of material ultimately analyzed is strongly dependent on leaching procedures, not all of which are standard. The relationships between the exposed bottom surface area and the true collection surface area are poorly understood and probably not consistent under field conditions. As the bucket shape is known to result in unusual turbulence over the sampler (Munn and Bolin, 1971), these data should not be used to estimate absolute mass transfer rates. The effect of such turbulence on deposition should be quantified under controlled conditions. The bucket is also subject to higher levels of contamination because of the need to leach the large surface area. Plastic buckets may not be compatible with extraction and analysis of collected particles for trace organics because of adsorption problems.

These problems are minimized by the use of inert discs, plates, or shallow dishes. They are smaller than the bucket with much lower side-to-bottom surface area ratios, can be readily employed at various orientations in the canopy, and are more amenable to ultraclean laboratory procedures (if Teflon) or extraction for trace organics (if glass). In addition, chemical speciation and elemental solubility can be determined for samples collected over relatively short time intervals (2–3 days; Davidson, 1977; Lindberg et al., 1979a). Work has been underway in our laboratory (and perhaps others) to develop a standard inert surface with a precipitation-activated mechanical cover to use in the forest canopy; more such research is needed.

Coating these surfaces with a sticky substance will increase the aerosol capture efficiency, but it has several disadvantages. A standard coating which is easy to apply and is trace-contaminant free does not exist. The problems of efficient recovery of the captured particles for analysis and analytical matrix effects must be solved. Samples collected on sticky surfaces are not amenable to elemental solubility or speciation studies. Finally, we are not aware of any compelling experimental evidence to indicate that a sticky surface is any better analog of foliage than a nonsticky surface.

Filter paper, a widely used collection surface in the United Kingdom, is reproducible, relatively inert and contaminant free (if carefully selected), and amenable to standard collection methods (in practice, placed 1–5 m above the ground beneath a fixed 1 m^2 plastic rain cover; see Cawse, 1974). This surface may be incompatible with chemical speciation or solubility studies, and in practice, requires relatively long collection periods (approximately 10 days).

The other surface methods in Table 3 (plastic foliage, sheets, or screens) have generally been applied only in small-scale, highly specialized studies of the chemistry of rain above and below the forest canopy. Because of the size of some of these surfaces, they may be subject to contamination problems, difficulties in efficient deposited-particle recovery and analysis, and incompatibility with typical monitoring studies.

There are several general dry deposition research needs which must be addressed in future studies. The most urgent is for both field-scale and environmental-chamber or wind-tunnel studies to establish the relationship between material collected by surrogate surfaces and foliage. There was considerable discussion at the EPA workshop that surrogate surfaces must be calibrated with vegetation for a thorough understanding of inert-surface, dry deposition data. A combination of the inert-surface methods in Group I (Table 3) with foliar-washing experiments holds some promise, although there are many problems regarding foliar washing which must be considered. Such an experiment has been attempted on a very small scale by a comparison of deposition rates of four trace elements and sulfate determined by leaching (washing) inert surfaces and foliage situated in the upper canopy during a 7-day dry period (Lindberg et al., 1979a). Dry deposition rates measured to each surface type agreed within a factor of approximately 2 for Cd, Mn, Zn, and sulfate, while the apparent deposition rate of Pb to the inert surface exceeded that measured to the leaf surface by an order of magnitude. Leaf surface adsorption of soluble Pb may have been a factor and is a phenomenon worthy of further investigation for its potential effects (Lindberg and Harriss, 1981).

Before larger scale comparative studies are attempted, several questions raised in earlier work (Tukey et al., 1965; Mecklenberg et al., 1966; Tukey, 1970; Lindberg et al., 1979a) must be addressed:

1. What fraction of the material washed from excised foliage is internally derived (i.e., by cellular leaching)?
2. Is there an optimum leaching medium (such as distilled water, dilute

acid, detergent) which is amenable to trace-level analytical techniques and is efficient at dislodging or solubilizing surface-deposited particles, but does not promote internal foliar leaching?

3. What is the optimum leaching duration; does surface-associated material dissolve before internal material?
5. Is there any physiological reason to analyze for the non-water-soluble fraction of an element in deposited particles?
5. If vegetation samples are collected and washed on a daily basis for determination of the net increase in surface deposition, can one assume that the internal leaching pool of an element is constant over that time interval?

We have measured order-of-magnitude variations in the surface area concentrations of some elements on adjacent chestnut oak leaves (*Quercus prinus*) on the same tree and similar variations in leaves collected from trees several kilometers apart (Lindberg et al., 1979a). This has serious implications if foliar washing data are to be used to estimate canopy-wide dry deposition rates (see further discussion on these points in Parker et al., 1980). Work is also needed to determine the means by which deposition rates measured to individual inert or foliar surfaces can be extrapolated to whole canopies. Multiplication of the single-surface deposition rate by the applicable leaf area index should provide an upper limit, but research is needed to more completely quantify this relationship.

Another question involves several recent reports of measured net upward fluxes over forested canopies (determined by eddy correlation methods) and over bare ground (determined by gradient methods) for several pollutant and nonpollutant elements (Hicks and Wesely, 1980; Droppo, 1980; Davidson and Elias, 1979). These observations indicate that a serious problem may exist in our interpretation of deposition data based on surface-accumulation measurements. The upward fluxes measured over bare ground were attributed to local, ground-level dust resuspension (Droppo, 1980; Davidson and Elias, 1979), while upward fluxes of S, measured over a pine forest, were tentatively attributed to foliar emissions (Hicks and Wesely, 1978b). Some grasses, several herbaceous plants, some crop plants, and pine trees are also known to release particulate forms of several trace metals (Beauford et al., 1975, 1977). These internal cycles of particles resuspended from soils or vegetation or released from within foliage can result in gross overestimates of true atmospheric deposition based on surface accumulation measurements.

We are obviously faced with a major challenge in this field to develop a standard dry deposition collection technique that meets the above requirements. This is best approached as a multidisciplinary task:

1. The effects researcher needs to define the optimum sampling period—or define a dry deposition event.

2. The meteorologist should provide guidance on the physical shape, orientation, and placement of the samplers in the field.
3. The plant physiologist needs to define the microstructure of a typical foliar surface.
4. The particle physicist should describe the mechanisms by which particles interact with various surfaces, including the relative importance of particle removal by sedimentation, impaction, and diffusion.
5. The atmospheric chemist needs to describe the gas–particle interactions that occur both in the air and on the surface of deposition which may influence the mobility or toxicity of a deposited element.
6. The analytical chemist needs to develop an inert, contaminant free surface which can be readily analyzed.
7. The geochemist must define the relative importance of internal element cycles to the captured particle pool.
8. The monitoring agency should prepare standard procedures for using the samplers.
9. The engineer must design the sampler with a precipitation-activated cover for reliable operation under field conditions.
10. The research community needs to consider a means of calibrating the instrument with living foliage.

4.2 Wet Deposition

The wet component of atmospheric deposition is much more addressable with current methodologies than the dry component, as indicated by recent workshops and symposia (Semonin and Beadle, 1977; Dochinger and Seliga, 1976; Hicks et al., 1980). As a result of the interest in precipitation chemistry generated by the acid rain phenomenon, there have been several recent studies concerned with development of precipitation sampling and analytical methodology for major and trace components (described in detail in Table 4).

It is our opinion that the most important research needs in this area concern the design of field experiments for collection of data appropriate for improving our understanding of plant–pollutant interactions. This topic is discussed in Section 5.2. However, there are several aspects of the physical collection of wet deposition which need to be addressed in future studies (some of these recommendations were drawn from Galloway et al., 1978):

1. Methods for contamination-free collection of precipitation for trace organic/inorganic analyses should be further refined.
2. Regional-scale multisite studies are needed to determine applicability of single-site measurements.

TABLE 4. Recent Developments in the Collection, Analysis, and Interpretation of Precipitation Chemistry Data

Need for collection of wetfall-only samples	Galloway and Likens, 1978
Need for event sampling and event definition	Galloway and Likens, 1978; Lindberg et al., 1979a
Design of wetfall-only automatic sampler	Volchok et al., 1974; Dana, 1977
Comparability of collectors	Galloway and Likens, 1976; Dana, 1979
Collection of representative samples	Galloway and Likens, 1976, 1978; Lindberg et al., 1977, 1979a
Preservation of samples in the field	Galloway and Likens, 1978; Lindberg et al., 1977
Laboratory storage of samples for chemical analysis	Peden and Skowron, 1978
Development of standard handling, analysis, and storage methods	Peden et al., 1979
Determination of precipitation and throughfall acidity by standard methods	Galloway et al., 1979; Hoffman et al., 1980
Collection procedures for trace inorganic analysis	Lindberg et al., 1977
Collection procedures for trace organic analysis	Lunde et al., 1977; Galloway, 1980
Filtration of rain samples	Kennedy et al., 1979; Lindberg et al., 1977
Treatment of suspended solids in rain	Kennedy et al., 1979; Lindberg et al., 1977
Stabilization of dissolved gases in collected rain	Dana, 1977
Analytical variability among several laboratories and techniques	Turner and Lindberg, 1976; Bogen et al., 1978
Procedural variability compared with natural field variations	Lindberg et al., 1977, 1979a
Temporal and spatial variability of precipitation chemistry	Semonin, 1976; Lindberg et al., 1977, 1979a; Slanina et al., 1979; Gatz, 1978
Frequency distribution of element concentrations in precipitation and rain volume for statistical analysis	Essenwanger, 1960; Lindberg et al., 1979a
Interpretation of rain volume–concentration relationships	Lindberg et al., 1979a; Kennedy et al., 1979; Parker et al., 1980

TABLE 4. (*Continued*)

Determination of relative importance of precipitation-scavenging mechanisms	Semonin and Beadle, 1977; Raynor, 1976; Lindberg et al., 1979a
Extrapolation of single-point precipitation volume measurements	Rogers and Zawadski, 1979
Urban influence on rain volume	Changnon, 1979
Optimum location of samplers in urban areas	Gatz, 1980
Use of wind trajectory analysis with precipitation data	Miller et al., 1976
Need for and utility of within-event sampling or sequential sampling	Raynor, 1976; Raynor and Haynes, 1977; Parker et al., 1980
Design of an automatic sequential sampler	Gascoyne, 1977; Ronneau et al., 1978; Raynor and McNeil, 1979
Development of standard methods for regional and national monitoring networks	Dana, 1977; Galloway and Cowling, 1978

3. Input from meteorologists is needed on the ancillary observations or measurements which should be made during precipitation collection.
4. Methods should be developed for routine collection of intercepted fog.
5. Elemental solubility studies in rain are needed to assess the relative importance of particulate components in precipitation.
6. Input from plant physiologists is needed on the most appropriate time scale for collection of precipitation data (subevent, event, weekly?) (discussed further in Section 5.2).
7. A wetfall-only sampler suitable for use below plant canopies (throughfall) needs to be developed.
8. Further work is needed to assess the effects of gas exchange on rain chemistry while samples are in the field prior to their return to the laboratory.
9. The wetness sensor head currently in use (Wong or HASL design) on wet/dry samplers needs to be redeveloped to provide more reliable operation and to assure collection of the entire precipitation event while minimizing exposure time of the sampler to dry deposition.
10. A sampler should be designed to filter precipitation reliably as it is collected.

11. Studies on the chemical stability of throughfall samples are needed.
12. Coordinated collection of cloud water and precipitation at several levels between the cloud base and ground is needed to increase our understanding of scavenging mechanisms and gas–particle interactions with precipitation.

5. UNDERSTANDING PLANT–POLLUTANT INTERACTIONS

All investigations of the effects of air pollutants on vegetation should be concerned with careful characterization and correlation of both pollutant dose and plant response. Fundamental problems confronting plant researchers in this regard are determining (1) how to measure pollutant dose so that principal phytotoxic components (both chemical species and exposure kinetics) are identified, and (2) how to characterize plant response in ways which can lead directly to quantifying actual or potential economic or ecological loss.

As the product of concentration and time, pollutant dose may be characterized in a wide variety of ways. The choice of an appropriate averaging time and the potential influence of peak concentrations during the exposure history will vary with both the biochemical reactivity and the mode of occurrence of the pollutant species. From measurements of SO_2 concentrations near smelters at Biersdorf, Germany, Stratmann (1963) identified several exposures which were considered important as indicators of pollutant stress. These included (1) the mean concentration (assumed to describe a single-exposure event), (2) the total dose (concentration–time integral), (3) the variability of concentrations, (4) the frequency of recovery periods between exposures, and (5) the frequency of concentration peaks. In our eastern regional environment, with a high density of urban and industrial sources, we must add a sixth factor—the concentration ratios and sequence of occurrence of additional pollutants such as ozone, nitrogen oxides, and PAN (peroxyacetyl nitrate).

In general rather minimal effort has been directed toward identifying the critical features of pollutant dose, particularly with respect to combinations of pollutants. With single pollutants such as SO_2 or HF which are released from point sources, localized effects may be envisioned to occur primarily as a result of well-defined exposure episodes determined by local meteorology or topography. Here, typical exposure kinetics are characterized by high peak-to-mean ratios and a high variability of exposure concentrations over time. This regime may be superimposed on regional-scale levels of background pollutants. Especially with pollutants such as SO_2 which may be metabolized and detoxified at low levels, these episodes achieve particular significance in influencing plant response to the total exposure dose.

With studies designed to evaluate chronic effects of regional-scale pollutants such as O_3, SO_2, and NO_x (see review by Reinert et al., 1975), conditions must be designed to approximate multiday exposures involving multiple pollutants under generally less-variable pollutant concentrations. With any exposure regime, the permutations of exposure episodes and relatively pollutant-free intervals can be considered to influence heavily plant physiological recovery from pollutant stress.

5.1 Controlled Exposure Studies

Important early conceptual contributions to the characterization of SO_2 dose were made by Zahn (1961, 1970) in a series of controlled laboratory studies. These studies emphasized the importance of the distribution of dose during the exposure interval, the duration of intervening recovery periods, and the role of low concentrations in amplifying visible injury following subsequent higher concentrations. With two species, wheat and larch, daytime pretreatment with subacute levels of SO_2 (0.4 ppm for 33–77 h) provided protection against visible injury by subsequent high doses (1–2 ppm for 2–3 h). More recently, McLaughlin et al. (1979) examined the influence of SO_2 dosage kinetics on photosynthesis, transpiration, and growth of kidney beans. Working at concentrations below the threshold for visible injury (3-h average $\leq$0.5 ppm), they found that SO_2-induced photosynthetic depression could be strongly influenced by varying the ratio of the peak to the mean concentration during exposure without changing the total SO_2 dose. No strong evidence was obtained to indicate that plants were sensitized to SO_2-induced suppression of photosynthesis by previous exposure. Plant growth appeared to be responding to a combination of exposure kinetics and exposure history in these experiments.

Exposure history has also been demonstrated to be important in plant responses to ozone (Johnston and Heagle, 1982). Here chronic exposure at levels below the visible injury threshold (4 pphm, 6 h/day for 10 days) predisposed plants to greater sensitivity to a subsequent acute O_3 dose (10 pphm for 3 h). Intervening periods without O_3 partially alleviated the sensitization, as did chronic exposure to levels $\geq$6 pphm, which produced visible injury.

The importance of exposure kinetics on plant uptake of pollutants has also been demonstrated for fluoride, a task made easier by the fact that fluoride is accumulated by foliage. Compared with continuous exposure, fluoride from the same total dose of HF was accumulated more readily with intermittent exposure at a higher peak/mean ratio (MacLean et al., 1969) and less readily by intermittent exposure at the same peak/mean ratio as in the continuous exposure case (MacLean and Schneider, 1973). These authors suggest that fluoride content of foliage be used as a basis for regulatory action, a possibility which is not feasible for most other pollutants, which are more readily metabolized in plant systems.

Attempts to generalize dose–response data into vegetation–response equations have been made since the time of O'Gara (1922). Heck and

Brandt (1977) have recently reviewed these equations for the major phytotoxic pollutants. One of the most intensive examinations of air quality data and the utility of dose–response relationships in air quality regulation was provided by Larsen and Heck (1976). In general, currently available dose–response equations are derived from acute exposures at continuous pollutant levels and use visible leaf injury as the dependent variable. The major need, at present, appears to be development of relationships which define plant response under chronic exposure regimes which include more than one pollutant and reflect the importance of the distribution of typical episodic stress sequences occurring under field conditions.

The definition of plant response represents the second critical feature of dose–response studies. Our obvious interest is in losses in productivity, stability, or quality of the plant system. The essential need is for suitable indices of these effects. For years, visible injury has been a primary measuring stick for assessing plant damage. This concept was based, in part, on experiments by Hill and Thomas (1933), which indicated that growth losses in alfalfa were directly proportional to visible injury of foliage. Evidence has accumulated rapidly in recent years from laboratory and field experiments to indicate that growth may be reduced in the absence of visible injury. Interest in subtle responses has led to a variety of mechanistic studies aimed at detecting cytological, enzymatic, or general physiological responses of plants to pollutants (see reviews by Ziegler, 1973; Mudd and Kozlowski, 1975). While both leaf injury and measures of physiological change offer a convenient short-term basis for defining plant response, neither is adequate alone. Plant physiological resilience, both in space and time, makes plant growth and yield the most reliable ultimate indicators of the effects of pollutant stress on plant systems and emphasizes the importance of measuring whole-plant responses. Recognition of the importance of whole-plant patterns of allocation of carbon (photosynthate), water, and nutrient resources in response to stress (McLaughlin and Shriner, 1980) implies that additional responses such as altered allocation between roots and shoots may be meaningful indicators of plant–environment (including pollutant and other stresses) interactions.

Ultimate development of dose–response relations in a way that can be meaningfully used in air quality regulation is obviously a tremendously complex task. At present, we have only a few of the pieces required to complete the task. To address it adequately will require generalization of the multiple interactions of plant species, growing environment, pollutant dose (kinetics, composition, and history), and measured plant processes. Deciding on the acceptable level of generalization is an important challenge to us all.

5.2 Field Research

Carefully designed field-level research on the effects of atmospheric deposition on vegetation is the logical extension of controlled exposure studies. These methods may provide the only means of validation of laboratory results, as well as providing realistic mixtures of wet and dry deposition and gas–particle interactions.

5.2.1 Quantifying Wet/Dry Deposition and Gas–Particle Interactions

A key need regarding field-level deposition research is the integration of mechanistic deposition studies with vegetation effects studies in the forest canopy. Dry deposition can no longer be neglected in studies of the role of the atmosphere in geochemical cycling or in research on the effects of atmospheric constituents on plants. This is particularly important if a constituent of dry deposition is delivered to the canopy in a relatively mobile (e.g., soluble) form, as recently demonstrated for Cd, Pb, Mn, Zn, and S in both suspended and deposited particles (Lindberg et al., 1979a).

The relative proportion of total atmospheric deposition attributed to dry processes has been largely ignored in field studies until quite recently. Table 5 (right-hand section) indicates the relative contribution of dry deposition to total deposition measured at the Walker Branch Experimental Watershed in eastern Tennessee over several time scales (from Lindberg et al., 1979a). The atmospheric input during the short-term experiments (periods W2, W3, and W6) was generally dominated by dry deposition; only during period W3 was the wetfall process of greater relative importance for Cd and SO_4^{2-}–S. Even over the longer time scales dry deposition constituted a significant fraction of the total atmospheric input of Cd (~20%), SO_4^{2-} (~35%), and Pb (~55%).

Particle deposition is not generally considered to be an episodic event, such as the sudden inundation of the leaf surface by precipitation, but rather a chronic, cumulative exposure of the vegetation to atmospheric constituents. However, certain conditions can combine to create an unusually harsh exposure of the vegetation to potentially toxic material. Wet deposition rates of Cd, Pb, and sulfate measured during events of short duration and low rainfall volume, and hence generally high concentrations, were considerably higher than any of the measured dry deposition rates when expressed on a comparative unit–time basis (center column of Table 5). The episodic wet deposition rates were calculated as the areal wetfall input divided by the duration of the precipitation event. Although the duration of the rain events during W2, W3, and W6 was short (10–30 min), the wet deposition rates during the events were approximately 2–3 orders of magnitude greater than the dry deposition rates measured for each period (when expressed on a per hour basis).

For the initial bioreceptor, this intense episodic flux of potentially toxic material can play an important role in physiological effects. Precipitation events of short duration, low volume, and high elemental concentrations were common during the growing season at Walker Branch Experimental Watershed. In many cases, these events followed relatively long (5–10 days)

TABLE 5. Comparison of Wet, Dry, and Total Deposition Measured at Walker Branch Watershed[a]

	Normalized Deposition Rates ($\mu g/m^2$ h)			Relative Contribution of Dry to Total Deposition (dry/total) · 100		
Period	Cd	Pb	SO_4^{2-}–S	Cd	Pb	SO_4^{2-}–S
W2[b]				82%	100%	70%
Wet	2.7	24.0	42,000			
Dry	0.01	0.62	50			
W3[c]				18	86	47
Wet	2.7	45.0	17,000			
Dry	0.0003	0.12	10			
W6[d]				67	100	80
Wet	3.1	25.0	23,000			
Dry	0.0001	0.24	17			
W2, 3, 6						
Mean wet/dry ratio	4100	170	1300			
Growing season[e]	—	—	—	26	56	34
Annual[f]	—	—	—	17	53	33

Data from Lindberg et al. (1979a).

[a]Wet and dry deposition rates to individual upper canopy surfaces are compared when normalized to a unit time basis in the center section of the table. The right-hand portion of the table indicates the relative contribution of dry deposition to the total deposited quantity of an element to the entire canopy over several time periods.

[b]5/16 to 5/20/77; wet event duration = 0.007 day, total period duration = 4.2 days.

[c]5/30 to 6/6/77; wet event duration = 0.02 day, total period duration = 7.0 days.

[d]7/12 to 7/18/77; wet event duration = 0.007 day, total period duration = 6.0 days.

[e]4/1 to 10/25/77; total period duration = 207 days.

[f]Calendar year 1977.

dry periods characterized by high air concentrations and dry deposition rates. When the subsequent precipitation event was very small in volume (~0.5–1.5 mm), much of the initial precipitation remained on the leaf surface, not being washed off or diluted by subsequent rainfall. The potential for physiological effects is enhanced when this solution contacts previously dry deposited material on the leaf surface. The high concentrations that develop under these conditions are further enhanced during the evaporation of droplets on the leaf surface. The event which occurred during period W2 provides an example of this phenomenon. By applying the dry deposition rate of the water-soluble fraction of elements measured to surfaces in the upper canopy, it is possible to estimate the approximate surface area concentrations of dry deposited material on a leaf prior to the precipitation event. These values are summarized in Table 6, including calculations of the total quantity of wet deposited material, dissolved concentrations in water droplets on the leaf surface, and total deposition to the leaf, expressed relative to the leaf internal content.

Precipitation of 1.3 mm falling on a 50-cm^2 leaf would deposit approximately 6 mL water. As this solution begins to evaporate, the leaf is exposed, although briefly, to extremely high concentrations resulting from the interactions between surface moisture and dry deposition. Hundred-fold evaporation of the surface moisture would result in concentrations of dissolved constituents several hundred to several thousand times higher than typical rain concentrations (compare the concentrations in Table 6 with average rain concentrations at Walker Branch Experimental Watershed of Cd = 0.44 μg/L, Pb = 6.8 μg/L, and SO_4^{2-}–S = 1200 μg/L). Knowledge of this situation is particularly important in relating precipitation chemistry to vegetation effects observed in the field; the need for both event sampling and wet/dry segregation is obvious. If wetfall had been sampled on a weekly basis during this same period, the occurrence of a larger storm following the 1.3-mm event would have further diluted the apparent concentrations to which the vegetation was exposed because of the inverse relationship between precipitation concentrations and rain volume (Lindberg et al., 1979a). The effect of weekly wetfall sampling in this situation would be to obliterate the details of the series of events which may or may not have resulted in an observed effect on the vegetation. Considerable effort should be directed toward understanding the combined effects of wet and dry deposition on vegetation.

The assumptions in the above calculations result in both over- and underestimates of concentrations. For example, rainfall of 1.3 mm may rapidly run off the leaf removing a considerable quantity of dissolved material. However, personal observations in the field indicate that a considerable quantity of rain can be retained by leaves if situated horizontally and under calm conditions (the interception storage of the forest in Walker Branch has been estimated at approximately 0.6 mm of precipitation, based on simulation studies, R. J. Luxmoore, personal communication). The calculations also assume a uniform surface concentration of dry deposited material, which is a poor assumption as indicated in Figure 1 (from Lindberg et al., 1979a). At isolated points on the leaf surface where deposited material accumulates, episodic concentrations in solution could be considerably higher than those estimated here. More work is needed on surface distribution of dry deposition on vegetation.

The physiological effects of surface-deposited metals on vegetation, either in particulate or dissolved form, require considerable study, as they are very poorly understood. There are several conflicting reports on toxicity in the literature (as reviewed by Krause and Kaiser, 1977; Zimdahl, 1976). However, the estimated pH of the resulting solution on the leaf surface ($\simeq$2, not indicated in table) can cause adverse effects in several plant

TABLE 6. Potential Concentrations of Several Elements in Solution on a Typical Upper-Canopy Chestnut Oak Leaf Surface (50 cm^2) Following a Brief Summer Shower Which was Preceded by a 10-Day Dry Period

		Water-Soluble Constituents		
Parameter	Units	Cd	Pb	SO_4^{2-}–S
Surface area concentration of previously dry deposited, water-soluble material[a]	ng/cm^2	0.1	2.6	1200
Mass of dry deposited, water-soluble material	ng	5.0	100	20×10^3
Concentration of soluble fraction dissolved by precipitation	μg/L	1.0	20.0	10×10^3
Concentration of elements in incident precipitation	μg/L	0.35	3.1	5×10^3
Estimated total concentration in solution on the leaf surface	μg/L	1.3	23.0	15×10^3
Potential concentration following evaporation	μg/L	130	2300	15×10^5
Total mass of element in solution on the leaf surface	μg	0.007	0.15	100
Total quantity of elements bound in leaf tissue	μg	0.065	0.22	2000
Total quantity of available fraction of elements delivered to the leaf surface during the growing season	μg	1.1	30.0	1700
Ratio of soluble element deposition during W2 to total leaf content[a]	—	0.11	0.68	0.05
Ratio of available element deposition during the growing season to total leaf content	—	17.0	140	0.8

Data from Lindberg et al. (1979a).

[a]Wet and dry deposition were separately sampled at Walker Branch Watershed from 5/16 to 5/20/77 (period W2). Although the rain event, which occurred on 5/18, was preceded by a 10-day dry period, dry deposited element concentrations were determined only for the ≃2.4-day sampling period preceding the rain. Dry deposited element concentrations are thus conservative estimates of total dry material on the leaf surfaces prior to the rain.

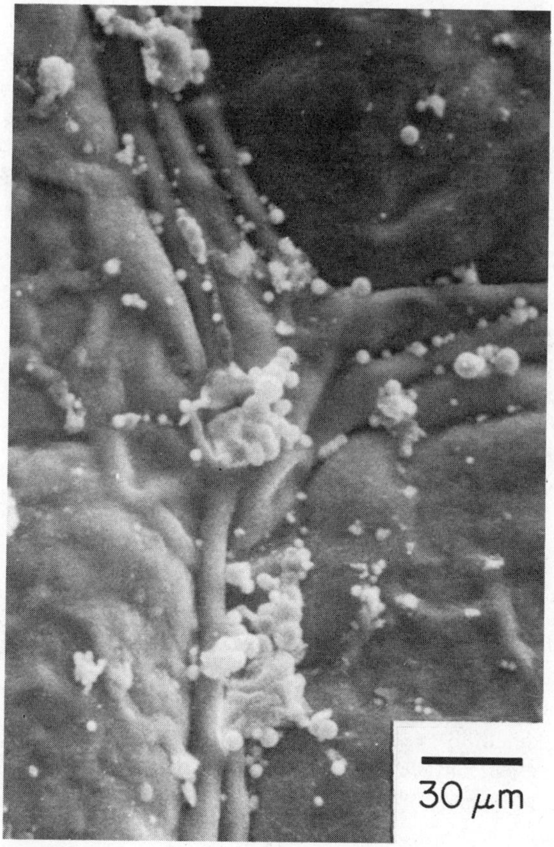

FIGURE 1. Scanning electron photomicrograph of a typical chestnut oak leaf collected in Walker Branch Watershed. Note the heterogeneous distribution of dry deposited particles, including combustion ash (spherical particles) (From Lindberg et al., 1979a.)

species (Shriner, 1976). Our calculation of acidity assumes no neutralizing capacity of the surface-deposited particles or of the leaf itself, a situation which is unusual but which has been documented (Lindberg et al., 1979a; Hoffman et al., 1980).

Table 6 also summarizes estimates of the magnitude of the deposition inputs during one event and over the growing season relative to the total leaf internal content of each element. The quantity of Pb estimated to be in solution on the leaf surface following the single event was nearly comparable to the total Pb content of the leaf. Lesser but still significant quantities of soluble Cd and SO_4^{2-} were deposited during this event. During the full growing season, the leaf surface was exposed to one to two orders of magnitude more dissolved Cd and Pb and nearly equal amounts of SO_4^{2-}

relative to the leaf content. While the effects of such exposures are unknown, the importance of atmospheric deposition in the cycling of these elements in the landscape is obvious and warrants continued research.

Another poorly understood and infrequently considered aspect of pollutant interactions in the field is the role of gas–particle reactions on the foliar surface. The role of trace metals, particularly Mn, in the catalytic oxidation of SO_2 in rain has been demonstrated in the laboratory; however, the application of these laboratory studies to the atmosphere is exceedingly complex (Barrie and Georgii, 1976; Barrie et al., 1977; Penkett et al., 1979). This phenomenon may play a significant role only under certain atmospheric conditions, such as in long-lived cloud droplets or urban fogs. A possibly more important situation is the absorption and oxidation of SO_2 by wetted foliar surfaces. The leaves of certain forest trees rapidly leach Mn into solution upon wetting, with concentrations in leaf moisture as high as 500 μg/L resulting from internally leached Mn and Mn solubilized from deposited particles (Lindberg et al., 1979a). Since the residence time of wetness (dew, rain) on leaf surfaces can easily approach several hours, this may represent a significant mechanism of increasing the deposition rate of SO_2 to vegetation. The ability of dew-covered surfaces to rapidly accumulate SO_4^{2-} was reported by Brimblecomb (1978). Deposition was controlled by the capacity of the solution to absorb SO_2; thus, presence of sufficient Mn to catalyze the oxidation should result in an increased rate of deposition. The relative importance of such interactions to deposition rates and to vegetation effects is largely unknown. The oxidation of SO_2 to SO_4^{2-} releases free H^+ to solution, which, if not neutralized by reaction with alkaline particle components, may result in damaging pH levels. Future studies should be directed at quantifying the role of gas–particle reactions in vegetation effects.

Continuing examination of the chemistry resulting from the rain–leaf interaction is both prudent and necessary to achieve a full understanding of internal and external mechanisms affecting leaf processes. The phenomenon of strong acid scavenging from rain by the forest canopy is a well-known but poorly understood phenomenon resulting in the release of plant-related weak acids and nutrients (Hoffman et al., 1980). The resulting pH of throughfall is often in the 5–6 range, not characteristic of acid rain. Future research in forests where this occurs (e.g., eastern deciduous forests) should concentrate both on canopy effects and on soil effects, particularly on the role of hydrogen uptake and subsequent organic carbon and nutrient loss from foliage.

Additional refinements in experimental design necessary for understanding plant–pollutant interactions should include studies of the following:

1. The influence of leaf surface microstructure and leaf exudates on particle capture, retention, and gas absorption.
2. Dose–response relationships for particle deposition.

3. Possible effects of shifting to a dominant direct atmospheric source of plant nutrients and toxins rather than a dominant soil-uptake route.
4. Episodic exposure of the canopy to high deposition rates of both wet and dry components.
5. Wet and dry deposited material distribution on individual leaves and within the canopy.
6. Deposited particle solubility under various moisture and chemical regimes.
7. Interaction of dew and intercepted fog or mist with deposited particles on vegetation and subsequent effects of the resulting solution.

5.2.2 Assessing Pollutant Effects on Plant Growth and Development

Quantifying pollutant impacts on plant growth and development under field conditions is a complex task for two primary reasons. First, variations in site quality between control and test sites can make differentiation of pollutant effects from experimental noise difficult. Secondly, particularly with perennial species, competition between plants within a terrestrial ecosystem is a potentially important modifier of individual and community-level responses. The task is somewhat easier when high levels of emissions from point sources are involved. Gordon and Gorham (1963), for example, were able to measure community-level changes along a 63-km gradient from an iron-sintering plant at Wawa, Ontario. Alteration of community composition under high pollutant burdens may be ascribed to differences in sensitivity between species which can result either in direct effects on those species or indirect effects due to altered competitive relationships within the plant community. In most present situations, however, such well-defined concentration gradients do not exist, and regionally elevated levels of pollutants from multiple sources are involved (Shriner et al., 1977). Here, quantifying pollutant impact requires evaluation of growth under comparable control conditions. Major progress in creating a basis of comparison has come with the development of the field chamber approach (Heagle et al., 1973; Mandl et al., 1973). With charcoal filtration providing the control condition and compared with growth of plants in unfiltered ambient plots, assessment of impacts of ambient air regimes on productivity of annual crops (see review by Heck and Brandt, 1977) has been greatly facilitated. Two main research needs can still be identified with this approach: (1) determining whether pollutant uptake, plant physiological condition, and growth are comparable between unfiltered chambers and field grown plants; and (2) providing adequate description of pollutant dose characteristics so critical features can be identified. Primary concerns in the first regard are that gas exchange characteristics including CO_2, water, and pollutants are comparable inside and outside the chambers. The increased likelihood of developing water deficits in the continuously stirred atmosphere of the chambers must be recognized, and supplemental irrigation must be provided when significant differences between chambers and field plots occur. With respect to pollutant monitoring, it is essential that the exposure history over the plant growth cycle be characterized both with respect to occurrence of primary pollutants and the sequence of principal stress episodes. Here, a variety of physiological processes (photosynthetic capacity, nutrient and water relations, carbon allocation, etc.) may serve as useful indicators of the sequence of events producing growth responses. Information of this type will be important in identifying which principal features of air quality data may be useful in estimating potential yield losses at other sites. Another recently developed field chamber technique which shows promise for assessing pollutant impacts under field conditions is the linear gradient system of Shinn (personal communication). This approach allows a predictable concentration gradient to be produced by linear dilution of a pollutant introduced into the incoming air stream. Dose–response data obtained by subsampling along the gradient at levels both below and above ambient concentrations may be a very cost-effective way to provide the response functions necessary to accurately predict potential impacts of future increases in concentrations of air pollutants on field-grown crops. Questions of physiological comparability with unchambered plants and dose distribution patterns should also be answered before this approach is put into wide-scale field use.

Field research on growth impacts on natural communities under chronic stress regimes is sorely needed. Here, both temporal and spatial characteristics of community structure greatly increase the complexity of the task. A good example of the challenges inherent in addressing ecosystem-level effects on a grassland ecosystem is the Colstrip Project (EPA, 1979). In this study, multitrophic-level studies are being conducted around an artificial source of SO_2 to assess potential effects of siting coal-fired power plants in the area. The fact that sensitivity to pollutant stress varies both within and between plant populations must be recognized. Work with natural populations of *Germanium carolinianum* L., a weedy species sensitive to SO_2, has demonstrated that genetic selection for resistance to air pollution stress may occur rather rapidly ($\leq$30 yr) (Taylor, 1978). Studies in England with perennial ryegrass have also indicated that populations growing in areas with high SO_2 concentrations (100–700 $\mu g/m^2$) evolved resistance to growth reduction by SO_2. The degree of evolution of tolerance was also shown to be directly related to the level of SO_2 under which they had grown. Additional work on the plasticity of genetic pools in response to SO_2 and other pollutants is necessary if we expect to predict responses of plant communities to chronic stress.

Because of their ubiquity, ecological, and economic importance, forest trees constitute a major focus of concern with respect to chronic air pollution stress on natural systems. Unfortunately, few data exist to permit adequate assessment of these effects. A number of approaches are being used to address this task. An intensive multidisciplinary approach to oxidant effects on the Coniferous San Bernardino National Forest near Los Angeles, California, has been described by Miller and McBride (1973). It includes both field sampling of responses at several trophic levels and mathematical simulations in an effort to assess alterations in system components. Another multidisciplinary effort (Legge et al., 1978, 1981; Legge, 1980) conducted in a coniferous forest ecosystem near a gas processing plant in northern Alberta, Canada, has demonstrated the utility of considering pollution impacts on the total plant–soil system. In this case, reduced growth of overstory lodgepole X jack pine growing within 1.5 km of the SO_2 source was attributed to reduced photosynthesis of foliage, an apparent consequence of reduced foliar retention of potassium and phosphorus, reduced pigment content, and lower levels of adenosine triphosphate. Growth losses appeared to be due more to sulfur-mediated changes in availability of soil nutrients than to chronic levels of sulfur uptake by foliage. A unique advantage available to these researchers was the distinctly different and constant $^{34}S:^{32}S$ ratios of SO_2 being introduced into the soil–plant system. Thus, impacts and fate of sulfur could be identified within the system components. More attention to research opportunities with radioisotopes, both stable and nonstable, may greatly assist future efforts to evaluate chronic pollutant impacts on ecosystem processes. In the eastern environment, McClenahen (1978) has used ecological indices of community diversity and structure to assess responses of deciduous forests along a 50-km gradient of air pollution levels along the heavily industrialized Ohio River Valley. Rosenberg et al. (1979) also found increases in species diversity and importance in a mixed oak forest in Pennsylvania as distance from a coal-burning power plant increased. They found that compositional changes were generally more sensitive indicators of pollution damage to overstory species than were growth assessments.

Two newly emerging tools, dendroecology and simulation modeling, offer promise for quantifying individual tree and community-level responses. Dendroecology is a discipline of dendrochronology, the science of dating growth rings of woody plants (Fritts, 1971). As a tool for studying air pollution effects, dendroecology relies on multivariate statistical techniques to separate effects of tree age and local climate from those induced by air pollution (Nash et al., 1974). These data may be particularly useful when coupled with forest simulation models (see Shugart et al., 1980, and Section 6) to examine long-term community-level changes. Of basic concern to this task are the mechanisms by which species' responses are integrated into system responses and the effects of these stresses over time. Additional data on growth responses at levels ranging from seedlings to mature trees are badly needed.

6. APPLYING POLLUTANT VEGETATION INTERACTION DATA TO INTEGRATED ASSESSMENT MODELS

There is increasing emphasis on evaluating ecosystem-level effects of developing energy technologies. Under our present air quality regimes, such studies imply assessment of changes within and between components of both soil and plant systems over multiyear time scales. In reality, almost all evaluations of air pollution effects under these conditions focus on only a few components, with inferences being made to unmeasured ecosystem-level changes at a larger scale. Nevertheless, emphasis on ecosystem changes and total impacts of developing technologies has defined the need as never before to integrate and focus our necessarily smaller scale experiments toward larger scale objectives. The essential challenge may be posed as a question: How do we evaluate present air pollution impacts to terrestrial systems in such a way that both present and projected future impacts may be quantified? This task may be approached in several ways.

One very fundamental approach would involve the development of empirical dose–response equations developed from both laboratory and field research with a variety of species. These data would then be used with existing air monitoring data for a region to predict the range and distribution of growth and yield alterations to be expected at those or projected future levels. Such an approach implies that experiments have identified the principal exposure characteristics which limit yield and that air quality data are reported in a manner such that these characteristics can be quantified. Neither condition is adequately satisfied at present. When these data are available, generalizations will have to be made with respect to dose–response equations and regions within which air quality data and modifying environmental variables can be considered homogeneous. This approach appears to have its greatest potential with monospecific agriculture or forest crops where interspecific competition is not a factor.

A second and very promising approach involves the use of an empirical dose–response function for a single site or selected group of sites to determine likely responses of vegetation over a range of other sites. An example of this approach was recently discussed by Heagle et al. (1979). In this study, yield reductions of corn, wheat, soybeans, and spinach measured with open-top field chambers at ambient and above-ambient levels of ozone were used to predict potential nationwide losses to these crops. In spite of the many assumptions implied in such an approach, it represents a very reasonable initiation point of large-scale estimates of crop loss. At present, the EPA is planning an expanded network of these chambers to quantify regional differences in response of several crop species. Thus, nationwide damage estimates may be possible in the future, based on much-needed regionalized response functions. Potential applicability of these results to other species should be broadened by comparative controlled fumigation or field studies involving those test species and other species of regional importance.

Another research approach which has evolved rapidly during the last decade is the use of mathematical models to integrate data from diverse sources to describe responses of complex whole systems. Modeling has been a necessary and integral part of developing an understanding of how ecosystems function and respond to stress. In air pollution studies, models have been restricted primarily to single-plant or single-species dose–response functions involving visible injury and dose or visible injury and yield (see review by Heck and Brandt, 1977). A much broader application of modeling is now gradually evolving and provides a framework within which we can address more comprehensive analyses. Such questions as the relationship of alterations in plant physiological processes to whole-plant function, and the relationship of community responses to individual species' responses can be addressed with simulation techniques.

The area in which models would appear to most effectively extend our understanding involves responses of forest ecosystems to chronic air pollution stress. Here, the complex structure and perennial growth habit of forest communities, particularly uneven-aged mixed deciduous forests, makes measurement and prediction of responses a difficult task. Botkin et al. (1972) made important early contributions to the understanding of forest community dynamics with the development of JABOWA, a 12-species model of growth and succession for a northern hardwood forest. With it, Botkin (1976) subsequently examined and emphasized the important role of species interactions in responses of forest ecosystems to environmental perturbation. More recently, Shugart and West (1977) developed and parameterized a 33-species model of southeastern forests (FORET), which was used to examine the influence of chestnut blight on growth and successional pattern of eastern forests. This model has subsequently been used by McLaughlin et al. (1978) and West et al. (1980) to examine the potential influence of chronic growth inhibition by air pollution on both species and stand-level biomass responses. In these studies, hypothetical levels of externally introduced stress on growth of each species were assumed, based on literature data on the species' relative sensitivity to visible injury and growth responses. A maximum growth stress of 0–20% was imposed, depending on a species' sensitivity ranking. A model of this type allows one to study interactions between stand age and composition at the time of stress initiation, level of stress, and time since stress was initiated on total stand biomass and its distribution between component species. In these studies, both competition and stand age were identified as important modifiers of forest responses to air pollution. Other types of forest models and their potential applicability to air pollution problems have been reviewed by Shugart et al. (1980).

At present, models of the type described above must be regarded as tools with which to address difficult tasks, integrate data, and identify needed research. Their output cannot provide final quantitative answers until we have better data on chronic stress effects on individual species. They emphasize the need for data on relative species' sensitivity to chronic stress and for better quantification of responses to stress of mature trees in mixed communities. The former may be addressed through use of field-chamber techniques now used primarily for agricultural crops. The second task is more difficult. However, the field of dendroecology has great potential for addressing this problem. To document impacts of air pollution on forest growth over the past 40 years, during which chronic regional-scale stress has increased sharply, will require techniques for assessing changes in annual pollutant levels. One potentially useful tool for obtaining histories of exposure to general air pollution stress is heavy-metal analysis in the individual rings (Lepp, 1975). This approach has been used in Sweden (Symeonides, 1979) to construct histories of heavy-metal pollution, although Tian and Leep (1977) caution that factors such as radial transport and soil uptake must be fully understood to use this technique accurately. In the Swedish study, both lead and copper showed little lateral movement and were useful in constructing a decade-level history of metal pollution at the study site. More recent developments coupling X-ray emission spectroscopy (Valkovic et al., 1979) with growth-ring analysis show promise for using a variety of trace elements for historical analyses. As these techniques are further developed, they may provide useful data for constructing historical indices of regional-scale chronic stress.

Ultimately, assessment of air pollution impacts over large spatial scales must rely on development of large-scale data bases defining the nature and distribution of the resources which may be sensitive to those impacts. Such a system depends on computer analysis of multiple variables for a large number of landscape units. The size and amount of detail entered for these units may vary depending on the nature of the task addressed. An example of such a system is the Geoecology Data Base developed by Olson et al. (1980). Formatted at a country level, it includes over 1000 variables on terrain, water resources, forestry and other vegetation, wildlife, agriculture, land use, climate, air quality, population, and energy. The utility of such a system in addressing regional-scale assessment of effects of acid precipitation on soils was illustrated by Klopatek et al. (1980). Here computer maps of soil pH, cation exchange capacity, and base saturation were overlain with similar maps of current levels of hydrogen ion loading from acid rain. Resultant analysis identified at a county level of resolution those areas in which soil pH was most likely to be impacted. This type of system has great diversity and utility for a wide range of ecological problems. While one of its great values is its ability to use generalized ecological relationships in large-scale analyses, it calls on us to identify and perform those experiments which will make generalizations as accurate as is reasonably possible.

ACKNOWLEDGMENTS

This research was sponsored by the Office of Health and Environmental Research, U.S. Department of Energy, under contract W-7405-eng-26 with Union Carbide Corporation; Publication No. 1525, Environmental Sciences Division, Oak Ridge National Laboratory, Oak Ridge, TN 37830.

REFERENCES

Altshuller, A. P. Analytical problems in air pollution control. In W. W. Meinke and J. K. Taylor (eds.) *Analytical Chemistry: Key to Progress on National Problems*. NBS Spec. Publ. No. 351, National Bureau of Standards, Washington, D.C., 1972.

Andren, A. W., and S. E. Lindberg. Atmospheric input and origin of selected elements in Walker Branch Watershed, Oak Ridge, Tennessee. *Water Air Soil Pollut*. **8**, 199–215 (1977).

Applequist, H., K. O. Jensen, T. Sevel, and C. Hammer. Mercury in the Greenland ice sheet. *Nature* **273**, 657–659 (1978).

Barrie, L. A., and H. W. Georgii. An experimental investigation of absorption of SO_2 by water drops containing heavy metal ions. *Atmos. Environ*. **10**, 743–749 (1976).

Barrie, L. A., and J. L. Walmsley. A study of sulphur dioxide deposition velocities to snow in N. Canada. *Atmos. Environ*. **12**, 2321–2332 (1978).

Barrie, L. A., S. Beilke, and H. W. Georgii. Sulfur dioxide removal by cloud and fog drops as affected by ammonia and heavy metals. In R. G. Semonin and R. W. Beadle (eds.) *Precipitation Scavenging (1974)*. ERDA Symposium Series No. 41, CONF-741003, NTIS, Springfield, Virginia, 1977.

Beauford, W., J. Barber, and A. R. Barringer. Heavy metal release from plants into the atmosphere. *Nature* **256**, 35–37 (1975).

Beauford, W., J. Barber, and A. R. Barringer. Release of particles containing metals from vegetation into the atmosphere. *Science* **195**, 571–573 (1977).

Bennett, J. H., and A. C. Hill. Absorption of gaseous air pollutants by a standardized plant canopy. *J. Air Pollut. Control Assoc*. **23**, 203–206 (1973).

Bennett, J. H., and A. C. Hill. Interactions of air pollutants with canopies of vegetation. In J. B. Mudd and T. Kozlowski (eds.) *Response of Plants to Air Pollution*. 383 pp. Academic, New York, 1975.

Bennett, J. H., A. C. Hill, and D. M. Gates. A model for gaseous pollutant uptake by leaves. *J. Air Pollut. Control Assoc*. **23**, 957–962 (1973).

Berg, W. W., Jr. Chlorine Chemistry in the Marine Atmosphere. Ph.D. Dissertation, Department of Oceanography Technical Report 2-76, Florida State University, Tallahassee, Florida (1976).

Bjorseth, A., G. Lunde, and A. Lindskog. Long-range transport of polycyclic aromatic hydrocarbons. *Atmos. Environ*. **13**, 45–53 (1979).

Bogen, D. C., S. J. Nagourney, and G. A. Welford. MAP3S Precipitation Chemistry Intercomparison Study. Environmental Measurements Laboratory Report No. EML-341, Department of Energy, New York (1978).

Botkin, D. B. The role of species interactions in the response of a forest ecosystem to environmental perturbation. In B. C. Patten (ed.) *Systems Analysis and Simulation in Ecology, Vol. IV*, pp. 147–173, Academic, New York, 1976.

Botkin, D. B., J. F. Janak, and J. R. Wallis. Some ecological consequences of a computer model of forest growth. *J. Ecol*. **60**, 849–872 (1972).

Boutron, C. Past and present day tropospheric fallout fluxes of Pb, Cd, Cu, Zn, and Ag in Antarctica and Greenland. *Geophys. Res. Lett.* **6**, 159–162 (1979).

Boutron, C., and C. Lorius. Historical record of trace metals in Antarctic snows since 1914. *Nature* **277**, 551–554 (1978).

Brimblecomb, P. Dew as a sink for SO_2. *Tellus* **30**, 151–157 (1978).

Brosset, C., et al. Report of the Expert Meeting on Dry Deposition, World Meteorological Organization, Gothenburg, Sweden, April 18–22 (1977).

Brown, D. K., M. P. Maskarinec, F. W. Larimer, and C. W. Francis. Mobility of organic compounds from hazardous waste. EPA-600/54-83-001, Environmental Monitoring System Laboratory, U.S. EPA, Las Vegas, Nevada (1983).

Carlson, R. W., F. A. Bazzaz, and J. J. Stukel. Physiological effects, wind reentrainment, and rainwash of Pb aerosol particulate deposited on plant leaves. *Environ. Sci. Technol.* **10**, 1139–1142 (1976).

Casey, A. W. Scanning Electron Microscopy Techniques for Determining Atmospheric Particle-size Distribution. UCRL-51895, Lawrence Livermore Laboratory, Livermore, California (1975).

Cawse, P. A. Survey of Atmospheric Trace Elements in the United Kingdom (1972–1973). R-7669, Atomic Energy Research Establishment, Harwell, England (1974).

Chamberlain, A. C. Aspects of Travel and Deposition of Aerosols and Vapor Cloud. MP/R 1261, Atomic Energy Research Establishment, Harwell, England (1955).

Chamberlain, A. C. Interception and retention of radioactive aerosols by vegetation. *Atmos. Environ.* **4**, 57–78 (1970).

Changnon, S. A., Jr. Rainfall changes in summer caused by St. Louis. *Science* **205**, 402–404 (1979).

Cheney, J. L., and J. B. Homolya. Sampling parameters for sulfate measurement and characterization. *Environ. Sci. Technol.* **13**, 584–588 (1979).

Cheng, Yung-Sung, and Hsu-Chi Yeh. Particle bounce in cascade impactors. *Environ. Sci. Technol.* **13**, 1392–1396 (1979).

Chrisp, C. E., G. L. Fisher, and J. Lammert. Mutagenicity of filtrates from respirable coal fly ash. *Science* **199**, 73–75 (1977).

Clough, W. S. Transport of particles to surfaces. *Aerosol Sci.* **4**, 227–234 (1973).

Conner, W. D. An inertial-type particle separator for collection of large samples. *J. Air Pollut. Control Assoc.* **16**, 35–38 (1966).

Coutant, R. W. Effect of environmental variables on collection of aerosol sulfate. *Environ. Sci. Technol.* **11**, 873–878 (1977).

Dana, M. T. The MAP3S Precipitation Chemistry Network: First Periodic Summary Report, PNL-2402. Pacific Northwest Laboratory, Richland, Washington (1977).

Dana, M. T. The MAP3S Precipitation Chemistry Network: Second Periodic Summary Report, PNL-2829. Pacific Northwest Laboratory, Richland, Washington (1979).

Davidson, C. I. Deposition of Trace-Metal-Containing Aerosols on Smooth Flat Surface and on Wild Oatgrass. Ph.D. Dissertation, California Institute of Technology, Los Angeles, California (1977).

Davidson, C. I., and R. W. Elias. Dry deposition of trace elements at a remote site in the high Sierra. Preprint of a paper presented at the *Meeting of the American Institute of Chemical Engineers, Department of Civil Engineering. Carnegie-Mellon Institute, Pittsburgh, Pennsylvania* (1979).

Dingle, A. N., and B. M. Joshi. Ammonium sulfate crystallization in Andersen cascade impactor samples. *Atmos. Environ.* **8**, 1119–1130 (1974).

Dochinger, L. S., and T. Seliga (eds.). *Proceedings of the First International Symposium on Acid*

Precipitation and the Forest Ecosystem. USDA Forest Service Technical Report NE-12 (1976).

Dovland, H., and A. Eliassen. Dry deposition on a snow surface. *Atmos. Environ.* **10**, 783–785 (1976).

Droppo, J. G. Experimental techniques for dry deposition measurements. In D. S. Shriner, C. R. Richmond, and S. E. Lindberg (eds.) *Atmospheric Sulfur Deposition: Environmental Impact and Health Effects*. Ann Arbor Science Publishers, Ann Arbor, Michigan, 1980.

Duce, R. A., G. Hoffman, and W. H. Zoller. Atmospheric trace metals at remote northern and southern hemisphere sites: Pollution or natural? *Science* **187**, 59–61 (1975).

Dzubay, T. G., and R. K. Stevens. Ambient air analysis with dichotomous sampler and x-ray fluorescence spectrometer. *Environ. Sci. Technol.* **9**, 663–668 (1975).

Dzubay, T. G., L. E. Hines, and R. K. Stevens. Particle bounce errors in cascade impactors. *Atmos. Environ.* **10**, 229–234 (1976).

Eisenbud, M., and T. Kneip. *Trace Metals in Urban Aerosols*. EPRI-117, Electric Power Research Institute, Palo Alto, California, 418 pp., 1975.

Elias, R. W., T. Hinkley, T. Hirao, and C. Patterson. Improved techniques for study of mass balance and fractionation among families of metals within the terrestrial ecosystem. *Geochem. Cosmochem. Acta.* **40**, 583–587 (1976).

Environmental Protection Agency (EPA). Bioenvironmental Impact of a Coal-Fired Power Plant, Fourth Interim Report, Colstrip, Montana. E. M. Preston and T. M. Gullet (eds.) EPA-600/3-79-044, Corvallis Environmental Research Laboratory, Corvallis, Oregon (1979).

Essenwanger, O. Frequency distribution of precipitation. In *Physics of Precipitation*. (Geophysical Monograph No. 5) NAS-NRC No. 746, National Academy of Sciences, American Geophysical Union, Washington, D.C., 1960.

Fan, Kuo-Chung, and J. W. Gentry. Clogging in nuclepore filters. *Environ. Sci. Technol.* **12**, 1289–1294 (1978).

Fisher, G. L., C. E. Chrisp, and O. G. Raabe. Physical factors of the mutagenicity of fly ash from a coal fired power plant. *Science* **204**, 879–881 (1979).

Fritts, H. C. Dendroclimatology and dendroecology. *Quat. Res. N.Y.* **1**, 419–449 (1971).

Fowler, E. B. (ed.) *Radioactive Fallout, Soils, Plants, Food, Man*, 317 pp. Elsevier, New York, 1965.

Galloway, J. N. (ed.) *Proceedings of a Workshop on Toxic Substances in Atmospheric Deposition, Jekyl Island, Georgia, November 1979*, U.S. Environmental Protection Agency, Washington, D.C. (1980).

Galloway, J. N., and E. B. Cowling. The effects of precipitation on aquatic and terrestrial ecosystems: A proposed precipitation chemistry network. *J. Air Pollut. Control Assoc.* **28**, 229–235 (1978).

Galloway, J. N., and G. E. Likens. Calibration of collection procedures for the determination of precipitation chemistry. In *Proceedings of the First International Symposium on Acid Precipitation and the Forest Ecosystem*. pp. 137–156. USDA Forest Service Tech. Report NE23, 1976.

Galloway, J. N., and G. E. Likens. Collection of precipitation for chemical analysis. *Tellus* **30**, 71–82 (1978).

Galloway, J. N., and G. Parker. Difficulties in measuring wet and dry deposition on forest canopies and soil surfaces. In T. C. Hutchinson and M. Havas (eds.) *Effects of Air Pollutants on the Terrestrial Ecosystem. Proceedings of NATO Advanced Research Institute on the Effects of Acid Rain on Terrestrial Ecosystems* (NATO Conf. Series 1: Ecology, Vol. 4) pp. 57–68. Plenum, New York, 1980.

Galloway, J. N., E. B. Cowling, E. Gorham, and W. W. McFee. A national program for assessing the problem of atmospheric deposition. Report to the Council on Environmental Quality, National Atmospheric Deposition Program, Natural Resource Ecology Laboratory, Colorado State University, Fort Collins, Colorado (1978).

Galloway, J. N., B. J. Crosby, Jr., and G. E. Likens. Acid precipitation: Measurement of pH and acidity. *Limnol. Oceanogr.* **24**, 1161–1165 (1979).

Garland, J. A. Dry deposition of SO_2 and other gases. In R. J. Engelmann and G. A. Sehmel (eds.) *Atmospheric–Surface Exchange of Particulate and Gaseous Pollutants (1974)*. ERDA CONF-740921, Technical Information Center, Oak Ridge, Tennessee, 1976.

Gascoyne, M. Design and operation of a simple sequential precipitation sampler with some preliminary results. *Atmos. Environ.* **11**, 397–400 (1977).

Gatz, D. F. An urban influence on deposition of sulfate and soluble metals in summer rains. In D. S. Shriner, C. R. Richmond and S. E. Lindberg (eds.) *Atmospheric Sulfur Deposition: Environmental Impacts and Health Effects*. Ann Arbor Science Publishers. Ann Arbor, Michigan, 1980.

Gatz, D. F. Spatial variability of atmospheric deposition. In R. G. Semonin, D. F. Gatz, M. E. Peden, G. J. Stensland (eds.) *Study of Atmospheric Pollution Scavenging*. COO-1199-59, Illinois State Water Survey, University of Illinois, Urbana, Illinois, 1978.

Gehrs, C. W., D. S. Shriner, S. E. Herbes, E. Salmon, and H. Perry. Environmental safety and health implications of increased coal utilization. In M. Elliott (ed.) *Chemistry of Coal Utilization*, Chapter 31. National Academy of Sciences, Washington, D.C., 1980.

Gibson, J. H., and C. V. Baker. First Data Report—July 1978 to February 1979. National Atmospheric Deposition Program, Natural Resource Ecology Laboratory, Colorado State University, Fort Collins, Colorado (1979).

Gifford, F. A. Atmospheric dispersion models for environmental pollution application. In *Lectures on Air Pollution and Environmental Impact Analysis*. pp. 35–38. American Meteorological Society, Boston, Massachusetts, 1975.

Gifford, F. A. Turbulent diffusion typing schemes: A review. *Nucl. Saf.* **17**, 68–86 (1976).

Gordon, A. G., and E. Gorham. Ecological aspects of air pollution from an iron sintering plant at Wawa, Ontario, Canada. *J. Bot.* **41**, 1063–1078 (1963).

Gordon, G. Study of the Emissions from Major Air Pollution Sources and their Atmospheric Interactions. Progress Report-75. University of Maryland, College Park, Maryland (1975).

Gordon, G. E., E. S. Gladney, J. M. Ondov, T. Conry, and W. H. Zoller. Problems with the use of cascade impactors. In W. Fulkerson, W. D. Shultz, and R. I. Van Hook (eds.) *First Annual NSF Trace Contaminants Conference*, pp. 138–145. CONF-730802, Oak Ridge National Laboratory, Oak Ridge, Tennessee, 1973.

Graustein, W. C., and R. L. Armstrong. Measurement of dust input to a forested watershed using $^{87}Sr/^{86}Sr$ ratios. *Geol. Soc. Am. Abstr.* **10**, 411 (1978).

Greist, W. H., and M. R. Guerrin. *Identification and Quantification of Polynuclear Organic Matter (POM) on Particulates from a Coal-Fired Power Plant*. EA-1092, Electric Power Research Institute, Palo Alto, California, 1979.

Hanna, S. R. Local and global transport and dispersion of airborne effluents. Presented at *Topical Meeting on Controlling Airborne Effluents from Fuel Cycle Plants, Sun Valley, Idaho, July 1–14, 1976*. ATDL Contribution No. 76/3, National Oceanic Atmospheric Administration, Air Resources Laboratory, Atmospheric Turbulence Diffusion Laboratory, Oak Ridge, Tennessee (1976).

Harris, F. S. *Atmospheric Aerosols: A Literature Summary of Their Physical Characteristics and Chemical Composition*. NASA Report CR2626, National Aeronautics and Space Administration, Washington, D.C., 1976.

Hart, G. E., and D. R. Parent. Chemistry of throughfall under Douglas-fir and Rocky Mountain juniper. *Am. Midl. Nat.* **92**, 191–201 (1974).

Haugen, D. A. (workshop coordinator). *Lectures on Air Pollution and Environmental Impact Analyses*. American Meteorological Society, Boston, Massachusetts, 1975.

Heagle, A. S., D. E. Body, and W. W. Heck. An open-top field chamber to assess the impact of air pollution on plants. *J. Environ. Qual.* **2**, 365–368 (1973).

Heagle, A. S., A. J. Riordan, and W. W. Heck. Field methods to assess the impact of air pollutants on crop yields. Paper 79-46.6. In *Proceedings 72nd Annual Meeting Air Pollution Control Association, Cincinnati, Ohio* (1979).

Heck, W. W., and C. S. Brandt. Air pollution effects on vegetation: Native crops and forests. In A. Stern (ed.) *Air Pollution, 3rd ed.* pp. 157–229. Academic, New York, 1977.

Heindryckx, R. Comparison of the mass size functions of the elements in the aerosols of the Ghent industrial district with data from other areas: Some physico-chemical implications. *Atmos. Environ.* **10**, 65–71 (1976).

Hering, S. V., S. C. Hagan, and S. K. Friedlander. Design and evaluation of new low-pressure impactor, 1. *Environ. Sci. Technol.* **12**, 667–673 (1978).

Hering, S. V., S. K. Friedlander, J. J. Collins, and L. W. Richards. Design and evaluation of a new low-pressure impactor, 2. *Environ. Sci. Technol.* **13**, 184–188 (1979).

Hicks, B. B., and M. L. Wesely. *An Examination of Some Micrometeorological Methods for Measuring Dry Deposition*. EPA-600/7-78-116, 18 pp. U.S. Environmental Protection Agency, Research Triangle Park, North Carolina, 1978a.

Hicks, B. B., and M. L. Wesely. *Recent Results for Particle Deposition Obtained by Eddy Correlation Methods*. ERC-78-12, 17 pp. Argonne National Laboratory, Argonne, Illinois, 1978b.

Hicks, B. B., and M. L. Wesely. Turbulent transfer processes in the canopy and at vegetation surfaces. In D. S. Shriner, C. R. Richmond, and S. E. Lindberg (eds.) *Atmospheric Sulfur Deposition: Environmental Impacts and Health Effects*. Ann Arbor Science Publishers, Ann Arbor, Michigan, 1980.

Hicks, B. B., M. L. Wesely, and J. L. Durham. Critique of methods to measure dry deposition: Workshop summary. Argonne National Laboratory, Argonne, Illinois, December 4–5, 1979. Sponsored by Environmental Science Research Laboratory, U.S. Environmental Protection Agency, Research Triangle Park, North Carolina—avail. National Technical Information Service, Springfield, Virginia, as PB81-126443 (1980).

Hill, A. C. Vegetation: A sink for atmospheric pollutants. *J. Air Pollut. Control Assoc.* **21**, 341 (1971).

Hill, A. C., and E. M. Chamberlain. The removal of water soluble gases from the atmosphere by vegetation. In R. Engelmann and G. Sehmel (eds.) *Atmosphere–Surface Exchange of Particulate and Gaseous Pollutants*. ERDA Symposium Series-38, CONF-740921, pp. 153–171. National Technical Information Service, Springfield. Virginia, 1976.

Hill, G. R., and M. D. Thomas. Influence of leaf destruction by sulphur dioxide and by clipping on yield of alfalfa. *Plant Physiol.* **8**, 223–245 (1933).

Hodge, V., S. R. Johnson, and E. Goldberg. Influence of atmospherically transported aerosols on surface ocean water composition. *Geochem. J.* **12**, 7–20 (1978).

Hoffman, W. A. Jr., S. E. Lindberg, and R. R. Turner. Precipitation acidity: The role of forest canopy in acid exchange. *J. Environ. Qual.* **9**, 95–100 (1980).

Hosker, R. P., Jr. Estimates of dry deposition and plume depletion over forests and grassland. In *The Physical Behavior of Radioactive Contaminants in the Atmosphere*, pp. 291–310. IAEA Symposium SM-181, Vienna, Austria, 1974.

Jernelov, A., and T. Wallin. Airborne mercury fallout on snow and around five Swedish chlor-alkali plants. *Atmos. Environ.* **7**, 209–214 (1973).

Johnston, J. W., and A. S. Heagle. The response of chronically ozonated soybean plants to an acute ozone exposure. *Phytopathology* **72**, 387–389 (1982).

Kellogg, W. W., R. D. Cadle, E. R. Allen, A. L. Lazrus, and E. A. Martell. The sulfur cycle. *Science* **175**, 587–596 (1972).

Kennedy, V. C., G. W. Zellwenger, and R. J. Avanzino. Variation of rain chemistry during storms at two sites in northern California. *Water Resour. Res.* **15**, 687–702 (1979).

Keyser, T. R., D. F. S. Natusch, C. A. Evans, Jr., and R. W. Linton. Characterizing the surfaces of environmental particles. *Environ. Sci. Technol.* **12**, 768–773 (1978).

Klement, A. W., Jr. Radioactive Fallout from Nuclear Weapon Tests. AEC Symposium No. 5, CONF-765, National Bureau of Standards, U.S. Dept. of Commerce, Springfield, Virginia (1965).

Klepper, B., and D. K. Craig. Deposition of airborne particulates onto plant leaves. *J. Environ. Qual.* **4**, 495–499 (1975).

Klopatek, J. M., W. F. Harris, and R. J. Olson. A regional ecological assessment approach to atmospheric deposition. In D. S. Shriner, C. R. Richmond, and S. E. Lindberg (eds.) *Atmospheric Sulfur Deposition: Environmental Impacts and Health Effects*. Ann Arbor Science Publishers, Ann Arbor, Michigan, 1980.

Krause, G. H. M., and H. Kaiser. Plant response to heavy metals and SO_2. *Environ. Pollut.* **12**, 63–71 (1977).

Larsen, R. I., and W. W. Heck. An air quality data analysis system for interrelating effects, standards, and needed source reductions: Part 3, vegetation injury. *J. Air Pollut. Control Assoc.* **26**, 325–333 (1976).

Lawson, D. R., and J. W. Winchester. Sulfur and crystal reference elements in nonurban aerosols from Squaw Mountain, Colorado. *Environ. Sci. Technol.* **12**, 716–721 (1978).

Legge, A. H. Primary productivity, sulphur dioxide, and the forest ecosystem: An overview of a case study. In *Proceedings of the Symposium on Effects of Air Pollutants on Mediterranean and Temperature Forest Ecosystems, Riverside, California, June 22–27, 1980.* Available as Gen. Tech. Rep. PSW-43, 256 pp. Pacific Southwest Forest and Range Exp. Stn., Forest Serv., U.S. Dept. Agric., Berkeley, California, 1980.

Legge, A. H., D. R. Jaques, G. W. Harvey, H. R. Krouse, H. M. Brown, E. C. Rhodes, M. Nosal, H. U. Schellhase, J. Mayo, A. P. Hartgerink, P. F. Lester, R. G. Amundson, and R. B. Walker. *Sulfur Gas Emissions in the Boreal Forest: The West Whitecourt Case Study* (Final report to the Whitecourt Environmental Study Group, Kananaskis Centre for Environmental Research) The University of Calgary, Calgary, Alberta, Canada (1978).

Legge, A. H., D. R. Jaques, G. W. Harvey, H. R. Krouse, H. M. Brown, E. C. Rhodes, M. Nosal, H. U. Schellhase, J. Mayo, A. P. Hartgerink, P. F. Lester, R. G. Amundson, and R. B. Walker. Sulphur gas emissions in the boreal forest. The West Whitecourt case study I: Executive Summary. *Water Air Soil Pollut.* **15**, 77–85 (1981).

Lepp, N. W. The potential of tree-ring analysis for monitoring heavy metal pollution patterns. *Environ. Pollut.* **9**, 49–61 (1975).

Lim, M. Y. Trace Elements from Coal Combustion-Atmospheric Emissions. Report No. ICTIS/TR05, International Energy Agency of Coal Research, London, 58 pp. (1979).

Lindberg, S. E. Mercury partitioning in a power plant plume and its influence on atmospheric removal mechanisms. *Atmos. Environ.* **14**, 227–231 (1980).

Lindberg, S. E., and R. C. Harriss. Emissions from coal combustion: Use of aerosol solubility in hazard assessment. In J. Singh and A. Deepak (eds.) *Environmental and Climatic Impact of Coal Utilization*. Academic, New York, 1980.

Lindberg, S. E., and R. C. Harriss. The role of atmospheric deposition in an Eastern U.S. deciduous forest. *Water Air Soil Pollut.* **15**, 13–31 (1981).

Lindberg, S. E., and R. R. Turner. Mercury emissions from chlorine production solid waste deposits. *Nature* **268**, 133–136 (1977).

Lindberg, S. E., R. R. Turner, N. M. Ferguson, and D. Matt. Walker Branch Watershed element cycling studies: Collection and analysis of wetfall for trace elements and sulfate. In D. L. Correll (ed.) *Watershed Research in Eastern North America*, pp. 135–150. Smithsonian Institute Press, Edgewater, Maryland, 1977.

Lindberg, S. E., R. C. Harriss, R. R. Turner, D. S. Shriner, and D. D. Huff. Mechanisms and Rates of Atmospheric Deposition of Selected Trace Elements and Sulfate to a Deciduous Forest Watershed. ORNL/TM-6674, Oak Ridge National Laboratory, Oak Ridge, Tennessee, 514 pp. (1979a).

Lindberg, S. E., D. R. Jackson, J. W. Huckabee, S. A. Janzen, M. J. Levin, and J. R. Lund. Atmospheric emission and plant uptake of mercury from agricultural soils near the Almaden Mercury Mine. *J. Environ. Qual.* **8**, 572–578 (1979b).

Lindeken, D. D., R. L. Morgin, and K. F. Petrock. Collection efficiency of Whatman-41 filter paper for submicron aerosols. *Health Phys.* **9**, 305–308 (1963).

Liu, B. Y. H., and K. W. Lee. Efficiency of membrane and Nucleopore filters for submicrometer aerosols. *Environ. Sci. Technol.* **10**, 345–350 (1976).

Lockhart, L. B., Jr., R. L. Patterson, and W. L. Anderson. Characteristics of Air Filter Media Used for Monitoring Airborne Radioactivity. Naval Research Laboratory Report 6054, Washington, D.C. (1963).

Lunde, G., J. Gether, N. Gjos, and M. B. S. Lande. Organic micropollutants in precipitation in Norway. *Atmos. Environ.* **11**, 1007–1014 (1977).

Lundgren, D. A., F. S. Harris, W. H. Marlow, M. Lippmann, W. E. Clark, and M. D. Durham. *Aerosol Measurement*, 720 pp. University Presses of Florida, Gainesville, Florida, 1979.

McClenahen, J. R. Community changes in a deciduous forest exposed to air pollution. *Can. J. For. Res.* **8**, 432–438 (1978).

McFarland, A. R., J. B. Wedding, and J. E. Cermak. Wind tunnel evaluation of a modified Andersen impactor and an all weather sampler inlet. *Atmos. Environ.* **11**, 535–539 (1977).

McLaughlin, S. B., and D. S. Shriner. Allocation of resources to defense and repair. Chapter 20, in J. G. Horsfall and E. B. Cowling (eds.) *Plant Disease: An Advanced Treatise*. Academic, New York, 1980.

McLaughlin, S. B., D. C. West, H. H. Shugart, and D. S. Shriner. Air pollution effects on forest growth and successions: Applications of a mathematical model. Paper 78-24.5, In *71st Meeting Air Pollution Control Association, Houston, Texas* (1978).

McLaughlin, S. B., D. S. Shriner, R. K. McConathy, and L. K. Mann. The effects of SO_2 dosage kinetics and exposure frequency on photosynthesis and transpiration of kidney beans (*Phaseolus vulgaris* L.). *Environ. Exp. Bot.* **19**, 179–191 (1979).

MacLean, D. C., and R. E. Schneider. Fluoride accumulation by foliage: Continuous vs. intermittent exposures to hydrogen fluoride. *J. Environ. Qual.* **2**, 501–503 (1973).

MacLean, D. C., R. E. Schneider, and L. H. Weinstein. Accumulation of fluoride by forage crops. *Contrib. Boyce Thompson Inst.* **24**, 165–166 (1969).

McMullen, T. B., and R. B. Faoro. Occurrence of eleven metals in airborne particulates and in surficial materials. *J. Air Pollut. Control Assoc.* **27**, 1198–1202 (1977).

Mandl, R. H., L. H. Weinstein, D. C. McCune, and M. Keveny. A cylindrical, open-top chamber for exposure of plants to air pollutants in the field. *J. Environ. Qual.* **2**, 371–376 (1973).

Mecklenberg, R. A., H. B. Tukey, and J. V. Morgan. A mechanism for the leaching of Ca from foliage. *Plant Physiol.* **41**, 610–613 (1966).

Melo, O. T., and C. R. Phillips. Aerosol-size spectra by means of Nuclepore filters. *Environ. Sci. Technol.* **9**, 560–564 (1975).

Miller, J. M., J. N. Galloway, and G. E. Likens. The use of ARL trajectories for the evaluation of precipitation chemistry data. In L. S. Dochinger and T. A. Seliga (eds.) *Proceedings of the First International Symposium on Acid Precipitation and the Forest Ecosystem*. USDA Forest Service Tech. Report NE-23 (1976).

Miller, P. R., and J. R. McBride. Oxidant Air Pollution Effects on a Western Coniferous Forest Ecosystem. Statewide Air Pollution Research Center, University of California, Riverside, California, 1973.

Morgan, J. J. Sampling and analysis of metals in air, water and waste products. In *International Conference on Environmental Sensing and Assessment, Vol. I*, Section 1-2. University of Nevada, Las Vegas, 1975.

Mudd, J. B., and T. T. Kozlowski. *Responses of Plants to Air Pollution*, 383 pp. Academic, New York, 1975.

Munn, R., and B. Bolin. Review: Global air pollution—meteorological aspects. *Atmos. Environ.* **5**, 363–402 (1971).

Murphy, B. D. The Influence of Ground Cover on the Dry Deposition Rate of Gaseous Material. UCCND/CSAD-19, Oak Ridge National Laboratory, Oak Ridge, Tennessee, 28 pp. (1976).

Nappo, C. J. The simulation of atmospheric transport using observed and estimated winds. Preprint of *Third Symposium on Atmospheric Turbulence, Diffusion, and Air Quality, Raleigh, North Carolina*, pp. 100–106, American Meteorological Society (1976).

Nash, T. H., H. C. Fritts, and M. A. Stokes. A Technique for examining nonclimatic variation in widths of annual tree rings with special reference to air pollution. *Tree-Ring Bull.* **35**, 14–24 (1974).

National Academy of Sciences (NAS). *Arsenic—Medical and Biological Effects of Environmental Pollutants*. National Academy of Sciences, Washington, D.C., 1977.

National Research Council (NRC). *Airborne Particles*. (Report by the Subcommittee on Airborne Particles, Committee on Medical and Biological Effects of Environmental Pollutants, Division of Medical Sciences) University Park Press, Baltimore, Maryland, 343 pp. (1979).

Natusch, D. F. S., and J. Wallace. Urban aerosol toxicity: Influence of particle size. *Science* **186**, 695–699 (1974).

Natusch, D. F. S., J. Wallace, and C. Evans. Toxic trace elements: Preferential concentration in respirable particles. *Science* **183**, 202–204 (1974).

Neustadter, H. E., S. M. Sidik, R. B. King, J. S. Fordyce, and J. C. Burr. The use of Whatman-41 filters for high volume air sampling. *Atmos. Environ.* **9**, 101–109 (1975).

Nihlgard, B. Precipitation, its chemical composition and effect on soil water in a beech and spruce forest in southern Sweden. *Oikos* **21**, 208–217 (1970).

Noggle, J. C. The role of gaseous sulfur in crop nutrition. In D. S. Shriner, C. R. Richmond, and S. E. Lindberg (eds.) *Atmospheric Sulfur Deposition: Environmental Impacts and Health Effects*. Ann Arbor Science Publishers, Ann Arbor, Michigan, 1980.

O'Gara, P. I. Sulfur dioxide fume problems and their solutions. *J. Ind. Eng. Chem.* **14**, 744 (1922).

Olson, R. J., J. M. Klopatek, and C. J. Emerson. Regional environmental studies: Application of the geoecology data base. In P. A. Moore (ed.) *Mapping Collection Harvard Library of Computer Graphics, Vol. V*. Laboratory for Computer Graphics and Spatial Analysis, Cambridge, Massachusetts (1980).

Owens, M. J., and A. W. Powell. Deposition velocity of SO_2 on land and water surfaces using a ^{35}S tracer method. *Atmos. Environ.* **8**, 63–67 (1974).

Parker, G. G., S. E. Lindberg, and J. M. Kelly. Atmosphere/canopy interactions of sulfur in the southeastern United States. In D. S. Shriner, C. R. Richmond, and S. E. Lindberg (eds.) *Atmospheric Sulfur Deposition: Environmental Impacts and Health Effects*. Ann Arbor Science Publishers, Ann Arbor, Michigan, 1980.

Parker, R. D., G. H. Buzzard, T. G. Dzubay, and J. P. Bell. A two-stage respirable aerosol sampler using Nuclepore filters in series. *Atmos. Environ.* **11**, 617–621 (1977).

Pattenden, N. J., and R. D. Wiffen. The particle size dependence of the collection efficiency of an environmental aerosol sampler. *Atmos. Environ.* **11**, 677–681 (1977).

Peden, M. E., and L. M. Skowron. Ionic stability of precipitation samples. *Atmos. Environ.* **12**, 2343–2349 (1978).

Peden, M. E., L. M. Skowron, and F. M. McGurk. Precipitation Sample Handling, Analysis, and Storage Procedures. Research Report 4, Illinois State Water Survey, Urbana, Illinois (1979).

Penkett, S. A., B. M. R. Jones, and R. E. J. Eggleton. A study of SO_2 oxidation in stored rainwater samples. *Atmos. Environ.* **13**, 139–147 (1979).

Pierce, R. C., and M. Katz. Dependency of polynuclear aromatic hydrocarbon content on size distribution of atmospheric aerosols. *Environ. Sci. Technol.* **9**, 347–353 (1975).

Raynor, G. S. Meteorological and Chemical Relationships from Sequential Precipitation Samples. CONF-761102, Brookhaven National Laboratory, Upton, New York (1976).

Raynor, G., and J. Haynes. Experimental Data from Analysis of Sequential Precipitation Samples at Brookhaven National Laboratory. BNL-50826, Atmospheric Sciences Division, Brookhaven National Laboratory, Upton, New York (1977).

Raynor, G. S., and J. P. McNeil. An automatic sequential precipitation sampler. *Atmos. Environ.* **13**, 149–155 (1979).

Reinert, R. A., A. S. Heagle, and W. W. Heck. Plant responses to pollutant combinations. In J. B. Mudd and T. T. Kozlowski (eds.) *Response of Plants to Air Pollution*. 383 pp. Academic, New York, 1975.

Rogers, H. H., H. E. Jeffries, E. P. Stahel, W. W. Heck, L. A. Ripperton, and A. M. Witherspoon. Measuring air pollutant uptake by plants: A direct kinetic technique. *J. Air Pollut. Control Assoc.* **27**, 1192–1197 (1977).

Rogers, R. R., and I. I. Zawadski. Rainfall statistics for application to plume rainout models. *Atmos. Environ.* **13**, 279–286 (1979).

Ronneau, C., J. Cara, J. L. Navarre, and P. Priest. An automatic sequential rain sampler. *Water Air Soil Pollut.* **9**, 171–176 (1978).

Rosenberg, C. R., R. J. Hutnik, and D. D. Davis. Forest composition at varying distances from a coal-burning power plant. *Environ. Pollut.* **13**, 307–317 (1979).

Schlesinger, W., W. Reiners, and D. Dropman. Deposition of water and cations on artificial foliar collectors in fir-Krummholz of New England mountains. *Ecology* **55**, 378–386 (1974).

Schroeder, H. A. Metals in the air. *Environment* **13**, 18–32 (1971).

Sehmel, G. A. A summary of dry deposition velocity data. In D. S. Shriner, C. R. Richmond, and S. E. Lindberg (eds.) *Atmospheric Sulfur Deposition: Environmental Impacts and Health Effects*. Ann Arbor Science Publishers, Ann Arbor, Michigan, 1980.

Sehmel, G. A., and W. Hodgson. Particle and Gaseous Removal in the Atmosphere by Dry Deposition. BNWL-SA-4941. Battelle Pacific Northwest Laboratory, Richland, Washington (1976).

Semonin, R. G. The variability of pH in convective storms. *Water Air Soil Pollut.* **6**, 395–406 (1976).

Semonin, R., and R. Beadle (eds.) Precipitation Scavenging. (1974) ERDS Symposium Series 41. Available from National Technical Information Service, Springfield, Virginia as CONF-741003 (1977).

Shinn, J. H. Estimation of aerosol plutonium transport by the dust-flux method: A perspective on application of detailed data. In M. G. White and P. B. Dunaway (eds.) *Transuranics in Natural Environments*. NVO-178, Nevada Applied Ecology Group, Las Vegas, Nevada, 1977.

Shriner, D. S. Effects of simulated rain acidified with sulfuric acid on host-parasite interactions. *Water Air Soil Pollut.* **8**, 9–14 (1976).

Shriner, D. S., S. B. McLaughlin, and C. F. Baes. Character and Transformation of Pollutants from Major Fossil Fuel Energy Sources, 39 pp. ORNL/TM-5919, Oak Ridge National Laboratory, Oak Ridge, Tennessee (1977).

Shriner, D. S., C. R. Richmond, and S. E. Lindberg (eds.) *Atmospheric Sulfur Deposition: Environmental Impacts and Health Effects.* (Proceedings of the Second Life Sciences Symposium, Potential Environmental and Health Consequences of Atmospheric Sulfur Deposition, Gatlinburg, Tennessee, Oct. 14–18, 1979) Ann Arbor Science Publishers, Ann Arbor, Michigan, 1980.

Shugart, H. H., and D. C. West. Development and application of an Appalachian deciduous forest succession model and its application to assessment of the impact of the chestnut blight. *J. Environ. Manage.* **5**, 161–179 (1977).

Shugart, H. H., S. B. McLaughlin, and D. C. West. Forest models: Their development and potential applications for air pollution effects research. In *Proceedings, Symposium on Effects of Air Pollutants on Temperate and Mediterranean Forest Ecosystems*, USDA Forest Service (1980).

Sievering, H., J. Dave, G. McCoy, and K. Walther. Cellulose filter high-volume cascade impactor aerosol collection efficiency, A technical note. *Environ. Sci. Technol.* **12**, 1435–1437 (1978).

Skogerboe, R. K., D. L. Dick, and P. L. Lamothe. Evaluation of filter inefficiencies for particulate collection under low loading conditions. *Atmos. Environ.* **11**, 243–249 (1977).

Slanina, J., J. G. Van Raaphorst, W. L. Zijp, A. J. Vermeulen, and C. A. Roet. An evaluation of the chemical composition of precipitation sampled with 21 identical collectors on a limited area. *Int. J. Environ. Anal. Chem.* **6**, 67–81 (1979).

Spurny, K. R., J. P. Lodge, Jr., D. C. Sheesley, and E. R. Frank. Aerosol filtration by means of Nuclepore filters—aerosol sampling and measurement. *Environ. Sci. Technol.* **3**, 453–463 (1969).

Spurny, K. R., J. Havlova, J. P. Lodge, Jr., E. R. Ackerman, D. C. Sheesley, and B. Wilder. Aerosol filtration by means of Nuclepore filters. *Environ. Sci. Technol.* **8**, 758–761 (1974).

Steinhardt, I., D. G. Fox, and W. E. Marlott. Modeling the uptake of SO_2 by vegetation. In *Proceedings, 4th National Conference on Fire and Forest Meteorology, St. Louis, Missouri, November 16–18, 1976*. USDA For. Serv. Gen. Tech. Rep. RM-32, 239 pp. (1977).

Stratmann, H. Field experiments to determine the effects of SO_2 on vegetation, Part II. (Measurement and evaluation of SO_2 ground level concentrations. Forschungsberichte des Landes Nordrhein–Westfalen Freiland versuche zur Ermittlung von schwefeldioxyd wirkung auf die vegetation, Part II) No. 1184, Landesanstalt für Immissions- und Bodennutzungsschutz des Landes Nordrhein–Westfalen, Essen, West Germany, 69 pp. (1963).

Symeonides, C. Tree-ring analysis for tracing the history of pollution: Application to a study in northern Sweden. *J. Environ. Qual.* **8**, 482–486 (1979).

Taylor, G. E. Genetic analysis of ecotype differentiation within an annual plant species, *Geranium carolinianum* L., in response to sulfur dioxide. *Bot. Gaz.* **139**, 362–368 (1978).

Thom, A. S. Momentum, mass, and heat exchange of plant communities. In J. L. Monteith (ed.) *Vegetation and the Atmosphere*, 439 pp. Academic, London, 1975.

Tian, T. K., and N. W. Lepp. Further studies on the development of tree-ring analysis as a method for constructing heavy metal pollution histories. Proceedings, University of Missouri Annual Conference on Trace Substances. *Environ. Health* **11**, 440–447 (1977).

Tolg, G. Elemental analysis with minute samples. In *Comprehensive Analytical Chemistry, III*. Elsevier, New York, 1975.

Tukey, H. B. The leaching of substances from plants. *Annu. Rev. Plant Physiol.* **21**, 305–324 (1970).

Tukey, H. B., R. A. Mecklenberg, and J. V. Morgan. Mechanisms for the leaching of metabolites from foliage. In *Isotopes and Radiation in Soil-Plant Nutrition*. IAEA Symposium Series 1965, International Atomic Energy Agency, Vienna, 1965.

Turner, R. R., and S. E. Lindberg. An Interlaboratory Comparison of Trace Analysis by Graphite Furnace AAS. ORNL/TM-5422, 21 pp. Oak Ridge National Laboratory, Oak Ridge, Tennessee (1976).

U.S. Atomic Energy Commission. Toxicity of Power Plant Chemicals to Aquatic Life. USAEC WASH-1249, Washington, D.C. (1973).

Valkovic, V., D. Rendic, E. K. Biegert, and E. Andrade. Trace element concentrations in tree rings as indicators of environmental pollution. *Environ. Int.* **2**, 27–32 (1979).

Van Hook, R. I., and D. Shultz (eds.) *Effects of Trace Contaminants from Coal Combustion, Proceedings of a Workshop*, 79 pp. ERDA 77-64, Energy Research and Development Administration, Washington, D.C., 1977.

Vaughan, B. E., K. H. Abel, D. A. Cataldo, J. M. Hales, C. E. Hane, L. A. Rancitelli, R. C. Routson, R. E. Wildung, and E. G. Wolf. Review of Potential Impact on Health and Environmental Quality from Metals Entering the Environment as a Result of Coal Utilization. A Battelle Energy Program Report, Battelle Memorial Institute, Richland, Washington, 75 pp. (1975).

Volchok, H. L., L. E. Toonkel, and M. Schonberg. Trace Metals—Fallout in New York City, II. U.S. Atomic Energy Commission Report HASL-281, Health and Safety Laboratory, New York (1974).

Wedding, J. B., R. W. Carlson, J. Stukel, and F. A. Bazzaz. Aerosol deposition on plant leaves. *Environ. Sci. Technol.* **9**(2), 151–154 (1975).

Wedding, J. B., R. W. Carlson, J. J. Stukel, and F. A. Bazzaz. Aerosol deposition on plant leaves. *Water Air Soil Pollut.* **7**, 545–550 (1977a).

Wedding, J. B., A. R. McFarland, and J. E. Cermak. Large particle collection characteristics of ambient aerosol samplers. *Environ. Sci. Technol.* **11**, 387–390 (1977b).

Weisburger, K. Natural carcinogenic products. *Environ. Sci. Technol.* **13**, 278–281 (1979).

Wesely, M. L., and B. B. Hicks. Some factors that affect the deposition rates of SO_2 and similar gases on vegetation. *J. Air Pollut. Control Assoc.* **27**, 1110–1116 (1977).

West, D. C., S. B. McLaughlin, and H. H. Shugart. Simulated forest response to chronic air pollution stress. *J. Environ. Qual.* **9**(1), 43–49 (1980).

Whelpdale, D. N. Atmospheric pathways of sulphur compounds: A general report. MARC Report No. 7, Chelsea College, University of London, London (1978).

White, E. J., and F. Turner. A method of estimating income of nutrients in a catch of airborne particles by a woodland canopy. *J. Appl. Ecol.* **7**, 441–460 (1970).

White, M. G., and P. B. Dunaway. Transuranics in Natural Environments. NVO-178, UC-2, Nevada Applied Ecology Group, Las Vegas, Nevada (1977).

Winchester, J. W., and G. G. Desaedleer. Applications of trace element analysis to studies of the atmospheric environment. In Amiel, S. (ed.) *Nondestructive Activation Analysis*. Elsevier, Amsterdam, 1975.

Winkler, P. Relative humidity and the adhesion of atmospheric particles to the plates of impactors. *Aero. Sci.* **5**, 235–240 (1974).

Winner, W. E., and H. A. Mooney. Ecology of SO_2 resistance: I. Effects of fumigators on gas exchange of deciduous and evergreen shrubs. *Oecologia* **44**, 290–295 (1980a).

Winner, W. E., and H. A. Mooney. Ecology of SO_2 resistance: II. Photosynthetic changes of shrubs in relation to SO_2 absorption and stomatal behavior. *Oecologia* **44**, 296–302 (1980b).

Zahn, R. Effects of sulfur-dioxide on vegetation. *Staub* **21**, 56–60 (1961).

Zahn, R. The effect on plants of a combination of subacute and toxic sulfur dioxide doses. *Staub* **30**, 20–23 (1970).

Ziegler, I. S. The effect of air polluting gases on plant metabolism. *Env. Qual. Safety* **2**, 182–208 (1973).

Zimdahl, R. Entry and movement in vegetation of lead. *J. Air Pollut. Control Assoc.* **26**, 655–660 (1976).

Zoller, W. H., E. S. Gladney, and R. A. Duce. Atmospheric concentration and sources of trace metals at the South Pole. *Science* **183**, 198–200 (1974).

Practical Application of Air Pollutant Deposition Models—Current Status, Data Requirements, and Research Needs

R. P. HOSKER, Jr.

Air Resources
Atmospheric Turbulence and Diffusion Laboratory
National Oceanic and Atmospheric Administration
Oak Ridge, Tennessee

1. INTRODUCTION

This paper largely grew from a requirement to model the dispersion of toxic contaminants in the atmosphere over a variety of weather conditions, sites and usage patterns, and travel distances. As a first step, a comprehensive literature survey was performed to review the types and suitability of various models available for this purpose. In particular, since delivery of effluent to the underlying landscape is a primary concern, methods to estimate both wet and dry plume scavenging were examined. Considerable effort was expended collecting either experimental data tabulations or computational relations for the major parameters required in these techniques. On the basis of this work, it appears to be possible to generate computer codes to plausibly simulate transport, diffusion, and removal processes over a range of distances and terrain types with a varying degree of success. At short ranges in simple terrain the results should be relatively good. Using empirically determined parameters, accuracy for a single event is probably order of magnitude, but the time-averaged results should be better—perhaps four or five times better. At longer distances and/or in complex terrain, however, single-event accuracy will be much worse, and the time-averaged estimates will probably be only order of magnitude.

The text first describes the most significant currently available air pollutant transport and diffusion models; this may be helpful for investigators who have not been active in this particular field. The presentation is biased toward simple models whenever feasible. Available parameterizations of wet and dry removal processes appropriate for use in these models are then discussed. To apply such techniques, information is needed about certain quantities, notably dry deposition velocity, wet scavenging rate, washout ratio, or (equivalently) wet deposition velocity. The report describes explicitly the computation of dry deposition velocity via the resistance analog for a variety of circumstances, and lists sources for empirical values of this quantity. Calculation of the wet scavenging rate for irreversibly captured particles and gases is described, along with removal estimation procedures for gases which are not permanently collected within hydrometeors. The method for estimating the washout ratio and/or wet deposition velocity is outlined; this technique is especially appropriate for long-term modeling. References for empirical wet removal data are also listed. Finally, some recommendations are made for model types according to the size of the region under study, and the data requirements and likely availability are discussed. Topics where additional research is needed are suggested.

Symbols are defined as they appear in the text. An attempt has been made to utilize notation common in the meteorological and air pollution literature without injuring too severely the sensibilities of workers in other disciplines. Actually the notation problem is quite serious, particularly in the strongly chemically oriented wet removal process. Hales (1972) suggested for this reason that the format of Bird et al. (1960) be used exclusively. Their symbolism, however, will generally be unfamiliar and somewhat confusing to meteorologists attempting to apply the concepts to air quality problems. There may be no good way to resolve this dilemma. In any event, the literature must be approached with great caution since errors of many orders of magnitude may arise from misinterpretation of symbols and definitions.

2. ATMOSPHERIC TRANSPORT AND REMOVAL MODELS IN CURRENT USE

2.1 Transport and Dispersion Model Types, Limitations, and Applicability

This is not the place for a comprehensive review of atmospheric transport and diffusion modeling; such an undertaking would require its own workshop. Instead, the major types of models in common use will be indicated, together with their theoretical bases and a few recent examples from the literature. Excellent summaries by Barr and Clements (1984), Drake et al. (1979), Gifford (1975), Johnson et al. (1976a), Pasquill (1975), and U.S. EPA (1978) should be consulted for further information. Table 1, based on Drake et al. (1979), illustrates some of the significant characteristics of various model types; the following discussion relies fairly heavily on this reference.

2.1.1 Rollback Models

These models are representative of a class of empirically calibrated relationships between atmospheric concentrations and effluent source strengths. The simplest version assumes proportionality and uses historical air quality data to calculate the appropriate constant. Once calibrated for a particular region, the model can be used to determine the reduction in emissions (the rollback) needed to achieve some given concentration level. No real understanding of the physical processes involved is required.

2.1.2 Statistical Models

The statistical models of Table 1 are a somewhat more elaborate way of establishing empirical relationships between concentrations and emissions. In these models, however, the dependence of air quality on a number of physical and meteorological variables is explored, and a certain amount of theoretical understanding of the underlying processes may be assumed. Many statistical tools have been utilized. Once the particular method is calibrated to a site, it can be used in a predictive mode, much like the rollback model. One advantage of these more complex models is that they may elucidate the roles of different sources (e.g., automobiles versus

TABLE 1. Characteristics of Transport and Dispersion Models

Model Type	Geographical Scale	Steady State or Time Dependent	Frame of Reference	Reaction Mechanisms	Removal Mechanisms	Treatment of Turbulence	Plume Additivity	Topography Treated
Rollback	Local, regional, or national	Steady state	Eulerian	Nonreactive	Implicit	Implicit	Not applicable	All terrain
Statistical	Local or regional	Steady state or time dependent	Eulerian	Nonreactive, reactive, or gas-to-particle	Implicit	Implicit	Not applicable	All terrain
Gaussian plume and puff	Local	Steady state or time dependent	Eulerian or Lagrangian	Nonreactive or reactive	Dry and wet	Diffusion coefficients	Yes and no	Homogeneous to simple terrain
Regional trajectory	Regional or national	Time dependent	Lagrangian, or mixed Lagrangian and Eulerian	Nonreactive or reactive	Dry and wet	Diffusion coefficients or eddy diffusivities	Yes	Nonhomogeneous to complex terrain
Box and multibox	Local or regional	Steady state or time dependent	Eulerian or Lagrangian	Nonreactive, reactive, or gas-to-particle	Dry and wet	Well-mixed or eddy diffusivities	Yes and no	Homogeneous to simple terrain
Grid	Local or regional	Steady state or time dependent	Eulerian	Nonreactive, reactive, or gas-to-particle	Dry and wet	Eddy diffusivities or complex formulation	Yes and no	All terrain
Particle	Local or regional	Time dependent	Mixed Lagrangian and Eulerian	Nonreactive, reactive, or gas-to-particle	Dry and wet	Eddy diffusivities	Yes and no	All terrain
Global	Global	Time dependent	Eulerian	Nonreactive or reactive	Dry and wet	Eddy diffusivities	Yes	All terrain
Physical	Local	Time dependent	Mixed Lagrangian and Eulerian	Nonreactive	None	Not applicable	Not applicable	All terrain

Modified from Drake et al. (1979).

industries), to facilitate control measures. In both cases, however, major changes in source strength and/or distribution will influence the model calibration, making further estimates unreliable.

2.1.3 Gaussian Plume

The best known model for diffusion of material emitted from a continuous point source is undoubtedly the Gaussian plume formulation:

$$\chi(x, y, z) = \frac{Q_0}{2\pi\sigma_y\sigma_z U} \exp\left[-\frac{y^2}{2\sigma_y^2}\right]\left\{\exp\left[-\frac{(z-h)^2}{2\sigma_z^2}\right] + \exp\left[-\frac{(z+h)^2}{2\sigma_z^2}\right]\right\} \quad (1)$$

which gives concentration χ at a receptor location (x, y, z) due to a continuous constant emission rate $\dot{Q}_0$ at $(0, 0, h)$. The model includes the effect of reflection of the plume at the ground. The dispersion parameters σ_y and σ_z describe crosswind and vertical dispersion of the plume; they are functions of distance downwind, atmospheric stability, and terrain roughness. Discussions of semiempirical curves or tabulated values describing these parameters are commonplace in the literature. A review by Gifford (1976) and recommendations of an American Meteorological Society workshop on the topic (Hanna et al., 1977) should be consulted for more recent guidance.

As pointed out by Gifford (1975), the Gaussian model rests on two distinct theoretical arguments (not quite the same as derivations!). The most easily understood development begins with the three-dimensional diffusion equation, assuming gradient transport of material in the atmosphere:

$$\frac{D\chi}{Dt} = \frac{\partial\chi}{\partial t} + \mathbf{v}\cdot\nabla\chi = \nabla\cdot(\mathbf{K}\cdot\nabla\chi) \quad (2)$$

where $\mathbf{v}$ is the vector wind and $\mathbf{K}$ is the eddy diffusivity tensor. If effects such as wind shear are insignificant so that the off-diagonal tensor components are negligible, and if the remaining terms are constant (though not necessarily equal), the equation reduces to the classical Fickian form:

$$\frac{D\chi}{Dt} = K_x\frac{\partial^2\chi}{\partial x^2} + K_y\frac{\partial^2\chi}{\partial y^2} + K_z\frac{\partial^2\chi}{\partial z^2} \quad (3)$$

This can be solved (Roberts, 1923) using the boundary conditions for an instantaneous point source, and then integrated over all time, assuming along-wind diffusion is negligible. A form of Eq. (1) results, using $\sigma_y^2 = 2K_y x/U$, $\sigma_z^2 = 2K_z x/U$. Application of the solution to the atmosphere unfortunately violates a number of the assumptions utilized in the derivation; in particular, the eddy diffusivities can be expected to be nonconstant. Neglect of along-wind diffusion is erroneous at low to zero wind speeds.

A second argument for the validity of Eq. (1) is grounded in the statistical theory of turbulence (Taylor, 1921). In this development, one appeals (Monin and Yaglom, 1971, 1975) to theoretical and experimental evidence that the probability density function for particles dispersing in homogeneous turbulence is Gaussian, both immediately after the process begins, and after a long time has elapsed. At intermediate times, the distribution is nearly Gaussian. The distribution can then simply be written down for dispersion relative to a fixed axis, and integrated in time to obtain a relation much like Eq. (1). In the atmosphere, however, turbulence is not homogeneous, particularly in the vertical direction; hence a rigorous application of the result is not possible.

In practice, however, the Gaussian formulation has been found to have considerable utility. At short travel distances, the assumption of homogeneity of turbulence may not be violated too badly, at least in the horizontal. For very large travel times, the statistical theory continues to predict that the Gaussian form is correct. At intermediate distances, one must be cautious, since effects such as wind shear may distort the plume. By using empirically derived dispersion parameters σ_y and σ_z, it appears that the model may be utilized out to distances of 10 km or so, at least in simple terrain. Techniques to modify the dispersion parameters to account for a variety of meteorological, terrain, and ground cover conditions have been widely discussed in the literature, further extending the applicability of the model.

2.1.4 Gaussian Puff

Gaussian puff models assume that a diffusing cloud is normally distributed as it is advected along some trajectory by the mean wind. The main advantage of the technique is that it permits the effluent path to vary with wind direction; this may be especially significant in regions of complex topography, where the mean wind is steered by the terrain. There are several disadvantages, however (see, e.g., Barr and Clements, 1984). Any real individual puff is not likely to possess a Gaussian concentration distribution, which can be expected to describe only an *ensemble* of puffs. Furthermore, the puff diffusion coefficients σ_{yp} and σ_{zp}, say, differ from those associated with plumes, since the dispersion processes are fundamentally different. The growth rate of a puff changes with time, since only eddies of a scale similar to the puff size contribute to the relative diffusion process, whereas eddies of all sizes influence the average diffusion of a plume (see Gifford, 1968). Finally, it must be recognized that an individual puff is transported by the wind field present at the time of release; this is not likely to coincide with the average wind field. Hence one must estimate the likely envelope of puff trajectories generated by fluctuations in wind direction about the mean.

The Gaussian plume and puff formulations have been used to generate line (e.g., highway) and area (e.g., urban) source models by superposition, assuming the effluents are not mutually reactive. A rather massive literature on this topic is available in the air pollution control field; Drake et al.

(1979) list some of the important models in current use, particularly by regulatory agencies.

2.1.5 Trajectory Models

Trajectory models are most commonly utilized for regional and larger travel distances. They simulate the behavior of a tagged air parcel moved about by a temporally and spatially varying wind field. The driving winds are generally specified externally, either from direct meteorological observations or as output data from a wind field prediction model. Pollutant dispersion occurs via continuous plumes or a series of moving puffs or boxes, which follow the computed trajectories. Many different assumptions can be made to account for the crosswind and vertical diffusion (e.g., uniform, Gaussian, or gradient-transfer-driven distributions), and for wet and dry removal processes.

The main advantage of these models is their ability to deal with changing wind and stability conditions. Such variations may be induced by any large-scale terrain or ground cover changes, or by the diurnal cycles experienced by any air parcel which travels large distances. Episodic events, such as encounters with rainstorms, can also be treated with such models. Drake et al. (1979) suggest that the disadvantages of trajectory models arise mainly from their sensitivity to errors in specifying the driving winds, and from problems associated with modeling effluent diffusion and removal processes along the trajectory. Difficulties may be expected in regions of multiple sources if the effluents are chemically interactive. Despite their drawbacks, trajectory models have become very popular for treating effluent dispersion over large distances and/or in complex terrain. The references provided by Drake et al. (1979) represent a reasonable cross section of versions in current use.

2.1.6 Box Models

Box, multibox, moving-cell, integral, and other such models are all based on mass conservation within some volume of interest. The volume may be approximated by a box, a set of boxes, a mixing region capped by a spatially varying upper boundary, or some other similar scheme. Pollutants within such subregions are generally supposed to be well mixed, or else obey some well-defined mixing process, such as gradient transport via *a priori* specified profiles of wind and eddy diffusivity. Topographical variations and the effect of elevated inversions can be treated; chemical reactions and effluent removal processes can be incorporated. Models of this type have been heavily utilized in the air pollution control field. Their complexity and the consequent computing problem naturally vary greatly with the situation addressed and the assumptions used. Agreement of the solutions with observed data can range anywhere between qualitatively and quantitatively good.

2.1.7 Grid Models

Grid models deal directly with the complete set of equations required to describe physical phenomena such as fluid motion, diffusion, chemical reactions, and assorted removal processes. Finite difference approximations to these equations are developed, a grid or net is superposed on the region of interest, and solutions are generated at the grid intersection points. Both two- and three-dimensional formulations are common. Very complex physical phenomena, terrain, or meteorological conditions can in principle be readily accommodated.

The major problems associated with such models are computational: instabilities may arise in the calculations if time or space increments are improperly selected, inaccuracies may appear because of the particular finite differencing scheme utilized, and the time and storage requirements may stretch or exceed computer capacities and budgets. In addition, physical processes taking place at subgrid scales usually must be parameterized so their effect on larger scale events can be included. This can be a formidable task. A comprehensive survey of grid-type air quality models is given by Drake et al. (1979), and should be consulted as a guide to both numerical techniques and specific physical approximations such as higher-order-closure turbulence theory, coupled chemical reactions, or aerosol formation.

2.1.8 Particle Models

So-called particle models represent the dispersion of a pollutant by allowing a very large number of particles to be transported through a network of grid cells by a pseudovelocity field composed of the actual velocity plus a "diffusion velocity" derived from gradient transport theory (e.g., Lange, 1978). Consider Eq. (2), and assume the wind field is not divergent, so that

$$\nabla \cdot \mathbf{v} = 0 \tag{4}$$

Then define the diffusion velocity $\mathbf{u}_D$ as

$$\mathbf{u}_D \equiv -\mathbf{K} \cdot \frac{\nabla \chi}{\chi} \tag{5}$$

Equation (2) can then be written as

$$\frac{\partial \chi}{\partial t} + \nabla \cdot (\mathbf{u}_P \chi) = 0 \tag{6}$$

where the pseudovelocity $\mathbf{u}_P \equiv \mathbf{v} + \mathbf{u}_D$. One solves Eq. (6) over a three-dimensional grid superposed on the region of interest. To begin, a large number of particles (generally 10^4 or more) are injected at the source location and distributed within the grid in some specified way—perhaps by a Gaussian plume for this first step. The subsequent number of particles in each cell represents the concentration in that cell. From this initial concentration field, the diffusion velocity is generated using Eq. (5) and a prescribed form for the eddy diffusivity tensor. The pseudovelocity $\mathbf{u}_P$ can then be calculated and used to transport the particles to new cellular locations, satisfying Eq. (6). A new value for $\mathbf{u}_D$ can then be found and the process repeated. Processes such as gravitational settling, wet removal, and radioactive decay can easily be incorporated.

The technique sidesteps many of the spurious effects associated with finite difference solutions of the diffusion equation. Accuracy and resolution are determined by the cell size and by the number of particles which can be followed—that is, by computer capacity. As in other methods, a vector wind field and a formalism for evaluating the eddy diffusivity tensor must be provided; in this case, however, the wind field must be constrained to obey Eq. (4). This may necessitate some "massaging" of observed winds to generate the requisite nondivergent flow.

2.1.9 Global Models

Global models, as defined by Drake et al. (1979) in Table 1, are the global circulation models originally developed for weather forecasting, and recently utilized for preliminary studies of global-scale pollution problems, particularly CO_2-induced climate changes (Smagorinsky, 1974, 1977). These models deal with the equations for conservation of mass, momentum, and energy on a rotating earth. They incorporate sinks and sources of these quantities by means such as drag at the earth's surface, a complete radiation balance, and allowance for evaporation, condensation, and precipitation. Many important phenomena occur on subgrid scales—convective activity leading to cloud formation, for example—and must be incorporated through parameterizations.

There are a number of problems associated with these models. First, they generally require very large computers, and take a great deal of time to run. Secondly, many of the effects which are significant to air pollutant studies occur on the subgrid scale and may or may not be adequately parameterized. Improvements in features such as biomass distributions, source and sink assessments, and atmospheric chemistry are needed. Finally, the time scales of existing models are keyed to the one- or two-week scale used in weather forecasting; they may be far too small for global scale modeling of pollutant dispersion.

On such large scales, somewhat simpler models may still be adequate. For example, one can argue (Gifford, 1975) that an assumption of constant eddy diffusivities at these scales should be quite good, so that Eq. (3) may be applied. Since the Gaussian plume is a particular solution of Eq. (3), it too can be used, perhaps in conjunction with a trajectory calculation (Heffter et al., 1975). Box models can be similarly utilized (Draxler, 1977). Features such as atmospheric stability changes and pollutant removal processes can be incorporated as well. The advantages and disadvantages of the various models can be expected to apply on global scales, but with an additional penalty of considerable computer storage and running time if much resolution is sought.

2.1.10 Physical Models

Physical modeling is another topic which has generated its own voluminous literature. Models of structures, terrain features, or even whole regions are constructed to scale and immersed in flowing air or liquid. If the modeling conditions are adjusted so a body of scaling or similarity parameters (e.g., Reynolds and Froude numbers) have the same values in the laboratory and in nature, then the simulation has been correctly performed, and the model results should be a good analog of the real phenomena. Many laboratory compromises are necessary, however, and the simulations are hardly ever exact (Snyder, 1972). Nevertheless, because of the degree of control one has over the experimental process, and because of the relative ease of changing configurations (like source locations), physical modeling has proven extremely useful for study of flow and dispersion in complex situations where mathematical models may be too difficult and/or field work too expensive. Scales of interest are usually fairly modest (a kilometre or two at best), although some attempts at regional scale modeling have been made. Data from these latter are probably best regarded as qualitative. Current physical models of geographic regions used to date have not been able to simulate any chemical or removal processes which are often important in real work problems; hence they will not be discussed further. Studies of individual receptors (such as leaves) in wind tunnels may, however, be extremely useful in quantifying pollutant and uptake mechanisms. Such methods have been profitably employed by many investigators (e.g., Chamberlain, 1974; Thom, 1968).

2.2 Parameterization of Removal Mechanisms

2.2.1 Rollback and Statistical Models

These models are not well suited to deposition modeling; they must be calibrated to a particular site using historical records and are limited to long-term results. Special techniques such as multivariate analysis may, however, be somewhat useful for understanding the dependence of removal rates on atmospheric conditions. Extrapolation of results to new or heavily altered old sites is not feasible.

2.2.2 Gaussian Plume

Since the most widely used dispersion model is the Gaussian plume, it is not surprising that it has been modified by many investigators to include wet and dry removal mechanisms. For many years, *dry deposition* was handled following Chamberlain (1953), who seems to have introduced the notion of an effective or residual source strength which depends on the amount of material removed from the plume, and consequently decreases with downwind distance. This modified source strength $\dot{Q}(x)$ replaces $\dot{Q}_0$ in Eq. (1). The method, known as the source depletion model, has been succinctly outlined by Van der Hoven (1968):

$$\frac{\dot{Q}(x)}{\dot{Q}_0} = \left[\exp\left(\int_0^x \frac{dx'}{\sigma_z \exp(h^2/2\sigma_z^2)}\right)\right]^{-(\sqrt{2/\pi})(v_d/U)} \tag{7}$$

where v_d is the effective dry deposition velocity. The technique is physically somewhat unrealistic, since material losses from the lower edge of the plume are implicitly distributed instantaneously throughout the vertical depth of the plume. Under stable atmospheric conditions in particular, this assumption of very rapid mixing must be poor. Miller et al. (1978) have examined the response of this model to changes or errors in v_d; they found that at short distances and with elevated sources, the results were insensitive to v_d. Ground level releases, however, were quite sensitive to such changes.

A somewhat different approach has been presented by Csanady (1955, 1957, 1958) and extended by Overcamp (1976) as the "partial reflection" model. In this instance, the "image term" (involving $z+h$) in Eq. (1) is preceded by a reflection coefficient α, which is thus a fraction of the strength of the real source. This coefficient is determined by setting the deposition flux equal to the difference in fluxes from the real and image terms. The plume is also allowed to tilt, to incorporate gravitational settling of large particles at terminal speed v_t:

$$\chi = \frac{\dot{Q}_0}{2\pi\sigma_y\sigma_z U}\exp\left[-\frac{y^2}{2\sigma_y^2}\right]\left\{\exp\left[-\frac{(z-h+v_t x/U)^2}{2\sigma_z^2}\right] + \alpha(x_G)\exp\left[-\frac{(z+h-v_t x/U)^2}{2\sigma_z^2}\right]\right\} \tag{8}$$

(Overcamp, 1976). The reflection coefficient $\alpha(x_G)$ can be computed by solving an implicit relation for x_G, and an expression for $\alpha(x)$ which depends on v_d, v_t, U, and the form of the vertical diffusion coefficient. The procedure is straightforward. In this model, dry deposition acts to remove material from the lower portions of the plume, which therefore begins to assume a non-Gaussian vertical concentration profile. This is a physically plausible result. Overcamp's (1976) comparison of the source depletion and partial reflection models suggests little difference during unstable conditions; this is reasonable, since atmospheric turbulence would act to keep the plume well mixed. In stable conditions, however, the fluxes to the surface computed by the two methods begin to differ near the downwind flux maximum and are significantly different far from the source. The flux estimated from the partial reflection model is less than that of the source depletion model, since the concentration just above the surface is also smaller in the former.

A somewhat similar though much more complex relation has been derived by Ermak (1977), who solved the steady-state version of Eq. (3) with a term added to account for gravitational settling, while the lower boundary condition explicitly allowed for nongravitational deposition. The vertical distribution of the plume changes with downwind distance in a plausible way. No direct comparison with other deposition models seems to be available. The original paper should be consulted for details of the solution, although an error, evidenced by dimensional inconsistencies, may be present.

Horst (1974, 1977) presented a quite different and physically appealing concept. In effect, the plume of effluent is gradually eaten away from below by "negative sources" (sinks) distributed continuously over the ground surface; these negative sources naturally emit negative plumes, which diffuse upward into the atmosphere like real plumes. The strength of each infinitesimal negative source is related to the deposition flux at its location. Superposition of the real plume and the integrated contributions of the distributed negative plumes yields the depleted plume concentration:

$$\chi = \dot{Q}_0 D(x, y, z; h) - \int_{-\infty}^{\infty} \int_0^x v_d \chi(\xi, \eta, z_d) D(x - \xi, y - \eta, z; 0)\, d\xi\, d\eta \quad (9)$$

where $D(x, y, z; h)$ is just Eq. (1) times $\dot{Q}_0^{-1}$, and z_d is the reference height where the dry deposition velocity is evaluated. The solution is computationally complex since an integral equation must be solved—usually numerically. Horst (1980) compared the results of his surface depletion model to those obtained from both the traditional source depletion method and the partial reflection model. He concluded that the partial reflection model is the easiest to use and, at short distances, is fairly close to his "exact" surface depletion model. Further downwind, however, the partial reflection model becomes quite inaccurate. The source depletion model is somewhat more difficult to use, and is less accurate close to the source, but does better than the partial reflection model far downwind. The differences between the source and surface depletion models diminish with decreasing deposition rate and increasing atmospheric instability. It should be noted that the source depletion and partial reflection models can be applied to cases where gravitational settling is important, while the surface depletion model in its current form cannot.

Horst (1980) introduced a modified source depletion model which utilized a factor to account for the dry-deposition-induced change in vertical concentration distribution as a function of downwind distances:

$$\chi = \dot{Q}(x) D(x, y, z; h) P(x, z) \quad (10)$$

where

$$\dot{Q}(x) = \dot{Q}_0 \exp\left[-\int_0^x v_d D(x', y, z_d; h) P(x', z_d)\, dx'\right] \quad (11)$$

The correction factor P is calculated on the basis of a simple gradient transfer model, and depends on the ratio v_d/U, mixing layer depth, and integrations involving σ_z; the original paper should be consulted for details. The calculated concentrations agree very closely with the surface depletion model, and the method is much easier to apply. Its predictions are least accurate very close to the source in stable conditions.

For routine calculations, the complexity of the above methods is obviously undesirable. Consequently, alternative approaches have been introduced which permit plumes to deplete in a simple way. For example, all the U.S. EPA (1978) recommended air quality models which deal with plume decay, do so via an exponential term $\exp(-t/\tau)$, where the characteristic time for dry deposition (or wet removal, or any other process) must be user-supplied. Similarly, Heffter et al. (1975) use $\exp(-2v_d t/\sqrt{\pi K_z t})$, while Wendell et al. (1976) use $\exp(-v_d t/\Delta z)$, to give plume concentration at time t. In the latter expression, $\Delta z = \sqrt{\pi/2}\,\sigma_z$ near the source, while Δz = depth of the mixing layer far downwind. Church (1976) describes the simple procedure used in WASH-1400 (U.S. NRC, 1975) to remove a fraction of plume material $v_d\,\Delta t/z_e$ at each time step Δt. The effective plume depth z_e is just $\int_0^\infty \chi(z)/\chi(0)\,dz$.

All these simple methods seem to be equivalent to a source depletion model (e.g., Eq. (7)). Their accuracy relative to the more detailed dry removal techniques does not seem to have been explored in the literature. A comparison would be useful to establish whether such simplifications of the expression for plume concentration are sufficiently accurate for practical purposes, and the conditions under which this is true.

As in the case of dry deposition, most attempts at *wet removal* modeling using the Gaussian plume can be traced back to the pioneering work of Chamberlain (1953). The theory, neatly summarized by Slinn (1984), is limited to scavenging of particles or gases which are captured irreversibly by raindrops or other hydrometeors falling through the effluent plume. In such conditions, the rate of removal of particles per unit volume should be proportional to their concentration: $\tilde{R} = \Lambda\chi$. The diffusion equation (2) can be amended to account for such removal:

$$\frac{D\chi}{Dt} = \nabla\cdot(\mathbf{K}\cdot\nabla\chi) - \tilde{R} \tag{12}$$

In general, the scavenging rate Λ is dependent on both position and time. Suppose, however, that Λ is a function only of time; then multiply Eq. (12) by $\exp[\int \Lambda\,dt]$ to get

$$\frac{D\chi'}{Dt} = \nabla\cdot(\mathbf{K}\cdot\nabla\chi')$$

which is just Eq. (2)—that is, wet removal is effectively eliminated from the problem in this special case. Given the solution of Eq. (2), as a Gaussian plume, for example, then the solution of Eq. (12), which involves wet removal, can simply be written

$$\chi(\mathbf{r}, t) = \chi'(\mathbf{r}, t) \exp\left[-\int \Lambda \, dt\right] \tag{13}$$

If Λ is conveniently independent of time as well as space, then the solution of Eq. (12) is just the solution of (2) multiplied by the factor $\exp(-t/\tau)$, where the characteristic decay time τ is just Λ^{-1}.

In practice, Gaussian-based wet removal models generally use Eq. (1) with the source strength $\dot{Q}$ allowed to decay exponentially with time or distance, through $t = x/U$ (see reviews by Drake et al., 1979; U.S. EPA, 1978). As with the basic Gaussian equation itself, the restrictions governing the derivation of Eq. (13) are sometimes relaxed during applications (e.g., Nieuwstadt et al., 1976) by allowing Λ to vary with position to account for precipitation changes over the region of interest. The basic problem in applying Eq. (13) is, of course, the actual evaluation of Λ, which depends in a very complicated way on the characteristics of the rainfall and scavenged effluent (Dana and Hales, 1976). This is discussed below.

As noted, the scavenging rate approach inherently assumes an *irreversible* collection process. Consequently it is suitable for gases only if they are extremely reactive (e.g., I_2). For less reactive gases, or those which merely form simple solutions in water, it is essential to account for possible desorption of gas from droplets as they fall from regions of high concentration toward the ground (Hales, 1972; Slinn, 1978). Hales' (1972) paper is an excellent guide to this whole question of gas scavenging.

Hales et al. (1973) have dealt with the soluble gas problem for a Gaussian plume under the simplifying assumptions of steady state, negligible chemical reactions, and vertical fall of raindrops. More complex cases are briefly considered below. Here the gas obeys Henry's law at equilibrium, $[\chi]_e = H[C]_e$, where $[\chi]$ and $[C]$ are the *molar* concentrations of gas in the ambient atmosphere and within the drops, respectively. Trace gases which do not react with water usually follow Henry's law quite closely (Hales, 1972). Analysis of mass transfer to and from a raindrop of radius R gives the expression

$$\frac{d[C]}{dz} = \frac{3\kappa}{v_t R}([\chi] - H[C]) \tag{14}$$

where κ is the overall mass transfer coefficient, and v_t is the drop's terminal velocity. The molar concentration within such droplets at any height z is then given by

$$[C(x, y, z; R)] = \frac{3\kappa}{v_t R} \exp\left(\frac{-3\kappa H z}{v_t R}\right) \int_\infty^z [\chi(x, y, z')] \exp\left(\frac{3\kappa H z'}{v_t R}\right) dz' \tag{15}$$

If gas absorption and desorption by falling drops have a negligible effect on the overall plume distribution, then a relation such as Eq. (1) can be used to generate a complex expression for [C]. Reference can be made to Hales et al. (1973) or Hales (1975) for details. To evaluate the molar flux of gas carried to the surface in the drops, an integration over the spectrum of droplet sizes is also necessary. Use of Eq. (15) requires expressions for κ and v_t. The evaluation of κ has been discussed in some detail by Hales (1972), and will be summarized below. Empirical expressions for v_t as a function of droplet radius are also mentioned later.

Central to the above calculation is the assumption that the gas concentration distribution was not affected by absorption/desorption in falling raindrops. Slinn (1974a) derived a Gaussian-like plume expression considering such reversible scavenging, but explicitly allowing the plume distribution to be coupled to the scavenging process. The steady-state modification of Eq. (12) is just

$$U\frac{\partial \chi}{\partial x} = K_y \frac{\partial^2 \chi}{\partial y^2} + K_z \frac{\partial^2 \chi}{\partial z^2} - \alpha\chi + \beta C \tag{16}$$

assuming that the raindrop size distribution can be characterized by some single drop radius R_m. The terms $\alpha\chi$ and βC represent absorption and desorption, respectively, of gas from the raindrops. An expression similar to Eq. (14) couples the in-drop concentration and gas phase concentration. Slinn (1974a) finds a complicated integral expression for χ, and evaluates it at small and large times $t = x/U$:

$$\chi \cong \frac{\dot{Q}}{[4\pi K_y t]^{1/2}[4\pi K_z t]^{1/2} U} \exp\left[-\frac{wt}{\delta}\right] \exp\left[-\frac{y^2}{4K_y t}\right] \exp\left[-\frac{(z-h)^2}{4K_z t}\right] \tag{17}$$

for small t, while

$$\chi \cong \frac{\dot{Q}}{[4\pi K_y t]^{1/2}[4\pi (K_z + w\delta) t]^{1/2} U} \exp\left[-\frac{y^2}{4K_y t}\right] \exp\left[-\frac{(z-(h-wt))^2}{4(K_z + w\delta)t}\right] \tag{18}$$

at large times. The term $w \equiv J_0 H$ is a "washdown" speed for the plume related to the precipitation rate J_0, while δ is a characteristic length related to the gas chemistry and droplet size: $\delta = (HR_m v_t)/(3\,\kappa)$. The effect of wet scavenging on the plume distribution is roughly as follows: Close to the source, the plume experiences only an exponential decay, much as in the case of particle or irreversible gas scavenging. Diffusion at these short distances is unaltered by the process. Far downwind, however, the plume is washed down from its initial height h, and the vertical diffusivity is also increased slightly. The solution for intermediate times is not yet published; Slinn (1974a) suggests using the sum of Eq. (17) plus $[1-\exp(-wt/\delta)]$ times Eq. (18) as an interim result. The value of w is apparently rather small (Slinn, 1974a; Barrie, 1978).

One other approach to quantifying the wet removal process can be traced back to Chamberlain (1953), who defined a wet deposition velocity v_w by analogy to the dry. Thus v_w = wet flux/concentration in air, at the surface. This can be estimated from empirically determined washout ratios

W_r as follows (Slinn, 1978): C is the contaminant (particle or gas) concentration within the hydrometeors, and χ is the concentration in air. Measurement at a convenient reference level near the ground gives

$$W_r = C_0/\chi_0 \tag{19}$$

Since the wet flux to the surface is just C_0J_0, where J_0 is the (rain-equivalent) precipitation rate,

$$v_w = W_rJ_0 \tag{20}$$

The wet deposition velocity can then be used analogously to the dry version to develop models for the wet scavenging process. A few authors (Heffter et al., 1975; Draxler, 1976) have used the washout ratio directly to give an exponential decay term for a plume: $\exp(-W_rJ_0t/\Delta z_w)$, where Δz_w is the thickness of the wetted plume layer.

The requisite washout ratios can be evaluated theoretically or measured directly, as discussed below. Use of the empirical values is generally recommended (Van Hook and Shults, 1976; Hoffman et al., 1977; Slinn et al., 1978) if the contaminant and precipitation characteristics of a previously studied case are similar to those of current interest. Fortunately the variability of measured values of washout ratio is surprisingly small (Engelmann, 1971; Slinn et al., 1978). The method is probably best suited to long-term estimates, where the variability induced by single storm events is integrated out.

The wet removal concepts above have been presented in terms of what is often called "washout", in which material is captured by hydrometeors falling through a plume located somewhere below cloud base. The same concepts of scavenging rate, washout ratio, and wet deposition velocity can be applied equally well to "rainout", where material at cloud height is scavenged by in-cloud processes such as nucleation, although the numerical values naturally are different. Slinn (1974b) has an interesting discussion of precipitation scavenging terminology and its pitfalls; in view of these, the common nomenclature has been sidestepped in this paper. The distinction between the operative physical processes should be borne in mind, however.

2.2.3 Puff Models

The Gaussian puff and its relatives have been used until recently mostly to estimate concentrations due to a single episodic release of material—e.g., a nuclear weapons test. In such work, at least in the United States, *dry deposition* has been recognized by using a depleting source term $Q(\chi)$ in the puff equation instead of the actual source strength Q_0; the form is quite like Eq. (7) (Fuquay, 1968). Knox et al. (1971) describe a model used in the U.S.S.R. for similar purposes, as reported by Izrael and Petrov (1970). The basic puff equation is modified by a complicated term involving deposition velocity, source height, and vertical diffusivity in exponential and complementary error function form.

As puffs have become more widely used in conjunction with trajectory models for regional-scale environmental assessment, simpler means for incorporating dry deposition have been introduced. For example, McClure (1976) depletes the vertically integrated concentration of puffs by a factor $\exp(-k_d t)$, where k_d = (dry deposition flux)/(vertically integrated concentration). A somewhat similar approach has been taken by Heffter and Ferber (1977), who remove a mass fraction $\chi_0 v_d \Delta t / \chi_0 z_m$ after each time step Δt of their trajectory/puff model; z_m is the effective depth of the mixed layer. Evidently after n steps, the concentration must be given by $\chi = \chi_0[1-(v_d \Delta t)/z_m]^n$, which is similar to an exponential decay process if $v_d \Delta t / z_m$ is small. Johnson et al. (1977) also use a simple technique; their puffs have a constant height (z_m, say), and expand horizontally with their radius proportional to $\sqrt{t}$. The mass/volume ratio is then calculated at any time t after allowing the mass to be depleted in the amount $k_d \int_0^t \chi V \, dt'$, where V is the puff volume at time t.

Wet removal of puff material is generally treated (Johnson et al., 1976b) via Eq. (13); that is, an exponential decay whose parameter depends on the characteristics of the effluent and the precipitation, as discussed below. The method should be restricted, strictly speaking, to irreversible scavenging of particles and highly reactive gases, since it is unable to describe the behavior of mildly reactive or merely soluble gases. A somewhat related treatment has been used by Heffter and Ferber (1977), who allow puff mass to deplete after each time step by the quantity $W_r J_0 \Delta t / z_w$, where z_w is the depth of the plume through which rain falls. As in their treatment of dry deposition, Johnson et al. (1977) use a simple mass–volume calculation for wet removal as a function of time; the mass removed from a puff at time t is just $\int_0^t k_w \chi V \, dt'$, where $k_w = \lambda J_0(t)$, and λ is a washout coefficient of appropriate units.

2.2.4 Trajectory Models

These generally incorporate a puff model of some sort (Gaussian, or expanding cylinder, for example), or a puff–box combination, or, for large scales, just a box model to handle the diffusion and removal processes. Hence the discussion above and below for these types of models is relevant here as well, and need not be restated.

2.2.5 Box Models

Many box-type models utilize stacked layers to permit variations with height of important meteorological parameters. *Dry deposition* can then be treated in a very natural way, allowing depletion to occur in the lowest layer such that the dry flux to the surface is given by the product of surface level dry deposition velocity and effluent concentration. For example, Draxler

(1976) computes the concentration in the lowest layer after a time step Δt as

$$\chi_1(t+\Delta t) = \chi_1(t)[1-(v_d\,\Delta t)/\Delta z_1] \qquad (21)$$

where Δz_1 is the depth of the model's lowest layer. The result is a kind of surface depletion model, where material is removed from the lower portion of the plume by dry deposition and replaced by material from above—depending on the intensity of atmospheric mixing present. Draxler and Elliott (1977) explicitly compared this approach with a simulated source depletion layered model, where deposition was distributed simultaneously throughout the plume depth. They found, as expected, that the surface depletion method left much more material aloft after several days of travel, particularly if v_d was rather high (≥ 3 cm/s). They also found deposition to be quite sensitive to the stability conditions at the time of effluent release.

Other box models utilize a sector or wedge whose height z_m is determined by mixing layer depth, and whose width at some distance x from the source depends on the crosswind spread of material as well as x (Scriven and Fisher, 1975). For well-mixed effluent in such a box, a simple exponential decay factor is often used: $\exp[-v_d x/Uz_m]$; this corresponds to a source depletion model, and probably works best in neutral to unstable conditions.

Wet deposition in box models seems to be nearly always treated with a scavenging rate decay factor $\exp[-\Lambda t]$ (e.g., McMahon et al., 1976), and is therefore appropriate mainly for irreversibly scavenged materials, as discussed earlier. Layered versions behave similarly, although the scavenging rate may vary with height because of different hydrometeor characteristics within the layers. It is also possible to treat wet removal in these models using empirical washout ratios W_r or, equivalently, the wet deposition velocity v_w. For example, in the ith layer after a time step Δt, Draxler (1977) computes the concentration $\chi_i(t+\Delta t) = \chi_i(t)[1-(W_r J_0\,\Delta t)/z_{wi}]$, where z_{wi} is the wetted depth of the effluent within layer i. The similarity to Eq. (21) is clear. This method has the potential, at least, of treating wet removal of soluble and moderately reactive gases as well as particles and highly reactive gases; it should be particularly useful when long-term assessments are required.

2.2.6 Grid Models

These generally use wet and dry deposition schemes similar to those already discussed. These models are especially well suited to use of the relation between mass flux, dry deposition velocity, and concentration at or just above the surface (Rao et al., 1976). Constant values for v_d are often utilized, probably for simplicity, although there is no reason why this quantity could not be allowed to vary in time and space to reflect changes in terrain, ground cover, and atmospheric conditions. Such modifications might well be significant in regional or global scale models. An interesting scheme for estimation of plume depletion and dry deposition has been suggested by the Nuclear Regulatory Commission (U.S. NRC, 1977; see also Markee, 1967; Markee et al., 1977). The two-dimensional form of Eq. (2) is used with empirically derived values for U and K_z to generate curves for the plume fraction remaining aloft and for the relative deposition rate for material transferred to the surface. The user's plume model is then compensated for dry deposition by simply applying these factors for the appropriate stability category. Miller and Hoffman (1979) have criticized this approach, however, because of apparent internal inconsistencies, and because this surface depletion model does not behave as it should when compared with a source depletion model—in particular, the NRC and source depletion models agree best in stable conditions, contrary to expectations and the results of other authors.

Wet removal is generally handled by an exponential decay term, $\exp(-\Lambda t)$, although some models simply assume *all* effluent is scavenged immediately when precipitation is encountered (Peterson and Crawford, 1970; Sheih, 1977). An interesting variation is due to Bolin and Persson (1975), who calculate the wet removal rate from $\beta \int_0^\infty \chi \, dz$. The coefficient β is a kind of "expected" overall scavenging rate, which takes into account the probability of rainfall, its likely duration and intensity, and the actual scavenging rate Λ expected for such precipitation (see Rodhe and Grandell, 1972, for details). Evidently β can vary with locale and season; the method seems best suited to long-term-average investigations. Wet deposition velocities and/or washout ratios do not seem to have been utilized in grid models to any extent, although they should be capable of dealing with gas scavenging problems that the more common scavenging rate concept cannot properly address.

Extremely complex numerical models dealing with precipitation scavenging and cloud dynamics have been described by Hane (1974, 1978), Molenkamp (1974), and others. These models deal with the equations of motion for cloud formation, precipitation formation, and the various scavenging phenomena which may apply. This work is still in a highly experimental stage, requires many parameterizations of only partially understood processes, and (like most deposition models) is still unvalidated. In short, such research, while intriguing and potentially extremely useful, is presently unsuitable for practical applications.

2.2.7 Particle Models

Particle models can deal with *dry deposition* in a particularly straightforward way. Near the ground, the deposition velocity is added to the pseudovelocity experienced by the particles (see Eq. (6)) so that the mass flux to the surface obeys the usual relation (Knox, 1974; Lange et al., 1976).

Wet removal is a bit more complicated (Lange and Knox, 1974). An individual "particle" actually corresponds to a "clump" of effluent, taken to be an aerosol. The scavenging rate depends (among other things) on the size of the aerosol particles, so that $\Lambda = \Lambda(a)$. If $m(a, 0)da$ is the mass of aerosol with radii between a and $a + da$ in the "particle", then at any time t the mass of the particle must be $M(t) = \int_0^\infty m(a, 0) \exp[-\Lambda(a)t]\, da$, for a given rainfall rate and size distribution. The mass fraction deposited to the surface from this particular "particle" during time step Δt is then just $[1 - M(t + \Delta t)/M(t)]$. The code keeps track of both the mass still aloft and that deposited for all the "particles" moving through the system. This approach cannot deal with gases. However, the wet deposition velocity could be introduced in such cases and treated in the same way as the dry deposition velocity.

2.2.8 Global Models

Incorporation of wet and dry removal processes into global scale models does not seem to require fundamental changes in methods already discussed. The concepts of surface depletion via a dry deposition velocity and wet removal by means of a scavenging rate or wet deposition velocity should be applicable on these scales. Other problems associated with parameterizations in these models probably represent a larger handicap to their application than do scavenging techniques.

3. DRY DEPOSITION MODELING

3.1 Deposition Velocity and the Resistance Analog

It is clear from the above that dry deposition modeling by any of the common techniques requires knowledge of the parameter v_d, the so-called deposition velocity. Hidden within this parameter are a host of variables which can influence the deposition rate. Table 2 (Sehmel, 1980) alphabetically lists some of these.

There are two main approaches to determine v_d. The first is a straightforward appeal to experiment; if the depositing effluent, underlying surfaces, and meteorological conditions of interest coincide with those previously considered, then empirical values of v_d can be used for predictive purposes. Some of the available sources for such data are discussed below. The second approach attempts to utilize our somewhat limited understanding of the physical processes involved to actually calculate v_d, given the necessary meteorological data from the site of interest. The usual scheme divides the atmosphere into layers or boxes, within each of which certain transport processes dominate, while others are negligible. Mass transfer is then held to be limited by a resistance characteristic of each layer, and the

TABLE 2. Some Factors Influencing Dry Deposition Removal Rates

Micrometeorological Variables	Characteristics of Particles	Characteristics of Gases	Surface Variables
Aerodynamic roughness	Agglomeration	Chemical activity	"Accommodation"
Mass transfer of	Diameter	Diffusion effects	Exudates
Particles	Diffusion effects	Brownian	Trichomes
Gases	Brownian	Eddy	Pubescence
Heat transfer	Eddy	Partial pressure in equilibrium with surface	Wax
Momentum transfer	Particle	Solubility	Biotic surfaces
Atmospheric stability	Momentum		Canopy growth stages
Diffusion, effect of	Heat		Dormant
Canopy	Effect of canopy		Expanding
Diurnal variation	Diffusiophoresis		Senescent
Fetch	Electrostatic effects		Canopy structure effects
Flow separation	Attraction		Areal density
Above canopy	Repulsion		Bark
Below canopy	Gravity settling		Bole
Friction velocity	Hygroscopicity		Leaves
Inversion layer	Impaction		Porosity
Pollutant concentration	Interception		Reproductive structure
Relative humidity	Momentum		Soils
Seasonal variation	Physical properties		Stem
Solar radiation	Resuspension		Type
Surface heating	Solubility		Electrostatic properties
Temperature	Thermophoresis		Leaf–vegetation effects
Terrain effects			Boundary layer
Uniform			Change at high winds
Nonuniform			Flutter
Turbulence			Stomatal resistance
Wind velocity			pH effects on
Zero plane displacement effect			Reaction
Mass transfer of			Solubility
Particles			Pollutant penetration and distribution in canopy
Gases			Prior deposition loading
Heat transfer			Water
Momentum transfer			

From Sehmel (1980).

overall deposition velocity (corresponding to mass transport from the free atmosphere to some final sink) is given as the reciprocal of the sum of these various resistances (Thom, 1975).

Slinn et al. (1978) summarize the formalism: let F_D be the dry deposition flux to the receptor surface from the atmosphere. The concentration gradient which is usually assumed to drive such fluxes is approximated by the concentration difference,

$$F_D = v_d(\chi_A - \chi_R) \tag{22}$$

where χ_A and χ_R are the effluent concentrations in the atmosphere and at the receptor surface. The term v_d is the overall deposition velocity or effective mass transfer velocity. For steady-state homogeneous conditions, the flux is continuous across the various layers; for a four-layer model, for example,

$$F_D = k_A(\chi_A - \chi_B) = k_B(\chi_B - \chi_C) = k_C(\chi_C - \chi_D) = k_D(\chi_D - \chi_R)$$

By expanding $\chi_A - \chi_R = \chi_A - \chi_B + \chi_B - \cdots + \chi_R$ in Eq. (22), F_D can be eliminated to show

$$v_d^{-1} = k_A^{-1} + k_B^{-1} + k_C^{-1} + k_D^{-1} \tag{23}$$

or, in terms of the Ohm's law analogy,

$$r_T = r_A + r_B + r_C + r_D \tag{24}$$

where the total mass transfer resistance ($\equiv v_d^{-1}$) is equal to the sum of the

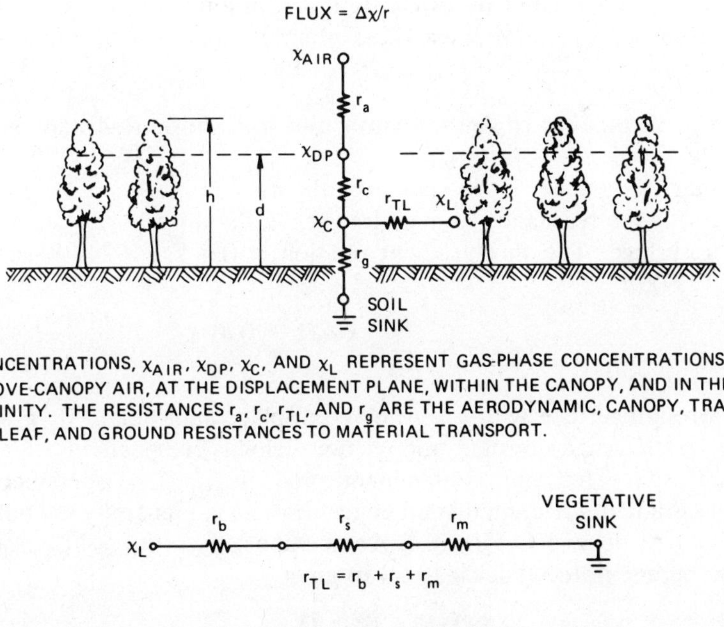

(a) CONCENTRATIONS, χ_{AIR}, χ_{DP}, χ_C, AND χ_L REPRESENT GAS-PHASE CONCENTRATIONS IN THE ABOVE-CANOPY AIR, AT THE DISPLACEMENT PLANE, WITHIN THE CANOPY, AND IN THE LEAF VICINITY. THE RESISTANCES r_a, r_c, r_{TL}, AND r_g ARE THE AERODYNAMIC, CANOPY, TRANSFER-TO-LEAF, AND GROUND RESISTANCES TO MATERIAL TRANSPORT.

(b) THE TRANSFER-TO-LEAF RESISTANCE, r_{TL}, IS COMPOSED OF THE LEAF BOUNDARY LAYER RESISTANCE, r_b, THE STOMATAL (AND SIMILAR) RESISTANCE, r_s, AND THE ASSIMILATION RESISTANCE r_m.

FIGURE 1. Electrical analog of transfer of materials to a forest. (Hosker and Lindberg, 1982; based on Thom, 1975.)

resistance to transfer in each of the layers. If any one of the layer resistances is much larger than the others, then the processes restricting mass transfer within that layer provide the rate-limiting step for the overall transfer process. Figure 1 shows a typical model for transfer from an elevated plume to a vegetative canopy. Note the presence of two possible in-canopy sinks, soil and leaves (others may well exist), so that the final resistance can be composed of several contributing ones; in particular, the transfer to leaf resistance includes the effects of the viscous boundary layer surrounding the leaf, the stomatal or other resistances restricting entrance to the internal portions of the leaf, and an assimilation resistance (Bennett et al., 1973).

The following sections examine some of the problems in computing resistances on a layer-by-layer basis. Turbulent transfer across a mixing layer above a potential receptor surface is assumed. The receptor may be water, bare soil, or perhaps a vegetative canopy. If deposition is to a canopy, then the resistance of the canopy to transfer must somehow be recognized. Finally, transport to the actual sink surface must occur and must consider effects such as the viscous boundary layer and possible biological resistances, as mentioned above. The discussion applies to gases and to relatively small particles (diameter less than 10 μm or so) which are transferred rather like gases. Large particles with significant settling velocities can be treated by explicitly incorporating this velocity into the overall transport model (e.g., a tilted plume) and the deposition velocity formulation.

3.2 Evaluation of Aerodynamic and In-Canopy Resistances

Planetary boundary layer similarity theory can be used to construct estimates for the resistance to mass transfer in the mixed layer above some receptor surface. The flux–gradient relation is (Hicks, 1974; Wesely and Hicks, 1977)

$$F_{\mathrm{D}} \equiv K_\chi \frac{\partial \chi}{\partial z} = \frac{k u_*(z-d)}{\Phi_\chi} \frac{\partial \chi}{\partial z} \tag{25}$$

above a rough surface with a zero plane displacement height d; k and u_* are the von Karman constant and friction velocity, respectively, K_χ is the eddy diffusivity for the contaminant, and Φ_χ is the nondimensional similarity gradient for contaminant concentration χ. Generally the pollutant is assumed to behave like heat or water vapor: $\Phi_\chi \cong \Phi_{\mathrm{H}} \cong \Phi_{\mathrm{W}}$. Equation (25) can be integrated (Hicks, 1974) to give

$$\chi(z) - \chi(d) = \frac{F_{\mathrm{D}}}{k u_*}\left[\ln\left(\frac{z-d}{z_0}\right) - \Psi_{\mathrm{H}} + \ln\left(\frac{z_0}{z_{0\chi}}\right)\right] \tag{26}$$

The term Ψ_{H} is obtained from integrating $d\Psi_{\mathrm{H}} \equiv (1-\Phi_{\mathrm{H}})\zeta^{-1}\, d\zeta$, where $\zeta \equiv (z-d)/L$ and L is the Monin–Obukhov length, which accounts for both latent and sensible heat transfer (Wesely and Hicks, 1977). The gradient Φ_{H}

is obtained from Dyer's (1974) review, so that (Wesely and Hicks, 1977)

$$\Psi_H = \begin{cases} -5\zeta, & 0<\zeta<1 \text{ (stable)} \\ 0, & \zeta=0 \text{ (neutral)} \\ \exp\{0.598+0.39\ln(-\zeta)-0.09[\ln(-\zeta)]^2\}, & -1<\zeta<0 \end{cases} \tag{27}$$

The term involving $z_0/z_{0\chi}$ is just the ratio of the well-known aerodynamic roughness length for momentum transfer to the less familiar characteristic roughness length for pollutant transfer.

Thus the effective resistance r_A to mass transfer between height z and the surface's displacement height d is

$$r_A = [\chi(z) - \chi(d)]/F_D = r_a + r_{z0} \tag{28}$$

where

$$r_a \equiv (ku_*)^{-1}[\ln((z-d)/z_0) - \Psi_H] \tag{29}$$

and

$$r_{z0} \equiv (ku_*)^{-1}\ln(z_0/z_{0\chi}) \tag{30}$$

The first term, r_a, is usually called the bulk aerodynamic resistance (Thom, 1972) or perhaps the eddy resistance (Hicks, 1974), while r_{z0} is a receptor surface transfer resistance associated with the differences in the way momentum and mass are transferred to the receptor (Thom, 1972, 1975); r_{z0} is the same as Monteith's (1973) and Thom's (1975) excess resistance r_b. Note that

$$r_A = \int_d^z \frac{dz}{K_\chi(z)} \tag{31}$$

which is a potentially convenient relation utilized later.

There seem to be several current schools of thought regarding the surface transfer resistance. First, it must be understood that the surface here is the *overall* receptor surface—for example, a canopy or an expanse of soil—and not the surface of an individual receptor element such as a leaf or sand grain. For the case of vegetative canopies or other such receptors where turbulent transport can also occur in the atmosphere below the displacement height, the physics of such transfer can either be implicitly incorporated into the surface transfer resistance, or can be explicitly treated separately, perhaps by integrating an eddy diffusivity within the canopy, much as in Eq. (31).

The implicit treatment is considered first. The overall receptor is essentially treated as a single receptor surface in this approach, and all the aerodynamic resistances for the receptor as a whole and for its individual elements are lumped together. Shuttleworth (1976) discusses some of the implications of this. The method proceeds as follows. The surface resistance r_{z0} often appears in the literature in the guise of the Stanton number B used by Owen and Thompson (1963) and many others. The relation is simply (Chamberlain, 1966)

$$ku_* r_{z0} = kB^{-1} = \ln(z_0/z_{0\chi}) \tag{32}$$

To determine B or $z_{0\chi}$, appeal is made to experiments. Brutsaert (1975) has summarized the many semiempirical expressions available to estimate $z_{0\chi}$; these generally depend on the Schmidt number $\mathrm{Sc} \equiv \nu/D_\chi$, where D_χ is the molecular diffusivity of the contaminant; and on the roughness Reynolds number $\mathrm{Re}_* \equiv u_* z_0/\nu$. Garratt and Hicks (1973), however, also examined the experimental data and found that relations of the sort recommended by Brutsaert (1975), Owen and Thompson (1963), and others seem to apply only to surfaces where the roughness elements form a regular array. This is characteristic mostly of wind tunnel studies and a few special agricultural cases such as vineyards. The data obtained over more common natural surfaces showed a nearly negligible dependence on Re_* over its range of three orders of magnitude. The surfaces in this instance were described as having randomly distributed fibrous roughness elements.

Garratt and Hick's (1973) recommendation is as follows: For $\mathrm{Re}_* < 50$ or so, the data for kB^{-1} tend to fall between the curves (Fig. 2)

$$kB^{-1} \cong \ln(k\,\mathrm{Re}_*\,\mathrm{Sc}) \tag{33}$$

and

$$kB^{-1} \cong 0.69\,\mathrm{Re}_*^{0.45}\,\mathrm{Sc}^{0.8} \tag{34}$$

The former expression rests on the assumption $z_{0\chi} = D_\chi/(ku_*)$ (Sheppard, 1958), while the latter is a best-fit adjustment of Owen and Thompson's

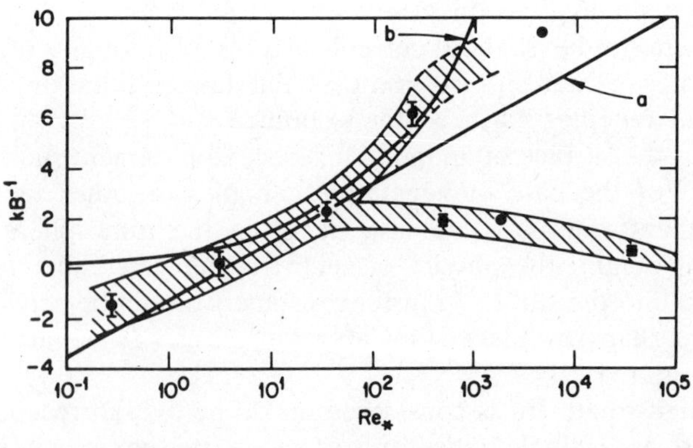

FIGURE 2. The overall behavior of kB^{-1}, compared with the relationship resulting from the assumption that $z_H = \kappa_t/ku_*$ (line a) and with a modified form of the Owen and Thomson (1963) relation $kB^{-1} = 0.69\mathrm{Re}_*^{0.45}\mathrm{Sc}^{0.8}$ (line b). The bands of data are drawn by eye to represent the probable 95% confidence limits of the observations. (From Garratt and Hicks, 1973, by permission.)

(1963) result. Equation (34) is similar in form to many of the expressions summarized by Brutsaert (1975).† For $Re_* > 50$, there are two possibilities. If the surface under consideration appears to be an orderly array of large bluff roughness elements, then Eq. (34) should be used. If, on the other hand, the surface is closely and randomly packed with fibrous individual elements, then kB^{-1} has very little dependence on Re_*. Shepherd (1974) suggests simply using $kB^{-1} \cong 2$ as a good approximation to the available data. Wesely and Hicks (1977) propose that this should be modified to show the effect of molecular diffusivity of the contaminant relative to thermal diffusivity κ_t; i.e.,

$$kB^{-1} \cong 2(\kappa_t/D_\chi)^{2/3} \tag{35}$$

Equation (35) may be applicable to most vegetated canopies of meteorological interest.

Garratt (1978), however, found a somewhat higher value of $\ln(z_0/z_{0T}) \cong 2.5 \pm 0.5$ for heat transfer over a rough ($z_0 = 0.4$ m) partially wooded site; he noted that the similarity profiles over this region did not apply below roughly 4.5 tree heights above the ground. Thom et al. (1975) observed similar departures from similarity theory just above a rather uniformly wooded site with $z_0 \cong 0.9$ m, and suggested that wakes behind individual tree tops act as a supplementary diffusing mechanism for heat and mass transport. It may be that forests differ enough from other receptors by virtue of their open trunk space and high roughness that the usual similarity theories for r_a and the semiempirical results for r_{z0} must be modified, but this is still a subject for research. In the interim, Eqs. (29), (32), and (34) or (35) may be the best available means for evaluating transfer resistance to vegetated regions, depending on Re_* and the canopy structure.

What if the surface is a body of water, rather than vegetation? Generally Re_* will be small, so Eq. (33) should apply. In strong winds, the presence of waves might be expected to modify this result. However, Hicks' (1975) analysis of data for heat and water vapor transfer over water suggests that the relation $z_{0\chi} = D_\chi/ku_*$ continues to hold up to wind speeds in excess of 14 m/s. Under the usual assumption that contaminants are transferred in a similar fashion, Eq. (33) should be useful for most nonstormy situations over bodies of water.

As indicated earlier, the method just outlined implicitly incorporates any overall canopy resistances into the surface resistance term. For surfaces such as bare soil or water, no real understanding seems to have been lost. For vegetation, however, all the turbulent transfer occurring within the

†Brutsaert's (1975) recommendation for $z_{0\chi}$ in meteorological applications (his Eq. (12)) leads to a curve kB^{-1} which lies above Eq. (34) by 50% or more for $10 < Re_* < 500$ or so, and which seems to be an upper bound on the data presented by Garratt and Hicks (1973). If the maximum likely value of r_{z0} is needed in a given application, then Brutsaert's paper should be consulted.

canopy region has been accounted for through the empirical parameterizations. The resistance calculated in this way may be an appropriate value, but little or no insight into the processes involved is provided. The approach is nevertheless useful and has been employed in many modeling efforts (Murphy, 1976; Wesely and Hicks, 1977). In all cases, the receptor element surface resistances must of course be added to $r_a + r_{z0}$.

An alternate approach to evaluating in-canopy resistance deals explicitly with the physical processes, at least to the extent of postulating eddy diffusivities within the canopy from which a resistance can be calculated (e.g., Eq. (31)). Waggoner and Reifsnyder (1968) and Shuttleworth (1976, 1978, 1979) have dealt extensively with canopies on a layer by layer basis. Various models have been used for the in-canopy flow and diffusivity. The so-called "ideal canopy" (Cionco et al., 1963; Cionco, 1971, 1972) has a uniform leaf area distribution with height and a uniform vertical distribution of Reynolds-number-independent drag coefficient of the canopy elements. The characteristic turbulent mixing length is constant with height, as is the turbulence intensity. Under these assumptions the velocity profile is exponential, $u(z)/u(h) = \exp[-a_1(1 - z/h)]$, where a_1 is an attenuation coefficient which depends on the canopy type. With a mixing length formulation for eddy diffusivity, K_χ has a similar dependence on height, $K_\chi = K_\chi(h)\exp[-a_2(1 - z/h)]$. The attenuation coefficient a_2 is allowed to differ from a_1 to permit the best fit to experimental data. Cionco (1972) discussed the values available in the literature for these coefficients, and Shreffler (1976, 1978) used this method to estimate transport within canopies using the diffusion equation.

The assumptions of the ideal canopy may not be valid for many stands of vegetation. Cionco (1971) described the difficulties in trying to reconcile such relations with data from nonuniform canopies like forests, although they may fit within the crown region. Oliver (1975) mentioned the "bulge" in the wind profile that often appears in forests, and attributed this to convective activity induced above the tree crowns during unstable conditions. Generally, one can fit some convenient expression to a given set of wind profile data, or make some assumptions about an appropriate form for eddy diffusivity, and derive a usable formula. For example, Murphy et al. (1977) allowed mixing length within the canopy to depend on the space between tree crowns, the area (at a particular height) of the crowns, and a characteristic eddy size for mixing between the branches, and deduced values for K_χ. It is probably also possible to estimate plausible eddy diffusivities from higher-order-closure models of canopy turbulence (Wilson and Shaw, 1977). All such possibilities, however, rest on rather shaky foundations due to our current lack of real understanding of forest aerodynamics. Bergen's (1976) summary is probably still valid:

> *There is currently no generally accepted model for . . . air flow in a forest canopy. The defects in existing models are fundamental. They reflect*

basically a lack of information on the flow near highly porous, bluff, and irregular bodies at appreciable speeds, as well as conceptual difficulties in superimposing turbulence fields of different macroscales and random secondary flow patterns.

In view of these uncertainties, the best approach to account for turbulent transfer processes within vegetation may still be the implicit method described earlier—*if* the distribution of materials deposited within the canopy is of no consequence. This may be the case if one is simply estimating aerial concentrations or doses at some receptor location downwind from the vegetation, so that only the amount removed by the canopy is of interest. For some purposes, however, the in-canopy distribution of contaminant may well be significant, and one must then attempt to utilize available descriptions of the flow. Caution is clearly advisable in these circumstances.

A few papers have dealt with in-canopy transfer processes in a manner quite different from any of the previously discussed methods. These papers tend to view a canopy as a filter. Slinn's (1974c, 1977) approach in particular is novel, in that material is transferred to and from the upper reaches of the vegetation by intermittent up and downdrafts. Some of the material inserted into the canopy is removed by direct turbulent transfer to individual receptor elements, while some is removed by filtration processes. Davidson and Friedlander (1978) and Bache (1979) are among those developing related methods. Chamberlain (1975) discusses some of the processes important to these computations. Such work offers some hope of additional understanding of the important physical processes governing in-canopy transport and removal.

3.3 Evaluation of Near-Receptor Resistances

The distinction between gases and small particles seems to become important mainly where transfer by molecular mechanisms becomes important—that is, close to individual receptor elements. In this region, molecular diffusion becomes significant for gases, while Brownian diffusion, gravitational settling, and viscous drag influence particle deposition. Expressions for estimating these final resistances to contaminant transfer are therefore discussed separately.

Within canopies or other bulk receptors composed of vertical arrays of individual receptor elements, the large number of deposition resistances acting essentially in parallel must also be considered. Because of the vertical variations in wind speed and turbulent diffusivity encountered in such situations, it is probably necessary to divide the canopy into layers within which terms such as $u(z)$ and $K_x(z)$ are fairly constant. An individual receptor resistance in a given layer is then estimated, and the overall resistance to deposition within the layer is determined by dividing the individual value by a weighting function such as the leaf area index for that layer. The total receptor resistance would then be the parallel sum of these various layer resistances. So-called shelter factors (Thom, 1971; Landsberg and Powell, 1973) may modify the individual resistances because of mutual aerodynamic interference. These factors are not well known for most vegetation types, but in the cases investigated seem to fall between 1 and 4 in value, increasing individual resistances by this amount. The degree of detail in the computations may be justifiable only when the sites of deposition must be known with some precision—in a food-chain analysis, for example.

3.3.1 Transfer of Gases to Receptor Elements

Extensive literature exists on the transfer of momentum, heat, and gases across boundary layers (Bird et al., 1960; Eckert and Drake, 1959). A number of authors have applied these techniques to individual leaves and other likely receptor surfaces (Chamberlain, 1974; Murphy and Knoerr, 1977; Parkhurst and Pearman, 1974; Pearman et al., 1972; Shreffler, 1978; Thom, 1968). Mass transfer to flat surfaces is generally found to follow the relation,

$$\mathrm{Sh} = c\mathrm{Re}^{1/2}\,\mathrm{Sc}^{1/3} \tag{36}$$

where Reynolds number $\mathrm{Re} \equiv \tilde{u}l/\nu$ and Schmidt number $\mathrm{Sc} \equiv \nu/D_\chi$. Here $\tilde{u}$ is the characteristic "outer" velocity for a viscous boundary layer over a surface of characteristic dimension l; D_χ is the molecular diffusivity of the gas in air. The Sherwood number Sh (corresponding to the Nusselt number of heat transfer) is related to the boundary layer resistance by

$$r_b = l/(D_\chi\,\mathrm{Sh}) \tag{37}$$

The constant c depends somewhat on the surface being tested; the usually cited value is $c = 0.664$, although values ranging up to about 1.1 have been reported for leaves (Chamberlain, 1974; Murphy and Knoerr, 1977; Pearman et al., 1972). Equation (37) gives the resistance to transfer for one surface of a receptor; it must be applied to both sides of a leaf. The angle of incidence of the wind relative to the surface can affect the transfer, especially if the wind is turbulent (Chamberlain, 1974; Parkhurst and Pearman, 1974; Thom, 1968). If the resistance for one gas is known, r_{b1}, say, then it can be calculated for other gases by

$$r_{b2}/r_{b1} = (D_{\chi1}/D_{\chi2})^{2/3} \tag{38}$$

The main difficulty in applying these results rests in the problem of estimating $\tilde{u}$, the velocity driving the receptor boundary layer. For ideal canopies, $\tilde{u} \sim u(z)$, which is modeled as noted earlier; shelter factors may, of course, alter the result. Nonideal cases are even less tractable.

3.3.2 Transfer of Particles to Receptor Elements

Particle deposition is an extremely complex process, coupling small-body aerodynamics and interparticle forces with a plethora of particle–surface interactions. Many of these factors are difficult to quantify, including effects of turbulence on particle motions, receptor movement (e.g., leaf flutter), and thermophoresis and diffusiophoresis when the receptors are sources or sinks of sensible and latent heat. Chamberlain's (1975) review discusses impaction and retention efficiency of particles in plant communities and should be consulted. He notes that bounceoff can be a significant factor in retention for particle diameters in the range of 5 to 50 μm; below 5 μm bounceoff is unlikely unless the incident velocity is very high and the receptor is smooth and hard. Depending on particle size, the microtopography of the receptor surface and its wetness, stickiness, and state of electric charge can strongly influence retention (see Holloway, 1971, for leaves). Important vegetation features include the number and type of epidermal protuberances; stomatal size, distribution, and number; the presence of exudates or wax platelets; surface moisture and wettability; and charge distribution and strength (Greene and Bukovac, 1974). For example, Wedding et al. (1975) found that retention of aerosols on rough pubescent leaves was about ten times greater than on smooth leaves. Even if a potential receptor such as dry vegetation turns out to be rather inefficient for particle retention, it nonetheless may serve to modify the local winds, enhancing deposition and retention on other receptors, such as soil.

Once a particle is deposited, a number of events may occur. If the particle is inert and insoluble and the receptor is not biologically active, the particle may simply sit on the surface, possibly being resuspended later. Wettable particles may be absorbed into bodies of water. Reactive particles may combine with other materials already present or arriving later on the receptor surface. Deliquescent particles may convert to the liquid phase in the presence of high humidity. Soluble particles contacting wet receptors may experience a number of fates—they may simply dissolve, solidifying later if the water evaporates. They may be transported in solution from one receptor to another by runoff. Or they may enter biologically active receptors through a number of pathways, possibly reacting chemically within the receptor, or even being assimilated into the biomass.

All of these possibilities for particle deposition and postdeposition transport or modification can be considered resistive links in the chain of fluxes. Little information exists for estimating resistances for many of these pathways for particle movement. However, models for the actual deposition process have been postulated by Davidson (1977), Sehmel and Hodgson (1974, 1978) and by Slinn (1974c, 1977).

Davidson (1977) dealt with the deposition of aerosols on to cylindrical receptors; his model was developed for grasses, but can probably be adapted to other related cases, such as conifer forests. Descriptions of the physics may be found in Fuchs (1964) and Friedlander (1977). For very small particles, Brownian diffusion is regarded as the dominant process, and the resistance at an individual receptor element is calculated like that for a gas, Eq. (37). Davidson used the cylindrical receptor diameter d_c as the characteristic dimension l, and followed Friedlander's (1977) recommendation for the Sherwood number:

$$\mathrm{Sh} = 0.683\mathrm{Re}^{0.466}\mathrm{Sc}^{0.333} \tag{39}$$

The Schmidt number here is based on the Brownian diffusivity D_B (e.g., calculated from Davies, 1966), and $\mathrm{Re} = \tilde{u}d_c/\nu$. For larger particles, impaction is considered the dominant mechanism; the resistance to deposition therefore depends on the collection/retention efficiency η_c:

$$r_b = \pi/(\tilde{u}\eta_c) \tag{40}$$

This efficiency is an empirical function of the Stokes number St:

$$\eta_c = \frac{\mathrm{St}^3}{\mathrm{St}^3 + 0.753\mathrm{St}^2 + 2.796\mathrm{St} - 0.202} \tag{41}$$

where

$$\mathrm{St} \equiv k_c \frac{(\rho_p - \rho_a)\tilde{u}d^2}{9\rho_a \nu d_c} \tag{42}$$

k_c is the Cunningham slip factor, d is the particle diameter, ρ_a and ρ_p are the air and particle densities.

Sehmel and Hodgson (1974, 1978), through a series of careful wind tunnel experiments, were able to develop an empirical relation between the individual receptor boundary layer resistance and a number of micrometeorological variables:

$$r_b = v_t^{-1}[1 - \exp(-\tilde{I}v_t/\tilde{u}_*)] \tag{43}$$

where v_t is the particle's terminal velocity and $\tilde{u}_*$ is the local friction velocity. $\tilde{I}$ is a complicated empirical function:

$$\begin{aligned}\tilde{I} = \exp\{&16.5 \ln(\nu/D_B) - 378.1 + [\ln \tau][0.3226 \ln(d/\tilde{z}_0) \\ &- 0.3385 \ln(D_B/\tilde{u}_* \tilde{z}_0) - 0.2863 \ln \tau - 11.82] - 12.8 \ln d\}\end{aligned} \tag{44}$$

where z_0 is the receptor's aerodynamic roughness and τ is the particle's so-called relaxation time such that $v_t = g\tau$; $\tau = \rho_p d^2/(18\rho_a \nu)$ (Chamberlain, 1975). Equation (44) was developed using only particles of density $\rho_p = 1.5\ \mathrm{g/cm^3}$; consequently particle density enters only indirectly through τ.

Slinn (1974c, 1977), on the other hand, utilizes a physical picture of the deposition process rooted in turbulent boundary layer concepts. Turbulent eddies are envisioned as intermittently advecting particles from the upper reaches of the receptor boundary layer down to a smooth surface. The resultant marching of eddies across the surface is described as "Joukowski's caterpillar tread" (Slinn, 1974c); this advective process is treated as a viscous jet impacting some fraction of the total surface with a velocity related to $\tilde{u}_*$. Particles are retained according to some characteristic jet collection efficiency E_j. The presence of water vapor flux modifies the process. Slinn's expression for monodisperse particle transfer resistance is then

$$r_b^{-1} = v_t + (\dot{m}_W \times 10^3) + \frac{\tilde{u}_*^2}{\beta \tilde{u}} E_j \tag{45}$$

where $\dot{m}_W$ is vapor flux (g cm^{-2} s^{-1}), positive when condensation occurs. Slinn took the unknown constant $\beta \cong$ the von Karman constant, and suggested

$$E_j \sim 10^{-(3\nu/\tilde{u}_*^2\tau)} + \frac{\beta}{\gamma}\left(\frac{\nu}{D_B}\right)^{-0.6} \tag{46}$$

With $\beta/\gamma \cong 1$, Slinn (1977) found reasonably good agreement with measured deposition velocities. For a polydisperse aerosol, Slinn (1977) noted that the concentration flux must be computed using the particle-size distribution as a weighting function, and suggested that a log–normal distribution is typical.

All of these resistance models may be difficult to apply in some situations, such as within canopies, where it is difficult to estimate quantities such as $\tilde{u}$, $\tilde{u}_*$, and $\tilde{z}_0$ for the individual receptor boundary layers. For ideal canopies, one can model $\tilde{u}(z)$ and $K_M(z)$ (Cionco, 1972; Murphy et al., 1977); a rough estimate of $\tilde{u}_*$ is then possible using $\tilde{u}_*^2 \sim K_M \, \partial \tilde{u}/\partial z$. Whether this approximation is appropriate close to individual receptors within densely packed canopies is entirely conjectural. In nonideal canopies the K-theory flow model is probably inadequate, and some other means of estimating the needed terms must be found.

3.3.3 Stomatal and Related Resistances

When the individual receptors for deposition are biologically active surfaces such as leaves, transport of the deposited material into the receptor and its subsequent interactions with internal cells can become limiting mechanisms for the overall transfer process. Figure 3 illustrates the complex linkage of pathways for pollutant uptake by a typical leaf. In particular, stomatal control of diffusion into a leaf is a major factor influencing the exchange of pollutants with internal tissues. The stomates are controlled by plant physiological processes; their closure increases resistance to such exchange. An important point (Hosker and Lindberg, 1982) is that both environmental and physiological factors can cause large changes in stomatal aperture. Other important pathways include (Rutter, 1975) the epidermal cells via the cuticle, and mesophyll cells by way of the substomatal cavities. Table 3 gives typical values for resistances to water vapor diffusion for some medium-sized leaves. Since cuticular resistances are generally very

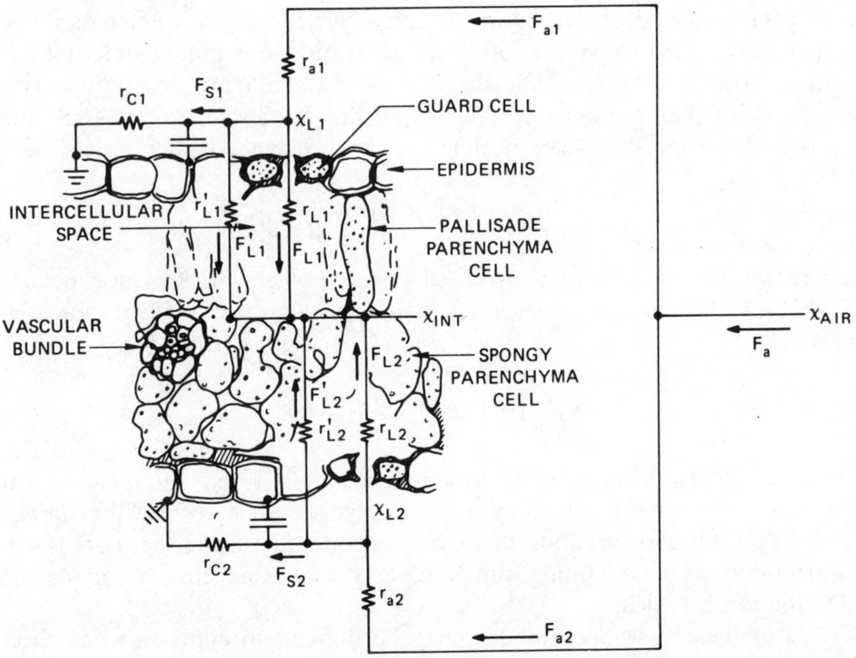

FIGURE 3. Electrical analog of pollutant exchange between leaf and surrounding air. The circuitry is superposed on a cross section of an amphistomatous leaf (stomates on both sides). χ_{AIR}, χ_{L1} and χ_{L2}, and χ_{INT} denote the gaseous pollutant concentration in well mixed surrounding air, concentrations at the upper and lower leaf surfaces, and average gas phase concentration within the leaf mesophyll, respectively. The (variable) resistances r_{a1} and r_{a2}, r_{L1} and r_{L2}, and r'_{L1} and r'_{L2} are the upper and lower boundary layer resistances, stomatal plus intercellular resistances, and cuticular plus internal resistances, respectively. Resistances r_{C1} and r_{C2} denote resistance to chemical reaction at the upper and lower leaf surfaces. The fluxes F_a, F_{a1}, and F_{a2}, F_{S1} and F_{S2}, F_{L1} and F_{L2}, and F'_{L1} and F'_{L2} denote the total flux to both surfaces ($F_a = F_{a1} + F_{a2}$), surface uptake at the upper and lower surfaces, and the fluxes through the upper and lower stomata and cuticles, respectively. The electrical symbols depicting grounds (⏚) represent the termination of fluxes due to reaction with leaf surface materials. The capacitance symbols (—| |—) represent surface adsorption or retention. (Based on Bennett et al., 1973.)

large, uptake of pollutants via this pathway can often be neglected. In fact, field studies (Fowler, 1978; Garland and Branson, 1977; Tan and Black, 1976) suggest that stomatal resistance may often be the controlling parameter in canopy deposition.

A number of investigators have reported the dependence of stomatal resistance on environmental factors. The text by Meidner and Mansfield (1968) is a good reference on stomate physiology. Low CO_2 level tends to cause stomatal opening, as does increasing temperature; these effects may be coupled in a complicated way through respiration and photosynthesis (Rutter, 1975). Stomatal opening also correlates strongly with incident light

TABLE 3. Typical Resistances (s/cm) to Water Vapor Diffusion for Medium Size (1 dm²) Leaves

Resistance	Range of Values
$r_{intercellular}$	0.1–0.5
$r_{stomatal}$ (for open stoma)	
cultivated crops	1–5
broad-leaf trees	3–10
xerophyte	5–20
$r_{cuticular}$	30–200
$r_{boundary\ layer}$	0.1–2

From Hosker and Lindberg (1982).

intensity. This latter fact has led to expressions for stomatal resistance which explicitly depend on local illumination intensity I. Shawcroft et al. (1973) suggested

$$r_s = r_{sm} + \frac{\beta_0}{I + I_0} \tag{47}$$

where r_{sm} is the (minimum) resistance at maximum irradiance for a given plant and stress condition, and β_0 and I_0 are equally specific empirical constants. This relation has been used by Lister and Lemon (1974), and by Murphy et al. (1977). Waggoner and Reifsnyder (1968) suggested, for an ideal canopy,

$$r_s = r_{sm}/\mathrm{erf}(\gamma_0 S) \tag{48}$$

where γ_0 is an empirical parameter and S is the radiation absorbed per square centimeter by the leaf. Relations such as Eq. (43) or (44) are especially useful when information on the penetration of sunlight into a vegetative canopy is available (Hutchison and Matt, 1977; Norman, 1971). Layer by layer estimation of the stomatal resistance of an average leaf can then be carried out. The stomatal resistance of a layer is due to the parallel resistance of all the leaves in that layer and can be approximated by dividing the average leaf resistance by the leaf area index for that level. The overall stomatal resistance of the canopy is then the parallel sum of the layer resistance.

If the known stomatal resistance of a leaf to a particular gas is r_{s1}, then, assuming a different gas is absorbed into the leaf through the same pathways, the resistance to this second gas, r_{s2}, is given by (Bennett et al.,

1973)

$$r_{s2}/r_{s1} = D_{\chi 1}/D_{\chi 2} \tag{49}$$

This is useful since stomatal resistances to water vapor and CO_2 have been widely studied and published (Monteith, 1976).

3.4 Experimental Data

Observations of dry deposition velocities or resistances have been widely reported. A comprehensive review paper by McMahon and Denison (1979) tabulates most of these data for studies published before March, 1978. Other references include Chamberlain (1974, 1980), Droppo (1980), Gash and Stewart (1975), Menzel (1967), Milne et al. (1979), Schwela (1979), and Smith and Jeffrey (1975). Chamberlain (this volume) provides additional information, and additional data are doubtless available.

Some caution must be used in applying these data in models. "Typical" deposition velocities must be selected with care, since they can vary over orders of magnitude, depending on the receptor and pollutant, as well as environmental conditions (see Table 2). Only results from experiments carefully matched to the case to be modeled are likely to be applicable.

Furthermore, one must determine exactly what was measured before using the numbers. For example, Hoffman et al. (1977) pointed out that deposition of radioactive iodine has often been evaluated by sampling only the vegetation at a particular site; material deposited on the underlying litter and soil is neglected. Overall deposition velocities based on such techniques are therefore underestimated. McMahon and Denison (1979) noted that hard-to-determine resistances such as r_b are often not explicitly evaluated, and are therefore implicitly included in measurements of surface resistance.

Experimental methods are subject to many constraints; the degree to which they were satisfied in any particular trial must be examined. Hicks and Wesely (1978) discussed some of these sources of error. For example, in flux-gradient measurements, very high accuracy and resolution are required of the instruments. Rapid instrument response is needed for direct eddy-correlation flux measurements. Vertical alignment within a small part of a degree is necessary for eddy-accumulation devices. Atmospheric stability, inaccuracies in estimates of zero displacement plane and roughness heights, and fetch inadequacies of size and homogeneity can also influence results. The observations must be made over a period long enough for all important turbulent fluctuations to contribute, but short enough for mean conditions to be stationary. If the receptors are biological, the influence of temperature, humidity, illumination, and water stress must be considered.

Finally, it should be recognized that, under some circumstances, negative deposition velocities may occur—that is, certain receptors may become pollutant sources because of resuspension or direct emission (Chamberlain,

1970; Droppo, 1980). This may influence long-term deposition experiments, and may be important for receptors further downwind. Chamberlain (1970) addressed the problem in terms of a field loss coefficient for a few cases, but the problem in general is still poorly understood.

4. WET DEPOSITION MODELING

4.1 Evaluation of Scavenging Rate

4.1.1 *Monodisperse Aerosols*

As stated earlier, the treatment of wet removal by means of a scavenging rate λ is appropriate only for aerosols and highly reactive gases which are irreversibly scavenged from the atmosphere by falling hydrometeors. Expressions for Λ are derived in many references (Chamberlain, 1953; Engelmann, 1968; Slinn, 1977, 1984). If the pollutant's gravitational settling speed is small relative to the fall velocity of the precipitation, then, for a monodisperse aerosol of radius a,

$$\Lambda(\mathbf{r}, t; a) = \int_0^\infty \epsilon E(a, l) A(a, l) v_t(l) N(\mathbf{r}, t; l)\, dl \tag{50}$$

where ϵ and E are the retention and particle and hydrometeor collection efficiencies, respectively, A is the effective cross-sectional area of a hydrometeor of characteristic dimension l and terminal speed v_t, and $N(\mathbf{r}, t; l)\, dl$ is the number density of hydrometeors with dimensions between l and $l + dl$ at position r and time t.

The retention efficiency ϵ is usually assumed to be unity (Hidy, 1970; Slinn, 1984). The collision efficiency depends in a very complex way on the details of the collision process (Slinn, 1974c). A suggested semiempirical result for raindrops of radius R is (Slinn, 1977)

$$E_r(a, R) \cong \frac{4[1 + 0.4\mathrm{Re}^{1/2}\mathrm{Sc}^{1/3}]}{\mathrm{ReSc}} + 4\left(\frac{a}{R}\right)\left[\left(\frac{a}{R}\right) + \frac{1 + 2(\mu_w a)/(\mu_a R)}{1 + (\mu_w/\mu_a)\mathrm{Re}^{-1/2}}\right] + \left[\frac{\mathrm{St} - \mathrm{St}_*}{\mathrm{St} - \mathrm{St}_* + 2/3}\right]^{3/2} \tag{51}$$

where $\mathrm{Re} \equiv v_t(R)R/\nu$, $\mathrm{Sc} \equiv \nu/D_B$, μ_a and μ_w are air and water viscosities, and $\mathrm{St} \equiv v_t(R)\tau(R)/R$, the Stokes number for a raindrop. $\tau(R)$ is the drop's relaxation time, while St_* is a critical value defined by

$$\mathrm{St}_* = \frac{1.2 + 0.0833 \ln(1 + \mathrm{Re})}{1 + \ln(1 + \mathrm{Re})} \tag{52}$$

E_r is shown in Figure 4, from Slinn (1977). Davenport and Peters (1978) and Pilat and Prem (1976) have also suggested expressions for E_r.

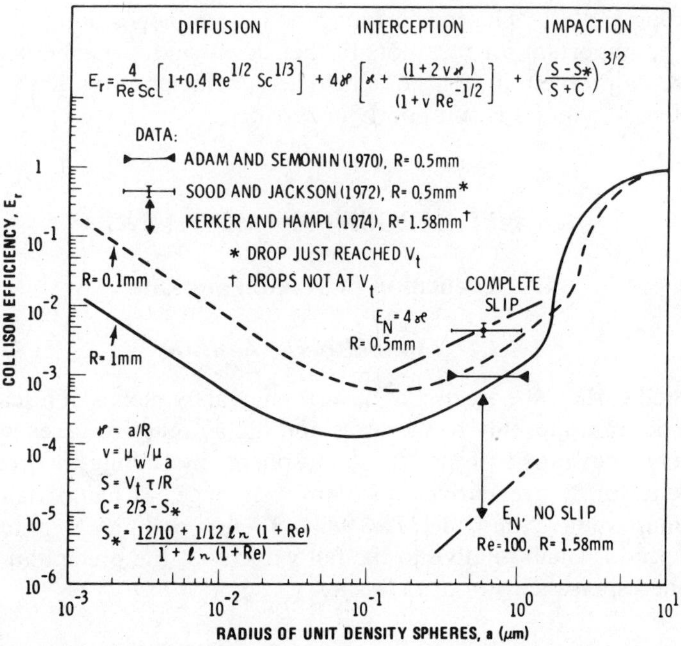

FIGURE 4. Slinn's proposed semiempirical expression for particle–raindrop collision efficiency E_r as a function of particle radius a, for two raindrop radii R. (From Slinn 1977, by permission.)

If the raindrops are approximately spherical, the effective cross section $A = \pi(a + R)^2 \cong \pi R^2$, since the particle radius is generally small compared with the drop size.

Evaluation of drop terminal velocity can be accomplished from several expressions (Best, 1950; Davenport and Peters, 1978; Gunn and Kinzer, 1949; Kessler, 1969). The recommendations of Davenport and Peters cover the droplet diameter range of 0.01–4.0 mm, and agree well with experimental data.

$$v_t = \begin{cases} \dfrac{2(\rho_w - \rho_a)gR^2}{9\mu_a}, & \text{Re} < 0.5 \\[2ex] \left[\dfrac{0.219(\rho_w - \rho_a)gR^{1.6}}{\rho_a^{0.4}\mu_a^{0.6}}\right]^{5/7}, & 0.5 \le \text{Re} \le 250 \\[2ex] 2.46\left[\dfrac{(\rho_w - \rho_a)gR}{\rho_a}\right]^{1/2}, & \text{Re} > 250 \end{cases} \tag{53a}$$

Hales et al. (1973) recommend the simpler empirical polynomial

$$v_t = 32184.5R^3 + 18029.7R^2 - 8710.9R \tag{53b}$$

The raindrop size distribution cannot be readily specified, since there is great variability from storm to storm and even within single storms. The distribution suggested by Marshall and Palmer (1948) is often used to characterize steady rains:

$$N(R) \cong N_0 \exp(-82RJ_0^{-0.21}) \tag{54}$$

where $N_0 \cong 0.08/\text{cm}^4$, and $N(R)\,dR$ is the concentration of drops (per cubic centimetre) with radii between R and $R+dR$; R is in centimetres, and the rainfall rate J_0 is in millimetres per hour. Waldvogel (1974) has shown, however, that N_0 not only varies from storm to storm, but also that it may jump in value within a given storm according to the cellular structure and convective activity within the system. Beard (1974) used the distribution proposed by Sekhon and Srivastava (1971) for a thunderstorm:

$$N(R) \cong 0.07J_0^{0.37} \exp(-76RJ_0^{-0.14}) \tag{55}$$

The uncertainties in evaluating v_t and N can be sidestepped by the following approximation to Eq. (50) (Slinn, 1974c). Multiply the integrand by R, and divide by the mean droplet radius R_m; assume the efficiency is determined fairly well by evaluating at this single value R_m. Thus

$$\Lambda_r \cong \frac{E_r(a, R_m)}{R_m} \int_0^\infty v_t(R) N(\mathbf{r}, t; R) \pi R^3 \, dR \tag{56}$$

Since the precipitation rate is

$$J_0(\mathbf{r}, t) = \int_0^\infty v_t(R) N(\mathbf{r}, t; R) \frac{4\pi}{3} R^3 \, dR \tag{57}$$

one can write

$$\Lambda_r(\mathbf{r}, t; a) \cong c \frac{J_0(\mathbf{r}, t) E_r(a, R_m)}{R_m} \tag{58}$$

where c is a constant $\cong 0.5$. Slinn takes R_m to be the volume-mean droplet radius, which is often related to the rainfall rate. For example, for continuous rain and J_0 in mm/h, $R_m \cong 0.035J_0^{0.25}$ (Mason, 1971), although Scott (1978) suggests $R_m \cong 0.045J_0^{0.21}$ for a Marshall–Palmer size distribution. For a convective thunderstorm, Sekhon and Srivastava (1971) found $R_m \cong 0.065J_0^{0.14}$. Evidently Λ_r should vary roughly as J_0^p, where p is between 0.7 and 0.9 or so; Slinn (1974c) points out that this is consistent with available data.

If the hydrometeors are snowflakes rather than raindrops, a similarly approximate result can be derived (Slinn, 1974c):

$$\Lambda_s(\mathbf{r}, t; a) \cong \frac{\rho_w g J_0(\mathbf{r}, t) E_s(a, l_s)}{\rho_a \langle v_t \rangle^2} \tag{59}$$

where E_s is the particle–snowflake collision efficiency for a flake of charac-

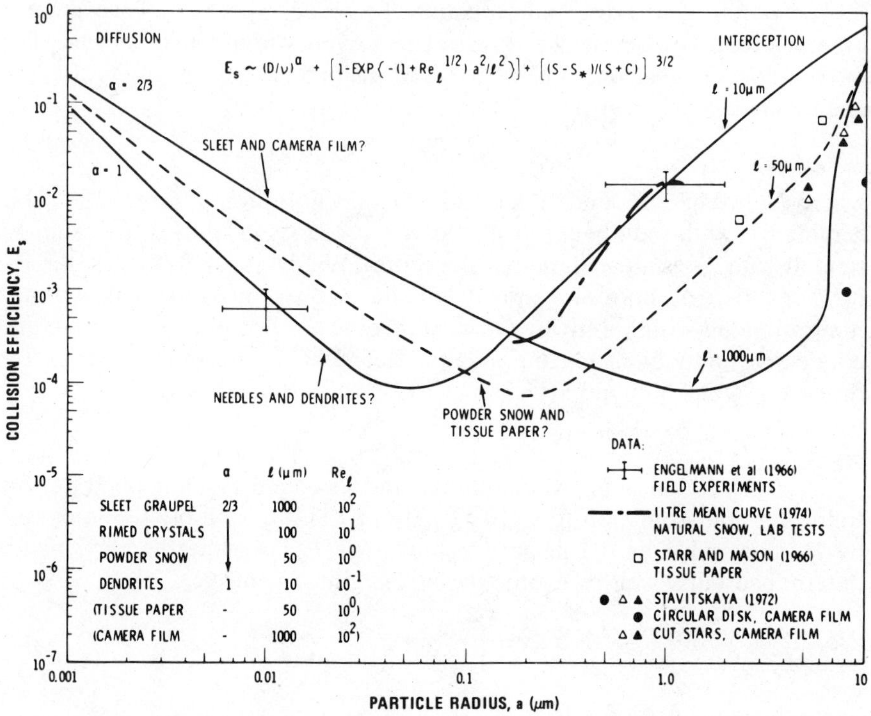

FIGURE 5. Slinn's proposed semiempirical expression for particle–snowflake collision efficiency E_s as a function of particle radius a, for three characteristic snowflake sizes l. (From Slinn, 1977, by permission.)

teristic dimension l_s, $\langle v_t \rangle$ is the average settling speed of the flakes, and J_0 is the rainfall-equivalent precipitation rate. Knutson et al. (1976), Redkin (1973), and Slinn (1974c, 1977) are among those who have tried to estimate E_s. Slinn's suggestion is shown in Figure 5. Zikmunda (1972) and Knutson et al. (1976) recommend $\langle v_t \rangle \cong 102 + 51 \log_{10} d_c$, where $\langle v_t \rangle$ is in centimetres per second and d_c is the diameter (in centimetres) of the circle circumscribed about the average snowflake. An expression for Λ_s very similar to Eq. (58) has recently been given by Slinn (1984), along with suggested values for its use.

Slinn (1974c, 1978) is very emphatic about the approximate nature of Eqs. (58) and (59). He states that use of removal rates such as these to describe precipitation scavenging is at best an order of magnitude estimation procedure. The work of Davenport and Peters (1978) seems to justify such conservatism, since their observed scavenging coefficients were generally one or two orders of magnitude larger than their calculated values. Slinn (1978) recommends that predictions using scavenging rates should be reserved for episodic single events, since the use of washout ratios, described below, provides much better long-term accuracy.

4.1.2 Polydisperse Aerosols

For most purposes it is usually necessary to determine a particle-averaged scavenging rate, which accounts for the particle size distribution. The nature of the property being removed—quantity of radioactivity, say, or mass scavenged—is the significant feature in such calculations. Suppose, for example, the mass remaining aloft in a cloud of particles must be evaluated. The mass of any single particle is just $4\pi\rho_p a^3/3$. Let $n(a)\,da$ be the number of particles with radii between a and $a+da$; the probability density function for the particle distribution is just $f(a)=n(a)/N_T$, where N_T is the total number of particles, $N_T=\int_0^\infty n(a)\,da$. The rate of removal of mass per unit volume due to precipitation scavenging of particles in the size range a to $a+da$ is $4\pi\rho_p a^3\Lambda(a)n(a)\,da/3$. Hence the rate of change of mass concentration χ is

$$\frac{d\chi}{dt}=-\frac{4\pi}{3}\rho_p N_T\int_0^\infty \Lambda(a)a^3 f(a)\,da \tag{60}$$

Define the nth order normalized particle-averaged scavenging rate by

$$\langle\Lambda_n\rangle\equiv\int_0^\infty \Lambda(a)a^n f(a)\,da\bigg/\int_0^\infty a^n f(a)\,da \tag{61}$$

Then Eq. (60) is just

$$\frac{d\chi}{dt}=-\langle\Lambda_3\rangle N_T\int_0^\infty \frac{4\pi}{3}\rho_p a^3 f(a)\,da \tag{62}$$

or

$$\frac{d\chi}{dt}=-\langle\Lambda_3\rangle\chi \tag{63}$$

which is solved in the usual way. If the quantity of interest is, on the other hand, radioactivity distributed on the particles according to their surface area, then the rate equation corresponding to Eq. (63) would involve the second order coefficient $\langle\Lambda_2\rangle$, which can also be calculated from Eq. (61).

Dana and Hales (1976) have examined particle-averaged scavenging coefficients under the common assumption of log–normal particle and raindrop size distributions:

$$f(x)\,dx=[\sqrt{2\pi}\ln\sigma_g]^{-1}\exp\left[\frac{-(\ln x-\ln x_g)^2}{2(\ln\sigma_g)^2}\right]d(\ln x) \tag{64}$$

where x_g is the geometric mean and σ_g the geometric standard deviation of the distribution, where x is either a or R. They found that the mass scavenging rate for a polydisperse aerosol is greater by up to a few orders of magnitude than the rate for a monodisperse aerosol of radius equal to the geometric mean radius a_g, at least for $0.1<a_g<10\ \mu m$. As the distribution for the polydisperse aerosol broadens (larger σ_{ag}), the rate increases, apparently because of the increased contribution of larger particles which are effectively scavenged. Dana and Hales discussed the example of a particle distribution characterized by $a_g = 0.1$ μm and $\sigma_{ag} = 2.0$; their calculations suggested that the particle-averaged scavenging rate in a 1 mm/h rain would be $\cong 0.0115$/h. If one only knew the geometric mean particle size, but not the breadth of the distribution (i.e., σ_{ag}), one might try to assume a monodisperse ($\sigma_{ag} = 1$) aerosol of similar size; however, the rate obtained under this assumption would be more than an order of magnitude too small. It is clear that a reasonably accurate picture of particle size distribution is essential for even order of magnitude accuracy in predicting scavenging rates using Eq. (61). Whitby (1980) suggests that $\sigma_{ag} = 2$ is a good rule of thumb for particle distributions, unless data are actually available.

Slinn (1977, 1984) describes a number of complex processes which can affect the scavenging rate. For example, multiple step removal can occur, whereby particles are collected by small droplets which are in turn scavenged by larger drops, which can also collect particles directly. Furthermore, particles can change their size and chemical characteristics during their residence in the atmosphere; these will in turn modify the scavenging process. If the particles are at high altitude, they may serve as droplet condensation nuclei within clouds. Such phenomena are too complicated to deal with in this review; the reader may refer to Slinn (1984) for further detail.

4.1.3 Irreversibly Scavenged Gases

Highly reactive gases can be irreversibly collected by raindrops; the resultant scavenging rate is given by Eq. (58), with the efficiency

$$E_r(R_m) \cong 4[1 + 0.4\mathrm{Re}^{1/2}\mathrm{Sc}^{1/3}]/(\mathrm{ReSc}) \qquad (65)$$

(Slinn, 1978, 1984). For less reactive gases, or those which only form simple solutions, gas can desorb from falling droplets as they enter regions of low effluent concentration. Equations (58) and (65) are not applicable to such cases; these are considered in the next section.

4.2 Overall Mass Transfer Coefficient for Gas Scavenging

Hales (1972) shows that the molar concentration [C] of a soluble gas in an average raindrop of radius R and terminal velocity v_t is

$$\frac{d[\mathrm{C}]}{dz} = \frac{3\kappa}{v_t R}([\chi] - h([\mathrm{C}])) \qquad (66)$$

where the function h depends on the equilibrium condition $[\chi]_e = h([\mathrm{C}])_e$; if the gas follows Henry's law, h is linear in [C], and Eq. (14) is obtained. It can also be shown (Bird et al., 1960; Hales, 1972) that the overall mass transfer velocity κ is related to the mass transfer velocities for the gas and liquid phases k_G and k_L

$$\kappa^{-1} = k_G^{-1} + mk_L^{-1} \tag{67}$$

where m is related to the slope of the equilibrium concentration curve. For systems obeying Henry's law, m is equal to H. In Slinn's (1984) notation, $m = 1/\alpha_S$, where α_S is the solubility coefficient, defined on a volume basis. Slinn provides convenient tables of α_S for some common gases, and discusses conversion of such coefficients from one system of units to another.

If κ is independent of [C], then Eq. (66) can be solved relatively easily; if κ does depend on concentration, the solution is more complex. In either case, [C] depends on both position and raindrop size. To find the mass flux delivered to the surface by rain, the molar concentration [C] (perhaps similar to that in Eq. (15)) is evaluated at ground level and integrated over the droplet size distribution (Hales et al., 1973):

$$F_W(x, y, 0, t) = \frac{4\pi M_G}{3} \int_0^\infty v_t(R)[C(x, y, 0; R)]R^3 N(x, y, 0, t; R)\, dR \tag{68}$$

where M_G is the molecular weight of the scavenged gas.

The gas phase mass transfer velocity k_G is an important part of κ; it can be calculated from the Frössling (1938) relation

$$k_G \cong \frac{D_x}{R}[1 + 0.42\mathrm{Re}^{1/2}\mathrm{Sc}^{1/3}] \tag{69a}$$

although Barrie (1978) and Beard (1974) recommend that this expression be limited to drops of radius $R > 0.6$ mm or so. For smaller drops, the result of Beard and Pruppacher (1971) may be slightly better:

$$k_G \cong \frac{D_x}{R}[0.78 + 0.43\mathrm{Re}^{1/2}\mathrm{Sc}^{1/3}] \tag{69b}$$

In both cases, D_x is the molecular diffusion coefficient of the effluent gas in air, $\mathrm{Re} \equiv v_t R/\nu$, and $\mathrm{Sc} \equiv \nu/D_x$, as before.

The liquid phase mass transfer velocity is more complicated. Hales (1972) related it to the molecular diffusivity of the gas in water, D_W, say. This may be appropriate in very small drops; however, for drops larger than about 0.1 mm, internal circulations develop within the drop. These are driven by viscous effects generated by the fall of the drop through the air; the induced velocities are typically a few percent of v_t (Slinn, 1984). Consequently mixing within a raindrop will usually be fairly rapid. Hales (1972) and Slinn (1984) consider some special cases.

Suppose the gas is very soluble in water (large α_S) and the internal circulation is active. Slinn (1984) suggests that k_L will range between 1 and

10 cm/s for raindrops between 0.1 and 1 mm. For many low molecular weight gases, k_G will be of order 10 cm/s also. Hence, since α_S is large, the resistance to transfer within the drop will be much smaller than the resistance in the atmosphere outside; that is, $\kappa \approx k_G$. On the other hand, if the solubility is small and/or the internal circulations are weak, then both k_G and k_L may contribute significantly to κ. This may also occur for high molecular weight, strongly soluble gases. Then κ depends not only on the raindrop–gas characteristics, but also on the concentration history of the drop, since a finite time is required for the falling drop to adjust to external concentration changes. Hales (1972) discussed this in some detail; his results have been used, for example, by Hales et al. (1973) to estimate pollutant removal from gaseous plumes.

The situation is even more complex when the pollutant gas is not only soluble in raindrops, but also enters into reactions within them. In this case, any of the atmospheric transfer, internal mixing, or chemical conversion processes can be rate-limiting. Hales (1972, 1978) and Slinn (1984) discuss some of the possible cases. Information on SO_2 scavenging is particularly common in the literature (Adamowicz, 1979; Barrie, 1978; Dana et al., 1975; Hales and Dana, 1979; Overton et al., 1979). The presence of several chemical species within a raindrop is an additional complicating factor.

4.3 Washout Ratios and Wet Deposition Velocities

As mentioned earlier, convenient means of quantifying wet deposition are available in the washout ratio W_r, as defined in Eq. (19), and its near relative, the wet deposition velocity v_w (Eq. (20)). These parameters are useful for both particle and gas scavenging, particularly when a long-term estimate of wet deposition is needed.

Suppose precipitation is falling through a particulate plume. Given a scavenging rate Λ_r or Λ_s (Eqs. (58) or (59)), and assuming the monodisperse particle concentration χ does not vary strongly in the horizontal, then the wet mass flux to the surface can be approximated by

$$F_W \cong \int_0^{z_m} \Lambda \chi \, dz' \tag{70}$$

where the integration along a typical hydrometeor path has been replaced by a vertical path, and z_m is the depth of the scavenged layer. The wet deposition velocity is just $v_w = F_W/\chi_0 = W_r J_0$ by Eq. (20), so

$$W_r \cong \frac{1}{\chi_0 J_0} \int_0^{z_m} \Lambda \chi \, dz' \tag{71}$$

Using Eq. (58) for rain scavenging,

$$W_r \cong \frac{1}{\chi_0 J_0} \int_0^{z_m} \frac{c J_0 E_r}{R_m} \chi \, dz' = \frac{c E_r}{\chi_0 R_m} \int_0^{z_m} \chi \, dz' \tag{72}$$

If χ is fairly uniform throughout the rain-scavenged layer, then (Slinn, 1977)

$$W_r \cong \frac{cE_r z_m}{R_m} \tag{73}$$

where $c \cong 0.5$. Notice that W_r does not depend explicitly on rainfall rate, although there is a weak implicit dependence on J_0 through the mean droplet radius R_m, as discussed above. The wet deposition velocity is just

$$v_w \cong \frac{cE_r J_0 z_m}{R_m} \tag{74}$$

If the layer depth $z_m \cong 500$ m and $R_m \cong 0.5$ mm, then $W_r \cong 0.5 \times 10^6 E_r$ and $v_w \cong 0.5 \times 10^6 J_0 E_r$; for a particle–raindrop collision efficiency of 0.2 and a precipitation rate of 1 mm/h, $W_r \cong 1 \times 10^5$ and $v_w \cong 2.8$ cm/s. These seem to be fairly typical values (Slinn et al., 1978); precipitation tends to concentrate particulate pollutants by a factor of more than one hundred thousand.

Scott (1978) has parameterized W_r for sulfate particulate present at cloud height in terms of rainfall rate J_0, sulfate density in hydrometeors at the top of the riming zone $M_s(0)$, and sulfate density just below the cloud base S_0. The result depends on storm type because of their different modes of precipitation formation. Scott's result for the washout ratio is

$$W_r \cong \frac{1.4 \times 10^4 M_s(0)}{S_0 J_0^{0.88}} + \frac{750[1 - 4.41 \times 10^{-2} J_0^{-0.88}]}{1.56 + 0.44 \ln J_0} \tag{75}$$

where

$$M_s(0) = \begin{cases} 0, & \text{cold storms where Bergeron ice growth process initiates precipitation} \\ 0.1 S_0, & \text{warm rain from synoptic-scale storms} \\ \tilde{f}(J_0) S_0, & \text{convective storms} \end{cases} \tag{76}$$

The term $\tilde{f}(J_0)$ is a measure of the importance of below-cloud particle scavenging relative to accretion. Figure 6 shows Scott's estimates together with some observations; for heavy rainfall rates (large J_0), the washout ratios become independent of storm type. Hales (this volume) notes that these curves are generated from some very specific assumptions about the nucleating capabilities of the pollutant aerosol being scavenged. Application to other particles must be made with caution, particularly if the aerosol is hydrophobic or is distributed as very small particles of limited nucleating ability. Competing mechanisms such as chemical reactions with other pollutants may also alter the process.

For polydisperse aerosols below the clouds, the mass flux to the surface is given by $F_W = \int_0^{z_m} \langle \Lambda_3 \rangle \chi \, dz'$, where $\langle \Lambda_3 \rangle$ is given by Eq. (61). The same approximations as above yield the same expressions (Eqs. (73) and (74)) for

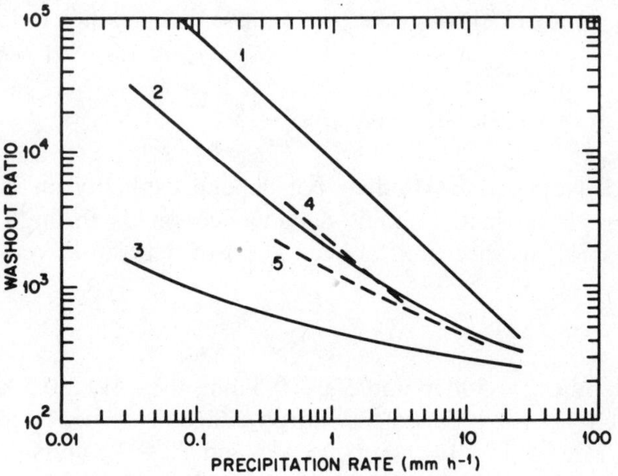

FIGURE 6. Sulfate washout ratio as a function of precipitation rate for three storm types. Curve 1 applies to intense convective storms, curve 2 to warm storms where rain develops without an ice growth state, and curve 3 to cold storms where ice growth initiates precipitation. Curves 4 and 5 are observations of ^{24}Na concentrations in rainwater. (From Scott, 1978, by permission.)

W_r and v_w, with E_r replaced by $\bar{E}_3$, where

$$\bar{E}_3 \equiv \int_0^\infty E_r(a, R_m)a^3 f(a)\,da \Big/ \int_0^\infty a^3 f(a)\,da \tag{77}$$

the particle-averaged third-order collision efficiency. As in the discussion of scavenging rate, information on both the mean particle size and the breadth of the size distribution is required for a correct evaluation of $\bar{E}_3$, W_r, and v_w.

For soluble gases which go to equilibrium quickly, the concentration in the precipitation can be predicted simply by knowing the concentration in the surrounding atmosphere. If Henry's law is obeyed, the concentration within droplets near the ground is just $C_0 = \alpha_S \chi_0$, where $\alpha_S(=1/H)$ is the solubility coefficient as tabulated, for example, by Slinn (1984). For such gases,

$$W_r = \alpha_S \tag{78}$$

and

$$v_w = \alpha_S J_0 \tag{79}$$

As an example (Slinn, 1984), α_s for SO_2 is about 3×10^3; if $J_0 \cong 1$ mm/h, $v_w \cong 0.08$ cm/s. This is much smaller than the typical value for particles, even though SO_2 is a highly soluble gas. Consequently the wet removal of gases is usually negligible compared with dry removal (Slinn et al., 1978). Nevertheless, gases can be very important in rain because they can be converted to less volatile species whose scavenging is important, or they can alter other chemical reactions including the scavenging ability of the precipitation.

For reactive gases, it is necessary to evaluate the possible reactions which may occur in the presence of other dissolved gases and trace substances (Slinn et al., 1978). Often it is possible to define an enhanced solubility coefficient α_{S*} (Postma, 1970; Hales, 1972). Then

$$W_r = \alpha_{S*} \tag{80}$$

and

$$v_w = \alpha_{S*} J_0 \tag{81}$$

Slinn et al. (1978) list α_{S*} for SO_2; the enhanced solubility depends on the atmospheric concentration χ_0 and the solution pH. This is probably true for other gases as well. The main difficulty with this approach is the fact that chemical equilibration for aqueous solutions in contact with gases at low concentrations in air is not always well characterized. Consequently empirical values for W_r should be utilized whenever feasible.

4.4 Experimental Data

Estimation of precipitation scavenging requires knowledge of many variables, such as raindrop size spectra, particle size distribution, gas solubility and reactivity, and precipitation collision retention efficiencies. In view of the liberties taken in deriving expressions such as Eq. (58) or (73), one is probably better off using measured values for scavenging rates or washout ratios, as long as the pollutants and meteorological conditions to be modeled are similar to those of the precursor experiments. Differences may occur depending on the location of the pollutant plume relative to the precipitating clouds.

Here again the review by McMahon and Denison (1979) is especially useful; values of $\Lambda_{r,s}$ and W_r are conveniently tabulated. Work by Davenport and Peters (1978), Engelmann (1968), and McMahon et al. (1976) may also be consulted for values of Λ_r. Articles by Knutson and Stockham (1974) and Summers (1974) may provide Λ_s. Washout ratios have been reported by Engelmann (1968, 1971), Krey and Toonkel (1974), Postma (1970), Slinn (1977, 1978, 1984), and Slinn et al. (1978); the last reference is particularly useful for tabulations of α_{S*}, the enhanced solubility coefficient for reactive gases.

If suitable empirical values for the removal parameters are unavailable, they can be estimated as outlined above. Detailed information on the precipitation and contaminant characteristics will be required. Some discussion of raindrop terminal velocity has already been presented; empirical relationships are often based on the data of Best (1950) or Gunn and Kinzer (1949). Zikmunda (1972) has suggested expressions for v_t for snow. Raindrop size distributions, other than those described above, have been presented by Carbone and Nelson (1978), Dana et al. (1975), Davenport and Peters (1978), Kessler (1969), and Waldvogel (1974), among others. Snowflake distributions have been reported by Gunn and Marshall (1958), Knutson and Stockham (1974), Knutson et al. (1976), and Passarelli (1978). Texts by Friedlander (1977), Fuchs (1964), and Junge (1963) are useful for characterizing particle distributions and behavior. Gas behavior largely depends on the individual chemistry of the situation to be modeled; information on SO_2, NH_4, CO_2, and a few other common gases is readily available in the meteorological literature. For more exotic substances, one must turn to chemistry references.

Slinn (1977) suggested that data are missing or deficient in the following areas: particle/raindrop collection efficiencies for particles in the 0.1–1 μm range; particle/snowflake collection efficiencies for all cases; growth rate of particles and the consequent effect on scavenging; and solubility and reaction rates for gases at low concentrations in polluted rainwater and soil water. Rainout and snowout data are often biased by simultaneous dry deposition, which must be accounted for. Finally, the in-cloud scavenging processes are not well understood; more work is needed on particle and hydrometeor circulations and interactions within storms of all types.

When dealing with reactive gases such as SO_2, a few additional caveats are appropriate. In particular, Hales and Dana (1979) suggest that typical precipitation-chemistry network data should not be utilized unless the particular network has been operated with valid sampling methods—many have not. They emphatically state that standard scavenging rate theory is not applicable to reactive gases such as SO_2; although one can usually fit an empirical scavenging rate coefficient to the data, the theoretical basis for this is incorrect, and the result is characteristic only of the data set from which it was derived. Hill and Adamowicz (1977) noted that calculations based on the concept of irreversible scavenging effectively assume that the initial rate of uptake by precipitation is maintained at all times. In actuality, the chemical process tries to go to equilibrium, so that the uptake drops off with time. Scavenging rate coefficients for reactive gases estimated on the basis of constant uptake may therefore be too large by orders of magnitude. Explicit consideration of the chemistry or use of measured washout ratios are necessary for such cases.

5. MODELING RECOMMENDATIONS AND RESEARCH NEEDS

5.1 Model Selection

The choice of a particular dispersion/deposition model depends on many factors, including the size and topography of the region under study, the length of time over which the pollutant source operates and/or concentration estimates are required, the characteristics of the effluent, and the availability of computer facilities, personnel, and funds for model development and execution. One way to begin the selection process is to determine the range over which calculations must be performed.

5.1.1 Short Range (<10 km)

In most instances the Gaussian plume is the model of choice at these distances. It is easily implemented, and a great deal of effort has been devoted to estimating appropriate dispersion coefficients for use in less than ideal terrain. Dry removal processes can be adequately handled by the partial reflection model (Eq. (8)) or by the modified source depletion method (Eqs. (10) and (11)). Ermak's (1977) model may also be suitable, although its performance does not seem to have been checked against the other models. Wet removal of irreversibly scavenged materials can be handled on an episodic basis by the usual exponential depletion term. For longer-term estimates and/or gas scavenging, the wet deposition velocity concept should probably be applied. The Gaussian transport and diffusion model alone is generally held to be accurate to a factor of 2 on an annual average basis, and only to a factor of 10 for individual case estimates. It is highly unlikely that the addition of more or less uncertain removal parameterizations can in any way enhance the accuracy of the results.

For hilly terrain, even at short distances, the Gaussian model is generally inadequate. The simplest approach is probably a trajectory model which handles dispersion via either a segmented plume or a sequence of puffs. The biggest obstacle to applying this method is inadequate information about the wind field. Egan (1975) offers some guidance for estimating such wind fields. Dry and wet deposition velocities, perhaps following the puff scavenging method of Heffter and Ferber (1977), may be the best way to implement scavenging.

If dispersion and deposition within vegetation must be treated, several models are extant. Cionco's (1971, 1972) ideal canopy is adequate for vegetation such as corn and some grains; Shreffler's (1976, 1978) example may be followed. In nonideal canopies like forests, modeling is still in its infancy. The K-theory model of Murphy et al. (1977) is available, although the entire notion of gradient transport in a forest rests on rather shaky ground. Higher-order-closure models are still under development, but probably represent the best method. A great deal of (probably unavailable) forest structure information as well as details of the biomass receptiveness to pollutant uptake will be required in such models. Deposition velocities can be calculated from the various resistances discussed earlier but are subject to a great deal of uncertainty.

5.1.2 Medium Range (10–50 km)

For simple terrain, the Gaussian plume model is probably still adequate, although there must be additional uncertainty about the dispersion coefficients at these distances. Gillani (1978) suggests that such steady-state models are acceptable, since variability of the planetary boundary layer and the removal processes is generally small. This is probably true in many cases, although the presence of, say, scattered afternoon thundershowers can introduce considerable variability into the wet removal process. Patchy vegetation may influence the dry removal rate. One can operate the models in a quasi-steady mode, where the meteorological inputs are periodically updated to reflect changing conditions. The scavenging methods suggested at shorter ranges should still be appropriate.

In complex terrain, including shoreline sites, mountains, or forested country, a trajectory model with segmented plume or puff dispersion is again a strong candidate. The wind field problem is worse, since synoptic conditions as well as locally dominated flows can contribute. National Weather Service data will generally be almost useless, since surface stations are roughly 200 km apart, while upper air soundings are reported only every 12 hr at a spacing of about 400 km. Separate models to estimate the wind field (Sherman, 1978) or a dense tower network for direct measurement may be needed.

5.1.3 Long Range (50–250 km)

The review of Bass (1980) may be consulted here to select models which are presently available and economically practical at these distances. Short-term-average models generally consist of puff or plume-segment dispersion superposed on trajectories, although a few grid models have been used. Puff models as well as a statistical trajectory model are recommended for long-term estimates. "Particle" models are useful, particularly in complex terrain, but are expensive and time-consuming to prepare and run. Dry and wet deposition velocities should be adjusted with time and position to reflect local conditions. The Heffter–Ferber (1977) technique is very straightforward. Specification of dispersion parameters is difficult at this scale. Nevertheless, models for use in this mesoscale range abound, largely because of the current regulatory interest in fossil fuel pollutant transport and effects over regional areas.

5.1.4 Very Long Range (>250 km)

Trajectory ensembles, box models, grid models with constant eddy diffusivity, and even Gaussian models have been used at these distances. On such scales many details of the transport and diffusion processes should become unimportant, especially if fairly long-term results are sought. The Heffter–Ferber (1977) depletion method seems suitable for both wet and dry removal. Grid or "particle" models may also be useful since deposition velocities can be adjusted from point to point to reflect local circumstances. On these scales the available synoptic wind fields may be adequate for operation of the model.

5.2 Model Weaknesses and Data Requirements

Transport and diffusion models, regardless of scale, usually require more meteorological information than is available. Data on the spatial and temporal variation of the wind field are especially crucial, since errors affect the accuracy of the whole transport simulation. The situation is aggravated in complex terrain where our understanding of flow processes is still quite limited, and our ability to predict local winds accurately is rather poor. At night the problem is compounded by the frequent occurrence of strong wind shear with height, often by up to 180° from the surface wind direction over a height of a few hundred metres or less. Data at plume height rather than at the surface are essential in such conditions. Wind shear can be a problem even by day for distances of more than 5–10 km; the Coriolis-induced turning of the wind vector with height introduces an additional cross-wind dispersion effect that is difficult to incorporate in models.

Diffusion parameterizations generally require information on atmospheric stability, mixing height, and surface characteristics such as roughness and displacement height. Data to adequately evaluate these parameters are hardly ever available, which accounts for the popularity of schemes such as Pasquill–Gifford–Turner (P-G-T) turbulence typing. Climatological estimates of mixing layer depth are available for at least the United States (Holzworth, 1972, 1974), and rules of thumb (e.g., displacement height is about 75% of vegetation height) exist for characterizing the ground cover. None of these methods will be highly accurate for any given site and time, but may be acceptable if only long-term averages are needed.

Dry deposition velocities can be estimated from the resistances described above if experimental data are available. However, many poorly quantified variables are needed for the evaluations. Above the receptor, the aerodynamic resistance to mass transfer requires the friction velocity u_* and the Monin–Obukhov length L. If the surface characteristic lengths d and z_0 can be estimated, u_* can be evaluated from a velocity measurement and the theoretical velocity profile (Dyer, 1974). Evaluation of L generally requires data on heat flux as well as air temperature and u_*. In the absence of this information, Golder's (1972) paper may be used to estimate a range of L values for a given P-G-T stability class and roughness length z_0. Moisture flux effects may be estimated through the Bowen ratio. The problem becomes very complex within canopies. An "ideal canopy" model requires estimates of the canopy attenuation coefficients (Cionco, 1971, 1972), which may or may not be available. Diffusivity within nonideal canopies like forests is still very poorly understood; K-theory models are available if mixing lengths can be specified, but these models are probably inappropriate anyway. Forest parameters such as biomass distribution with height are needed to weight individual receptor resistances. The latter require knowledge of surface boundary layer characteristics and biologically dominated resistances for the particular vegetation type. Stomatal resistance may be crucial, and it depends at least on local temperature, CO_2 concentration, illumination, and plant water stress. Fortunately empirically determined stomatal resistances to water vapor and CO_2 transport are often available (Monteith, 1976) and can be scaled to gaseous pollutants through Eq. (49).

Estimates of wet removal parameters are at least as complicated. The scavenging rate coefficient for irreversibly collected monodisperse particles and gases requires knowledge of the collection/retention efficiency as a function of hydrometeor size and particle size, as well as the raindrop size distribution. Approximate versions require only the mean droplet size R_m, say, and the precipitation rate. Both of these probably vary from storm to storm, and within storms. If the aerosol is polydisperse, the particle size distribution is also required; approximations to it will require at least the geometric mean and standard deviation of the distribution. Scavenging of reactive gases depends on the overall mass transfer velocity κ, which may be hard to evaluate, and the molar concentration within the precipitation [C]; this latter variable is coupled to the airborne concentration $[\chi]$ as calculated from the plume transport and diffusion model. The washout ratio method demands the least: collision/retention efficiency, mean droplet size, rainfall rate, and the depth of the rain-scavenged layer. Information about cloud and plume height and thickness is necessary to estimate the latter. If gases are involved, only the precipitation rate and the solubility (or enhanced solubility) coefficient are needed; the latter probably will depend, however, on raindrop pH and the presence of other (possibly catalyzing) trace contaminants.

At the present state of the art, many of the required terms will be only poorly known. Educated guesses will be necessary in many cases. Modeling accuracy will probably be order-of-magnitude at best. Data of sufficient quality to validate such models are still rather sparse, although recent field work has improved the outlook.

5.3 Research Needs

Research must continue on the development of transport and diffusion models at all ranges, particularly in complex terrain. Methods of determining wind fields from rather limited field data are needed. The parameterization of dispersion, especially in puff models, must be further investigated. As additional data become available, models must be tested and improved. Inexpensive techniques to measure variables such as heat flux, Bowen ratio, and friction velocity would be useful, since this would permit direct calculation of Monin length L. This parameter is useful both in the transport model and in estimating the aerodynamic dry deposition resistance. Further tabulations and improvements of estimates for displacement height and roughness length would be helpful.

Canopy flow models must be further developed, particularly in nonideal cases such as forests. Higher-order-closure turbulence modeling seems appropriate here, although it is presently difficult to deduce reasonable closure parameterizations. Methods of incorporating direct or inferred (from, say, solar radiation penetration data) canopy structure parameters like biomass distribution and aerodynamic drag should be pursued. It may be necessary to include aeroelastic effects. Detailed field studies of in-canopy turbulence are also needed; high-order covariance measurements might be quite useful in evolving proper parameterization and closure schemes. Deposition resistances related to receptor biological mechanisms require further study. In particular, simple parameterizations (e.g., Eq. (47)) of stomatal resistance to various substances for common plant species and micrometeorological conditions would be especially useful to modelers, since this may often be a controlling factor.

Additional detailed pollutant–precipitation interaction studies are needed to verify terms such as the collection/retention efficiencies that have been suggested for rain and snow. The within-cloud scavenging processes need further study. Characterizations of hydrometeor size distributions and their variability for different types of storms should continue. Relations between such distributions and easily measured quantities such as rainfall rate are useful. Tabulations of particle size distributions and better methods to evaluate these in the field are needed. Additional tabulations of gas solubility or enhanced solubility coefficients should be provided; the experiments should be conducted so as to simulate conditions occurring in the atmospheric scavenging process. The effects on these coefficients of other materials in the precipitation should be examined.

ACKNOWLEDGMENTS

This work was accomplished under agreements among the Environmental Protection Agency, Oak Ridge National Laboratory, the National Oceanic and Atmospheric Administration, and the Department of Energy. The author is grateful to B. B. Hicks and G. A. Sehmel for helpful discussions and to M. J. Van Hoesen for assistance in the literature search.

REFERENCES

Adamowicz, R. F. A model for the reversible washout of sulfur dioxide, ammonia, and carbon dioxide from a polluted atmosphere and the production of sulfates in raindrops. *Atmos. Environ.* **13**(1), 105–121 (1979).

Bache, D. H. Particle transport within plant canopies—I. A framework for analysis. *Atmos. Environ.* **13**(9), 1257–1262 (1979).

Barr, S., and W. E. Clements. Diffusion modeling: Principles and applications. In D. Randerson (ed.) *Atmospheric Sciences and Power Production*. Chapter 14. U.S. Dept. of Energy DOE/TIC-27601. Available NTIS DE84005177, Springfield, Virginia (1984).

Barrie, L. A. An improved model of reversible SO_2-washout by rain. *Atmos. Environ.* **12**(1–3), 407–412 (1978).

Bass, A. Modeling long range transport and diffusion. In *Proc. of AMS/APCA 2nd Joint Conf. on Applications of Air Pollution Meteorology, New Orleans, March 24–27.* pp. 193–215. American Meteorological Society, Boston, 1980.

Beard, K. V. Rain scavenging of particles by electrostatic-inertial impaction and Brownian diffusion. In *Precipitation Scavenging (1974)*, pp. 183–194. U.S. ERDA Tech. Info. Ctr., Oak Ridge, Tennessee, 1977; avail. NTIS, Springfield, Virginia, as CONF-741003, 1974.

Beard, K. V., and H. R. Pruppacher. A wind tunnel investigation of the rate of evaporation of small water drops falling at terminal velocity in air. *J. Atmos. Sci.* **28**(8), 1455–1464 (1971).

Bennett, J. H., A. C. Hill, and D. M. Gates. A model for gaseous pollutant sorption by leaves. *J. Air Poll. Cont. Assoc.* **23**(11), 957–962 (1973).

Bergen, J. D. Airflow in the forest canopy—a review of current research on modeling the momentum balance. In *Proc. of 4th Nat. Conf. on Fire and Forest Meteorol., St. Louis, Missouri, Nov. 16–18* (1976).

Best, A. C. Empirical formulae for the terminal velocity of water drops falling through the atmosphere. *Quart. J. Roy. Meteorol. Soc.* **76**(329), 302–311 (1950).

Bird, R. B., W. E. Stewart, and E. N. Lightfoot. *Transport Phenomena.* 780 pp. Wiley, New York, 1960.

Bolin, B., and C. Persson. Regional dispersion and deposition of atmospheric pollutants with particular application to sulfur pollution over Western Europe. *Tellus* **27**(3), 281–310 (1975).

Brutsaert, W. The roughness length for water vapor, sensible heat, and other scalars. *J. Atmos. Sci.* **32**(10), 2028–2031 (1975).

Carbone, R. E., and L. D. Nelson. The evolution of raindrop spectra in warm-based convective storms as observed and numerically modeled. *J. Atmos. Sci.* **35**(12), 2302–2314 (1978).

Chamberlain, A. C. Aspects of travel and deposition of aerosol and vapour clouds. A.E.R.E. report H.P./1261. 32 pp. Atomic Energy Research Estab., Harwell, Berks., England (1953).

Chamberlain, A. C. Transport of gases to and from grass and grass-like surfaces. *Proc. Roy. Soc. Lond. A* **290**, 236–265 (1966).

Chamberlain, A. C. Transport of gases to and from surfaces with bluff and wavelike roughness elements. *Quart. J. Roy. Meteorol. Soc.* **94**(401), 318–332 (1968).

Chamberlain, A. C. Interception and retention of radioactive aerosols by vegetation. *Atmos. Environ.* **4**(1), 57–78 (1970).

Chamberlain, A. C. Mass transfer to bean leaves. *Bound.-Layer Meteorol.* **6**(3–4), 477–486 (1974).

Chamberlain, A. C. The movement of particles in plant communities. In J. L. Monteith (ed.) *Vegetation and the Atmosphere*, Vol. 1, pp. 155–203. Academic, London, 1975.

Chamberlain, A. C. Dry deposition of sulfur dioxide. In D. S. Shriner, C. R. Richmond, and S. E. Lindberg (eds.) *Atmospheric Sulfur Deposition: Environmental Impact and Health Effects*, pp. 185–197. Ann Arbor Sci. Publishers, Ann Arbor, Michigan, 1980.

Church, H. W. The atmospheric dispersion model as used in the reactor safety study, WASH-1400. In *Preprints of 3rd Symp. on Atmospheric Turbulence, Diffusion, and Air Quality, Raleigh, NC, Oct. 19–22.* pp. 188–193. American Meteorological Society, Boston, 1976.

Cionco, R. M. Application of the Ideal Canopy Flow Concept to Natural and Artificial

Roughness Elements. U.S. Army Electronics Command tech. report ECOM-5372, OSD-1366. Avail. NTIS, Springfield, Virginia, AD730-638. 63 pp. (1971).

Cionco, R. M. A wind-profile index for canopy flow. *Bound.-Layer Meteorol.* **3**(2), 255–263 (1972).

Cionco, R. M., W. D. Ohmstede, and J. F. Appleby. A Model for Air Flow in an Idealized Vegetative Canopy. U.S. Army Electronics Research and Development Activity Meteorological Research Note no. 5. Avail. NTIS, Springfield, Virginia, AD 422705 (1963).

Csanady, G. T. Dispersal of dust particles from elevated sources. *Australian J. Phys.* **8**, 545–550 (1955).

Csanady, G. T. Dispersal of dust particles from elevated sources. II. Limitations of the approximate theory. *Australian J. Phys.* **10**, 558–564 (1957).

Csanady, G. T. Deposition of dust particles from industrial stacks. *Australian J. Appl. Sci.* **9**, 1–8 (1958).

Dana, M. T., and J. M. Hales. Statistical aspects of the washout of polydisperse aerosols. *Atmos. Environ.* **10**(1), 45–50 (1976).

Dana, M. T., J. M. Hales, and M. A. Wolf. Rain scavenging of SO_2 and sulfate from power plant plumes. *J. Geophys. Res.* **80**(30), 4119–4129 (1975).

Davenport, H. M., and L. K. Peters. Field studies of atmospheric particulate concentration changes during precipitation. *Atmos. Environ.* **12**(5), 997–1008 (1978).

Davidson, C. I. Deposition of Trace-Metal-Containing Aerosols on Smooth, Flat Surfaces and on Wild Oat Grass (*Avena fatua*). Ph.D. thesis, Calif. Inst. Tech., Univ. Microfilms, Ann Arbor, Michigan, 161 pp. (1977).

Davidson, C. I., and S. K. Friedlander. A filtration model for aerosol dry deposition: Application to trace metal deposition from the atmosphere. *J. Geophys. Res.* **83**(C5), 3243–3252 (1978).

Davies, C. N. *Aerosol Science*, 408 pp. Academic, New York, 1966.

Drake, R. L., D. J. McNaughton, and C. Huang. Available Air Quality Models. *Mathematical Models for Atmospheric Pollutants*, (Appendix D) EPRI EA-1131, project 805. Electric Power Research Institute, Palo Alto, California, 1979.

Draxler, R. R. A Diffusion-Deposition Scheme for Use Within the ARL Trajectory Model. NOAA tech. memo. ERL-ARL-63. NOAA/Air Resources Lab., Silver Spring, Maryland, 19 pp. (1976).

Draxler, R. R. A Mesoscale Transport and Diffusion Model. NOAA Tech. memo. ERL-ARL-64. NOAA/Air Resources Lab., Silver Spring, Maryland, 31 pp. (1977).

Draxler, R. R., and W. P. Elliott. Long-range travel of airborne material subjected to dry deposition. *Atmos. Environ.* **11**(1), 35–40 (1977).

Droppo, J. G. Dry deposition processes on vegetation canopies. In *Atmosphere-Surface Exchange of Particulate and Gaseous Pollutants (1974)*, pp. 104–111. Avail. NTIS, Springfield, Virginia as CONF-740921, 1980.

Droppo, J. G. Experimental techniques for dry deposition measurements. In D. S. Shriner, C. R. Richmond, and S. E. Lindberg (eds.) *Atmospheric Sulfur Deposition: Environmental Impact and Health Effects*, pp. 209–221. Ann Arbor Science Publishers, Ann Arbor, Michigan, 1980.

Dyer, A. J. A review of flux-profile relationships. *Bound.-Layer Meteorol.* **7**(3), 363–372 (1974).

Eckert, E. R. G., and R. M. Drake, Jr. *Heat and Mass Transfer*, 2nd ed., 530 pp. McGraw-Hill, New York, 1959.

Egan, B. A. Turbulent diffusion in complex terrain. In D. A. Haugen (coord.) *Lectures on Air*

Pollution and Environmental Impact Analyses, pp. 112–135. American Meteorological Society, Boston, 1975.

Engelmann, R. J. The calculation of precipitation scavenging. In D. H. Slade (ed.) *Meteorology and Atomic Energy 1968*. pp. 208–221. Avail. NTIS, Springfield, Virginia, as TID-24190 (1968).

Engelmann, R. J. Scavenging prediction using ratios of concentrations in air and precipitation. *J. Appl. Meteorol.* **10**(3), 493–497 (1971).

Ermak, D. L. An analytical model for air pollutant transport and deposition from a point source. *Atmos. Environ.* **11**(3), 231–237 (1977).

Fowler, D. Dry deposition of SO_2 on agricultural crops. *Atmos. Environ.* **12**(1–3), 369–373 (1978).

Friedlander, S. K. *Smoke, Dust, and Haze: Fundamentals of Aerosol Behavior*. Wiley, New York, 1977.

Frössling, N. Uber die Verdunstung fallender Tropfen. *Gerlands Beitraege Geophysik* **52**, 170–216 (1938).

Fuchs, N. A. *Mechanics of Aerosols*, 408 pp. Pergamon Press, Oxford; MacMillan, New York, 1964.

Fuquay, J. J. Environmental safety analysis. In D. H. Slade (ed.) *Meteorology and Atomic Energy 1968*. pp. 379–399. Avail. NTIS, Springfield, Virginia as TID-24190 (1968).

Garland, J. A., and J. R. Branson. The deposition of sulphur dioxide to pine forest assessed by a radioactive tracer method. *Tellus* **29**(5), 445–454 (1977).

Garratt, J. R. Transfer characteristics for a heterogeneous surface of large aerodynamic roughness. *Quart. J. Roy. Meteorol. Soc.* **104**(440), 491–502 (1978).

Garratt, J. R., and B. B. Hicks. Momentum, heat, and water vapour transfer to and from natural and artificial surfaces. *Quart. J. Roy Meteorol. Soc.* **99**(442), 680–687 (1973).

Gash, J. H. C., and J. B. Stewart. The average surface resistance of a pine forest derived from Bowen ratio measurements. *Bound.-Layer Meteorol* **8**(3–4), 453–464 (1975).

Gifford, F. A., Jr. An outline of theories of diffusion in the lower layers of the atmosphere. In D. H. Slade (ed.) *Meteorology and Atomic Energy 1968*, pp. 65–116. Avail. NTIS, Springfield, Virginia as TID-24190 (1968).

Gifford, F. A., Jr. Atmospheric dispersion models for environmental pollution applications. In D. A. Haugen (coord.) *Lectures on Air Pollution and Environmental Impact Analyses*, pp. 35–58. American Meteorological Society, Boston, 1975.

Gifford, F. A., Jr. Turbulent diffusion-typing schemes: a review. *Nucl. Safety* **17**(1), 68–86 (1976).

Gillani, N. V. Project MISTT: Mesoscale plume modeling of the dispersion, transformation, and ground removal of SO_2. *Atmos. Environ.* **12**(1–3), 569–588 (1978).

Golder, D. Relations among stability parameters in the surface layer. *Bound.-Layer Meteorol* **3**(1), 47–58 (1972).

Greene, D. W., and M. J. Bukovac. Stomatal pentration: Effect of surfactants and role in foliar absorption. *Am. J. Bot.* **61**(1), 100–106 (1974).

Gunn, K. L. S., and J. S. Marshall. The distribution with size of aggregate snowflakes. *J. Meteorol.* **15**, 452–461 (1958).

Gunn, R., and G. D. Kinzer. The terminal velocity of fall for water droplets in stagnant air. *J. Meteorol.* **6**, 243–248 (1949).

Hales, J. M. Fundamentals of the theory of gas scavenging by rain. *Atmos. Environ.* **6**(9), 635–659 (1972).

Hales, J. M. Atmospheric transformations of pollutants. In D. A. Haugen (coord.) *Lectures on*

Air Pollution and Environmental Impact Analsyses, pp. 228–242. American Meteorological Society, Boston, 1975.

Hales, J. M. Wet removal of sulfur compounds from the atmosphere. *Atmos. Environ.* **12**(1–3), 389–399 (1978).

Hales, J. M., and M. T. Dana. Regional-scale deposition of sulfur dioxide by precipitation scavenging. *Atmos. Environ.* **13**(8), 1121–1132 (1979).

Hales, J. M., M. A. Wolf, and M. T. Dana. A linear model for predicting the washout of pollutant gases from industrial plumes. *A.I.Ch.E. J.* **19**(2), 292–297 (1973).

Hane, C. E. Precipitation scavenging in a squall line: Numerical experimentation. In R. G. Semonin and R. W. Beadle (coord.) *Precipitation Scavenging (1974)*, Avail. NTIS, Springfield, Virginia as CONF-741003. pp. 759–768 (1974).

Hane, C. E. Scavenging of urban pollutants by thunderstorm rainfall: numerical experimentation. *J. Appl. Meteorol.* **17**(5), 699–710 (1978).

Hanna, S. R., G. A. Briggs, J. Deardorff, B. A. Egan, F. A. Gifford, and F. Pasquill. AMS workshop on stability classification schemes and sigma curves—summary of recommendations. *Bull. A.M.S.* **58**(12), 1305–1309 (1977).

Heffter, J. L., and G. J. Ferber. Development and verification of the ARL regional-continental transport and dispersion model. In *Preprints of AMS/APCA Joint Conf. on Applications of Air Pollution Meteorology, Salt Lake City, Nov. 29–Dec. 2*, pp. 400–407. American Meteorological Society, Boston, 1977.

Heffter, J. L., A. D. Taylor, and G. J. Ferber. A Regional-Continental Scale Transport, Diffusion, and Deposition Model. NOAA tech. memo. ERL-ARL-50 NOAA/Air Resources Lab., Silver Spring, Maryland, 28 pp. (1975).

Hicks, B. B. Some micrometeorological aspects of pollutant deposition rates near the surface. In R. J. Engelmann and G. A. Sehmel (coord.) *Atmosphere–Surface Exchange of Particulate and Gaseous Pollutants (1974)*, pp. 434–449. Avail. NTIS, Springfield, Virginia as CONF-740921 (1974).

Hicks, B. B. A procedure for the formulation of bulk transfer coefficients over water. *Bound.-Layer Meteorol.* **8**(3–4), 515–524 (1975).

Hicks, B. B. and M. L. Wesely. An Examination of Some Micrometeorological Methods for Measuring Dry Deposition. U.S. EPA report EPA-600/7-78-116. U.S. EPA, Research Triangle Park, North Carolina, 19 pp. (1978).

Hidy, G. M. Theory of diffusive and impactive scavenging. In R. J. Engelmann and W. G. N. Slinn (coord.). *Precipitation Scavenging (1970)*, pp. 355–371. Avail. NTIS, Springfield, Virginia as CONF-700601 (1970).

Hill, F. B., and R. F. Adamowicz. A model for rain composition and the washout of sulfur dioxide. *Atmos. Environ.* **11**(10), 917–927 (1977).

Hoffman, F. O. A reassessment of the deposition velocity in the prediction of the environmental transport of radioiodine from air to milk. *Health Phys.* **32** (May), 437–441 (1977).

Hoffman, F. O., D. L. Shaeffer, C. W. Miller, and C. T. Garten, Jr. (eds.) *Proc. of Workshop on the Evaluation of Models Used for the Environmental Assessment of Radionuclide Releases, Gatlinburg, Tennessee, Sept. 6–9, 1977.* Oak Ridge National Lab., Oak Ridge, Tennessee. Avail. NTIS, Springfield, Virginia as CONF-770901. 131 pp. (1977).

Holloway, P. J. The chemical and physical characteristics of leaf surfaces. In T. F. Preece and C. H. Dickinson (eds.) *Ecology of Leaf Surface Microorganisms*. pp. 39–53. Academic, New York, 1971.

Holzworth, G. C. Mixing Heights, Wind Speeds, and Potential for Urban Air Pollution Throughout the Contiguous United States. U.S. EPA publication AP-101. U.S. EPA, Research Triangle Park, North Carolina. 118 pp. (1972).

Holzworth, G. C. Climatological data on atmospheric stability in the United States. Presented at *Symp. on Atmospheric Diffusion and Air Pollution, Santa Barbara, California, Sept. 9–13*. Avail. U.S. EPA, Research Triangle Park, North Carolina (1974).

Horst, T. W. A surface depletion model for deposition from a Gaussian plume. In R. J. Engelmann and G. A. Sehmel (coord.) *Atmosphere–Surface Exhange of Particulate and Gaseous Pollutants (1974)*, pp. 423–433. Avail. NTIS, Springfield, Virginia as CONF-740921 (1974).

Horst, T. W. A surface depletion model for deposition from a Gaussian plume. *Atmos. Environ.* **11**(1), 41–46 (1977).

Horst, T. W. A review of Gaussian diffusion-deposition models. In D. S. Shriner, C. R. Richmond, and S. E. Lindberg (eds.) *Atmospheric Sulfur Deposition: Environmental Impact and Health Effects*. pp. 275–283. Ann Arbor Science Publishers, Ann Arbor, Michigan, 1980.

Hosker, R. P., and S. E. Lindberg, Review: atmospheric deposition and plant assimilation of gases and particles. *Atmos. Environ.* **16**(5), 899–910 (1982).

Hutchison, B. A., and D. R. Matt. The distribution of solar radiation within a deciduous forest. *Ecolog. Monographs* **47**(2), 185–207 (1977).

Izrael, Yu. A. and V. N. Petrov. Spread of underground-nuclear-explosion radioactive products over long distances. In I. D. Morokhov (ed.) *Nuclear Explosions for Peaceful Purposes*, transl. by R. Addis, pp. 130–141. Atomizdat, Moscow, 1970. (Cited by Knox et al., 1971).

Johnson, W. B., R. C. Sklarew, and D. B. Turner. Urban air quality modeling. In A. C. Stern (ed.) *Air Pollution, Vol. I, Air Pollutants, Their Transformation and Transport*, 3rd ed. pp. 504–562. Academic, New York, 1976a.

Johnson, W. B., D. E. Wolf, and R. L. Mancuso. The European regional model of air pollution (EURMAP) and its application to transfrontier air pollution. In *Proc. of 7th Internal. Tech. Meeting on Air Poll. Modeling and its Application, Airlie, Virginia, Sept. 7–10*, pp. 292–371. Avail. U.S. EPA Tech. Info. Staff, Cincinnati, Ohio (1976b).

Johnson, W. B., D. E. Wolf, and R. L. Mancuso. Seasonal and annual patterns and international exchanges of SO_2 and SO_4^{2-} concentrations and deposition in Europe as simulated by the EURMAP model. In *Preprints of AMS/APCA Joint Conf. on Applications of Air Pollution Meteorology, Salt Lake City, Nov. 29–Dec. 2*, pp. 52–63. American Meteorological Society, Boston, 1977.

Junge, C. E. *Air Chemistry and Radioactivity*, 382 pp. Academic, New York, 1963.

Kessler, E. On the distribution and continuity of water substance in atmospheric circulations. *Meteorol. Monographs*, Vol. 10 (32), 84 pp. American Meteorological Society, Boston, 1969.

Knox, J. B. Numerical modeling of the transport, diffusion, and deposition of pollutants for regions and extended scales. *J. Air Poll. Cont. Assoc.* **24**(7), 660–664 (1974).

Knox, J. B., T. V. Crawford, K. R. Peterson, and W. K. Crandall. Comparison of U.S. and U.S.S.R. Methods of Calculating the Transport, Diffusion, and Deposition of Radioactivity. Lawrence Radiation Laboratory report UCRL-51054. Avail. NTIS, Springfield, VA, TID-4500, UC-35. 50 pp. (1971).

Knutson, E. O., and J. D. Stockham. Aerosol scavenging by snow: comparison of single-flake and entire-snowfall results. In R. G. Semonin and R. W. Beadle (coord.) *Precipitation Scavenging (1974)*, pp. 195–207. Avail. NTIS, Springfield, Virginia as CONF-741003 (1974).

Knutson, E. O., S. K. Sood, and J. D. Stockham. Aerosol collection by snow and ice crystals. *Atmos. Environ.* **10**(5), 395–402 (1976).

Krey, P. W., and L. E. Toonkel. Scavenging ratios. In R. G. Semonin and R. W. Beadle (coord.) *Precipitation Scavenging (1974)*, pp. 61–70. Avail. NTIS, Springfield, Virginia as CONF-741003 (1974).

Landsberg, J. J., and D. B. B. Powell. Surface exchange characteristics of leaves subject to mutual interference. *Agricult. Meteorol.* **12**(2), 169–184 (1973).

Lange, R. ADPIC—a three-dimensional particle-in-cell model for the dispersal of atmospheric pollutants and its comparison to regional tracer studies. *J. Appl. Meteorol.* **17**(3), 320–329 (1978).

Lange, R., and J. B. Knox. Adaptation of a three-dimensional atmospheric transport-diffusion model to rainout assessments. In R. G. Semonin and R. W. Beadle (coord.) *Precipitation Scavenging (1974)*, pp. 732–758. Avail. NTIS, Springfield, Virginia as CONF-741003 (1974).

Lange, R., M. A. Dickerson, K. R. Peterson, C. A. Sherman, and T. J. Sullivan. Particle-In-Cell vs. Straight-line Airflow Gaussian Calculations of Concentration and Deposition of Airborne Emissions out to 70 km for Two Sites of Differing Meteorological and Topographical Character. Lawrence Livermore Laboratory Report UCRL-52133. Avail. NTIS, Springfield, Virginia, 67 pp. (1976).

Lister, R., and E. Lemon. Interactions of atmospheric carbon dioxide, diffuse light, plant productivity and climate processes—model predictions. In R. J. Engelmann and G. A. Sehmel (coord.) *Atmosphere–Surface Exchange of Particulate and Gaseous Pollutants (1974)*, pp. 112–135. Avail. NTIS, Springfield , Virginia as CONF-740921 (1974).

McClure, V. E. Transport of heavy chlorinated hydrocarbons in the atmosphere. *Environ. Sci. and Tech.* **10**(13), 1223–1229 (1976).

McMahon, T. A., and P. J. Denison. Empirical atmospheric deposition parameters—a survey. *Atmos. Environ.* **13**(5), 571–585 (1979).

McMahon, T. A., P. J. Denison, and R. Fleming. A long-distance air pollution transportation model incorporating washout and dry deposition components. *Atmos. Environ.* **10**(9), 751–761 (1976).

Markee, E. J., Jr. A parametric study of gaseous plume depletion by ground surface adsorption. In C. A. Mawson (ed.) *Proc. of U.S. AEC Meteorol. Info. Meeting, Chalk River Nucl. Labs., Sept. 11–14*, pp. 601–614. Atomic Energy of Canada, Ltd., Chalk River, Ontario, Canada, Report no. AECL-2787 (1967).

Markee, E. J., Jr., L. Andrews, and R. Jubach. On the behavior of dry gaseous effluent deposition and attendant plume depletion. In *Preprints of AMS/APCA Joint Conf. on Applications of Air Poll. Meteorol., Salt Lake City, Nov. 29–Dec. 2*, pp. 408–413. American Meteorological Society, Boston, 1977).

Marshall, J. S., and W. McK. Palmer. The distribution of raindrops with size. *J. Meteorol.* **5**, 165–166 (1948).

Mason, B. J. *The Physics of Clouds*, 2nd ed., 671 pp. Oxford University Press, Clarendon, 1971.

Meidner, H., and T. A. Mansfield. *Physiology of Stomata*, 178 pp. McGraw-Hill, New York, 1968.

Menzel, R. G. Airborne radionuclides and plants. In N. C. Brady (ed.) *Agriculture and the Quality of Our Environment*, pp. 57–75. (AAAS Pub. No. 85 (1967)). American Association for the Advancement of Science, Washington, D.C., 1967.

Miller, C. W., and F. O. Hoffman. An analysis of NRC methods for estimating the effects of dry deposition in environmental radiological assessments. *Nuc. Safety* **20**(4), 458–467 (1979).

Miller, C. W., F. O. Hoffman, and D. L. Shaeffer. The importance of variations in the deposition velocity assumed for the assessment of airborne radionuclide releases. *Health Phys.* **34**(June), 730–734 (1978).

Milne, J. W., D. B. Roberts, and D. J. Williams. The dry deposition of sulphur dioxide—field measurements with a stirred chamber. *Atmos. Environ.* **13**(3), 373–379 (1979).

Molenkamp, C. R. Numerical modeling of precipitation scavenging by convective clouds. In R. G. Semonin and R. W. Beadle (coord.) *Precipitation Scavenging (1974)*, pp. 769–793. Avail. NTIS, Springfield, Virginia as CONF-741003 (1974).

Monin, A. S., and A. M. Yaglom. *Statistical Fluid Mechanics*, vols. 1 and 2, 769 and 874 pp. J. L. Lumley (ed.). MIT Press, Cambridge, Massachusetts, 1971, 1975.

Monteith, J. L. *Principles of Environmental Physics*, 241 pp. Elsevier, New York, 1973.

Monteith, J. L. (ed.). *Vegetation and the Atmosphere, Vol. 2, Case Studies*, 439 pp. Academic, London, 1976.

Murphy, B. D. Deposition of SO_2 on ground cover. In *Preprints of 3rd Symp. on Atmos. Turb., Diff., and Air Qual., Raleigh, North Carolina, Oct. 19–22.* pp. 500–505. American Meteorological Society, Boston, 1976.

Murphy, C. E., Jr., and K. R. Knoerr. Simultaneous determinations of the sensible and latent heat transfer coefficients for tree leaves. *Bound.-Layer Meteorol.* **11**(2), 223–241 (1977).

Murphy, C. E., Jr., T. R. Sinclair, and K. R. Knoerr. An assessment of the use of forests as sinks for the removal of atmospheric sulfur dioxide. *J. Environ. Quality* **6**(4), 388–396 (1977).

Nieuwstadt, F. T. M., C. M. Verheul, and J. Addicks. The validation of the Gaussian dispersion model for long-term average ground level concentration in the Rijnmond region. In *Proc. of 7th Internat. Tech. Meeting on Air Poll. Modeling and its Application, Airlie, Virginia, Sept. 7–10*, pp. 169–218. Avail. U.S. EPA Tech. Info. Staff, Cincinnati, Ohio (1976).

Norman, J. M. Sunfleck-Size and Light-Intensity Distributions in Plant Canopies. Ph.D. Thesis, Univ. of Wisconsin. Univ. Microfilms, Ann Arbor, Michigan, 1971.

Oliver, H. R. Ventilation in a forest. *Agricult. Meteorol.* **14**(3), 347–355 (1975).

Overcamp, T. J. A general Gaussian diffusion-deposition model for elevated point sources. *J. Appl. Meteorol.* **15**(11), 1167–1171 (1976).

Overton, J. H., Jr., V. P. Aneja, and J. L. Durham. Production of sulfate in rain and raindrops in polluted atmospheres. *Atmos. Environ.* **13**(3), 355–367 (1979).

Owen, P. R., and W. R. Thompson. Heat transfer across rough surfaces. *J. Fluid Mech.* **15**(part 3), 321–334 (1963).

Parkhurst, D. F., and G. I. Pearman. Convective heat transfer from a semi-infinite flat plate to periodic flow at various angles of incidence. *Agricult. Meteorol.* **13**(3), 383–393 (1974).

Pasquill, F. The dispersion of material in the atmospheric boundary layer—the basis for generalization. In D. A. Haugen (coord.) *Lectures on Air Pollution and Environmental Impact Analysis*, pp. 1–34. American Meteorological Society, Boston, 1975.

Passarelli, R. E., Jr. Theoretical and observational study of snowsize spectra and snowflake aggregation efficiencies. *J. Atmos. Sci.* **35**(5), 882–889 (1978).

Pearman, G. I., H. L. Weaver, and C. B. Tanner. Boundary layer heat transfer coefficients under field conditions. *Agricult. Meteorol.* **10**(1–2), 83–92 (1972).

Peterson, K. R., and T. V. Crawford. Precipitation scavenging in a large-cloud diffusion code. In R. J. Engelmann and W. G. N. Slinn (coord.) *Precipitation Scavenging (1970)*, pp. 425–431. Avail. NTIS, Springfield, Virginia as CONF-700601 (1970).

Pilat, M. J., and A. Prem. Calculated particle collection efficiencies of single droplets including inertial impaction, Brownian diffusion, diffusiophoresis, and thermophoresis. *Atmos. Environ.* **10**(1), 13–19 (1976).

Postma, A. K. Effect of solubilities of gases on their scavenging by raindrops. In R. J. Engelmann and W. G. N. Slinn (coord.) *Precipitation Scavenging (1970)*, pp. 247–259. Avail. NTIS, Springfield, Virginia as CONF-700601 (1970).

Rao, K. S., I. Thomson, and B. A. Egan. Regional transport model of atmospheric sulfates. Presented at 69th Annual Meeting of Air Poll. Cont. Assoc., Portland, Oregon, June 27–July 1. Avail. NOAA/ARDL, Oak Ridge, Tennessee ATDL Contrib. File No. 76/13. 9 pp. (1976).

Redkin, Yu. N. Washout of atmospheric aerosols by snow in the surface layer. In V. M. Voloshchuk and Yu. S. Sedunov (eds.) *Hydrodynamics and Thermodynamics of Aerosols*. pp. 226–238. Wiley, New York; and Israel Prog. Scientific Transl., Jerusalem, 1973.

Roberts, O. F. T. The theoretical scattering of smoke in a turbulent atmosphere. *Proc. Roy. Soc. Lond. A.* **104**, 640–654 (1923).

Rodhe, H., and J. Grandell. On the removal time of aerosol particles from the atmosphere by precipitation scavenging. *Tellus* **24**(5), 442–454 (1972).

Rutter, A. J. The hydrological cycle in vegetation. In J. L. Monteith (ed.) *Vegetation and the Atmosphere, Vol. 1*, pp. 111–154. Academic, London, 1975.

Schwela, D. H. An estimate of deposition velocities of several air pollutants on grass. *Ecotoxicology and Environ. Safety* **3**(2), 174–189 (1979).

Scott, B. C. Parameterization of sulfate removal by precipitation. *J. Appl. Meteorol.* **17**(9), 1375–1389 (1978).

Scriven, R. A., and B. E. A. Fisher. The long range transport of airborne material and its removal by deposition and washout—I. General considerations. *Atmos. Environ.* **9**(1), 49–58 (1975).

Sehmel, G. A. Particle and gas dry deposition: A review. *Atmos. Environ.* **14**(9), 983–1011 (1980).

Sehmel, G. A., and W. H. Hodgson. Predicted dry deposition velocities. In R. J. Engelmann and G. A. Sehmel (coord.). *Atmosphere–Surface Exchange of Particulate and Gaseous Pollutants (1974)*, pp. 399–422. Avail. NTIS, Springfield, Virginia as CONF-740921 (1974).

Sehmel, G. A., and W. H. Hodgson. A Model for Predicting Dry Deposition of Particles and Gases to Environmental Surfaces. Battelle-Pacific Northwest Labs, report PNL-SA-6721. Avail. PNL, Richland, Washington, 42 pp. (1978).

Sekhon, R. S., and R. D. Srivastava. Doppler radar observations of drop-size distributions in a thunderstorm. *J. Atmos. Sci.* **28**(6), 983–994 (1971).

Shawcroft, R. W., E. R. Lemon, and D. W. Stewart. Estimation of internal crop water status from meteorological and plant parameters. In *Plant Response to Climatic Factors*, pp. 449–459. UNESCO, Paris, (1973).

Sheih, C. M. Application of a statistical trajectory model to the simulation of sulfur pollution over northwestern United States. *Atmos. Environ.* **11**(2), 173–178 (1977).

Shepherd, J. C. Measurements of the direct deposition of sulphur dioxide onto grass and water by the profile method. *Atmos. Environ.* **8**(1), 69–74 (1974).

Sheppard, P. A. Transfer across the Earth's surface and through the air above. *Quart. J. Roy. Meteorol. Soc.* **84**, 205–224 (1958).

Sherman, C. A. A mass-consistent model for wind fields over complex terrain. *J. Appl. Meteorol.* **17**(3), 312–319 (1978).

Shreffler, J. H. A model for the transfer of gaseous pollutants to a vegetational surface. *J. Appl. Meteorol.* **15**(3), 744–746 (1976).

Shreffler, J. H. Factors affecting dry deposition of SO_2 on forests and grasslands. *Atmos. Environ.* **12**(6–7), 1497–1503 (1978).

Shuttleworth, W. J. A one-dimensional theoretical description of the vegetation–atmosphere interaction. *Bound.-Layer Meteorol.* **10**(3), 273–302 (1976).

Shuttleworth, W. J. A simplified one-dimensional theoretical description of the vegetation–atmosphere interaction. *Bound.-Layer Meteorol.* **14**(1), 3–27 (1978).

Shuttleworth, W. J. Below-canopy fluxes in a simplified one-dimensional theoretical description of the vegetation–atmosphere interaction. *Bound.-Layer Meteorol.* **17**(3), 315–331 (1979).

Slinn, W. G. N. The redistribution of a gas plume caused by reversible washout. *Atmos. Environ.* **8**(3), 233–239 (1974a).

Slinn, W. G. N. Proposed terminology for precipitation scavenging. In R. G. Semonin and R. W. Beadle (coord.) *Precipitation Scavenging (1974)*, pp. 813–818. Avail. NTIS, Springfield, Virginia as CONF-741003 (1974b).

Slinn, W. G. N. Dry deposition and resuspension of aerosol particles—a new look at some old problems. In R. J. Engelmann and G. A. Sehmel (coord.) *Atmosphere–Surface Exchange of Particulate and Gaseous Pollutants (1974)*, pp. 1–40. Avail. NTIS, Springfield, Virginia as CONF-740921 (1974c).

Slinn, W. G. N. Some approximations for the wet and dry removal of particles and gases from the atmosphere. *Water, Air, and Soil Poll.* **7**(4), 513–543 (1977).

Slinn, W. G. N. Parameterizations for resuspension and for wet and dry deposition of particles and gases for use in radiation dose calculations. *Nucl. Safety* **19**(2), 205–219 (1978).

Slinn, W. G. N. Precipitation scavenging. In D. Randerson (ed.) *Atmospheric Sciences and Power Production*, Chapter 11, pp. 466–532. U.S. Dept. of Energy, DOE/TIC-27601. Available NTIS DE 84005177, Springfield, Virginia (1984).

Slinn, W. G. N., L. Hasse, B. B. Hicks, A. W. Hogan, D. Lal, P. S. Liss, K. O. Munnich, G. A. Sehmel, and O. Vittori. Some aspects of the transfer of atmospheric trace constituents past the air–sea interface. *Atmos. Environ.* **12**(11), 2055–2087 (1978).

Smagorinsky, J. Global atmospheric modeling and the numerical simulation of climate. In W. N. Hess (ed.) *Weather and Climate Modifications*, pp. 633–686. Wiley, New York, 1974.

Smagorinsky, J. Modeling and predictability. In *Energy and Climate*, pp. 133–139. National Academy of Sciences, Washington, D.C., 1977.

Smith, F. B., and G. H. Jeffrey. Airborne transport of sulphur dioxide from the U.K. *Atmos. Environ.* **9**(6–7), 643–659 (1975).

Snyder, W. H. Similarity criteria for the application of fluid models to the study of air pollution meteorology. *Bound.-Layer Meteorol.* **3**(1), 113–134 (1972).

Summers, P. W. Note on SO_2 scavenging in relation to precipitation type. In R. G. Semonin and R. W. Beadle (coord.) *Precipitation Scavenging (1974)*, pp. 88–94. Avail. NTIS, Springfield, Virginia as CONF-741003 (1974).

Tan, C. S., and T. A. Black. Factors affecting the canopy resistance of a Douglas-fir forest. *Bound.-Layer Meteorol.* **10**(4), 475–488 (1976).

Taylor, G. I. Diffusion by continuous movements. *Proc. Lond. Math. Soc.* **20**, 196–212 (1921).

Thom, A. S. The exchange of momentum, mass, and heat between an artificial leaf and the airflow in a wind-tunnel. *Quart. J. Roy. Meteorol. Soc.* **94**(399), 44–55 (1968).

Thom, A. S. Momentum absorption by vegetation. *Quart. J. Roy. Meteorol. Soc.* **97**(414), 414–428 (1971).

Thom, A. S. Momentum, mass, and heat exchange of vegetation. *Quart. J. Roy. Meteorol. Soc.* **98**(415), 124–134 (1972).

Thom, A. S. Momentum, mass, and heat exchange of plant communities. In J. L. Monteith (ed.) *Vegetation and the Atmosphere, Vol. 1*, pp. 57–109. Academic, London, 1975.

Thom, A. S., J. B. Stewart, H. R. Oliver, and J. H. C. Gash. Comparison of aerodynamic and energy budget estimates of fluxes over a pine forest. *Quart. J. Roy. Meteorol. Soc.* **101**(427), 93–105 (1975).

U.S. EPA. Guideline on Air Quality Models, OAQPS Guideline Series no. 1.2-080, EPA report no. EPA-450/2-78-027. EPA Office of Air Qual. Plan and Std., Research Triangle Park, North Carolina, 84 pp. (1978).

U.S. NRC. Calculation of Reactor Accident Consequences. Appendix VI to Reactor Safety Study. U.S. NRC report no. WASH-1400, NUREG-75/014. Avail. NTIS, Springfield, Virginia (1975).

U.S. NRC. Methods for Estimating Atmospheric Transport and Dispersion of Gaseous Effluents in Routine Releases from Light-Water-Cooled Reactors. U.S. NRC Regulatory Guide 1.111, Revision 1 (July). NRC Div. Document Cont., Washington, D.C. (1977).

Van der Hoven, I. Deposition of particles and gases. In D. H. Slade (ed.) *Meteorology and Atomic Energy 1968*, pp. 202–208. U.S. AEC Div. Tech. Infor., Oak Ridge, TN. Avail. NTIS as TID-24190 (1968).

Van Hook, R. I., and W. D. Shults. Effects of Trace Contaminants From Coal Combustion, *Proc. of Workshop, Knoxville, Tennessee, Aug. 2–6*. U.S. ERDA report 77-64, Avail. NTIS, Springfield, Virginia, 79 pp. (1976).

Waggoner, P. E., and W. E. Reifsnyder. Simulation of the temperature, humidity, and evaporation profiles in a leaf canopy. *J. Appl. Meteorol.* **7**(3), 400–409 (1968).

Waldvogel, A. The N_0 jump of raindrop spectra. *J. Atmos. Sci.* **31**(4), 1067–1078 (1974).

Wedding, J. B., R. W. Carlson, J. J. Stukel, and F. A. Bazzaz. Aerosol deposition on plant leaves. *Environ. Sci. and Tech.* **9**(2), 151–153 (1975).

Wendell, L. L., D. C. Powell, and R. L. Drake. A regional scale model for computing deposition and ground level air concentration of SO_2 and sulfates from elevated and ground sources. pp. 318–324. In *Preprints of 3rd Symp. on Atmos. Turb., Diff., and Air Qual., Raleigh, North Carolina, Oct. 19–22.* American Meteorological Society, Boston, 1976.

Wesely, M. L., and B. B. Hicks. Some factors that affect the deposition rates of sulfur dioxide and similar gases on vegetation. *J. Air Poll. Cont. Assoc.* **27**(11), 1110–1116 (1977).

Whitby, K. T. Personal Communication at *International Conference on the Terrestrial Ecosystem, Banff, Alberta, Canada, May 10–17*, 1980.

Wilson, N. R., and R. H. Shaw. A higher order closure model for canopy flow. *J. Appl. Meteorol.* **16**(11), 1197–1205 (1977).

Zikmunda, J. Fall velocities of spatial crystals and aggregates. *J. Atmos. Sci.* **29**(8), 1511–1515 (1972).

SYNTHESIS

DATA ACQUISITION, INTERPRETATION, AND APPLICATION IN VEGETATION–AIR POLLUTANT INTERACTION STUDIES

J. Shinn
Lawrence Livermore National Laboratory
Livermore, California

M. Unsworth
University of Nottingham
Sutton Bonington, Loughborough
England

BACKGROUND AND DISCUSSION

Through the process of atmospheric deposition of gaseous and particulate pollutants, the relationship between a given ambient air concen-

tration and a particular cellular receptor becomes exceedingly obscure and indirect. For example, it will be argued that for a complex, natural forest ecosystem the wet and dry deposition processes require greater theoretical and experimental research so that a rational strategy can be developed to protect the forests from deterioration. One may therefore infer that the present body of literature on the criteria for air quality regulation, which relates observed effects to controlled levels of air pollutant exposures of single plants, may be too simple a representation for a complex forest community. Evidence for this has become apparent in the concern for acidic rain effects, where the scientific approach has shown that the problem in forests may be assessed only with proper consideration for the cation–anion fluxes and balances and a host of other considerations that are greatly complicated by the soil–plant continuum of surface and solution chemistries. These factors are intimately related to the atmospheric wet and dry deposition processes.

We shall review the theoretical and experimental quantification of the important factors relating to wet and dry deposition and, as the title suggests, the principal questions arise when it becomes necessary to acquire data, to interpret data, and to apply the data in a predictive, general manner. We shall assess the current limitations to measurement, collection, and interpretation of data, first by reviewing the theoretical basis of pollutant transfer and deposition to show where weaknesses lie and data are required. The term model in this context represents a mathematical statement of physical fact based on first principles and does not become predictive without certain key data and parameterizations. Second, we shall review the measurement problems that have been discovered in recent field studies and we shall identify several research goals to improve both the measurement technique and interpretation in the context of complicated chemistry and physics. Also in the latter review we shall discuss some case studies where applications have identified research needs. In particular, the resolution of the problem of predicting chronic, low-level, air pollution stress in forest ecosystems has been attempted by combining mechanistic and ecological concepts in predictive succession models spanning 500 years. Such attempts will be more realistic with better resolution of the deposition and vegetation interactions. The application of improved deposition theory and experimental methods, and their inclusion in succession models, offer one approach to a rational air pollution economic assessment and development of new air pollution control strategies.

A. On Application of Air Pollutant Deposition Models

There are no easy means of reviewing the principles and requirements of air pollution deposition models, which ". . . cross the boundaries of many disciplines." In this review many atmospheric transport and dis-

persion models are listed, but with some simplifications: atmospheric chemical transformations are treated lightly, the soil as an in-canopy sink is only mentioned, other in-canopy transfers are omitted, such as the components of wet transfer (throughfall, interception, stemflow, and the like) and resuspension of dry particles, and the canopy stomatal diffusion resistance is expressed only as a function of light intensity (it depends on other environmental and biologic variables).

Assuming that one has at least an idea of how atmospheric transport models are applied, it is possible to discuss the research needs without considerable reference to equations. (Earlier in the conference, Dr. Warren Johnson showed how the Stanford Research Institute model, EURMAP, is used to predict ambient concentrations crossing national boundaries of Eastern Europe and how interpreted regional meteorological and air quality monitoring data are included.)

Although the various features and requirements of the important atmospheric transport and diffusion models are reviewed, by restricting the discussion to Gaussian-type models nearly all the features of deposition removal may be critically examined. Dry deposition is incorporated into the classical Gaussian solution of the diffusion equation to account for loss of pollutant material (plume depletion) as the plume disperses downwind.

The concept of plume depletion was a correction for source strength in early models, which worked well at long distances, and later was included as a term for partial reflection from the surface, which works better at short distances. Computationally, more complex models consider sinks (negative sources) for pollutant materials distributed over the ground surface. Two points must be pondered:

1. *As development of transport codes evolved, the deposition process was perceived as a correction to make the calculated concentrations more realistic and not as a means of calculating pollutant flux to vegetation and soil.*
2. *The models were constructed on the basis of their physical plausibility more than by comparison with real deposition measurements, which were largely unavailable. A key point for improvement for air pollution effects application therefore is to make the deposition terms more realistic by including parameterization to functional forms of the deposition velocity, which are derived from observations and include biological as well as environmental variables of parameterization.*

There are other technical difficulties related to dry deposition calculation in models that compute transport from the source. Near the pollutant source, model resolution and meaningful physical definitions are limited by the scale of the grid size chosen for numerical solution:

thus the regions likely to have severe impact by deposition of trace elements from smelters and combustion plants will be poorly predicted. Far from the pollutant source, there doesn't seem to be enough data available to select any of the particular models over the others. Realism in the model predictions must be improved by representing dry depletion occurring from ground level upward and altering the vertical distributions of material as a function of distance from the source. Very little of such data are available.

Wet deposition predictions are most successful when considering a particular episode, such as a given rainfall event where drop size, fall rates, and ambient gas or particle concentrations are known. One reviewed approach, the scavenging rate, is appropriate for irreversible processes only, where the pollutant will undergo capture by falling drops and not be desorbed. Collection efficiencies, however, are difficult to determine, especially for airborne particles, resulting in, at best, an order of magnitude estimate of wet deposition. An alternative method, the washout ratio, is advantageous for a long-term estimate of wet deposition, but again needs an accurate representation of the type of storm, rainfall rates, and the specific pollutant concentrations during events. Again, the review shows that experimental data are sufficient only to the extent of providing empirical values rather than a thorough description of mechanisms which the models need.

Having reviewed the above-ground factors of transport and deposition, the influence of ground-level interception is considered. Characterizing the overall ground surface and the distribution of vegetation (canopy) is difficult, and calculation of deposition resistances or particle collection efficiencies is complex and inaccurate. Flow fields within the canopy are dependent upon canopy structure. The functions for diffusion by leaf stomatal pores depend on local leaf temperature, solar radiation, and moisture stress. Leaf morphology and area distributions influence collection efficiencies. In spite of the complexities, some rational approaches are feasible. Most notably lacking are the coordinated field exercises required to combine the talents to obtain data for both the model requirements and the measurement of deposition for a determination of biologic effects. In some vegetation such as crops, predictive success may be better than, for example, in North American forests, where the degree of complexity is difficult due to multiaged, multispecies interactions with the transport and deposition process. In the case of forests, therefore, where prediction of air pollution effects are least certain, we find that there is less data on deposition, less understanding of the deposition mechanisms, and perhaps less capability and resources from responsible agencies to conduct such experiments.

We should conclude from this review of the theoretical processes that these are surmountable complexities nonetheless. The expression of theoretical concepts as models of transport and deposition serves several useful purposes, not the least of which is the ability to interpret data, as contrasted with their predictive role. Another practical use of models, which has been forced upon us for better or for worse, is their application in enforcing air quality standards. Industry and governments alike may misuse models, especially if their uncertainties are not explicitly stated. In the discussions there were repeated calls for application of models to assess vegetation impacts by users such as planners and control strategists, especially since they are already used (or misused) extensively. There are those who feel that site specific modeling is the only rational strategy for air pollution control. But there were also calls to consider the uncertainties carefully, as for example, when one asks for precise mechanistic measurements in the face of a factor of two or worse in predicting pollutant distributions. If models would be made more modular in nature, then they should be more easily validated, albeit piecemeal. Scales must be reconsidered also: Plants may respond to dry deposition in tens of minutes, whereas for wet deposition the scale of days is appropriate. At the risk of oversimplifying the review, we may summarize:

1. *Gaseous dry deposition is possible to predict and to measure (but hard to monitor).*
2. *Particulate deposition is both hard to predict and hard to measure.*
3. *Wet deposition is hard to predict but possible to measure and monitor.*

B. Data Acquisition and Interpretation

The analysis and interpretation of the pathway of air pollutants from the atmosphere to vegetation and soils, and the subsequent actions of pollutants on plants, require first, appropriate measurements, and then interpretation of the measurements and responses. Initially it may be possible only to evolve empirical expressions useful for predicting responses of similar plants in similar environments, but an ultimate aim for the scientist must be to develop an understanding by experiment which enables the prediction of responses in diverse circumstances. This section considers the present position and future needs in data acquisition and interpretation.

1. MEASUREMENTS OF ATMOSPHERIC CONCENTRATIONS

1.1. Gases. *Instrumentation for monitoring concentrations of the major air pollutant gases in the atmosphere may be adequate. The* *biologist may need information concerning short-term peak concentrations over periods of typically ten minutes, so monitoring networks should achieve such records.*

1.2 Particles. *The monitoring of airborne particles is unsatisfactory. Deposition on vegetation depends critically on size distribution, and such information is necessary, along with analyses of chemical composition and speciation.*

Size analysis would best be accomplished by in situ samples, whereas chemical analysis generally calls for a collection technique. Analytical methods for trace species are usually adequate if sufficient particulate material is collected. In analyzing for chemical species it is important to determine whether there are speciation changes during dissolution of particles. The time resolution necessary in particulate monitoring is not yet clear, but from the viewpoint of particle effects on plants, it seems unlikely to be shorter than about ten minutes.

2. MEASUREMENT AND INTERPRETATION OF DRY DEPOSITION TO VEGETATION

2.1 Gases. *For experimental studies there are suitable micrometeorological methods for field measurements of fluxes of SO_2 and O_3 to extensive canopies of vegetation. Methods for NO_2 and NH_3 should be developed.*

Interpretation of fluxes in terms of resistance analogs is a useful tool for predicting how deposition depends on atmospheric, bulk physiological, leaf surface, and soil properties. With suitable analysis it is possible to partition the total flux into that entering the canopy through stomata and that deposited on leaf surfaces and soil. The main limitation of the approach for modeling lies in predicting stomatal responses to light and water stress, and in evaluating uptake by soil and undercanopy vegetation. Information on stomatal responses from other environmental physiology projects is useful to attack these problems. When there is water lying on leaves in canopies, surface uptake of gases may be an important sink. Further studies of chemical properties of wet films on leaves and of gas solubility in such films are necessary.

In the laboratory, measurements of fluxes of pollutant gases to leaves of plants are possible in several ways, and these methods help to evaluate the roles of stomatal uptake (independently on upper and lower leaf surfaces) and cuticular deposition. Such measurements show that uptake is strongly influenced by air movement, stomatal responses (to pollutants and other environmental factors), and reactivity on the leaf surface. Extension of such techniques to the field would be useful, although there are difficulties in scaling measurements on components of a plant to predict deposition on a canopy. More needs to be known of the reactivity of liquid and cell walls inside leaves to pollutant gases before the stomatal uptake path can be modeled accurately.

Micrometeorological methods for determining the distribution of sinks for pollutants within a canopy (e.g., as a series of leaf layers) are theoretically questionable and the method cannot be recommended. Direct measurements (see later) are preferable but still present major problems.

Knowledge of the response time of stomata to pollutant gases would be useful for incorporation into models of pollutant deposition.

2.2 Particles. *The measurement of fluxes of particulate matter to canopies in the field by micrometeorological methods is extremely difficult at present because of the lack of appropriate instrumentation and/or analytical sensitivity, especially for sulfate and ammonium.*

Even if particulate fluxes were measurable, their interpretation is more difficult than for gases; particle size, wind speed, and leaf surface properties are most significant factors determining uptake. Further knowledge is necessary concerning (1) resuspension of particles within canopies, (2) sedimentation of particles in canopies where turbulence may be low, and (3) size fractionation as aerosols diffuse down through a canopy.

3. *MONITORING DRY DEPOSITION*

Because direct measurements of rates of dry deposition are very difficult, monitoring with either artificial collectors or by analyzing vegetation is an attractive concept. Bruce Hicks summarized the findings of a workshop devoted to this topic, (see Hicks et al., 1980) with particular reference to the problem of measuring the flux to a canopy:

1. *Artificial collection surfaces cannot simulate natural conditions.*
2. *Micrometeorological methods are all sufficiently complicated that monitoring use is presently impractical.*
3. *The major need is to evaluate surface resistances associated with different circumstances—in other words, to parameterize.*
4. *There is a need to promote laboratory and wind tunnel experiments to study specific parts of the deposition process.*
5. *There is a need to develop new methods for field measurement, whether these be budget methods, plume methods, or micrometeorological variance methods. The intent of these field studies, when they are conducted, is indeed, to relate the flux of a selected pollutant to air concentrations, as well as to surface characteristics and meteorological properties.*

The workshop recommended that, for routine monitoring use at this time, concentrations be measured. These would need to be interpreted by the use of deposition velocities or some appropriate resistance model.

For measurements within a canopy, use of real leaves or surrogate surfaces appears to be the only available approach. When using real leaves there may be leaching of elements from within the leaf. Washing methods seldom remove all particles on leaf surfaces. Collection and analysis of rainfall above and below a canopy after a period of dry weather is unlikely to be a useful technique of assessing dry deposition because of transformations and exchange on leaf surfaces, but further investigations for a range of elements are desirable.

Surrogate surfaces avoid the leaching problems of leaves, but their exposure and uptake and capture efficiency for gases and particulates is artificial. Whether leaves or surrogate surfaces are used as collectors, there are problems in scaling up measurements to estimate deposition on the canopy.

Further studies are necessary of liquid on leaf surfaces, especially of ion exchange, action on particulates (solubility as a function of particle size), and uptake of gases.

The impact of throughfall and stemflow in canopies also requires study—for example, the spatial variability of rainfall distribution in a forest, its influence on soils, soil animals and microorganisms, understory vegetation, and regenerating plants.

4. ACTION OF POLLUTANTS

4.1 Gases. *Measurements in the laboratory in cuvettes or chambers have been discussed (Section 2.1) in relation to dry deposition. Such methods also allow detailed studies of responses to pollutants and interactions with environmental and edaphic factors. Current designs of field chambers and other exposure systems allow less control of the environment and have mainly been used to study growth and yield of agricultural crops. Studies linking short-term physiological responses to long-term yield experiments are necessary. Action of gaseous pollutants on leaf cuticle also requires study.*

4.2 Particles. *Participants were divided in their opinion of the significance of dry-deposited particles on vegetation. Submicron particles can enter stomata, but stomata occupy only 1–5% of leaf surface area. Solubility of particles in the moist stomatal pore may be significant for some elements. It is not known whether the microscale of particle deposition on leaves (picometeorology) has any significance in terms of biological action. The interaction of particles with moisture on leaves, either macroscale or trapped in crevices of leaf cuticular waxes, may be important for the action of pollutants.*

There are few experimental systems designed for exposing plants to submicron particles, and further developments are necessary. In general, care is necessary in experimental design in the laboratory to ensure that the environment neither restricts realistic rates of pollutant uptake by insufficient air movement nor masks action of pollutants by uncontrolled fluctuations of environmental factors.

5. PATHWAYS FOR POLLUTANTS

Studies of pathways of pollutants through ecosystems received little attention in this session or elsewhere but are an obvious extension of studies of transport and deposition. For relatively restricted systems (e.g., agricultural crops) lysimeters have been used to study the sulfur cycle. On a larger scale, catchment studies are a useful approach, especially with pollutants which may act as nutrients.

6. MODELING INTERACTIONS BETWEEN POLLUTANTS AND COMPLEX ECOSYSTEMS

Ecosystems have evolved by complex interactions between species (discussed in detail in the next section), especially competition for light, space, water, and nutrients and differential sensitivity to stresses. Although our knowledge is, at present, inadequate to apply to such systems to estimate the response to pollutant stress, it is useful to consider the potential (see the chapter by Lindberg and McLaughlin in this volume). The method could be extended to allow interactions between stresses and to handle pollutant action that occurred infrequently. Many more biological measurements are necessary before such models could be used reliably as predictors, but present models may be useful to test ecosystem sensitivity. A general problem in assessing the impact of pollutants on complex ecosystems is the lack of information on productivity before pollution. Measurement of tree rings, dendroecology, may provide such information, and tree-ring analysis can also give a measure of pollutant concentrations in the past.

Assessment of the potential for pollution damage may be possible by using regional ecology data bases, but it is difficult to know which factors should contribute to such bases until fundamental information on biological action is obtained.

C. Progress and Priorities

Substantial progress has been made in the last decade in air pollution studies—for example, in quantifying effects on agricultural yields, in identifying differences in sensitivity, in understanding interactions with other environmental factors, in discovering the physiological and biochemical mechanisms of pollutant action, and in describing the transfer of pollutants from the atmosphere to vegetation. We can model dry deposition of gases rather well but that of particles only poorly. Wet deposition is difficult to model but relatively easy to measure.

To make further progress, a combination of studies of applied problems and fundamental processes is necessary. Without fundamental understanding, a different study will be necessary for each new problem—an inefficient use of money, resources, and manpower. Current criteria may be more successful in protecting agricultural crops than they are for protecting natural ecosystems such as forests and grasslands. In such ecosystems there are substantial physical differences in pollutant exposure compared with agricultural monoculture. For example, the distribution of dry deposition and throughfall in the canopy leads to different responses between species. Similarly, the biological complexity of the system—for example, competition, evolution of stress resistance, regeneration, and decay—indicates the responses to pollutants are unlikely to be as simple as in annual crops. Consequently, studies of ecosystem responses to pollutants are likely to take many years to complete. A corollary is that monitoring of dry and wet deposition should continue for a period appropriate for such studies.

In simpler systems the most important priorities for applied and fundamental studies in the near future are likely to be the responses to multiple pollutants and the varying concentrations.

The major need for further interdisciplinary study concerns the atmosphere–leaf interface where detailed investigations of the physics, chemistry, and biology of the interactions between material on leaf surfaces and plant tissue are required.

REFERENCE

Hicks, B. B., M. L. Wesely, and J. L. Durham. Critique of methods to measure dry deposition: Workshop summary. Argonne National Laboratory, Argonne, Illinois, December 4–5, 1979. Sponsored by Environmental Science Reseach Laboratory, U.S. Environmental Protection Agency, Research Triangle Park, North Carolina—avail. National Technical Information Service, Springfield, Virginia as PB81–126443 (1980).

PART NINE

TERRESTRIAL ECOSYSTEM IMPACTS BY AIR POLLUTANTS: STATE OF THE ART OF THE CURRENT STUDIES AND FUTURE NEEDS

One of the largest single studies attempted on terrestrial ecosystem impacts by air pollutants relates to the San Bernardino Mountain study. Scientists have evaluated, by and large, regional and point source emissions on agricultural and forest ecosystems. Few studies have been successful in integrating the physical and chemical aspects of the atmosphere with the ecosystem flux. Also of concern is the influence of the terrestrial ecosystem on the aquatic ecosystem and vice-versa. Existing approaches and models of ecosystem response analysis and derived predictions of ecosystem change are highly complex. In the process, the fine tuning of assessment of the individual components is lost. The integration of the disciplinary efforts of the analyses require further attention.

This section reviews the current state of the art of available examples of terrestrial ecosystem studies, critically evaluates the existing deficiencies, and discusses the current and future needs in the improvement of such studies.

Gaseous Air Pollutants

PAUL R. MILLER
Forest Service
U.S. Department of Agriculture
Riverside, California

RONALD N. KICKERT
Division of Biological Control
University of California
Berkeley, California

1. INTRODUCTION

The consequences of chronic low-level pollutant effects on ecological systems are mostly cumulative and indirect; these effects are often not foreseeable or are discounted because they are uncertain. When the time comes that effects are unmistakable, the recognized cause is already the result of development over several years, and its reduction or elimination is extremely difficult and often impossible. Discounting effects because they are not foreseeable occurs, in part, because ecological systems cannot easily be described in terms of their value to humans. We may be able to estimate economic losses resulting from the suppression or near elimination of particular species, but it is more difficult to interpret the significance of systems level changes in function and structure in terms of welfare to humans.

The Wilderness Act of 1964, the establishment of UNESCO Biosphere Reserves (Doherty, 1977) on a worldwide scale, and the Endangered Species Act of 1973 all suggest that substantial value is placed on maintaining the integrity of ecological systems and on protecting populations of component species. These acts of Congress and UNESCO, Man and the Biosphere movement, provide a source of justification for carrying out ecosystem-level analyses of pollutant stress.

1.1 General Guidelines for Studying Ecological Systems

It is a complex task to accomplish a holistic interpretation of the effects of air pollutant stress on an ecosystem, that is, on the totality of interactions of populations with one another and with their physical environment. Such a task requires that the effects on each ecosystem component be evaluated. The elements of time, space, and biological complexity associated with analyzing effects on ecological systems are illustrated in Table 1. The difficulty of either measuring or predicting effects with increasing time, space, and level of biological organization is related to increasing cost, greater time required, and more uncertainty of predictions. Because a holistic approach is difficult and only a few ecosystem components can be measured, the intuition of the researchers is an irreplaceable resource for selecting those ecosystem components and processes that will be studied. The results of most ecosystem level studies pertain to selected subsystems, and it is necessary to recognize this limitation.

Kickert and Miller (1979) discuss the distinction between studies of chronic air pollution effects by determining whether the stated objectives reach one or all of three levels of achievement, namely, description, analysis, and forecasting. Even when the forecasting level has been accomplished, it is mainly in relation to producer processes—for example, photosynthesis and growth. Consumer and decomposer processes have received less attention. Kickert and Miller (1979) suggest that designing an

TABLE 1. Time, Space, and Biological Complexity Gradients Associated with Analyzing Effects on Ecological Systems

Moving from	*through*	*to*
Short-term ⟶	Time ⟶	Long-term
Near-field ⟶	Space ⟶	Far-field
Individual ⟶	Level of biological organization ⟶	Ecosystem

Gradient of Increasing Difficulty of Either Measuring or Predicting Effects ⟶

	Due to	
Low ⟶	Cost ⟶	High
Short ⟶	Time required ⟶	Long
Relatively small ⟶	Uncertainty ⟶	Relatively great

From Van Winkle et al. (1976).

integrated study of pollutant stress in a quantitative, analytical manner requires that the first 1 to 2 years be used for assembling working hypotheses and appropriate computer simulation models so that data collection can be guided by the models. Further, it is necessary to study selected portions of all major categories of ecosystem components for example, producers, consumers, decomposers, and the physical environment.

Many of the same principles have been reported as guidelines for the study of the effects of pollutants or other stress-causing factors on ecosystems. As stated by Barrett et al. (1976), these guidelines are:

1. Inclusion of both structural and functional ecosystem parameters.
2. Studies of ecosystems representing three levels of magnitude, namely, microcosm, seminatural, and natural.
3. Establishment of standardized structural and functional indexes for future stress investigations.
4. Design studies that provide an integrative predictive model as a major end result.
5. Employment of an interdisciplinary problem solving approach.

The guidelines provide useful points for examining the current state of the science to analyze the chronic, low-concentration effects of gaseous pollutants on terrestrial ecosystems. Item 4, however, neglects to emphasize that models are useful as tools for investigation. Broad scale changes, such as temperature changes, which may result from increased concentrations of carbon dioxide, will not be considered here. Effects of gaseous pollutants on temperate and Mediterranean forest ecosystems will be the major focus of discussion.

1.2 Generalizations about Chronic Low-Concentration Pollutant Effects on Ecological Systems

Several literature reviews have attempted to synthesize a holistic view of chronic air pollution effects on forest ecosystems (Miller, 1973; Miller and McBride, 1975; Knabe, 1976; Miller and White, 1977; Smith, 1981). Information is available mainly from isolated studies of producer species; effects at the consumer and decomposer levels are less known.

1.2.1 Effects on Producer Populations

Woodwell (1970) concluded that severe chronic pollutant damage to a forest community results in this ordering of events: (1) loss of sensitive species, (2) loss of the tree canopy, and (3) maintenance of a residual cover of pollutant tolerant herbs or shrubs that are recognized as seral or successional species. The results of continuous growth suppression of selected species by agents other than air pollutants tend to reinforce Woodwell's conclusions about forest community behavior. In the central hardwood forest types, for example, 20 yr of treatment with 2,4,5-T resulted in a mosaic of shrub communities and a less stable herb cover that remained after cessation of treatment (Niering and Goodwin, 1974). Other examples of the inability of stressed systems to return to their former species composition are known from instances of intensive grazing in the Southwest and deforestation in the tropics. Glendenning (1952) described how the elimination of grazing during a 17 yr experiment did not result in reestablishment of grassland, because grazing had encouraged the establishment of mesquite and cactus; these species had reached a density and size that monopolized the water supply and thereby discouraged grass establishment. Gomez-Pompa et al. (1972) suggest that even a one-time event, namely, the large-scale clearing of tropical forests, may result in permanent deforestation because of severe nutrient losses and the poor seed dispersal capability of tropical trees. This example may be somewhat analogous to the result of chronic, high-concentration pollutant damage, for example, the formation of barrens near point sources (Balsillie et al., 1978).

1.2.2 Effects on Consumer Populations

Direct effects of air pollution on free-ranging large vertebrates may be almost impossible to document because they can escape exposure where the available habitat is large and it is possible to spend much time in relatively unpolluted areas. Certain species such as the desert big-horn sheep (*Ovis canadensis nelsoni*), whose summer range is confined to high elevations in southern California, may be exceptions. A portion of their habitat, in the San Gabriel Mountains about 40 km from the Los Angeles metropolitan area, is in a downwind path in summer. The older desert big-horn sheep, living in the severely oxidant exposed Lytle Creek from the east fork of the

San Gabriel River watersheds, suffer from a cataract-like eye disease and a lungworm–pneumonia complex. The development of these diseases may be influenced by chronic oxidant exposure, but the evidence is only circumstantial (Taylor, 1973). Injurious effects of fluoride accumulation by animals have been well documented, particularly for domestic livestock (National Academy of Sciences, 1971).

Small vertebrates that are unable to migrate to distant unpolluted areas may be expected to be affected directly by pollutant exposure and affected indirectly by changes in food abundance from plants that provide a major part of their diet. In the San Bernardino Mountains a trapping program at vegetation plots variously affected by chronic oxidant exposure indicated that the same species were present as were found in similar trappings 70 yr ago (Kolb and White, 1974). Population numbers, however, were lower compared with other similar forest regions (Miller et al., 1977). Evidence suggests that the size and frequency of acorn crops from California black oak (*Quercus kelloggii* Newb.) may be less in areas receiving the greatest seasonal oxidant dose (Miller et al., 1979).

Invertebrate consumer populations appear to be most subject to the influence of air pollution on their habitat or host. Dewey (1972) described the accumulation of fluoride by four major groups of insects—foliage feeders, cambial region feeders, pollinators, and predators. Although no adverse effects were observed, the presence of relatively high levels of fluoride in predators suggests the possibility of fluoride transfer through this food chain.

The weakening of ponderosa pine by chronic exposure to ozone makes them more vulnerable to successful infestation by pine bark beetles (*Dendroctonus brevicomis* Lec. and *D. ponderosae* Hopk.) (Stark and Cobb, 1969; Dahlsten, 1978). Two insect enemies of lodgepole pine, a needle miner (*Ocnerostoma strobivorum* Zeller) and the pine sheath needle miner (*Zellaria haimbachi* Busck.) appeared to be more abundant on trees experiencing chronic injury by fluorides from an aluminium smelter (Carlson et al., 1974).

1.2.3 Effects on Decomposer Populations

In a mixed-conifer forest ecosystem, one effect of chronic ozone exposure is the increase in needle litter fall (Miller et al., 1977). The increased bulk of litter and its associated nutrients may enhance the populations of decomposer organisms if other environmental factors are not limiting. Ruhling and Tyler (1972) have preliminary evidence that the litter decomposition rate in a spruce forest was limited by an increased concentration of heavy metal ions, but only when water and temperature were not limiting.

In three independent studies, the diversity of invertebrate decomposers showed insignificant changes in population parameters, but disturbances were detected in nutrient cycling (Auerbach, 1978). These results suggest that the sampling of populations of decomposers does not provide a sensitive indicator of ecosystem-level change, that is, nutrient cycling. Part of the problem is that "...communities contain a comparatively large number of species that are rare at any given locus in time or space..." (Odum, 1963).

2. CURRENT UNDERSTANDING OF THE STRUCTURAL AND FUNCTIONAL CHARACTERISTICS OF ECOLOGICAL SYSTEMS

The structural properties that may be examined include species diversity, species dominance, and trophic food web patterns. In selecting the structural properties to be measured in a study of stress effects, it may be helpful to use the analogy that not all the links connecting the coils of a bedspring can be expected to wear out at the same time. Holling (1978) emphasizes that because everything is not intimately connected to everything else, a need exists to assess those connections that are significant, and further states that "...structural features (size, distribution, age, who connects to whom) are more important to measure than numbers."

Functional aspects that may be measured are vegetational and animal productivity, energy flows, moisture flows, and nutrient cycling. The status of such studies is discussed in the paragraphs that follow.

2.1 Progress Toward a Holistic Analysis of Forest Ecosystems

Studies done at the Hubbard Brook Experimental Forest in the White Mountain National Forest of New Hampshire collectively have provided an outstanding example of ecosystem analysis (Likens et al., 1978; Bormann and Likens, 1979). The flow of energy into primary production, and subsequently to consumer food webs, and then to decomposition processes on the forest floor has been carefully analyzed for this hardwood forest (Gosz et al., 1978). About 75% of the net annual production of the stands studied enters directly into the detritus system on the forest floor. In some years, a species of caterpillar (*Heterocampa guttivita* Walker) may consume as much as 40% of the total leaf tissue. This is an interesting quantification of a severe stress affecting the rate and amount of energy fixation. Other valuable features of the Hubbard Brook ecosystem analysis are the inclusion of above- and below-ground biomass and the meterological inputs of carbon and its output through the watershed drainage. These studies are a landmark in the area of understanding both structure and function of a forest ecosystem through energy flow analysis.

Nutrient cycling between plant tissues and soil at Hubbard Brook has received major attention (Whittaker et al., 1979). Changes in nutrient cycling after experimental deforestation have been a major focal point of the studies (Likens et al., 1978). During a 4-yr deforested period, the net export of nitrogen, potassium, and calcium from the watershed peaked during the second year and declined thereafter, suggesting that appreciable losses occurred before nutrients could be taken up by vegetation during the recovery (reorganization) phase. The usefulness of nutrient status as an index of ecosystem stress is discussed in Section 4.3.

A final example of work based on the Hubbard Brook project is an analysis of forest recovery after major exogenous disturbances such as fire, windthrow, and clear cutting (Bormann and Likens, 1979). A shifting-mosaic, steady-state hypothesis, based to a large extent on forest growth simulations with the JABOWA model (Botkin et al., 1972a, b), is proposed to describe events during a 500-yr period after a major disturbance. The shifting-mosaic steadystate is ". . . an array of irregular patches composed of vegetation of different ages." Patches pass from one age state to another during four phases after disturbance: reorganization, aggradation, transition, and steady state. Instead of a simple asymptotic model that assumes a gradual increase in biomass until steady state is achieved, the Bormann and Likens (1979) model ". . . delineates four phases of biomass decline and recovery, with a sharp drop in the reorganization phase and maximum accumulation of biomass occurring prior to the steady state." The time span includes several hundred years for the four phases to be completed.

Air pollution effects on forest stands are not as immediately devastating as fire and wind events, but it is necessary to acknowledge that chronic pollutant effects may be superimposed on the results of these catastrophic events. The holistic approach of the Hubbard Brook study provides valuable insights into normal structure and function of this hardwood forest ecosystem, and therefore can aid in the analysis and interpretation of pollutant stress on similar stands. A combination of long-term research support by participating agencies and prolific productivity by participating scientists provides a guiding example of what it takes to analyze ecological systems.

3. RECENT INTERDISCIPLINARY STUDIES LEADING TO INTEGRATIVE PREDICTIVE MODELS OF STRESS EFFECTS

Studies of pollutant stress effects that incorporate the essential requirements listed by Barrett et al. (1976) are few. This is particularly true for an interdisciplinary problem solving approach leading to a simulation model, either to guide research or as a major result for predictive purposes. An additional quality of current integrated studies is that they are focused on a single large area, for example, the San Bernardino mountains of southern California (Miller et al., 1977), and the Colstrip, Montana, area (Preston and Lewis, 1978) where selected data are collected regarding all trophic levels. One major difference between these studies is that at Colstrip an attempt was made to determine the bioenvironmental effects during the first few years of pollutant stress, but at San Bernardino, where the damage began in the early 1940s, the study was not begun until 1972 at sites along a gradient of decreasing pollutant dose. Although the West Whitecourt study (Legge, 1980) of sulfur gas emissions in a boreal forest is site specific, the data pertained mainly to primary production and nutrient cycling; modeling was not attempted. Modeling was a primary focus for a study of heavy metal effects on oak–pine stands at the Crooked Creek Watershed in southeast Missouri, but only primary production and litter decomposition were included (Dixon et al., 1978a, b).

3.1 Modeling Approaches in the San Bernardino Mountain Study

In the San Bernardino study of oxidant air pollution effect on a mixed conifer forest ecosystem, the highest priority processes that were selected by the team of investigators included moisture flow, leaf injury, leaf litter fall and litter decay, seed production, seedling establishment, tree growth, and tree mortality especially resulting from bark beetles and root disease. To predict changes in population dynamics and succession resulting from the foregoing process was one of the most essential objectives. The processes were to be linked in a particular simulation sequence to achieve this result (Kickert and Miller, 1979). The entire simulation sequence was not accomplished during the planned life of the project. The attempt to model selected subsystems was supported by extensive data gathered at 18 vegetation plots and 6 semipermanent meterological and air pollution monitoring sites; all data were available from a computer databank. Recent progress reports include Miller et al. (1977, 1982) and Taylor (1977).

Several organizational and managerial lessons have been learned from problems experienced during the course of the project:

1. Frequent communication was difficult because personnel were broadly scattered on both the Berkeley and Riverside campuses of the University of California. The development of a data management system was attempted at the Lawrence Livermore Laboratory (Barbieri, 1977) on a computer system that could not provide reliable remote access to users. The data management system had to be redesigned for a more accessible computer system.
2. Field study design had been virtually completed and much data collected before an ecological systems analyst was hired. Models could not be formulated and tested soon enough to provide a guide for selecting the types of data needed.
3. Budget reductions threatened the survival of the entire project just as data analysis was beginning and finally resulted in the loss of

valuable personnel at a critical time; additional sources of funding had not been developed to help guard against this event.

3.2 The Design and the Modeling Approach of the Colstrip Study

The ecological impact assessment of a coal-fired power plant on the structure and function of producers, consumers, and decomposers in the grassland ecosystem of southeastern Montana was a 6-yr project terminated in 1980. Preston and Lewis (1978) defined the following characteristics for the study:

> *. . . plant and animal community structure, primary production, invertebrate animal consumers, and decomposers; plant and animal diseases; both beneficial and harmful insects; indicators and predictors of pollution (e.g., lichens and honeybees); physiological responses of plants and vertebrate animals; insect behavior (mainly honeybees) and production; and the behavior, reproduction, and development, population biology, health and condition of vertebrate animals.*

The program also included monitoring of aerosols and a number of field and laboratory sulfur dioxide exposure experiments.

One of the primary objectives was to develop methods for predicting the effects of coal-fired power plants; these predictive methods are needed to aid in the siting process. The conceptual framework of the project had an autecological perspective. Effects on a target population are envisioned as derived from the parallel influences of direct toxic effects (including food chain concentration) and indirect effects (altered primary production and altered environmental quality) on competition and predation among the several species populations of which the target population is a component (Lewis et al., 1978).

The modeling activities were directed toward adapting the ELM (grassland simulation model) to the Colstrip sites, developing the capability to simulate direct effects of sulfur dioxide on ecosystem components, and developing a submodel of the sulfur cycle for ELM (Dodd et al., 1978). The sulfur-cycling model is structured so that primary production will be influenced by sulfur dioxide in the atmosphere as well as by the distribution of sulfur in the plant and soil systems. Principal investigators have developed a working knowledge of the ecology of grasslands near Colstrip, Montana, leading to predictive capability.

4. PROGRESS TOWARD ESTABLISHING STRUCTURAL AND FUNCTIONAL INDEXES OF STRESS EFFECTS

In the foregoing sections the concept of a holistic approach to studying pollutant effects on ecological systems was stressed in an attempt to exercise the imagination about the kinds of effects that may occur. In this section, we shall discuss other approaches to the investigation of stress effects. A pioneering group in this area is the Institute of Animal Resource Ecology at the University of British Columbia. In the forward to *Adaptive Environmental Assessment and Management* (Holling, 1978), Martin Holdgate observes:

> *. . . some of the ideas about ecosystems and their methods of characterization have led us astray, because they have not been based on sound understanding. In consequence, much effort has been devoted to the wrong kind of analysis and to the collection of unnecessarily large quantities of data that have given rise to undue expectations and unsatisfactory predictions.*

Holling (1978) emphasizes four properties of ecosystem that should be acknowledged, namely, organized connection between parts, spatial heterogeneity, resilience, and dynamic variability. He also lists several lessons relating to these properties that appear to have application to the analysis of pollutant effects.

1. Since everything is not intimately connected to everything else, there is no need to measure everything. There is a need, however, to determine the significant connections.
2. Structural features (size, distribution, age, who connects to whom) are more important to measure than numbers.
3. Changes in one variable (e.g., a population) can have unexpected impacts on variables at the same place but several connections away.
4. Events at one place can re-emerge as impacts at distant places.
5. Monitoring of the wrong variable can seem to indicate no change, even when drastic change is imminent.
6. Impacts are not necessarily immediate or gradual; they can appear abruptly some time after the event.
7. Variability of ecological systems, including occasional major disruptions, provides a kind of self-monitoring system that maintains resilience. Policies that reduce variability in space or time, even in an effort to improve environmental quality, should always be questioned.
8. Many existing impact assessment methods (e.g., cost–benefit analysis, input–output, cross-impact matrices, linear models, discounting) assume that none of the above occurs or, at least, that none is important.

Items 1 through 5 are particularly cogent to the discussion of the selection and validity of various indicators for describing ecosystem level responses to pollutant stress. One of the questions being addressed is whether there is an ecosystem equivalent of adrenaline that can serve as a reliable index of pollutant stress (Auerbach, 1978).

4.1 Identifying Chronic Pollutant Effects Using Species Diversity Indexes and Path Analysis

McClenahen (1978) employed calculated values of the Shannon diversity index (Pielou, 1975), evenness (Williams, 1977), and coefficients of community (Pielou, 1975) to describe overstory, subcanopy, shrub, and herb layer composition in 7 deciduous forest stands along a 50-km portion of the upper Ohio River Valley. These stands were subject to chloride, sulfur dioxide, and fluoride along a dose gradient. Species richness, evenness, and the Shannon diversity index were depressed within the overstory, subcanopy, and herb layers. Similarity in species composition, as indicated by coefficient of community, decreased with increasing chronic pollutant exposure. McClenahen (1978) concludes that these are adequate qualitative estimates of the environmental effects of industrial development. In addition, it seems that the zone of influence of pollutants could be identified as the first step in designing more detailed studies of productivity and related processes.

Path analysis (Heise, 1975) was used by Westman (1979) to determine the most likely route of causation of the decline in cover in the California coastal sage scrub community in certain California sites. Sixty-seven sites were sampled throughout the range of coastal sage dominants (*Salvia*, *Eriogonum*, and *Encelia*) from San Francisco to Baja California. Foliar cover was recorded for all species at each site, and data of 43 habitat variables were obtained for each site. These variables included community structure, topographic position, substrate, climate, fire and grazing history, and mean annual air pollution concentration. The percentage total cover at each site was tested against the 43 habitat variables to obtain Pearson correlations. Mean annual oxidant concentration, elevation, and mean air temperature during the warmest month were negatively correlated with cover. The path analysis model suggests that the single strongest direct cause of decline in native cover is the increasing mean annual concentration of oxidants. This hypothesis was bolstered by Whittaker's log-cycle equitability index (Whittaker, 1972) and concentration of dominance as measured by Simpson's index (Simpson, 1949). In conclusion, Westman (1979) states that this kind of methodology can only suggest likely causal routes of the decline of total cover; additional experimentation is required to confirm the cause and to quantify the effects.

These two studies offer useful examples of the kinds of synecological methods that can provide indications for the need for more detailed analyses of pollutant effects.

4.2 Applying Simulation Models for Tree Growth and Stand Succession as Indicators of Long-Term Effects

Selected examples of pollutant-induced depression of tree radial growth include studies of eastern white pine (*Pinus strobus* L.) (Phillips et al., 1977a, b), ponderosa pine (*Pinus ponderosa* Law) (McBride et al., 1975) and California black oak (Miller et al., 1979; Gemmill, 1980). This kind of database is essential input for models that simulate forest stand dynamics for fairly long time spans.

The JABOWA forest growth simulation model (Botkin et al., 1972a, b) "grows" trees as functions of tree age, stand biomass per unit area, leaf area, moisture and temperature status, and percent of full sunlight available in an eastern deciduous forest. A modification of this model called FORET developed by Shugart and West (1977) has been applied to the study of air pollutant effects by McLaughlin et al. (1978) and West et al. (1980). The model considers 32 tree species native to the southern Appalachian region; it simulates growth of individual trees on a circular $\frac{1}{12}$-ha plot. Pollutant effects were imposed in terms of a 0, 10, and 20% (high) or 0, 5, and 10% (low) level of induced growth reduction of individual species within a simulated seedling forest on both 50- and 400-yr-old forests. Growth reductions were expressed as 10–20% for sensitive, 5–10% for intermediate, and 0% for tolerant species. The species composition of a mixture of 32 species was expressed as percent of total biomass with and without pollution-induced stress. The simulation results suggest that some species may show greater growth in spite of pollutant stress because of greater suppression of the growth of other species with which they interact in the successional process. Competition was identified as a significant modifier of individual plant response in a mixed-species community. The most important limitation of this approach at present is the quality of the data available for describing expected growth decreases because of SO_2 exposure. In particular, the assumption that the relative sensitivity to growth loss is proportional to the relative leaf injury reported in the literature is an uncertainty that can be remedied only by obtaining suitable quantitative measures of growth loss for each species. These simulations, nevertheless, provide useful insights and encouragement for improving the data required for quantifying pollutant effects on growth.

Emanuel et al. (1978) by using a spectral analysis approach examined the simulation results of FORET when successional change was perturbed by three separate factors: chestnut blight, a change in average annual temperature related to CO_2 buildup, and suppression of selected species by SO_2 exposure. Cyclic behavior in the total biomass of the stand with periods greater than 200 yr was evident in the control and in all three perturbations. Spectral analysis provided a means for further definition of differences in the time series of the three perturbations. Each significant peak in the power spectral density corresponds to a cyclic phenomenon in the model forest stand dynamics. The calculated power spectral densities for total biomass in each situation showed a decrease in spectral richness for both the temperature and SO_2 perturbation.

CERES is a high resolution model of forest stand biomass dynamics for predicting trace contaminant, nutrient, and water effects. Hourly, daily, weekly, monthly, and longer term simulation results were obtained (Dixon et al., 1978a). CERES incorporates sugar transport along concentration gradients between plant organs to predict effects of trace contaminants, nutrients, or both when coupled with a soil chemistry model (SCHEM) and models of solute uptake with the acronyms DIFMAS and DRYADS. CERES can be linked with a soil–plant–atmosphere model (PROSPER) to study the relationship of growth and solute transport to tree water status.

In a simulation of heavy metal effects on the Crooked Creek watershed in southeast Missouri, the parameter values for CERES were either (1) estimated, (2) obtained from the literature, or (3) calculated from Crooked Creek or Walker Branch watershed data (Dixon et al., 1978b). During the 6-yr application test, the simulation of heavy metal uptake by the vegetation was accompanied by a steady annual input of plant litter and a reduced decomposition rate that increased litter mass by nearly 50%. The authors suggest that new data associated with solute transport mechanisms are needed to improve model definition.

A growth simulator called SUCSIM developed for coniferous forests by Reed and Clark (1979) was proposed for the study of oxidant air pollutant effects in the San Bernardino Mountain Study (Kickert and Miller, 1979). SUCSIM is much faster than the JABOWA-based FORET model because it "grows" cohorts of trees, not single trees. It differs also in that the birth–death routines are probabilistic and environment is more directly defined in terms of light, temperature, and moisture. The SUCSIM application to the San Bernardino study would have been well supported by tree growth data from the study area, but data strings of similar length for weather and air pollution dose were not always available for the tree growth study plots.

Kercher and Axelrod (1981) have reported on another modification of JABOWA intended to describe SO_2 effects on a Sierra Nevada mixed-conifer forest. A major feature of this model is the incorporation of fire ecology for this forested area. The sulfur dioxide effect on growth is based mainly on values reported in the literature as in the modified FORET model (West et al., 1980). Application of the model is limited by the quality of the data used for choosing and defining the relationships in the model.

4.3 Nutrient Cycling as a Monitoring Point for Measuring Pollutant Stress

The effect of deforestation on nutrient loss has been well documented at Hubbard Brook (Likens et al., 1978); nitrogen loss is considered to be particularly critical. Studies at other sites in the United States showed also that nitrogen losses can be excessive following disturbances (Vitousek et al., 1979). Because a large number of interacting components are involved in nutrient cycling, several investigators have suggested that changes in total system dynamics may be measured in nutrient cycling (Auerbach, 1978).

In the West Whitecourt study of SO_2 effects in a boreal forest Legge (1980; Legge et al., 1981) found a direct relationship between lowered soil pH and the elevated levels of foliar manganese in lodgepole × jack pine (*Pinus contorta* Dougl. × *Pinus banksiana* Lamb.). Foliar manganese was proposed as an indicator of change in the forest ecosystem induced by SO_2 exposure. In addition, mineral nutrient cycling was affected both directly and indirectly.

4.4 Using Indicator Species to Monitor Injury to Forest Stands

Norway spruce (*Picea abies* (L.) Karst.) has been proposed as an indicator of sulfur dioxide and hydrogen fluoride damage in central and northern Europe because its level of sensitivity to pollution is high enough to provide satisfactory protection for most associated species (International Union of Forest Research Organizations, 1979). In the eastern United States, white pine (*Pinus strobus* L.) has frequently been used as an indicator of ozone injury (Skelly et al., 1977). Use of indicator species, however, provides little insight into ecosystem level effects of pollutants, but it can be an effective and inexpensive monitor of air quality changes if the data gathered are sufficiently quantitative.

5. EVALUATING CURRENT DEFICIENCIES IN ANALYSIS OF ECOSYSTEM LEVEL EFFECTS OF POLLUTANT STRESS

We have concluded that the reported study designs discussed earlier have generally indicated that a holistic study design has merit but is difficult to implement in short time spans, that is, 3–5 yr. The remaining questions about study design focus mainly on the issues of which ecosystem properties should be selected for measurement, how much data is sufficient, how long should measurements be continued, or both. In most situations, the availability of funding is the major factor controlling these issues. We are reminded that the difficulty of measuring or predicting effects increases as time and space increase and as the level of biological organization becomes more complex (see Table 1). Without sound quantitative data from the population level gained through appropriate field and laboratory studies, the ability to predict effects at the ecosystem level will be limited. Also parallel development of models is required as a means of showing who is connected to whom and how strongly. All of the successional models described in Section 4.2 are limited in application by the quality of data inputs, yet such models are needed to guide the collection of new data.

Auerbach (1978) warns that assessment of environmental effects cannot be reduced to a set of standard procedures by regulatory agencies because each problem is unique. A set of guidelines of the sort listed by Barrett et al. (1976) (Section 1.1), however, could be expanded to provide a set of minimum criteria for studying stress effects on ecological systems. Such criteria may encourage individual researchers working at the population level to collect the kinds of quantitative data required to improve and update existing models. The *Handbook of Methodology for the Assessment of Air Pollution Effects on Vegetation* (Heck et al., 1979) provides a compendium that serves as a departure point and not a cookbook of standard procedures. *Adaptive Environmental Assessment and Management* (Holling, 1978) provides a conceptual framework for approaching complex problems: It emphasizes the futility of the uncritical collection of information.

Another serious impediment to adequate assessment of environmental effects is the incremental nature of the funding available for ecological effects research. A framework for a program of Long-Term Ecological Research (LTER) was formulated under sponsorship of the National Science Foundation (1979). The program defined a set of core research questions relating mainly to stresses or disturbances in ecological systems. It provided a managerial plan, outlined measurement needs, standardized methods, and proposed study sites where extended observation of particular populations, communities, or ecosystems can be made. The LTER program called for the participation and cooperation of numerous federal agencies that may sponsor research. If the LTER program were implemented, then the combined sources of funding might help to prevent the dismantling of productive research projects just because of a change in the objectives of a single agency.

6. FUTURE RESEARCH NEEDS

The high costs of interdisciplinary research carried out over long time spans require justification. The economic returns expected from a single species, for example the harvest of timber, may not be adequate to justify the costs of such research. Westman (1977) has used examples to illustrate that social benefits are derived for ecosystem functioning in addition to the esthetic values championed by various conservation or environmental organizations. These benefits include ground water storage, water purification, erosion control, conservation of nutrients, pollutant absorption, and stabilization of climate. There is no social agreement about the use of monetary units to express the value of these ecosystem functions. The problem may be complicated further by the concept that the concern may be for the benefits to be derived by future generations where the object of concern is not only material welfare for humans but also the dignity of humans.

In practice, damage to ecosystems is not repaired in the sense that the original structure and function are restored. The repair bill cannot represent the total cost. Balsillie et al. (1978) described the variety of techniques that are being tested in an effort to revegetate the Sudbury, Ontario barrens. No single formula is appropriate for all sites, and in many situations more than one trial will be required to reestablish vegetation. Even if succession occurs, the resulting climax community will not compare favorably with the original vegetation on the Sudbury area. The economic feasibility of the task is a matter of concern.

Prevention of damage to ecosystems may be much less costly than the cure. Cooper (1975) supports the idea that ecosystem models are of potentially great value for guiding and in forming incremental decisions regarding environmental policy that attempts to inhibit environmental degradation. The United States Environmental Protection Agency (1979) listed a set of questions regarding pollutant effects on ecosystems. The answers may be approached through the improvement of ecosystem models. The questions are:

1. Are there systems-level structural properties, for example, food webs, which are critical to ecosystem functions?
2. Are these structural properties or functions more sensitive to stress than individual components of the system?
3. How can ecosystems be described in terms of their value to man? For example, are there characteristics of ecosystems that can be evaluated to determine if, in response to stress, they contribute to different conditions, but of equal value to man?
4. Are there inherently stable or unstable states for ecosystems relating to their value for man?
5. Assuming there are sets of conditions to which ecosystems return when stressed, are there limits beyond which they can be stressed and not return? What is their rate of return to their original state?
6. How can an ecosystem be most simply and economically characterized to determine whether or not it is under stress and, if so, how severe and how persistent a stress?
7. What is the significance of variations in the severity or timing, for example, continuous versus intermittent, of the stress?
8. Can an ecosystem's condition be usefully estimated by describing only its physical and chemical characteristics?
9. What systems-level processes or phenomena are significant for the transport and fate of toxic substances?
10. Are there certain components of ecosystems which are more useful or accurate in predicting toxic exposure levels than others?

In summary, future research activities should be guided by criteria such as those proposed by the LTER program (National Science Foundation, 1979). The basic purpose of such criteria would be to improve the quality of the quantitative database required by models of ecological systems. The time and space resolution of models should be geared to the needs of environmental policy decision makers. One of the primary goals should be to provide for the transfer of simulation technology to the appropriate managerial levels of agencies and private enterprises concerned with the control of chronic effects of pollutants of ecosystems. Communication between researchers and decision makers must be encouraged during all phases of data collection and model development if the products of research are to have maximum usefulness.

ACKNOWLEDGMENT

This study was supported in part by the United States Environmental Protection Agency Grant R805410-03 through its Corvallis Environmental Research Laboratory, Corvallis, Oregon.

REFERENCES

Auerbach, S. I. Current perceptions and applicability of ecosystem analysis to impact assessment. *Ohio J. Sci.* **78**, 163–174 (1978).

Balsillie, D., K. Winterhalder, and W. D. McIlveen. Problems of regeneration of stressed ecosystems. Paper 78-44.6, 71st Annual Mtg. Air Pollut. Control Assoc. (1978).

Barbieri, J. F. EPA–San Bernardino National Forest information system—final report. UCRL-52309. Lawrence Livermore Laboratory, Livermore, California, 66 pp. (1977).

Barrett, G. W., G. M. Van Dyne, and E. P. Odum. Stress ecology. *Bioscience* **26**, 192–194 (1976).

Bormann, F. H., and G. E. Likens. Catastrophic disturbance and the steady state in northern hardwood forests. *Amer. Scientist* **67**, 660–669 (1979).

Botkin, D. B., J. F. Janak, and J. R. Wallis. Rationale, limitations and assumptions of a northeastern forest growth simulator. *IBM J. Res. Devel.* **16**, 101–116 (1972a).

Botkin, D. B., J. F. Janak, and J. R. Wallis. Some ecological consequences of a computer model of forest growth. *J. Ecol.* **60**, 849–872 (1972b).

Carlson, C. E., W. E. Bousfield, and M. D. McGregor. The relationship of an insect infestation on lodgepole pine to fluorides emitted from a nearby aluminum plant in Montana. *USDA Forest Service, North. Reg., Insect-Disease Rep.* 74–14, 21 pp. (1974).

Cooper, C. F. Ecosystem models and environmental policy, Ch. 6, pp. 57–62 In G. S. Innis (ed.) *New Directions in the Analysis of Ecological Systems, Part* 1, 132 pp. Simulation Council Proc. Ser., LaJolla, California, 1975.

Dahlsten, D. L. The role of bark beetles in oxidant stressed forests. In *Proc. Simulation Modeling of Oxidant Air Pollution Effects of Mixed Conifer Forests and the Possible Role of Models in Timber Management Planning for Southern California National Forests*, 29 pp. Statewide Air Pollution Research Center, University of California, Riverside, 1978.

Dewey, J. E. Accumulation of fluorides by insects near an emission source in western Montana. *Environ. Entomol.* **2**, 179–182 (1972).

Dixon, K. R., R. J. Luxmoore, and C. L. Begovich. CERES—A model of forest stand biomass dynamics for predicting trace element contaminant, nutrient, and water effects. I. Model Description. *Ecol. Model.* **5**, 17–38 (1978a).

Dixon, K. R., R. J. Luxmoore, and C. L. Begovich. CERES—A model of forest stand biomass dynamics for predicting trace contaminant, nutrient, and water effects. II. Model Application. *Ecol. Model.* **5**, 93–114 (1978b).

Dodd, J. L., M. Coughenour, and W. K. Lauenroth. Section 16. Progress in modeling the effects of SO_2 fumigation on an eastern Montana grassland. In E. M. Preston and R. A. Lewis (eds.) *The Bioenvironmental Impact of a Coal-Fired Power Plant, Third Interim Report*, pp. 511–520. Available National Technical Information Service, Springfield Virginia as EPA-600/3-78-021 (1978).

Doherty, J. UNESCO's Biosphere Reserve Program. *International Wildlife* **7**, 24–28 (1977).

Emanuel, W. R., H. H. Shugart, Jr., and D. C. West. Spectral analysis and forest dynamics: The effects of perturbations on long-term dynamics. In H. H. Shugart, Jr. (ed.) *Time Series and Ecological Processes*, pp. 193–207. Proc. SIAM Instit. for Math. and Soc., Philadelphia, Pennsylvania (1978).

Environmental Protection Agency. EPA and the academic community—partners in research. EPA-600/8-79-032. 15 pp. (1979).

Gemmill, B. Radial growth of California black oak in the San Bernardino mountains. In *Proc. Symposium on Ecology, Management and Utilization of California Oaks*, pp. 128–135. USDA Forest Service, Gen. Tech. Report PSW 44. 368 pp. (1980).

Glendenning, G. Some quantitative data on the increase of mesquite and cactus on a desert grassland range in southern Arizona. *Ecology* **33**, 318–328 (1952).

Gomez-Pompa, A., C. Vazques-Yames, and S. Guevara. The tropical rain forest: a nonrenewable resource. *Science* **177**, 762–765 (1972).

Gosz, J. R., R. T. Holmes, G. E. Likens, and F. H. Bormann. The flow of energy in a forest ecosystem. *Sci. Amer.* **238**(3) 92–102 (1978).

Heck, W. W., S. V. Krupa, and S. N. Linzon (eds.) *Methodology for the Assessment of Air Pollution Effects on Vegetation, a Handbook*. Air Pollut. Control Assoc., Pittsburgh, Pennsylvania, 1979.

Heise, D. R. *Causal Analysis*. Wiley, New York, 1975.

Holling, C. S. (ed.) *Adaptive Environmental Assessment and Management*. 377 pp. Wiley, New York, 1978.

International Union of Forestry Research Organizations. Subject Group S 2.-09-00. Resolution on air quality standards for the protection of forests. *IUFRO News*, No. 25, 2 pp. (1979).

Kercher, J. R., and M. C. Axelrod. A model for forecasting the effects of SO_2 pollution on growth and succession in a western conifer forest. Interim Report. Lawrence Livermore Laboratory Internal Document, available NTIS, Springfield, Virginia as UCID-18537 (1981).

Kickert, R. H., and P. R. Miller. Chapter 14. Responses of ecological systems. In W. W. Heck, S. V. Krupa, and S. N. Linzon (eds.) *Methodology for the Assessment of Air Pollution Effects on Vegetation, a Handbook*. Air Pollut. Control Assoc., Pittsburgh, Pennsylvania 1979.

Knabe, W. Effects of sulfur dioxide on terrestrial vegetation. *Ambio* **5**, 213–218 (1976).

Kolb, J. A., and M. White. Small mammals of the San Bernardino Mountains. *California Southwest National* **19**, 112–113 (1974).

Legge, A. H. Primary productivity, sulphur dioxide, and the forest ecosystem: An overview of a case study. In *Proceedings of the Symposium on Effects of Air Pollutants on Mediterranean and Temperate Forest Ecosystems*, pp. 51–62. USDA, Forest Service Tech. Rep. PSW-43, 256 pp. (1980).

Legge, A. H., D. R. Jaques, G. W. Harvey, H. R. Krouse, H. M. Brown, E. C. Rhodes, M. Nosal, H. U. Schellhase, J. Mayo, A. P. Hartgerink, P. F. Lester, R. G. Amundson, and R. B. Walker. Sulphur Gas Emissions in the Boreal Forest: The West Whitecourt Case Study I. Executive Summary. *Water, Air and Soil Pollution* 15: 77–85 (1981).

Lewis, R. A., E. M. Preston, and N. R. Glass. Assessment of ecological impact from the operation of a coal-fired power plant in the North Central Great Plains. In E. M. Preston, and R. A. Lewis (eds.). *The Bioenvironmental Impact of a Coal-Fired Power Plant. Third Interim Report*. Available National Technical Information Service, Springfield, Virginia as EPA-600/3-78-021. 571 pp. (1978).

Likens, G. E., F. H. Bormann, R. S. Pierce, and W. A. Reiners. Recovery of a deforested ecosystem. *Science* **199**, 492–496 (1978).

McBride, J. R., V. P. Semonin, and P. R. Miller. Impact of air pollution on the growth of ponderosa pine. *Calif. Agric.* **29**(12) 8–9 (1975).

McClenahen, J. R. Community changes in a deciduous forest exposed to air pollution. *Canad. J. Forest Res.* **8**, 432–438 (1978).

McLaughlin, S. B., D. C. West, H. H. Shugart, and D. S. Shriner. Air pollution effects on forest growth and succession: Applications of a mathematical model. *Proc. 71st Ann. Meeting Air Pollut. Control Assoc., Houston, Texas, June* 25–30, paper 78-24.5. APCA, Pittsburgh, Pennsylvania (1978).

Miller, P. R. Oxidant-induced community change in a mixed conifer forest. In J. A. Naegele (ed.). *Air Pollution Damage to Vegetation*. pp. 101–117. American Chemical Society, Washington D.C. 1973.

Miller, P. R., and J. R. McBride. Effects of air pollution on forests. In J. T. Mudd and T. T. Kozlowski (eds.) *Response of Plants to Air Pollution*, pp. 195–235. Academic, New York, 1975.

Miller, P. R., and M. White. Ecosystems, In *Ozone and Other Photochemical Oxidants*, pp. 586–642. National Academy of Sciences, Washington, D.C., 1977.

Miller, P. R., R. N. Kickert, O. C. Taylor, R. J. Arkley, F. W. Cobb, Jr., D. L. Dahlsten, P. J. Gersper, R. F. Luck, J. R. McBride, J. R. Parmeter, Jr., J. M. Wenz, M. White, and W. W. Wilcox. Photochemical oxidant air pollution effects on a mixed conifer forest ecosystem. A progress report. EPA 600/3-77-104. 338 pp. (1977).

Miller, P. R., G. J. Longbotham, R. E. Van Doren, and M. A. Thomas. Effect of chronic oxidant air pollution exposure on California black oak in the San Bernardino mountains. In *Proc. Symposium on Ecology, Management and Utilization of California Oak*, pp. 220–229. USDA, Forest Service Gen. Tech. Rep. PSW-44, 368 pp. (1979).

Miller, P. R., O. C. Taylor, and R. G. Wilhour. Oxidant air pollution effects on a western coniferous forest ecosystem. Environmental Research Brief EPA-600/D-82-276. 10 pp. (1982).

National Academy of Sciences. Fluorides—biologic effects of atmospheric pollutants. NAS, Washington, D. C. 195 pp. (1971).

National Science Foundation. *Long-Term Ecological Research-Concept Statement and Measurement Needs. Workshop Summary*, 27 pp. The Institute of Ecology, Indianapolis, Indiana 1979.

Niering, W. A., and R. H. Goodwin. Creation of relatively stable shrublands with herbicides: arresting "succession" on rights-of-way and pasture land. *Ecology* **55**, 784–795 (1974).

Odum, E. P. *Ecology*, 152 pp. Holt, Rinehart and Winston, New York, 1963.

Phillips, S. O., J. M. Skelly, and H. E. Burkhart. Eastern white pine exhibits growth retardation by fluctuating air pollution levels: interaction of rainfall, age and symptom expression. *Phytopathology* **67**, 721–725 (1977a).

Phillips, S. O., J. M. Skelly, and H. E. Burkhart. Growth fluctuations of loblolly pine due to periodic air pollution levels: interaction of rainfall and age. *Phytopathology* **67**, 716–720 (1977b).

Pielou, E. C. *Ecological Diversity*. Wiley, New York, 1975.

Preston, E. M., and R. A. Lewis (eds.) The Bioenvironmental Impact of a Coal-Fired Power Plant. Third interim report. Available National Technical Information Service, Springfield, Virginia as EPA-600/3-78-021. 571 pp. (1978).

Reed, K. L., and S. G. Clark. SUCcession SIMulator: A coniferous forest simulator. Model documentation. Bulletin No. 11. Coniferous forest biome. Ecosystem analysis studies. 96 pp. (1979).

Ruhling, A., and G. Tyler. Effects of heavy metal pollution of the decomposition of spruce needle litter, 48 pp. Dept. Plant Ecology, University of Lund, Sweden. (1972).

Shugart, H. H., Jr., and D. C. West. Development and application of an Appalachian deciduous forest succession model. *J. Environ. Manage.* **5**, 161–179 (1977).

Simpson, E. H. Measurement of diversity. *Nature* **163**, 688 (1949).

Skelly, J. M., C. F. Crogan, and E. M. Hayes. Oxidant levels in remote mountainous areas of southwestern Virginia and their effects on native white pine (*Pinus strobus* L.). *Proc. International Conf. Photochem. Oxidant Pollut. and its Control*, pp. 611–620 (1977).

Smith, W. H. *Air Pollution and Forests*, 379 pp, Springer-Verlag, New York, 1981.

Stark, R. W., and R. W. Cobb, Jr. Smog injury, root disease and bark beetle damage in ponderosa pine. *Calif. Agric.* **23**(9), 13–15 (1969).

Taylor, O. C. Oxidant air pollutant effects on a mixed conifer forest ecosystem. Task B Report: Historical background and proposed systems study of the San Bernardino mountain area, 136 pp. Statewide Air Pollution Research Center, University of California, Riverside (1973).

Taylor, O. C. (ed.) Photochemical oxidant air pollution effects on a mixed conifer forest ecosystem, Final Report, Contract 68-03-2442, 194 pp. University of California, Statewide Air Pollution Research Center (1977).

Van Winkle, W. S., S. W. Christensen, and J. S. Mattice. The two roles of ecologists in defining and determining the acceptability of environmental impacts. *Intern. J. Environ. Stud.* **9**, 247–254 (1976).

Vitousek, P. M., J. R. Gosz, C. C. Grier, J. M. Melillo, W. A. Reiners, and R. L. Todd. Nitrate losses from disturbed ecosystems. *Science* **204**, 469–474 (1979).

West, D. C., S. B. McLaughlin, and H. H. Shugart. Simulated forest response to chronic air pollution stress. *J. Environ. Qual.* **9**, 43–49 (1980).

Westman, W. E. How much are nature's services worth? *Science* **197**, 960–964 (1977).

Westman, W. E. Oxidant effects on California coastal sage scrub. *Science* **205**, 1001–1003 (1979).

Whittaker, R. H. Evolution and measurement of species diversity. *Taxon*. **21**, 213–251 (1972).

Whittaker, R. H., G. E. Likens, F. H. Bormann, J. S. Easton, and T. G. Siccama. The Hubbard Brook ecosystem study: Forest nutrient cycling and element behavior. *Ecology* **60**, 203–220 (1979).

Williams, F. M. Model-free evenness: An alternative to diversity measures. Satellite Progrm. Statist. Ecol., Intern. Statist. Ecol. Progrm. The Pennsylvania State University, University Park, Pennsylvania (1977).

Woodwell, G. M. Effects of pollution on the structure and physiology of ecosystems. *Science* **168**,429–433 (1970).

Techniques for Assessing Ecosystem Impacts of Air Pollutants

CARL W. CHEN
Systech Engineering, Inc.
Lafayette, California

ROBERT A. GOLDSTEIN
Electric Power Research Institute
Palo Alto, California

1. INTRODUCTION

There exists a hirearchy of biological levels of organization starting at the subcellular and progressing through the ecosystem. Three of the levels are considered to encompass ecological phenomena: the organism, the population, and the ecosystem. The objective of this paper is to review approaches to assessing terrestrial effects of air pollutants at the ecosystem level. At the ecosystem level of organization, aquatic systems can be strongly coupled with terrestrial systems, since terrestrial components of an ecosystem (e.g., vegetation and soil) can affect through multiple processes the quality of water moving through the system and hence the biota of the streams, lakes, and rivers into which the water eventually enters.

In reviewing ecosystem assessment approaches, the literature on effects of atmospheric deposition of acids will be used as a source for examples of applications of the approaches, because analyses of acid deposition effects tend to stress the interaction of ecosystem components and the coupling of terrestrial and aquatic systems. This paper should not be considered a review of current understanding of acid rain effects, since references were not chosen with that objective in mind.

Approaches to analysis of ecosystem effects include:

1. Extrapolation of results of exposure of single ecosystem components (e.g., an individual species or a single soil layer) under controlled or partially controlled environmental conditions to specified levels of pollutants.
2. Extrapolation of results of exposure of multiple ecosystem components (microcosms) under controlled or partially controlled conditions to specified levels of pollutants.
3. Observation of ecosystem responses in the field under noncontrolled levels of pollutants.

Each of these three major categories can be subdivided into numerous variations, some of which will be discussed in the text.

In general, ease of control of environmental conditions and experimental replicability are inversely related to the complexity of the experimental system (Figure 1). Hence exposure of single ecosystem components such as in bioassay and environmental chamber experiments are the easiest to control and replicate, while observations under natural conditions are the most difficult.

Regardless of experimental approach, collected data need to be analyzed, interpreted, and reported. Available methods include averaging, plotting, mapping, correlation, input–output analysis, and mathematical modeling. Special attention must be given to the reporting of the result. Misleading generalizations can occur by not fully describing conditions under which data are obtained.

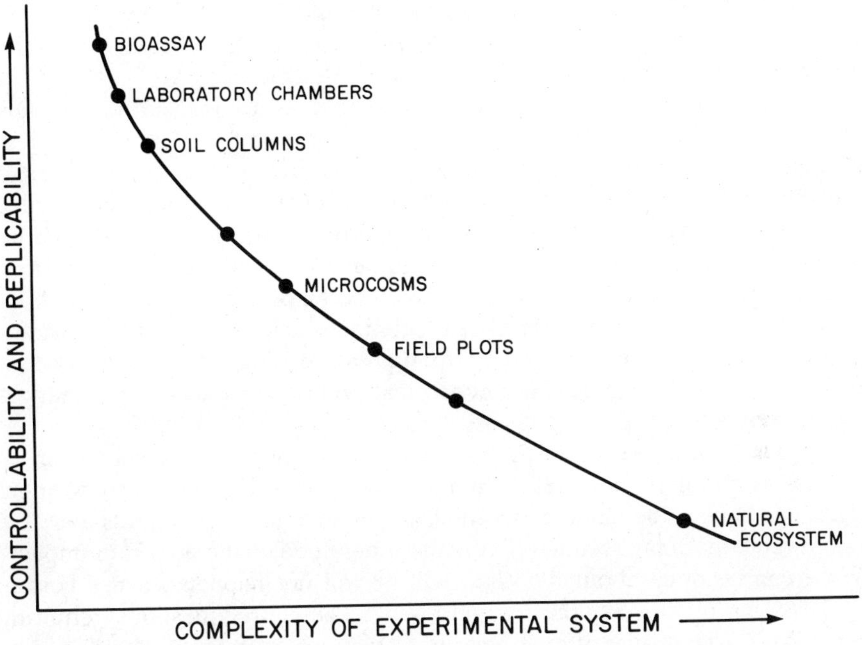

FIGURE 1. Relationship between complexity and controllability and replicability of experimental systems for assessing ecosystem effects.

Several gaseous and particulate air pollutants are hypothesized precursors to the formation of strong acids. These acids are deposited onto the land in rain, snow, and aerosols of liquid and particulate matter. The influx of acids, termed acid rain, sets into motion a myriad of biogeochemical reactions, first in the terrestrial then in the aquatic ecosystem. Depending on the magnitude, frequency, and duration of the acid rain and the characteristics of the receiving ecosystem, ecological impacts of various kinds and degrees may be observed.

The scientific community (e.g., Glass, 1978) has suggested many potential ecological impacts of acid rain, for example, direct injury to plant leaves (i.e., necrosis), leaching of soil nutrients, slowdown of nitrification process, decrease of forest productivity, and acidification of lakes and streams resulting in loss of fishes.

Some of the acid rain impacts are postulated as a result of laboratory experiments. Some are advanced through hypothesized simple cause–effect relations that are deemed plausible. However, not all potential impacts are proven and observed in the field. Possible reasons for this are that exposures in the field may be less severe than in the laboratory, the greater complexity of the ecosystem may result in reduction or delay of effects, biota in the field may develop increased tolerance to pollutants as a result of exposure over long periods of time; and greater variability of environmental conditions in the real system induces greater variability in the system's behavior and hence can make effects more difficult to detect.

Investigators have applied various techniques to measure ecosystem effects of acid rain. Regional surveys have been used to assess a real extent of acid rain damage or susceptibility to such damage. To understand mechanisms and processes involved with acid rain effects, controlled experiments have been performed on laboratory soil columns, in environmental chambers, on experimental greenhouse and field plots, and on forest microcosms where input conditions can be varied by design. Field investigations such as small watershed techniques and integrated lake–watershed studies have been designed to monitor ecological impacts as they occur in the real system where ecological processes may mediate changes initiated by acid rain.

Because of the energy crisis, a strong push is under way to shift from oil to coal as an energy source. Associated with this shift, many management decisions have to be made concerning the control of air pollutants and acid rain effects that may result. What is the magnitude of the acid rain impact? What can be done about it? What will or will not happen when a certain level of air quality standard is imposed? These are questions requiring technically sound quantitative answers. Many of the scientific studies conducted to date have not been designed to answer management questions. They do, however, provide bits and pieces of information that can be pieced together to help design more comprehensive, mechanistic, and integrated study efforts. It is hoped that this review of approaches, which includes laboratory experiments, field investigations, and methods of data interpretation, may point to some future directions to be taken.

2. REGIONAL SURVEYS

Regional surveys are usually conducted by taking samples from many places (e.g., lakes) in a region. Under the constraint of a limited budget, the survey is generally reconnaissance in nature. It covers a lot of ground and only a very few samples can be taken and analyzed for each site. In fact, a one-time-and-one-sample-only survey is a common practice.

Woods (1978) performed a one-time-only survey for 57 lakes in the Adirondack region of New York. Samples were taken from surfaces and bottoms of lakes and analyzed for major cations and anions. Presence or absence of fish in each lake was determined from secondary sources with undetermined reliability. Among the 57 lakes, 32 were reported to contain fish, 20 contained no fish, and for the remaining five there was no information. Table 1 summarizes the environmental characteristics of fish and fishless lakes. There are, of course, questions regarding the representativeness of one-time samples; however, the data seem to convey certain

TABLE 1. Comparison of Lakes With and Without Fish in the Adirondacks; Based on a Survey of 57 Lakes During the Summer of 1975

Parameter	Lakes With Fish	Lakes Without Fish	Statistically Significant at Indicated Level
Number	32.0	20.0	
Location	Eastern Adirondacks	Western Adirondacks	
Size of Basins	3.7×10^6 m^2	1.7×10^6 m^2	$P < .05$
Hydraulic Flash Factor[a]	5	13	$P < .01$
pH	4.3–6.8	3.9–4.7	$P < .01$
Ca^{2+} (meq/L)	0.09	0.06	$P < .01$
Na^+	0.026	0.023	—
NH_4^+ (meq/L)	0.0009	0.0065	$P < .001$
Cd (μg/L)	0.79	0.66	—
Pb (μg/L)	1.7	2.5	$P < .05$
Chlorophyll-a (μg/L)	5.1	2.8	—

From Woods (1978).
[a]Storage ratio Q/V (per year).

information. Small lakes situated in the western Adirondacks with high hydraulic flashing tend to have low pH and Ca^{2+} in their waters. These lakes are more likely to have no fish.

A regional survey of 155 lakes was conducted in southern Norway (Wright et al., 1977). Samples were collected at the lake surface and bottom in October 1974. A selected number of lakes were resampled in March 1975 and March 1976. Major cations and anions were measured. Phytoplankton and zooplankton concentrations were counted, but no fish data were collected in the field. Bedrock geology and basin vegetation were surveyed.

Figure 2 presents the pH distribution of the surface waters of the 155 lakes surveyed in October 1974. Generally speaking, pH of the lake water at the southernmost end is lowest, say 4.5. The pH increases in the northward direction to reach 7 near 63°N latitude. On the west coast there is also an area with low pH. The majority of the area has a pH above 5.6, which is the pH of distilled water equilibrated with CO_2 in the atmosphere.

The investigators hypothesize that the low pH of the lakes in southern Norway was a result of acid rain caused by the long distance transport of sulfate from central Europe. The study, however, did not consider local sources of pollutants (e.g., pulp and paper mills), potential acidification due to application of nitrogen fertilizer to forests, or other causes.

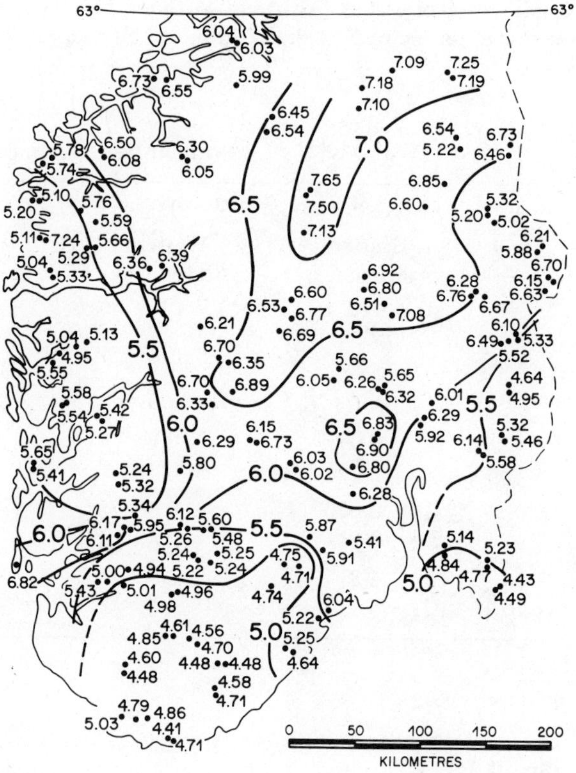

FIGURE 2. pH levels in surface waters of 155 lakes in southern Norway, October 1974. (After Wright et al., 1977.)

Figure 3 shows the weighted average pH of precipitation during the period July 1974 to June 1975. The pH of precipitation changed from 4.3 in the south to 4.7 or higher toward 63°N latitude. It is clear from comparing Figure 2 and 3 that precipitation cannot be the sole factor governing lake acidity, since the hydrogen ion concentration of the lakes ranges over two orders of magnitude while the rain ranges only over a factor of 2.5. Other factors that must be considered are vegetation, soils, geology, and morphology of catchments. The bedrock geology mapped in Figure 4 suggests that high pH lakes occur in areas of calcareous geology while low pH lakes occur in areas of granite geology.

Based on regional survey data, Henriksen (1980) has developed a nomograph to predict lake pH based on pH of precipitation and calcium concentration of the lake (Figure 5). This nomograph is derived by correlation analysis of data obtained in southern Norway. The relationship formulated by Henriksen describes an equilibrium and not a dynamic situation; hence, although it is intended to give the final response of a lake

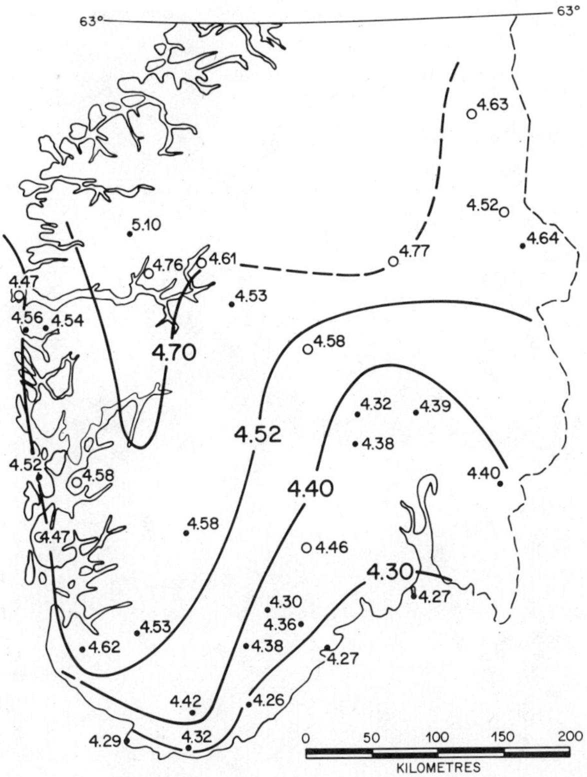

FIGURE 3. Weighted average pH of precipitation during the period July 1974–June 1975. Solid circles, daily sampling; open circles, weekly sampling. (After Wright et al., 1977.)

to a continual precipitation at a given average pH, it does not incorporate a basis for predicting how long the lake will take to reach its final state. The general validity and applicability of this model need to be tested.

Since fish data were not collected in the Norwegian regional survey, they cannot be analyzed together with the water quality data collected. Quotations of evidence observed previously within the region and outside the study area were used to argue that fish disappeared because of low pH (Hendrey and Wright, 1976).

Other regional lake surveys have been conducted in Sweden (Dickson, 1975, cited by Wright et al., 1977) in the Adirondacks (Schofield, 1976) and in Canada (Beamish and Harvey, 1972; Beamish, 1976). In the latter study, the fish and water quality data obtained from the LaCloche Mountain lakes were used to establish the pH levels at which different fish stopped reproduction (Table 2). Loss to the fishery was attributed to the cessation of reproduction rather than to the toxicity of metals or scarcity of foods.

Use of regional surveys to establish environmental tolerance ranges of

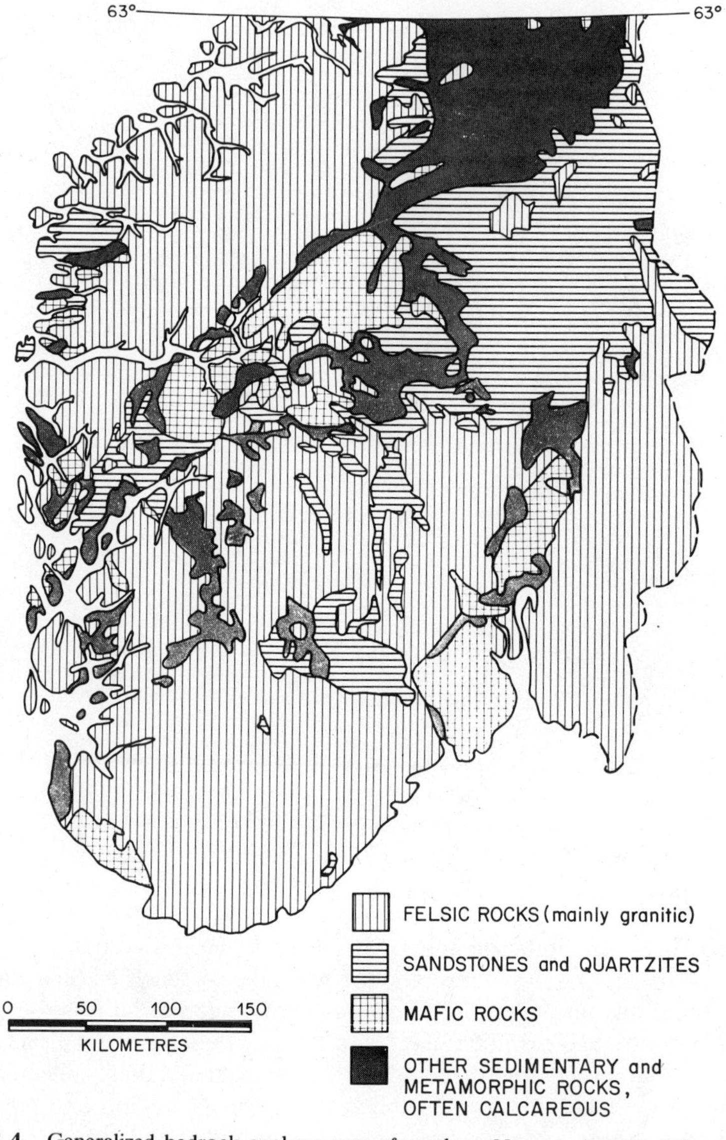

FIGURE 4. Generalized bedrock geology map of southern Norway. (Modified from Oftedahl, 1974, in Wright et al., 1977.)

biota is a good approach that has also been applied to assess acid rain effects on vegetation. Tree-ring analyses of Norway spruce and Scots pine have been conducted in regions having different assumed levels of acidic precipitation (Abrahamsen et al., 1976). However, results have been inconclusive. Some data showed no effect and other data indicated high growth with increasing acidification.

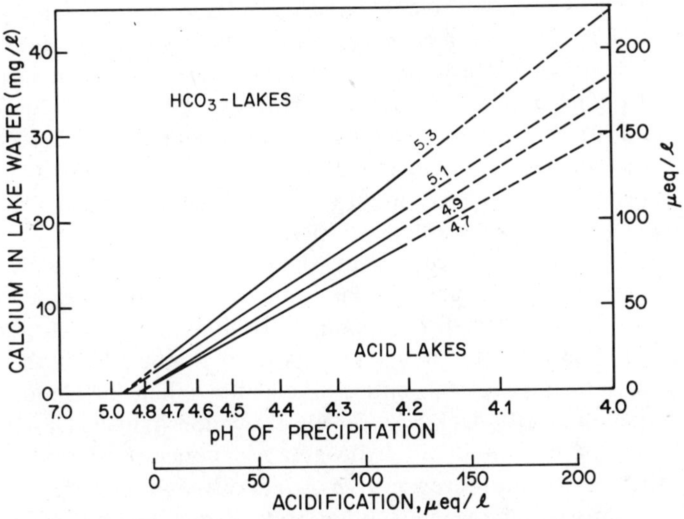

FIGURE 5. Nomograph for predicting lake pH. (After Henriksen, 1980.)

Unlike one-time, infrequent, or short-term surveys of surface water quality and biota, surveys of tree cores provide the potential opportunity to reconstruct long-term historical trends. Surveys of lake sediment cores have the same potential (Davis et al., 1980). Sediment core analysis can reveal past changes in the terrestrial as well as the aquatic system (Davis and Norton, 1978) as a result of the coupling between the systems.

TABLE 2. Approximate pH at Which Fish in the La Cloche Mountain Lakes Stopped Reproduction

pH	Species	Family
6.0–5.5	Smallmouth bass *Micropterus dolomieu*	Centrarchidae
	Walleye *Stizostedjon vitreum*	Percidae
	Burbot *Lota lota*	Gadidae
5.5–5.2	Lake trout *Salvelinus namaycush*	Salmonidae
	Troutperch *Percopsis omiscomaycus*	Percopsidae
5.2–4.7	Brown bullhead *Ictalurus nebulosus*	Ictaluridae
	White sucker *Catostomus commersoni*	Catostomidae
	Rock bass *Ambloplites rupestris*	Centrarchidae
4.7–4.5	Lake herring *Coregonus artedii*	Salmonidae
	Yellow perch *Perca flavescens*	Percidae
	Lake chub *Couesius plumbeus*	Cyprinidae

After Beamish (1976).

The major strength of a regional survey is its potential to identify quickly major ecological effects on a regional scale. Sets of survey data lend themselves to correlative analyses, such as among lake acidity, bedrock geology, and precipitation acidity, or tree growth, soil type, and precipitation acidity. Significant correlations suggest mechanisms by which effects occur and hence management strategies to reduce effects. Hypotheses regarding the nature of the mechanisms should be tested experimentally with one or more of the other approaches that will be discussed.

Generalizations of regional survey data must be made with caution because the general sparseness of data with respect to seasonal, year-to-year, long-term, and near-field spatial variation can lead to errors in interpretation. For instance, if a lake is stratified, the pH at the surface might not be indicative of the entire lake. Data taken by Hendrey et al. (1980) for a small Adirondack lake during snowmelt indicate differences of over an order of magnitude in hydrogen ion concentration with depth. Similarly, there exists a difference of 2 orders of magnitude for the surface water of this same lake between summer and snowmelt season (Galloway et al., 1980).

Another limitation of the regional survey approach is that at best it only suggests the mechanisms by which effects are brought about. Mechanistic understanding is usually required to predict quantitatively how the system will respond with time to different levels of pollutant concentrations. Such knowledge is important for making management decisions. Hence, although regional surveys may establish a link between acid precipitation and increased acidity of lakes, they have not provided a basis for predicting future rates of change in lake acidity for present or changed levels of atmospheric acidity. Thus we cannot predict what the magnitude of the problem will be in a decade if acidity of precipitation remains at its current levels nor to what degree the effects would be ameliorated by any specific decreases in the acidity of precipitation.

3. CONTROLLED EXPERIMENTS

Since ecosystems are complex, it is logical for scientists to isolate one or a few specific components and perform experiments to determine acid rain effects. Types of experimental systems that can be used are many—for example, bioassay, laboratory chambers, laboratory soil columns, experimental greenhouses and field plots, and microcosms.

In many of these experiments, severe acid rain conditions are imposed to produce measurable effects. This is accomplished either by the application of synthetic rain water stronger than acid rain, or by the compression of experimental time with high application rate. Extension of experimental results to field conditions is not straightforward. If a certain effect is observed under experimental conditions at pH 3.0, one should not simply report that acid rain causes such and such effects unless rain pH tends to be pH 3.0 and rainfall intensity and duration in the field is equal to that applied in the experiment.

There are reasons to use in controlled experiments pollution exposures more severe than those observed in the field. These reasons include determining what will happen if pollution increases and also determining threshold levels for responses. However, there is need for caution in extrapolating the results of such experiments to less severe field exposures.

Since there are practically infinite combinations of experimental variables possible, controlled studies may be perceived as a dream world for researchers to perform experiments and to publish original papers. Primary purposes of controlled experiments, however, are to determine biological tolerance to pollutants and to determine mechanisms and their rates. Experimental designs of controlled experiments should be standardized and coordinated so results can be synthesized into generalized principles.

3.1 Bioassay

For aquatic organisms, bioassay experiments are conducted with aquaria to which are added acid rain or synthetic rain water adjusted to different pH with sulfuric acid. Other variables include cation concentrations (e.g., Ca^{2+}, Mg^{2+}, Al^{3+}). This technique has been used to test the tolerance range of various organisms (e.g., phytoplankton, zooplankton, fish) and the hatching success of fish eggs.

Lakes affected by acid rain are most likely oligotrophic. Moss (1973, as cited by Giddings and Galloway, 1976) found that oligotrophic species of algae usually grow better at low pH. Below pH 4.5, however, their growth may stop. The organisms will survive at pH 3.6.

Bell (1971, cited by Giddings and Galloway, 1976) conducted experiments to examine the effect of pH on aquatic insects. Most insect larvae will survive pH 4.0. Emergence is affected at pH 5.0. The CO_2 concentration in the experiments, however, was very high (i.e., 20 mg/L).

A large volume of literature exists on the effect of pH on freshwater fish (EIFAC, 1969). A general conclusion is that adult lake trout can tolerate pH 4.0 but eggs will not hatch below 4.5.

3.2 Laboratory Chambers

Laboratory chambers are boxes of soil planted with seeds or young seedlings. Acid rain or synthetic acid rains of different pH are applied at a specific rate. For an enclosed chamber, acid mists can also be used.

Crowther and Ruston (1911) performed such experiments on grass as early as 1911. More recent works are on seedlings of different trees, mosses, and lichens (e.g., Wood and Bormann, 1976).

Direct injury to the young leaves of most plants can be demonstrated at pH below 3.0. Higher than pH 3.0, some increased growth by simulated acid rain has been observed, probably due to the fertilizing values of nitrate and sulfate (Wood and Bormann, 1976).

3.3 Laboratory Soil Columns

Soil samples collected in the field are packed in a column to recreate soil horizons. From each horizon, a lysimeter can be installed to extract leachate. Synthetic acid rains of different pH are applied on the surface. Leachate and final soil samples can be analyzed for acid rain effects. Rippon (1980) has described such an experimental design.

Typical responses of lysimeter experiments are that acid rain is neutralized by the soil column. H^+ is retained in the soil and Ca^{2+}, Mg^{2+}, K^+, and other cations are leached out. The response, however, follows a characteristic breakthrough curve for reactions that involve equilibria. It usually takes a long while before the acid rain effect can be detected in the leachate of the lower soil horizon.

A special case of lysimeter experiments is to measure acid rain effects on soil microbial activity, for example, ammonification, and nitrification. For such experiments, the rate of watering and the ambient temperature become important control variables (Wood, 1978).

The procedure of enriching soil with a substrate to follow its degradation by organisms should be made with caution. Alexander (1980) has shown that the procedure may change the pathway of microbial reaction. With synthetic acid rain of pH 4.1 and no enrichment, the nitrification process was not impaired. When the soil was enriched with NH_4^+, however, the nitrification process was inhibited.

3.4 Microcosms

A microcosm can be defined as a biological–physical model of an ecosystem that includes more than one component. Microcosm experiments permit the study of interactions between and integrated responses of ecosystem components under controlled conditions.

Lee and Weber (1976) and Kelly (personal communication) have established forest microcosms to study effects of acid rain on forest ecosystems. The microcosms consist of boxes containing reconstituted litter and soil layers. Each layer is mixed to minimize interplot differences. Seedlings of various trees are planted in the boxes. Simulated acid rain treatments are applied. A major difference between the facilities of Lee and Weber and that of Kelly is that that of Lee and Weber is in an open sided greenhouse and that of Kelly is outdoors. Also, any given treatment of Lee and Weber is always applied at the same pH, while Kelly will vary the pH around a specified mean.

Lee and Weber have used simulated acid rain treatments of pH 5.6, 4.0,

3.5, and 3.0 and have monitored tree growth, leaf production, nutrient uptake, and litter decomposition to estimate effects on forest productivity and nutrient cycling. Preliminary data (Lee and Weber, 1976) indicate that the leaching of Ca^{2+}, Mg^{2+}, and other cations from the soil is similar to that observed in the laboratory soil columns and field plots. Direct injury to leaves was observed at pH 3. Also at pH 3 germination of tulip poplar and sumac was inhibited, while that of red cedar, white pine, and birch was stimulated.

Tonnessen and Harte (1980) are using 50-L aquatic microcosms to study response of plankton and trace metal concentrations to additions of simulated acid rain.

3.5 Experimental Greenhouse and Field Plots

Experimental field plots function similarly to laboratory soil columns. The applied acid rain, however, can travel not only vertically but also laterally to a ditch. Findings of Norwegian field studies (Abrahamsen and Dollard, 1979) have been essentially similar to those of the soil column experiments described earlier.

Other experimental plot studies consist of exposing agricultural crops in the greenhouse and field to different levels of simulated acid rain (Jacobson, 1980). For simulated precipitation with values between pH 3 and 4, positive, negative, and no effects on growth and yield have been observed. Jacobson (1980) and Shriner (in this volume) discuss evidence that suggests greenhouse grown crops are more susceptible to acid rain injury than field grown crops.

Since 1974, the Norwegian SNSF Project has exposed field plots of Scots pine saplings to different treatments of simulated acid rain. Results for the early years of the study (Abrahamsen and Dollard, 1979) indicate enhanced growth in the acid treated plots. Abrahamsen (1980) points out that it is conceivable that short term effects can be positive as a result of increased nitrogen availability, but that longer term effects could be negative as a result of increased cation leaching from the soil.

Experimental plots can also be established in aquatic systems. Müller (1980) has placed tubes 10 m in diameter and 2–2.5 m in length in a lake in the Canadian Experimental Lakes Area to study effects of acidification on periphyton growth.

3.6 Overview

The major strength of the approaches described in this section is that they allow the testing of cause–effect hypotheses under fully or partially controlled environmental conditions and pollutant exposures. However, to obtain this control, the researcher needs to sacrifice complexity in the experimental system and hence some degree of realism. Complexity is reduced by omitting ecosystem components, physically reducing the size of the system, and, in terrestrial microcosms and laboratory soil columns, mixing each of the soil horizons.

Bioassay and environmental chamber studies allow maximum experimental control. They focus on the direct response of an organism, plant, fish, or plankton to a specific acid rain exposure. However, response in the real system can be different as a result of mediation by other components.

Microcosm (including soil column) studies incorporate component interactions and integrated responses. As the complexity of the microcosm experiment is increased, either by adding more system components or by making the rate of pollutant input as complicated as exists in the real system, replicability and controllability of the experiments are decreased. The results obtained become highly variable and are as difficult to interpret as the field observations under natural conditions.

Limitations of laboratory microcosms are that they cannot hold large organisms (e.g., fish and trees bigger than seedlings); they cannot model processes that cover large areas (e.g., migration of seed dispersal); and their behavior diverges from the real system as the time of the experiment increases, hence they are not applicable to measuring long-term effects. To consider large organisms, greater areas or volumes, and longer periods of time, the researcher must go to experimental field plots. The major limitation of the experimental field plot is inability to control weather conditions and difficulty in controlling other environmental conditions, for example, choosing a set of plots with identical soil conditions. To incorporate the full complexity of the real system, the researcher must turn to the real system as described in the next section.

4. TOTAL ECOSYSTEM APPROACHES

4.1 Small Watershed Approach (Input–Output Analyses)

Small watershed studies conduct detailed input–output analyses of biogeochemical fluxes on a basic ecosystem unit. Based on quantitative information on vegetation and geology of the ecosystem, it is possible to estimate internal cycling of materials between biotic and abiotic components (Likens et al., 1977).

The basic ecosystem unit comprises a small watershed and a small brook that drains it. The best known site for small watershed studies in Hubbard Brook in the White Mountains of New Hampshire (Likens et al., 1977).

In Hubbard Brook, six contiguous watersheds, ranging in size from 12 to 53 ha, were selected as the basic ecosystem units. These watersheds are located at the headwaters of two tributaries to Hubbard Brook. They are watertight, similar in vegetation and geology, and subject to the same climatic conditions. As a part of the experimental forest program of the

U.S. Forest Service, one watershed was cleared of woody vegetation in 1965 to determine the effects of clear cutting. Another watershed was cut in alternating strips in 1980. Remaining watersheds are undisturbed.

The inputs to the ecosystem are measured at 11 weather stations. The outputs are monitored at the lowest point (or the exit) of each watershed. Input and output samples are collected weekly and are analyzed for the quantity of water and its chemical concentrations in both dissolved and particulate forms.

The output fluxes are calculated by multiplying the volumetric flow rate by its chemical concentrations. They are normalized by the watershed area for interwatershed comparison and statistical calculations of averages and standard deviations.

To facilitate input–output analysis, a time period must be selected for calculating the annual budget. If the amount of water storage within the system is not monitored, then it is desirable to select a time when internal water storage would be approximately the same every year. Otherwise, it would not be possible to determine what portion of the difference between annual input and annual output for a specific material was attributable to a direct net uptake or release by ecosystem components and what portion was attributable to an annual change in the volume of water being held in the system.

In general, the internal storage must be measured to minimize errors in input–output analysis. This is particularly true when internal storage is very large. For a complex system where internal storage exists in numerous compartments, quantification of total storage may be difficult.

The starting and ending date of a period selected for one parameter may not be suitable for another. A good rule of thumb is a starting date when the system is quiescent. If one chooses a starting date when major activity is taking place, then, depending on the phenology for a given year, events may be included in the budget to a greater or lesser extent, thus increasing the interyear variance in the input–output budgets.

Considering the above criteria for places where there is no snow accumulation, a midwinter starting date would appear to be optimum, assuming that the vegetation is dormant and that the system's water storage capacity is fully recharged by fall and early winter precipitation. Such an approach has been taken for watersheds in the southeastern United States (Goldstein and Mankin, 1972). However, in climates where there is snow accumulation, the situation is more complex because the amount of snow present at a given date will vary from year to year and snow does not release the materials it contains at a uniform rate. Elevated concentrations of impurities are released during the early stages of thawing. Both high volume of water and variable release of impurities during snow thaw can lead to rapid large changes in water quality. Galloway et al. (1980) have observed changes in lakes of 2 orders of magnitude in hydrogen ion concentration during snow thaw.

In the Hubbard Brook study Likens et al. (1977) chose the period from June 1 to May 31 to calculate their annual budgets. For this period annual precipitation and streamflow were found to have the highest correlation, about .99. It was reasoned that the system's water storage capacity would be fully charged at this date as a result of snow thaw during the spring. While this date avoids the snow thaw period, it also starts the annual budget calculation during the start of the plant growth period when rapid changes in internal fluxes of materials between ecosystem components might be expected.

The water storage capacity of ecosystems in areas drier than the northeast and southeast United States might not be regularly recharged on an annual basis. For such areas, the monitoring of water storage would be necessary. This necessity would also apply to the northeast and southeast United States during extended dry periods.

Input–output analyses of the Hubbard Brook watersheds indicate that there is a net absorption of acidity and net loss of magnesium, calcium, potassium, sodium, and aluminum ions (Likens et al., 1977).

In calculating and analyzing materials budgets, one must be very careful in handling free hydrogen ion concentrations of pH. As water takes part in different reactions, free hydrogen ions will be created or consumed to satisfy chemical equilibria. If two solutions were to be mixed, their resultant pH would not necessarily be the average of their individual free hydrogen ion concentrations. Hence when fluxes and budgets of acidity are being calculated and analyzed, it is preferable to work with alkalinity, which is conserved, as the key variable.

Data collected in the small watershed approach can be used for interpretive techniques other than input–output analysis. Johnson et al. (1969) formulated a working model for the variation in stream water chemistry data collected at Hubbard Brook.

The model is based on a simple assumption that the stream water at a given point at any time is some mixture of ground water and surface water. Ground water is added to the stream at a constant rate. Surface water which comes from rain or snowmelt varied day to day and season to season. Stream water quality can therefore be calculated as the weighted average concentration of ground water and surface water.

The assumption leads to an equation for the calculation of stream water quality C as follows:

$$C = \frac{C_0 - C_x}{1 + BD} + C_x$$

Where C_0 is the concentration of ground water, C_x the concentration of rainwater or snowmelt, and D stream discharge. B is an empirical coefficient obtained by adjusting its value in the plot of C versus $1/(1 + DB)$ until a straight line relationship is established.

A limitation of input–output analysis is that it treats the ecosystem as a black box and thus, as does the regional survey, only suggests mechanisms by which effects are brought about. However, small watersheds at which input–output studies are being undertaken are excellent locations for experimental plot studies and sources of soils for laboratory experiments to test hypotheses about mechanisms. In such situations, the results of the laboratory and field plot studies can be related to the total integrated response of the real system.

Other limitations include lack of control of weather conditions and atmospheric inputs. There is potential to exert some control on inputs. Schindler et al. (1980) have undertaken an ecosystem study of a lake where they are artificially acidifying it over many years by adding sulfuric acid directly to the lake. A small watershed could conceivably be treated with different exposures of lime to determine how this would alter the acidity of surface waters.

4.2 Integrated Small Watershed Approach

The integrated approach attempts to look at watershed processes mechanistically within the context of an entire ecosystem. The Integrated Lake–Watershed Acidification Study (ILWAS) was designed to determine the ecological effects of acid rain under natural conditions (EPRI, 1979; Goldstein et al., 1980). Since the most widely reported effect of acid rain has been the acidification of lake water leading to elimination of fish, it is of interest to learn how and why ecosystems (i.e., watershed and lake) become acidified by acid rain.

The watershed ecosystem is envisioned to comprise a cascade of basic compartments: atmosphere, canopy, snow pack, litter layer, soil layers, bogs, stream segment, and lake. These are the compartments that the acid rain water must pass through before it reaches the lake outlet. As it passes through each compartment, biogeochemical processes acting in series and in parallel will produce or consume acids and will release chemicals that shift the pH and other chemical equilibria.

Three forested watersheds (Panther, Woods, and Sagamore) within 30 km of each other in the Adirondack Park region of New York State were selected for field investigation. Each watershed has different configurations and characteristics for the basic ecosystem compartments. The principal hypothesis of the study is that these differences produce different pH dynamics—that is, Panther Lake is alkaline, Woods Lake acidic, and Sagamore Lake in between.

Field surveys were conducted to characterize the properties of the basic ecosystem compartments in each watershed. At selected locations, measurements were made for ambient air quality and the quantity and quality of waters as they move through the system from tree top to lake output (Figure 6). Data were collected monthly, weekly, or only once,

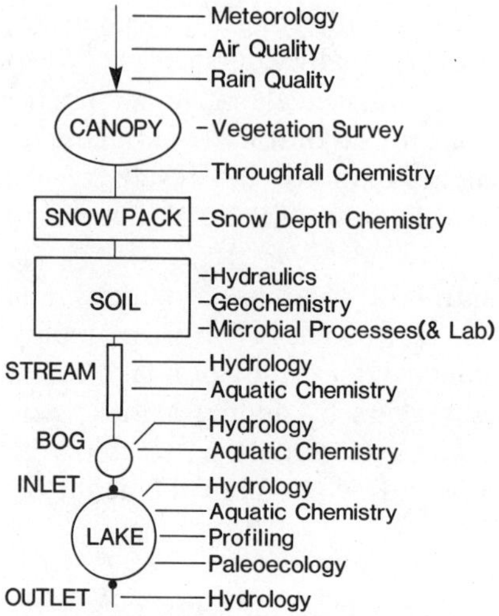

FIGURE 6. Field program of the integrated lake–watershed acidification study.

depending on the parameter and its temporal variability. The field program began in 1977 and was completed in December 1981.

In conjunction with the field program, a mathematical model was developed to calculate dry deposition as a function of ambient air quality, and to simulate the quantity and quality of water in throughfall, soil horizons, bogs, stream, lake, and lake outlet (Chen et al., 1979). The role of the model is to organize the measurement of lake–watershed acidification processes into an integrated theoretical framework (Goldstein et al., 1980). The model also serves as a vehicle to check the consistency of theory and data from rainfall quantity to lake outlet quality. Acidification processes and rate coefficients are based on theory established through controlled experiments. Calculated results are compared against data observed in the field. Since the model is conceived on the basis of fundamental biogeochemical ecosystem processes, it is expected that the model will be applicable to geographic areas other than the Adirondack Parks region where data were collected. The potential capability of the model to predict the dynamic response of surface waters to different levels of inputs of atmospheric acidity is also derived from its mechanistic underpinnings.

Preliminary data from ILWAS indicate that deposition rates of H^+ and SO_4^{2-} vary seasonally and are approximately the same for all three watersheds (see Figures 7 and 8). The deposition rates of various ions in acid rain as measured at the seven ILWAS stations correlate reasonably well with those measured at the nearby MAP3S station (Table 3). This finding is significant because it suggests that acid rain data from regional monitoring stations may be used to perform preliminary calculations of acid rain effects for nonmonitored sites.

Preliminary analysis of pH data shows that pH at Panther Lake inlet is

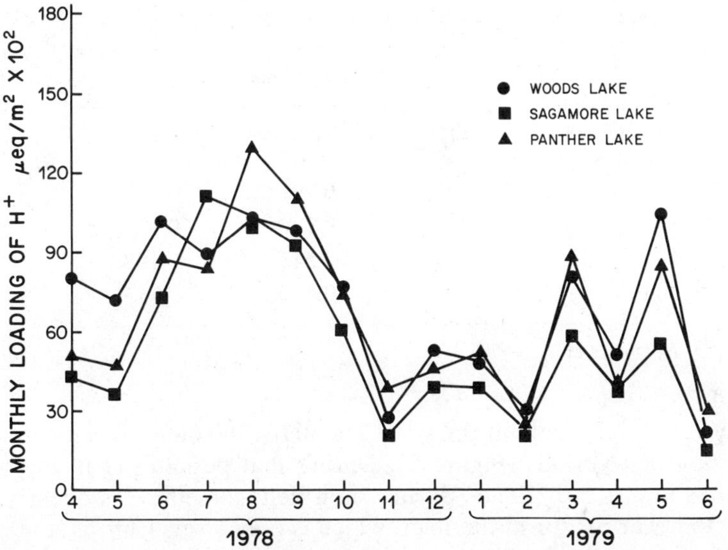

FIGURE 7. Precipitation input of H^+ at Woods, Sagamore, and Panther Lake Basins. (Johannes and Altwicker, 1980.)

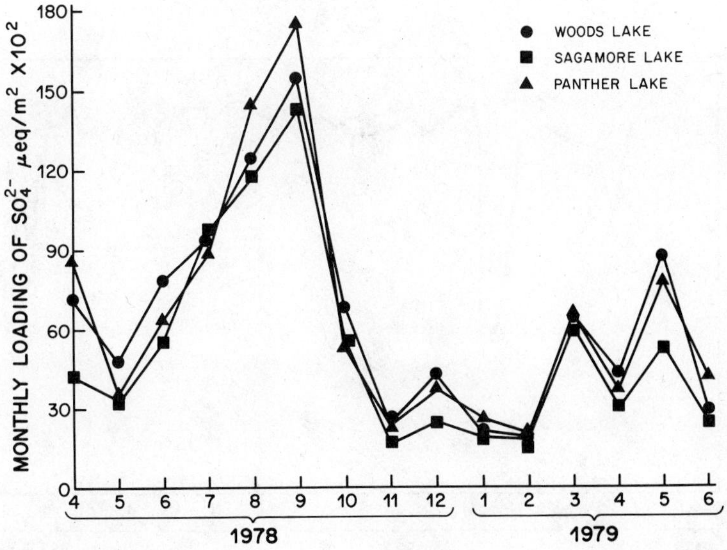

FIGURE 8. Precipitation input of SO_4^{2-} at Woods, Sagamore, and Panther Lake Basins. (Johannes and Altwicker, 1980.)

TABLE 3. Correlation on the Results from ILWAS Stations (7 sites) With Those From MAP3S Stations

	Correlation Coefficient (*R*)		
Parameters	Ithaca	Whiteface	Penn State
H^+	0.778	0.791	0.906
SO_4^{2-}	0.702	0.760	0.915
NO_3^-	0.510	0.543	0.468
NH_4^+	0.911	0.690	0.713

Altwicker et al. (1979).

normally 7.3–7.5 throughout the year. The pH at the outlet is similar except during snowmelt periods (Figure 9). During that period, pH in the outlet drops to as low as 5.0. Hence during snowmelt, something is occurring in the lake to lower pH, or in the inlet, which is only a small spring and is not representative of all inflows.

The pH profiles measured in the lake show that only surface water is acidified during the period of snowmelt (Figure 10). What is the source of H^+ ions that acidify the lake surface?

To attempt to resolve the puzzle, the model was used to help trace the sources of water at the outlet. The model is first calibrated to the Panther Lake Basin (Figure 11). After that, precipitation falling directly on the lake

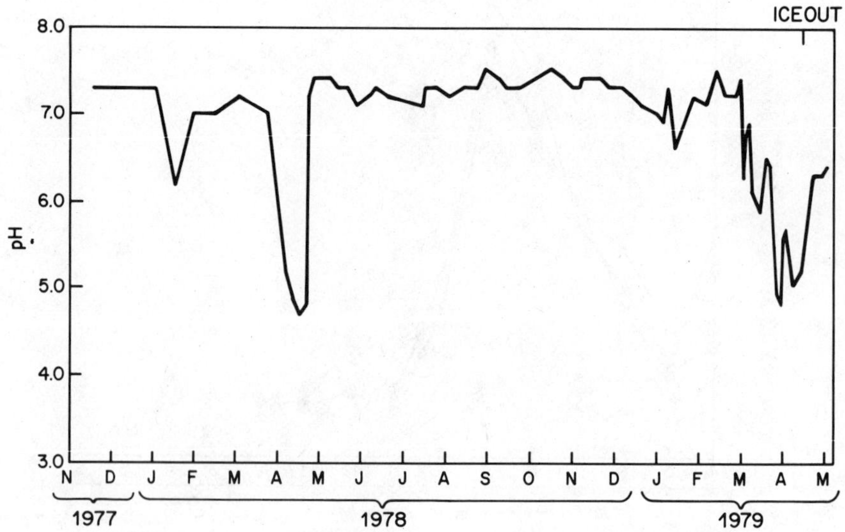

FIGURE 9. The pH at Panther Lake outlet. (Galloway et al., 1980.)

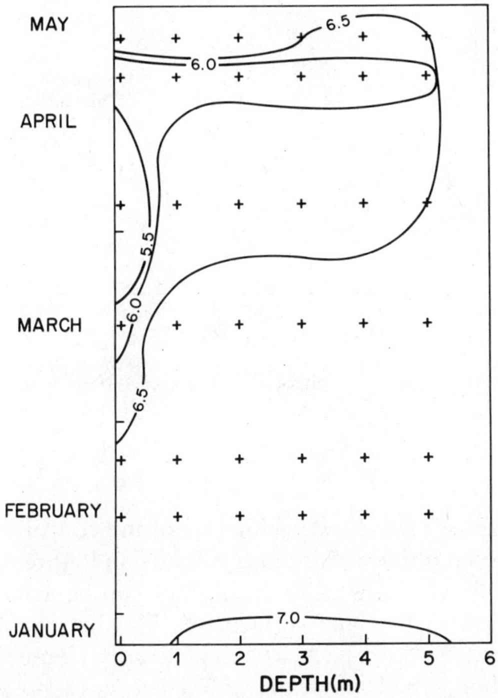

FIGURE 10. The pH profiles in Panther Lake. (Hendrey et al., 1980.)

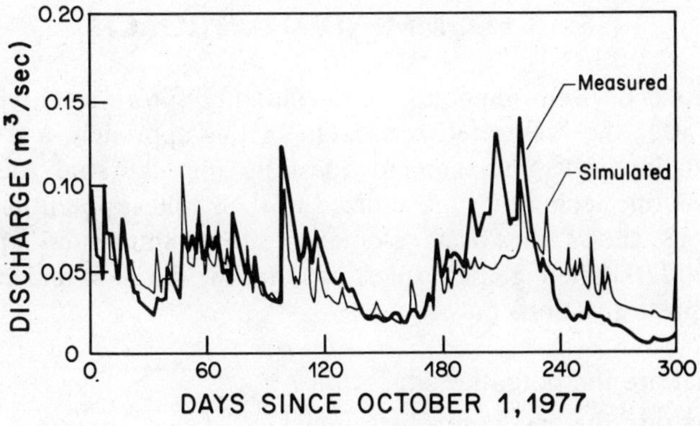

FIGURE 11. Comparison of calculated and observed hydrograph at Panther Lake outlet. (Goldstein et al., 1980.)

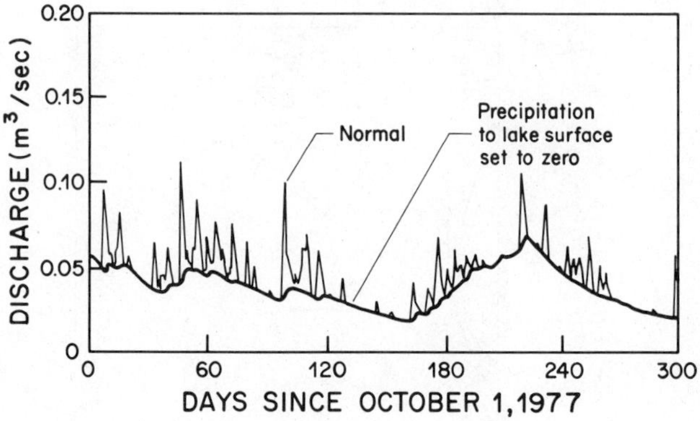

FIGURE 12. Effects of direct precipitation on the lake surface to the outflow of Panther Lake. (Goldstein et al., 1980.)

surface is set to zero. This allows estimation of the contribution of this input to the total observed outflow. Results are shown in Figure 12. It can be seen that approximately two-fifths of the peak flow can be accounted for by the direct precipitation to the lake surface. During that period, lake water is inversely stratified with respect to temperature (Hendrey et al., 1980). Direct precipitation, which has a pH of 3.8–4.7, quite possibly is deposited right on the surface to acidify the lake surface water (pH 5.0). Other possible explanations are surface and shallow subsurface runoff resulting from snowmelt (Galloway et al., 1980).

The significance of the capability to manipulate the model to examine the effect of a single process on the integrated response of the ecosystem should not be overlooked. The interactions between the model and the field research influence and strengthen each other.

5. RECOMMENDED APPROACH

To assess ecosystem impacts of air pollutants, future research should be patterned after the integrated approach. In this approach, a framework is provided to integrate management questions into the study design, when various scientific techniques are utilized to their fullest potential.

Figure 13 shows the basic elements of the integrated approach. It proceeds with the management questions that the decision makers may wish to ask. Typical questions may include

1. What are the potential effects?
2. What are the best control strategies?
3. What will the control do and not do?

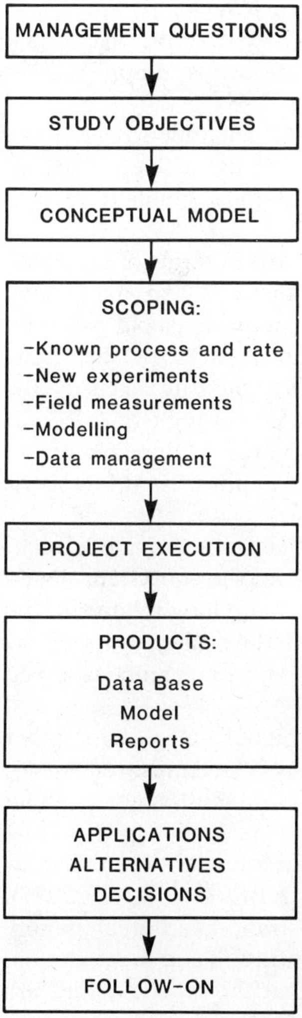

FIGURE 13. Framework of an integrated study.

Based on these broad questions, scientists begin to formulate the study objectives that will address the management questions. This, however, is easier said than done. Everyone has a personal bias about what is important. It is difficult for a scientist to accept that his or her project idea may not be relevant to the question at hand. For that reason, a system analyst with a broad background on the subject matter may be consulted. This individual will not be involved in any specific component of the research, other than modeling and integration efforts.

A conceptual model should then be developed. The conceptual model must include important physical, chemical, and biological processes operating in the ecosystem. It must accept the exposure level of pollutant and other inputs to the system and produce outputs that describe the system response. The variables used by the model must be as simple and as routinely measured as possible.

Based on the conceptual model, the scope of the scientific research can be delineated. Required knowledge of the process is first defined. Some of this can be obtained through literature. The rest must come from new experiments.

Results of controlled experiments will be expressed in terms of processes and rates useful to the model. For toxicity experiments, Chen and Selleck (1969) have formulated a kinetic model of fish toxicity threshold based on the toxification and detoxification rates for organisms exposed to different combinations of chemical. When the toxification rate exceeds the detoxification rate, the difference may be integrated through time to calculate biological response. Such a concept can help prevent the difficult and often unfruitful search for effects which frequently do not exist at low levels of exposure.

The field program is designed to produce a data base for model calibration and verification. The input to the responses of the ecosystem should be monitored concurrently for periods of time which are long relative to the reaction times of the key processes being studied. The early years of the data set can be used to calibrate the model. The later years can be reserved for model verification.

The data base can be used not only for the model but also for other interpretive techniques, including correlation analysis. One must remember, however, a positive correlation does not necessarily establish a cause–effect relationship.

For example, a near-perfect correlation has been found between increasing annual H^+ and NO_3^- deposition at Hubbard Brook (Likens, 1977). Annual H^+ deposition is the product of H^+ concentration and precipitation. The NO_3^- deposition is the product of NO_3^- concentration and precipitation. Hence, the good correlation may simply mean that the two variables are related through a common denominator (precipitation) rather than between H^+ and NO_3^- concentrations.

Conversely, a zero correlation does not mean the absence of a cause–effect relationship (Figure 14). At low concentration, a positive correlation exists between dose and response because the pollutant can be a nutrient required for biological growth. Beyond a certain limit and below the environmental tolerance limit, the increase of chemical dose may not produce any appreciable changes in biological responses—that is, a zero correlation.

When the chemical dose exceeds the toxicity threshold, a negative correlation is established between the dose and the response. Thus, the correlation can change from positive to negative depending on the level of pollutant exposure. If data covering the whole range of conditions are used

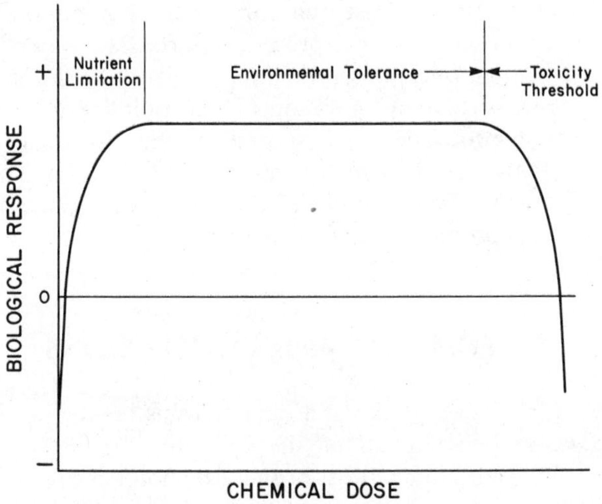

FIGURE 14. Generalized relationship between chemical dose and biological response.

in a correlational analysis, a zero correlation will result even though a dose–effect relationship actually exists.

Execution of the integrated research should include five major tasks:

1. Program Management.
2. Modeling.
3. Field Research.
4. Controlled Experiments.
5. Data Management.

The relationships among these tasks are depicted in Figure 15. As shown, modeling, field observations, and laboratory experiments should work hand in hand. Field data can lead to changes in model formulation. Model simulations can, in turn, suggest different sampling designs. All study components should be coordinated through program management, which permits modifications in accordance with scientific merit.

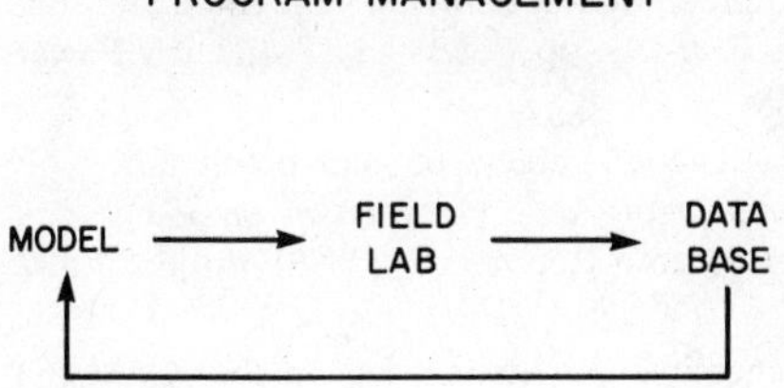

FIGURE 15. Major tasks of the Integrated Research Program.

Products of an integrated research study will be a comprehensive data base, a model, and a series of interpretive reports. The model can be used to answer management questions, which may help make decisions.

The study, however, should not stop at that point. Based on an understanding of system properties, one should attempt to identify a few key parameters that can characterize the response of the ecosystem. A long-term but less intensive measurement of these parameters should be carried out as a follow-up program. Such a program will furnish data to check against model predictions.

6. SUMMARY AND CONCLUSIONS

A number of laboratory, field, and data analysis techniques are available for assessing ecosystem impacts of air pollutants. Each technique has its strong points and weak points. Most of these techniques are complementary to each other.

Applications of these methodologies in the past have produced a wealth of qualitative and quantitative knowledge of acid rain effects on plants, soils, waters, and fishes. At present, the most apparent effect of acid rain is the acidification of lake waters.

Not all lakes are acidified to the same degree. This affords the opportunity to investigate how a lake is acidified to different levels under natural conditions. The integrated study technique couples field investigation with theoretical modeling. The model may provide scientific answers to such management questions as what will and will not happen if a certain air quality standard is imposed, and if the acidity of precipitation is to increase or decrease.

What is needed in the future is for atmospheric pollution effects researchers to develop comprehensive, integrated programs that combine the strengths of multiple techniques to answer complex questions regarding ecosystem responses.

REFERENCES

Abrahamsen, G. Acid precipitation in plant nutrients and forest growth. In *Proceedings of International Conference on the Ecological Impact of Acid Precipitation, Sandefjord, Norway, March* 11–14, pp. 58–63 (1980).

Abrahamsen, G., and G. J. Dollard. Effects of acid precipitation on forest vegetation and soil. In *Ecological Effects of Acid Precipitation*, EA-79-6-LD, pp. 4.2.1–4.2.17. Electric Power Research Institute, Palo Alto, California, 1979.

Abrahamsen, G., K. Bjor, R. Horntvedt, and B. Tveite. Effects of acid precipitation on coniferous forest. In F. H. Braekke (ed.) *Impact of Acid Precipitation on Forest and Freshwater Ecosystems in Norway*, pp. 37–63. Research Report 6/76, SNSF project, 1432 Aas-NLH, Norway (1976).

Alexander, M. Effects of acid precipitation on biochemical activities in soil. In *Proceedings on International Conference on the Ecological Impact of Acid Precipitation, Sandefjord, Norway, March* 11–14 (1980).

Altwicker, E. R., A. H. Johannes, C. R. Barnes, E. G. Chapman, and N. L. Clesceri. Atmospheric inputs into remote areas. *I. Proceedings of Fourth International Conference of the Commission on Atmospheric Chemistry and Global Pollution, Boulder, Colorado, August* (1979).

Beamish, R. J. Acidification of lakes in Canada by acid precipitation and resulting effects on fishes. *Proceedings of the First International Symposium on Acid Precipitation and the Forest Ecosystem, Columbus, Ohio*. Springfield, Virginia, PB-258-645, 479–498, August (1976).

Beamish, R. J., and H. H. Harvey. Acidification of the La Cloche Mountain Lakes, Ontario, and resulting fish mortalities. *J. Fish. Res. Board Can.* **29**(8), 1131–1143 (1972).

Bell, H. L. Effects of low pH on the survival and emergence of aquatic insects. *Water Research* **5**, 313–315 (1971).

Chen, C. W., and R. E. Selleck. A kinetic model of fish toxicity threshold. *Journal Water Pollution Control Federation* **41**(8), Part 2, R294–R308 (1969).

Chen, C. W., S. A. Gherini, and R. A. Goldstein. Modeling the lake acidification process. In M. G. Wood (ed.) *Ecological Effects of Acid Precipitation, Report of a Workshop, Gallway, United Kingdom, September* 4–8, 1978. EA-79-6-LD Electric Power Research Institute. Palo Alto, California, 1979.

Crowther, C., and A. G. Ruston. The nature, distribution, and effects upon vegetation of atmospheric impurities in and near an industrial town. *J. Ag. Sci.* **4**, 25–55 (1911).

Davis, R. B., and S. A. Norton. Paleolimnologic studies of human impact on lakes in the United States with emphasis on recent research in New England. *Pol. Arch. Hydrobiol.* **25**, 99–115 (1978).

Davis, R. B., S. A. Norton, D. F. Braekke, F. Berge, and C. T. Hess. Atmospheric deposition in Norway during the last 300 years as recorded in SNSF lake sediments: Part IV, synthesis and comparison with New England. In *Proceedings of International Conference on the Ecological Impact of Acid Precipitation Sandefjord, Norway, March* 11–14, 274–275 (1980).

Dickson, W. The Acidification of Swedish lakes. Institute Freshwater Research Drottningholm Report 54, pp. 8–20 (1975).

EPRI. Technical Work Statement for the Integrated Lake–Watershed Acidification Study RP 1109, Electric Power Research Institute, Palo Alto, California (1979).

European Inland Fisheries Advisory Committee (EIFAC). Water quality criteria for European freshwater fish. Report on extreme pH values and inland fisheries. *Water Research* **3**, 593–611 (1969).

Galloway, J. N., C. L. Schofield, G. R. Hendrey, and A. H. Johannes. Sources of acidity in three lakes acidified during snowmelt. In *Proceedings of International Conference on the Ecological Impact of Acid Precipitation, Sandefjord, Norway, March* 11–14, pp. 264–265 (1980).

Giddings, J., and J. N. Galloway. Effects of acid precipitation on aquatic and terrestrial ecosystem. A literature review. In *Literature Reviews on Acid Precipitation*, The Centre for Environmental Quality Management and the Water Resources and Marine Sciences Center, Cornell University, Ithaca, New York, 1976.

Glass, N. R. (ed.) Environmental Effects of Increased Cost Utilizations and Ecological Effects of Gaseous Emissions from Coal Combustion. EPA Report 600/7-78-108, U.S. EPA, Office of Health and Ecological Effects and Office of Research and Development, Washington, D.C. (1978).

Goldstein, R. A., and J. B. Mankin. Prosper: A model of atmosphere–soil–plant water flow. pp. 1181–1187. In *Proceedings of the* 1972 *Summer Computer Simulation Conference, San Diego*. AFIPS, Arlington, Virginia 1972.

Goldstein, R. A., C. W. Chen, S. A. Gherini, and J. D. Dean. A framework for lake watershed acidification study. In *Proceedings of International Conference on the Ecological Impact of Acid Precipitation, Sandefjord, Norway, March* 11–14, pp. 252–253 (1980).

Hendrey, G. R., and R. F. Wright. Acid precipitation in Norway: Effects on aquatic fauna. *J. Great Lakes Res.* **2**(supplement 1), 192–207 (1976).

Hendrey, G. R., J. N. Galloway, and C. L. Schofield. Temporal and spatial trends in the chemistry of acidified lakes under ice cover. In *Proceedings of International Conference on the Ecological Impact of Acid Precipitation, Sandefjord, Norway, March* 11–14, pp. 266–267 (1980).

Henricksen, A. Acidification of freshwaters—a large scale titration. In *Proceedings of International Conference on the Ecological Impact of Acid Precipitation, Sandefjord, Norway, March* 11–14, pp. 68–74 (1980).

Jacobson, J. S. The influence of rainfall composition on the yield and quality of agriculture crops. In *Proceedings of International Conference on the Ecological Impact of Acid Precipitation, Sandefjord, Norway, March* 11–14, pp. 41–46 (1980).

Johannes, A. H., and E. R. Altwicker. Atmospheric inputs to three Adirondack lake watersheds. In *Proceedings of International Conference on the Ecological Impact of Acid Precipitation, Sandefjord, Norway, March* 11–14, pp. 256–257 (1980).

Johnson, N. M., G. E. Likens, F. H. Bormann, D. W. Fisher, and R. S. Pierce. A working model for the variation in stream water chemistry at Hubbard Brook Experimental Forest, New Hampshire. *Water Resources Research* **5**(6), 1353–1363 (1969).

Lee, J. J., and D. E. Weber. A study of the effects of acid rain on model forest ecosystems. EPA Report 76-25-5, U.S. Environmental Protection Agency, Corvallis, Oregon (1976).

Likens, G. E., F. H. Bormann, R. S. Pierce, J. S. Eaton, and N. M. Johnson. *Biogeochemistry of a Forested Ecosystem*. Springer-Verlag, New York, 1977.

Moss, B. The influence of environmental factors on the distribution of freshwater algae. II. The role of pH and carbon dioxide–biocarbonate system. *J. Ecology* **61**, 157–177 (1973).

Müller, P. Effects of artificial acidification on the growth of periphyton. *Canadian J. Fisheries and Aquatic Sci.* **37**, 355–363 (1980).

Rippon, J. G. Studies of acid rain on soils and catchments. In T. C. Hutchinson and M. Havas (eds.) *Effect of Acid Precipitation on Terrestrial Ecosystems*, pp. 499–524. Plenum, New York, 1980.

Schindler, D W., R. Wagemann, R. B. Cook, T. Ruszczynski, and J. Prokopowich. Experimental acidification of lake 223, experimental lakes area: background data and first three years of acidification. *Can. J. Fisheries and Aquatic Sci.* **37**, 342–354 (1980).

Schofield, C. L. Acid precipitation: effects on fish. *Ambio* **5**, 228–230 (1976).

Tonnessen, K., and J. Harte. The potential for acid precipitation damage to aquatic ecosystems of the Sierra Nevada, California. In *Proceedings of International Conference on the Ecological Input of Acid Precipitation, Sandefjord, Norway, March* 11–14, pp. 338–339 (1980).

Wood, M. J. Effect of pH on nitrification rates in soil. Research report of Central Research Laboratories, Leatherhead, Surrey, England (1978).

Wood, T., and F. H. Bormann. Short term effects on an artificial acid rain upon the growth and nutrient relations of *Pinus strobus* L. *Proceedings of the International Symposium on Acid Precipitation and the Forest Ecosystem, Columbus, Ohio, May* 12–15, 1975. NTIS Springfield, Virginia, PB-258-645, August (1976).

Woods, L. W. Limnology of remote lakes in the Adirondack region of New York State with emphasis on acidification problems. Env. Health Report No. 4, New York State Department of Health (1978).

Wright, R. F., T. Dale, A. Henriksen, G. R. Hendrey, E. T. Gjessing, M. Johannessen, C. Lysholm, and E. Storen. Regional Surveys of Small Norwegian Lakes October 1974, March 1975, March 1976, and March 1977. SNSF Project IR 33/77 Olso, Norway, October (1977).

SYNTHESIS

ECOSYSTEM ANALYSIS OF AIR POLLUTION EFFECTS

Robert A. Goldstein
Electric Power Research Institute
Palo Alto, California

Allan H. Legge
Kananaskis Centre for Environmental Research
The University of Calgary
Calgary, Alberta, Canada

INTRODUCTION

Atmospheric deposition can affect ecosystems as well as ecosystem components and processes in multiple ways. Air pollution effects can be considered detrimental, beneficial, or innocuous. Effects can be direct—for example, sulfur dioxide could be absorbed into leaves and photosynthesis reduced; they can also be indirect—for example, deposition of strong acids could increase nitrogen and sulfur availability and increase primary production, or increase rate of leaching of essential cations from soil and decrease primary production. Ecosystem components interact through a host of processes, operating in series and in parallel. An effect *on one component can result in other changes throughout the system or, considering the homeostatic nature of ecosystems, result in a negative feedback that tends to dampen the original effect. Interaction between terrestrial and aquatic components is an important consideration.*

Ecosystem analysis, as a means of assessing air pollution impacts, stresses an integrated view of ecosystem behavior. Because multiple aspects of ecosystem dynamics are concerned, an interdisciplinary approach is required. The conceptualization of material cycling through the ecosystem is one potential pathway for integrating different ecosystem components. The papers presented in this section discussed the philosophy of and motivation for application of ecosystem analysis to assessment of air pollution effects, reviewed strengths and weaknesses of the approach in the context of past and present applications, and recommended future directions to be taken.

Although the philosophies and attitudes expressed in the two papers presented are very similar, there is a difference in perspective. The paper by Miller and Kickert justifies the need for ecosystem analysis on the basis of protecting the integrity of entire ecosystems. The paper by Chen and Goldstein justifies the need for ecosystem analysis because, to understand how any single ecosystem component is being affected—for instance, the vegetation—the component must be placed into an ecosystem context so the integration of all direct and indirect effects can be calculated.

Synthesis and Discussion

The results of an ecosystem analysis from India substantiate the importance of understanding linkages between different ecosystem components and performing ecosystem analyses. Ecosystem effects of atmospheric emissions of fluorides from an aluminum plant are being assessed. Effects have been observed in animals, vegetation, and soils in the vicinity of the plant. Cows and buffaloes develop fluorosis from grazing on grasses and feed that have accumulated fluorides. Species composition of the plant communities in the area is changing, hence implying long-term ecosystem structural change.

The value of remote sensing in assessing ecosystem effects and vegetation damage was discussed. It was clear that although remote sensing techniques—such as false-color infrared photography, thermal line scan imagery, and satellite imagery—are analytical tools with immense promise, these tools have definite scales of resolution and, hence, limitations. These limitations must be considered in the context of the research or application question posed. There was general agreement, however, that these techniques showed promise for the future and should be used.

It was suggested that detrimental effects of air pollution, resulting from industrial activity, should be evaluated within a context that also includes recognition of the socioeconomic benefits of the industry. Socioeconomic benefits connected with the protection of ecosystem integrity should also be considered. Evaluation of socioeconomic benefits and costs is not a component of ecosystem analysis but falls into a broader subject area commonly referred to as integrated assessment.

The difficulty of synthesizing observations made in different systems exposed to different pollutants is illustrated by the following. Miller and Kickert, in their paper, hypothesized that in the San Bernardino Forest in California decomposer populations may be enhanced as a result of increased needle fall caused by chronic ozone exposure. This observation elicited the comment that in the Sudbury, Ontario area near nickel smelters the bacterial populations were inhibited. This led to declines in rates of litter decomposition and wood decay and in population levels of bacterial pathogens. The disease blister-rust of white pine was practically nonexistent in the fumigated area but was present normally outside the area.

The general monitoring of ecosystem processes may not be adequate to identify effects produced by air pollution unless the monitoring program was specifically designed to test for the existence of such an effect. Thus the fact that the Hubbard Brook study had detected no effects of acidic deposition in the terrestrial ecosystem may be either a result of the nonexistence of any effects or a failure to consider such effects in the original experimental design.

In the response to a question about the general applicability of the Integrated Lake–Watershed Acidification Study (ILWAS) model, it was pointed out that models based on fundamental mechanistic biogeochemical processes should have widespread applicability. Current plans include testing the model at Swedish experimental sites and at an experimental watershed being studied at the West Point United States Military Academy.

It was asked if the ILWAS model had been validated. The hydrological component of the model has been compared with data from the experimental sites. Agreement in general is good, although there are some systematic discrepancies that are being investigated. The plant canopy and soil biogeochemistry parts of the model have only recently been programmed and are currently being tested. The experimental program is still in progress, and since the final testing of the model at the study sites cannot be made until the experimental program is completed, testing will not be completed for at least another two years. However, partial testing is expected to be conducted continuously using data from the study sites and other sources. (The current state of the ILWAS program has been summarized by Goldstein et al. (1984)).

There was also a question concerning how particle deposition and throughfall were treated in the ILWAS model. Deposition is calculated using empirical deposition velocities for different weather conditions. Deposition velocities are derived from the meteorological data collected at the sites. The measured ambient air quality and leaf area index are also used in the calculation of dry deposition. The pool of collected chemicals on the leaf surface is washed off as a function of rainfall intensity and duration.

A question was raised concerning how to define the boundaries of an experimental ecosystem. From the perspective of vegetation communities, boundaries are usually not distinct. Experimental ecosystems usually use surface drainage relationships to define boundaries. Flow budgets and groundwater measurements can be used to test if there are no groundwater inflows to or outflows from the experimental catchment. It is not clear how one would define analogous natural experimental ecosystems in geographical areas where small (ten to several hundred hectare) watertight catchments do not occur.

It was pointed out that mathematical models should be flexible enough to allow for simulation response to different air pollution exposure scenarios.

CONCLUSION

The ecological assessment of air pollution effects involves addressing many complex scientific problems. There exists no panacea for solving these problems. Each analytical approach, whether it be a controlled experiment, regional survey, or ecosystem analysis, has strengths and weaknesses. We need to make use of the strengths of all of these approaches. This statement, however, should not be considered as a license for each researcher to pursue his own isolated project.

When considering ecosystem analysis as a method for assessing air pollution effects, it is essential to distinguish between a holistic or total ecosystem approach and an integrated ecosystem approach because this determines the scope of the research. The holistic approach to ecosystem research indiscriminately examines structural and functional ecosystem responses. Since all ecosystem components are considered to be interrelated, all ecosystem components are accepted as relevant. A holistic ecosystem study is multidisciplinary with a minimum of interaction of experienced researchers from the diverse disciplines. The integrated approach, in contrast to the holistic approach, stresses the establishment of a specific question or objective and then examines only those facets of ecosystem structure and function that are strongly related to the specific question or objective. Scope and experimental design therefore are defined prior to the start of data collection. The integrated approach is interdisciplinary with experienced researchers from diverse disciplines working as an interactive research team. It is the integrated approach to ecosystem analysis which is considered to be the preferred method for air pollution effects research on ecosystems.

Modeling is an important element in ecosystem analysis. It is important to recognize, however, that when modeling ecosystem dynamics there exists a hierarchy of processes operating over a broad range of spatial and temporal scales. It is not practical to attempt to include all of the hierarchical levels within a single model. One should develop a suite of models, with each model designed for a specific spatial–temporal level of resolution. These models should be developed with the objectives of the ecosystem research program constantly in mind.

Because of the natural complexity of ecosystems, it is important to recognize at the outset that ecosystem analysis research requires a long-term commitment in the sense of time, personnel, and financial support. A five- to seven-year time period is recommended for an ecosystem analysis research program. Experience has shown that a major commitment by a research team leader and a core of scientific professionals is essential to provide continuity in the research. The number and disciplines of the professionals are a function of the scope of the problem. A 75% time allocation is recommended for the research team leader, while a 50% time allocation is recommended for core professionals. Student involvement should be minor and supportative of the research program. From a financial viewpoint, adequate funding must be assured so research goals can be met. It is recommended that the research program be reviewed by the funding agencies in concert with an independent interdisciplinary scientific advisory group at least on a yearly basis to assure adequate progress toward program goals and to provide constructive scientific support to the research team.

To be successful, an ecosystem analysis should

1. *Have clearly defined objectives and hypotheses that answer major management questions presented by government, industry, and the scientific community.*
2. *Formulate a conceptual model of the ecosystem at the start of the study.*
3. *Have a long-term funding commitment.*
4. *Have an interdisciplinary team of experienced researchers.*
5. *Strive to develop a predictive quantitative capability which is based on a mechanistic understanding of the ecosystem processes involved.*
6. *Have continuous interaction and synthesis between modeling and experimental investigations to provide perspective.*
7. *Have continuous review and evaluation to provide for evolution in the research.*
8. *Have a final synthesis with recommendations which not only relate*

to the management questions originally asked but also clearly state the increased understanding of ecosystem structure and function.

It was pointed out a number of times during the course of this conference that the costs associated with comprehensive air pollution effects research programs are relatively inexpensive when compared with the costs of most air pollution abatement control strategies. Environmental management decisions must be based upon facts, not fiction. The courtroom approach to issues is for lawyers, not scientists! As has been successfully shown in Alberta, Canada, industry and government can and should work together to fund ecosystem analysis research programs cooperatively. These programs must be firmly based upon guidance from industry, government, and the scientific community. With this approach, and through interdisciplinary cooperation, there is potential to have support during the next 20 years for the type of technically sound, exciting, and necessary ecosystem research for which the participants of this conference have indicated their enthusiasm.

REFERENCE

Goldstein, R. H., S. A. Gherini, C. W. Chen, L. Mok, and R. J. M. Hudson. Integrated acidification study (ILWAS): a mechanistic ecosystem analysis. Phil. Trans. R. Soc. Lond. **B305**, *409–425 (1984).*

AUTHOR INDEX

SUBJECT INDEX